Human-Computer Interaction

Designing for Diverse Users and Domains

Human Factors and Ergonomics

Series Editor

Gavriel Salvendy

Human-Computer Interaction

Designing for Diverse Users and Domains

Edited by

Andrew Sears
Julie A. Jacko

CRC Press
Taylor & Francis Group
Boca Raton London New York

CRC Press is an imprint of the
Taylor & Francis Group, an **informa** business

CRC Press
Taylor & Francis Group
6000 Broken Sound Parkway NW, Suite 300
Boca Raton, FL 33487-2742

First issued in paperback 2017

© 2009 by Taylor & Francis Group, LLC
CRC Press is an imprint of Taylor & Francis Group, an Informa business

No claim to original U.S. Government works

ISBN 13: 978-1-138-11505-7 (pbk)
ISBN 13: 978-1-4200-8887-8 (hbk)

Library of Congress Cataloging-in-Publication Data

Human-computer interaction. Designing for diverse users and domains / editors, Andrew Sears, Julie A. Jacko.
    p. cm. -- (Human factors and ergonomics)
  "Select set of chapters from the second edition of The Human computer interaction handbook"--Pref.
  Includes bibliographical references and index.
  ISBN 978-1-4200-8887-8 (hardcover : alk. paper)
  1. Human-computer interaction. I. Sears, Andrew. II. Jacko, Julie A. III. Human-computer interaction handbook. IV. Title: Designing for diverse users and domains. V. Series.

QA76.9.H85H85643 2009
004.01'9--dc22
                                                              2008050947

Visit the Taylor & Francis Web site at
http://www.taylorandfrancis.com

and the CRC Press Web site at
http://www.crcpress.com

For Beth, Nicole, Kristen, François, and Nicolas.

CONTENTS

CONTRIBUTORS

Norman Alm
School of Computing, University of Dundee, Scotland

Alisa Bandlow
College of Computing, Georgia Institute of Technology, USA

Bridget C. Booske
Population Health Institute, School of Medicine and Public Health, University of Wisconsin, USA

Amy Bruckman
College of Computing, Georgia Institute of Technology, USA

Alex Carmichael
Department of Psychology, Manchester University, UK

Joel Cooper
Department of Psychology, Princeton University, USA

Sara J. Czaja
Department of Psychiatry and Behavioral Sciences, Center on Aging, University of Miami Miller School of Medicine, USA

Paula J. Edwards
Health Systems Institute, Georgia Institute of Technology, USA

Jinjuan Feng
Department of Computer and Information Sciences, Towson University, USA

Andrea Forte
College of Computing, Georgia Institute of Technology, USA

Thomas Fuller
Microsoft Game Studios, Microsoft Corporation, USA

Paul Green
Human Factors Division, University of Michigan, Transportation Research Institute, USA

Peter Gregor
School of Computing, University of Dundee, Scotland

William M. Gribbons
Bentley College, USA

Vicki L. Hanson
IBM T.J. Watson Research Center, USA

Julie A. Jacko
Institute for Health Informatics, University of Minnesota, USA

Kevin Keeker
Microsoft Game Studios, Microsoft Corporation, USA

Matthew B. Kugler
Department of Psychology, Princeton University, USA

Steven J. Landry
Department of Industrial Engineering, Purdue University, USA

Nicole Lazzaro
XEODesign, Inc., USA

Chin Chin Lee
Center on Aging, University of Miami, USA

V. Kathlene Leonard
Alucid Solution, Inc., USA

Alan F. Newell
School of Computing, University of Dundee, Scotland

Randy J. Pagulayan
Microsoft Game Studios, Microsoft Corporation, USA

Ramon L. Romero
Microsoft Game Studios, Microsoft Corporation, USA

François Sainfort
Health Systems Institute, Georgia Institute of Technology, USA

Ingrid U. Scott
Department of Ophthalmology, Penn State College of Medicine, USA

Andrew Sears
Interactive Systems Research Center, Information Systems Department, UMBC, USA

Annalu Waller
School of Computing, University of Dundee, Scotland

Dennis Wixon
Microsoft Game Studios, Microsoft Corporation, USA

Mark A. Young
The Maryland Rehabilitation Center and Workforce Technology Center, State of Maryland Department of Education, USA

ADVISORY BOARD

PREFACE

We are pleased to offer access to a select set of chapters from the second edition of *The Human–Computer Interaction Handbook.* Each of the four books in the set comprises select chapters that focus on specific issues including fundamentals which serve as the foundation for human–computer interactions, design issues, issues involved in designing solutions for diverse users, and the development process.

While human–computer interaction (HCI) may have emerged from within computing, significant contributions have come from a variety of fields including industrial engineering, psychology, education, and graphic design. The resulting interdisciplinary research has produced important outcomes including an improved understanding of the relationship between people and technology as well as more effective processes for utilizing this knowledge in the design and development of solutions that can increase productivity, quality of life, and competitiveness. HCI now has a home in every application, environment, and device, and is routinely used as a tool for inclusion. HCI is no longer just an area of specialization within more traditional academic disciplines, but has developed such that both undergraduate and graduate degrees are available that focus explicitly on the subject.

The HCI Handbook provides practitioners, researchers, students, and academicians with access to 67 chapters and nearly 2000 pages covering a vast array of issues that are important to the HCI community. Through four smaller books, readers can access select chapters from the Handbook. The first book, *Human–Computer Interaction: Fundamentals,* comprises 16 chapters that discuss fundamental issues about the technology involved in human–computer interactions as well as the users themselves. Examples include human information processing, motivation, emotion in HCI, sensor-based input solutions, and wearable computing. The second book, *Human–Computer Interaction: Design Issues,* also includes 16 chapters that address a variety of issues involved when designing the interactions between users and computing technologies. Example topics include adaptive interfaces, tangible interfaces, information visualization, designing for the web, and computer-supported cooperative work. The third book, *Human–Computer Interaction: Designing for Diverse Users and Domains,* includes eight chapters that address issues involved in designing solutions for diverse users including children, older adults, and individuals with physical, cognitive, visual, or hearing impairments. Five additional chapters discuss HCI in the context of specific domains including health care, games, and the aerospace industry. The final book, *Human–Computer Interaction: The Development Process,* includes fifteen chapters that address requirements specification, design and development, and testing and evaluation activities. Sample chapters address task analysis, contextual design, personas, scenario-based design, participatory design, and a variety of evaluation techniques including usability testing, inspection-based techniques, and survey design.

Andrew Sears and Julie A. Jacko

March 2008

ABOUT THE EDITORS

Andrew Sears is a Professor of Information Systems and the Chair of the Information Systems Department at UMBC. He is also the director of UMBC's Interactive Systems Research Center. Dr. Sears' research explores issues related to human-centered computing with an emphasis on accessibility. His current projects focus on accessibility, broadly defined, including the needs of individuals with physical disabilities and older users of information technologies as well as mobile computing, speech recognition, and the difficulties information technology users experience as a result of the environment in which they are working or the tasks in which they are engaged. His research projects have been supported by numerous corporations (e.g., IBM Corporation, Intel Corporation, Microsoft Corporation, Motorola), foundations (e.g., the Verizon Foundation), and government agencies (e.g., NASA, the National Institute on Disability and Rehabilitation Research, the National Science Foundation, and the State of Maryland). Dr. Sears is the author or co-author of numerous research publications including journal articles, books, book chapters, and conference proceedings. He is the Founding Co-Editor-in-Chief of the *ACM Transactions on Accessible Computing,* and serves on the editorial boards of the *International Journal of Human–Computer Studies*, the *International Journal of Human–Computer Interaction,* the *International Journal of Mobil Human–Computer Interaction,* and *Universal Access in the Information Society*, and the advisory board of the upcoming *Universal Access Handbook*. He has served on a variety of conference committees including as Conference and Technical Program Co-Chair of the Association for Computing Machinery's Conference on Human Factors in Computing Systems (CHI 2001), Conference Chair of the ACM Conference on Accessible Computing (Assets 2005), and Program Chair for Asset 2004. He is currently Vice Chair of the ACM Special Interest Group on Accessible Computing. He earned his BS in Computer Science from Rensselaer Polytechnic Institute and his Ph.D. in Computer Science with an emphasis on Human–Computer Interaction from the University of Maryland—College Park.

Julie A. Jacko is Director of the Institute for Health Informatics at the University of Minnesota as well as a Professor in the School of Public Health and the School of Nursing. She is the author or co-author of over 120 research publications including journal articles, books, book chapters, and conference proceedings. Dr. Jacko's research activities focus on human–computer interaction, human aspects of computing, universal access to electronic information technologies, and health informatics. Her externally funded research has been supported by the Intel Corporation, Microsoft Corporation, the National Science Foundation, NASA, the Agency for Health Care Research and Quality (AHRQ), and the National Institute on Disability and Rehabilitation Research. Dr. Jacko received a National Science Foundation CAREER Award for her research titled, "Universal Access to the Graphical User Interface: Design For The Partially Sighted," and the National Science Foundation's Presidential Early Career Award for Scientists and Engineers, which is the highest honor bestowed on young scientists and engineers by the US government. She is Editor-in-Chief of the *International Journal of Human–Computer Interaction* and she is Associate Editor for the *International Journal of Human Computer Studies*. In 2001 she served as Conference and Technical Program Co-Chair for the ACM Conference on Human Factors in Computing Systems (CHI 2001). She also served as Program Chair for the Fifth ACM SIGCAPH Conference on Assistive Technologies (ASSETS 2002), and as General Conference Chair of ASSETS 2004. In 2006, Dr. Jacko was elected to serve a three-year term as President of SIGCHI. Dr. Jacko routinely provides expert consultancy for organizations and corporations on systems usability and accessibility, emphasizing human aspects of interactive systems design. She earned her Ph.D. in Industrial Engineering from Purdue University.

Human-Computer Interaction

Designing for Diverse
Users and Domains

DESIGNING FOR DIVERSITY

THE DIGITAL DIVIDE: THE ROLE OF GENDER IN HUMAN–COMPUTER INTERACTION

Joel Cooper and Matthew B. Kugler
Princeton University

INTRODUCTION

This is the age of the computer. Whether we use a personal computer to balance our checkbook, operate a mainframe that sorts signal intercepts for our government, or work the cash register at the local Burger King, the steady hum of computer technology permeates our daily existence. Comparison shopping can now be done in seconds online with a few mouse clicks, rather than in days of real-world driving. University bookstores now compete not only with local equivalents, but also with online merchants on other continents often with favorable consequences to the wallet and purse. The cost of distributing ideas has reached an all-time low, forever altering political dialogue, and websites have brought together far-flung communities dedicated to everything from Japanese cartoons to satanic cults. According to the latest available figures from the U.S. Census, 61.8% of households in the United States have home computers and 54.7% have Internet access (U.S. Census, 2005).

This prevalence masks a persistent problem, however. While the 61.8% figure is impressive from the perspective of merely a decade ago, it is nonetheless true that the increase in computers has not been uniform across every subgroup in society nor has it affected all groups in the same way. To the contrary, the computer revolution has left some groups behind. A person with a bachelor's degree is 30% more likely to own a computer than a person with only a high-school education. A household with an income of $75,000–$99,999 has a 90% chance of owning a computer; one with an income of $25,000–$49,999 has a 67% chance. There are also racial differences in computer ownership. White and Asian Americans are over 20% more likely to own a computer than Black and Hispanic Americans (U.S. Census, 2005). Moreover, in the last decade of the 20th century, the gap in computer ownership between African Americans and Whites widened. These differences persist even when controlling for income. It has been shown that owning a computer leads to dramatic advantages on academic test scores. It is particularly interesting that, controlling for the number of computers in a particular household, wealthy Americans and White Americans gained even more of an advantage than poor and minority students (Atwell & Battle, 1999).

A divide also exists between men and women, with women not enjoying the benefits of the technological revolution on par with men (Cooper & Weaver, 2003). The difficulties women face while using computers are sweeping. They are underrepresented in their use and ownership of computers (Pinkard, 2005; Wilson, Wallin, & Reiser, 2003; Yelland & Lloyd, 2001), take fewer technology classes in high school and college (Pinkard, 2005), are far less likely to graduate college with degrees in IT fields and, most significantly, enjoy interacting with computers much less than do men (Mitra, Lenzmeier, Steffensmeier, Avon, Qu, & Hazen, 2000).

Computers are becoming central to more jobs every year. Current estimates suggested that by 2010, 25% of all new jobs in the public and private sectors will be technologically oriented (AAUW, 2000). However, even more important, computers play a role in all of the basic activities of life from banking, to shopping, to—increasingly—voting. Decades ago, computer innovation was driven by the space program, the cold war, and military technology. Now, a new car's computer technology is more than 1,000 times more powerful than what guided the Apollo moon missions (Alliance of Automobile Manufacturers, 2006). Computers are inescapable. With all of this in mind, it is a societal problem that the path to computer efficacy is more difficult for the poor, ethnic minorities, and women (Wilson, Wallin, & Reiser, 2003).

THE DIGITAL DIVIDE

Discrimination against women, at least in certain domains, has deep and complex roots. Understanding the basis of such discrimination is important and complex. The roots of the digital divide share some commonalities with discrimination that women have faced in employment and professional advancement but also have their own distinct origins. The use of computers in the home, classroom, and workplace is only a few decades old, which affords us the opportunity to gain a glimpse at the genesis of the particular problem of the gender divide in information technology.

In the late 1970s, computers began to replace television as the technological innovation in the classroom. By the 1980s they were ubiquitous in education and on their way to becoming a fixture in most households. In this context, Wilder, Mackie, and Cooper (1985) surveyed school children to assess their attitudes toward computers. They found a large difference in the degree to which boys and girls were attracted to the computer. As early as kindergarten, boys indicated more positive attitudes about computer technology than girls. These small attitudinal differences became dramatic in the fifth grade and continued to grow through the middle- and high-school years (Wilder, Mackie, & Cooper, 1985). Computer use had just begun to spread into the mainstream of public life, and numerous explanations for the difference were considered. Wilder et al. hoped that the gender differences in regards to computers were either an artifact of the particular geographic area studied in the investigation or something that would diminish as technology became more widely accessible. It was easy to hope that the problem would fix itself in those days; public education is a great equalizer.

This was not to be the case. Disturbing effects discovered in the 1980s persisted into the 1990s. The clearest data was not on the question of usage, but on anxiety. In a host of domains, both young girls and older women reported that computers are not creators of fun and amusement but rather the source of apprehension. Weil, Rosen and Sears (1987) reported that about 1 in 3 adults in the United States experienced what they called "computerphobia"—adverse anxiety reactions to the use of computers. Dembrot and her colleagues were among the first to investigate the imbalance in computer anxiety as a function of gender. They found that female college students expressed considerably more anxiety about computers than did their male counterparts (Dambrot, Watkins-Malek, Silling, Marshall, & Garver, 1985; see also Temple & Lips, 1989). This finding was replicated frequently throughout the 1990s (i.e., Colley, Gale, & Harris, 1994; Todman & Dick, 1993). In the late 1990s, these differences between males and females were as ubiquitous as they were in the 1980s, with females from elementary school grades to university graduates expressing greater anxiety and negative attitudes (Brosnan, 1998; Whitley, 1997).

Now, in the first decade of the 21st century, there are some promising signs that the gender gap in computer technology may be weakening. The U.S. Census showed marked increase in computer use by women, especially in the use of the Internet and e-mail and in the workplace. Nonetheless, despite the increased use, women continue to lag behind men in feelings of competence with the computer. They also continue to suffer greater anxiety about using information technology and have fewer positive attitudes about working and playing with the computer than do men (Colley & Comber, 2003; Schumacher & Morahan-Martin, 2001). Surveying school-age children and comparing their responses to those collected more than a decade ago, Colley and Comber (2003) found that girls' interest in computer applications improved, but that girls continue to like the computer less than boys do. When given a chance to use computers in the voluntary world outside of school, girls use the computers less frequently than do boys. Similarly, Mucherah (2003) recently reported that teenage girls feel far less involved with computers and enjoy them less than boys of comparable ages. At Princeton University, researchers asked incoming college students about their reactions to computers (Cooper & Weaver, 2003). Despite having a highly capable and academically accomplished sample, they found that the young women were far less confident of their ability with computers than were the young men. The incoming female undergraduates reported feeling significantly less comfortable with computers than the men did, even though most of them had taken computer classes in their high schools and more than 80% of them had taken higher-level mathematics, including calculus. That any differences were seen in such circumstances is very discouraging, and the effects were not small.

Those same researchers also asked incoming students to imagine that they were going to take a course in psychological statistics. They presented the following question: Suppose that you were asked to complete a statistics homework assignment on the computer. How comfortable would you feel in doing that assignment? These highly capable students again differed based on gender. Men felt that they would be comfortable completing the assignment while women felt uncomfortable (Cooper & Weaver, 2003). Therefore, the lack of confidence just noted is not merely an abstract concept. Even in the context of a specific example, women were just not as sure of their abilities as men were. We can easily imagine that the difference might have been even greater had 4 out of 5 of the women not already completed courses in calculus.

The digital divide is a worldwide problem. Much of the previous research was conducted in the United States. Other studies from Western Europe and other highly developed countries show similar effects. Data have been gathered in Great Britain (Colley, Gale, & Harris, 1994), Australia (Okebukola & Woda, 1993), Canada (Temple & Lips, 1989), and Spain (Farina, Arce, Sobral, & Carames 1991), always with the same result. In a review of this literature for the International Association for the Evaluation of Educational Achievement, Reinen and Plomp (1997) concluded that, "concern about gender equity is right. . . . Females know less about information technology, enjoy using the computer less than male students and perceive more problems with . . . activities carried out with computers in schools" (p. 65).

As interest in this issue has intensified, additional international data have led to the same conclusion. Recent data reported from Romania (Dundell & Haag, 2002), Egypt (Abdelhamid, 2002) and Italy (Favio & Antonietti, 2002), for example, continue to show the persistence of the digital divide in a wide array of educational systems around the globe. Although there are some exceptions, (i.e., Solvberg, 2002, in Norway) gender differences have been remarkably durable.

What the Digital Divide Is Not

The gender gap is less about total hours using a computer than about using a computer voluntarily for enjoyment and comfort with information technology. In schools, it is not about total number hours spent in front of the computer screen, but rather the interference of computer anxiety with the ability and excitement to learn. In the workplace, the digital divide is not about the magnitude of use, but rather about women's reactions to the technology with which they interact. It is about their comfort, attitudes, and levels of anxiety. Women use computers at their jobs more than men do. The use of the computer as typewriter and cash register, for example, necessarily requires human–computer interaction (HCI) in the workplace, and with women holding far more service and administrative support jobs than men, their computer use is relatively high. In fact, in 2003, 63% of women used computers for their jobs, whereas only 51% of men did so.

Exposure in the workplace and in the school has not ended the disparity between men and women in terms of their levels of comfort using a computer, attitudes about computers, and willingness to use computers in contexts in which computer use is not required. Especially in educational settings, anxiety with using computers cannot only result in a feeling of discomfort, but also can lead to less-adequate performance with the computer and the material that was supposed to be learned more enjoyably and efficaciously with computer technology.

UNDERSTANDING THE ROOTS OF THE DIGITAL DIVIDE

The digital divide is not caused by lack of use, nor is it due to differences in economic status, social class, or heredity. We also assume that differences based on biological sex play, at most, a negligible role in accounting for the differences. Rather, we see the different reactions to information technology to be rooted in the socialization of boys and girls as they learn to cope with the social constructions that form the norms, rules, and expectations for their gender. As a heuristic guide to understanding the digital divide, we propose a model that described a series of factors whose result is differential attitudes and differential comfort levels with the use of computers in contemporary society.

The model, which we will describe in more detail in the following sections, takes as its starting point the idea that there exist in our social world entrenched stereotypes of the behaviors and attitudes that are appropriate for children and adults of each gender. Boys are supposed to be more eager to play with computers than girls. The most important consequence of this stereotype is that girls will experience more anxiety when playing with, or learning from, computers, thus making it difficult for them to have pleasant and successful computer interactions. This will happen whether or not girls accept the stereotype as valid. Girls who accept the stereotype as valid will be harmed by what is referred to as the "self-fulfilling prophecy" (Merton, 1948; Rosenthal & Jacobson, 1968). Ironically, girls who do not believe that the stereotype is true will nonetheless experience anxiety with computers because of the phenomenon known as "stereotype threat" (Steele & Aronson, 1995). The mere knowledge that the stereotype exists and other members of society believe it sets in motion processes that lead to confirmation of the stereotype. As the model showed, a girl who knows that there is a stereotype predicting poor computer competency on her part will experience more computer anxiety and, in the end, poorer performance and more negative attitudes about computers. This, in turn, will lead to anxiety and a greater chance of failure.

Our model also shows that different attributional patterns for boys and girls contribute to the cycle that perpetuates the digital divide. Because of the different interpretations that boys and girls are taught with regard to success in achievement domains, the stereotype about the relation of gender to computer use may become reinforced and more resistant to change. As Fig. 1.1 suggests, the dilemma is a self-reinforcing cycle in which boys, typically to their advantage, and girls, to their disadvantage, become enveloped in the veil of the gender stereotype for computing.

In the Beginning

Undeniably, gender stereotypes abound. Like most stereotypes, they were created by society over an extended time, and even though they are now undesirable, they are reluctant to be dismantled. Regardless of whether or not they are true, stereotypes have dramatic impact on behavior. For example, in most western societies, we share common societal expectations about the toys boys and girls are supposed to play with. We do not expect to see the war characters in our favorite toy store

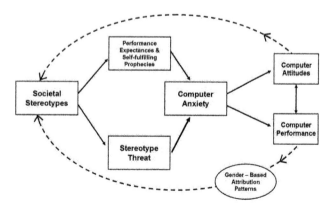

FIGURE 1.1. A logical model of digital divide.

sharing shelf space with dress-up dolls and doll strollers. To the contrary, we expect that the warriors will be near the cars, trains, and space heroes. The dolls will be near the carriages, play houses, and play schools. Boys will find their toys in the former section, girls in the latter. In reality, of course, there is gender overlap such that some boys find their favorite toy in the section with dolls and carriages. And many girls are in toy-heaven when confronted by the cars and trucks. However, in general, there is a strong effect for gender stereotyping of toys, based on and reinforced by adult expectations of what is expected to be interesting and pleasing to boys and girls.

Gender stereotypes abound in the classroom as well. Regardless of whether or not the stereotypes are true, we see mathematics, science, and technology as the province of boys more than girls. Girls write well and are interested in literature and poetry. Computers are the bedrock of information technology and, as we have seen from surveys of children and adults described above, in established democracies and developing nations, we have a similar stereotype about who enjoys and benefits from computers. It is not immediately apparent why gender stereotypes developed for computers become nearly identical to stereotypes about science and mathematics. Although the algorithms that comprise computer software are complex and mathematically sophisticated, and computers burst into our consciousness in large-scale space and science ventures, most computer users do not interact with computers at that level. A screen, keyboard, and mouse pad form the basis of the interfaces that most people have with computers. Why did the use of computers become associated with gender?

The answer to that question is multidetermined and a full analysis is beyond the scope of the current chapter. However, we can isolate one of the causes of the gender stereotype in the introduction of the computer into the educational system. The classroom is a ubiquitous melting pot and its influence on children's attitudes is profound. When educators first looked to computers to supplement their normal educational methods, they made an understandable, though fundamental, error. They drew their inspiration from the world of the video game and video arcade. The best examples of popular computer games in the 1980s were not the increasingly rich variety currently available, but the far less diverse sampling of the video arcade and the early Nintendo and Sega gaming systems. While these programs drew their contexts from a multiplicity of domains, everything from medieval combat to futuristic space adventures, what most had in common was an emphasis on competitive responding. As games grew more elaborate, story lines were increasingly incorporated to keep children's interest in space adventures, sports, and battles.

Educators have always searched for ways to make learning more efficient and more enjoyable. That computers can give students an interactive experience makes them an obvious and attractive addition to the classroom. Computer-software manufacturers turned out hundreds of programs designed to assist teachers in delivering instruction in every discipline from art to zoology. They most likely contemplated how to design such programs and wondered what children wanted. One thing that was obvious at this time was that childred would rush to finish—or ignore entirely—their homework for a chance to hit the arcade. Video-game designers were posting large profits and the growth of their industry seemed to be limited only by rapidly disappearing technical issues. If children would voluntarily spend hours navigating the story line of *Kings Quest*, it seemed fair to assume that instruction delivered in the same game-like format would be popular.

What went unnoticed was predominantly boys visited arcades and spent hours playing their favorite video games. Turning classrooms into video arcades by adopting software that resembles video games might have been attractive to most young boys, but it presented a problem for girls. Every benefit that was gained among video-game players by making learning software the image of video games was a deficit to those who did not relate to or enjoy such games.

One of the earliest analyses of the issue of computer technology in the classroom was that of Lepper and Malone (1987). They asked girls and boys what they liked about computers. The responses of boys matched intuitions. Boys liked activities that were in the form of games. Story-lines featuring sports, war, and space were popular. Boys liked eye-hand coordination and competition. They preferred their feedback in the form of flashing lights, blaring sounds, and explosions. Girls were different. They disliked competitive programs, the videogame emphasis on reflexes, and the dramatic end games. What they wanted was better learning tools. Frequent and clear feedback, preferably not in the form of explosions, was also favored. Girls appreciated computers as learning tools, but found the game focus very disheartening. With computer-assisted instruction (CAI) programs like *Word Invasion, Demolition Division,* and *Slam Dunk Math,* it is clear that boys were more precisely targeted than girls. Chappel (1996) reported that in the real worlds of education and business, programs overwhelmingly favor male identification and male interests. Yet such programs have been one of the major ways in which children are introduced to computers. The grand attempt to "make learning fun" has been premised on an unfortunate, gender-biased definition of the word "fun" that is not widely applicable.

A Designer's Bias

Why is there such a strange disconnect between program designers and female users? (See Lynn, Raphael, Olefsky, & Bachen, 2003 for a discussion.) There are many possibilities. Men are more interested in computer programming than women, and one could say that male programmers are more likely to produce male-focused software, but that is both uncharitable and hard to test. Of far greater interest is the possibility that software designers implicitly assume that their users will be male and tailor their work accordingly. In everyday communication, we often adapt our tone to our audience. If we think our audience is hostile, we are likely to act in a more hostile manner ourselves (Snyder & Swann, 1978). Similarly, if a male college student believes that the female he is speaking to is attractive, he is likely to act in a friendly and inviting manner (Snyder, Tanke, & Berscheid, 1977). In HCI, a software designer is interacting with a communication partner, albeit a strange one. In the mind's eye of the programmer is an eventual user. Someone is sitting at the front end of the computer screen, answering questions, and interacting with the program. But who is

that user? It is possible that the software designer uses stereotypical beliefs to portray the most likely consumer for whom he or she is preparing an educational or entertainment product. In this view, computer programmers write software as though they are communicating with boys because their automatic representation of the gender of the user is male.

Huff and Cooper (1987) investigated the possibility that the gender of the typical user influences the communication process and, accordingly, the characteristics of the software that is produced. They asked teachers in the New Jersey public schools to design software to help seventh-grade children learn the appropriate use of commas. The teachers were given one of three different instructions. One group of teachers was asked to design the software for seventh-grade boys, another for seventh-grade girls, and the third, most interestingly of all, were just told to tailor their work for seventh-grade "students." These teachers were all well versed in the likes and dislikes of seventh graders. Their designs were always fascinating and the two gender-specific conditions produced the kind of program concepts that Lepper and Malone (1987) would have expected. In describing her boy-directed idea, one teacher wrote:

Here is an opportunity to enjoy the world, do sports, and learn English grammar at the same time. Your child will enjoy shooting cannons and competing for the highest score. After playing with this program, your child will use commas in a natural and correct manner.

The group writing for girls had no difficulty in guessing the right features to include. A typical response was expressed by one teacher when she described her program as, "Two girls go on a shopping trip to a record shop to find music for a dance being given at school. They converse with each other and make decisions about what to buy. The use of commas and rules involved are taught through this trip. Reinforcement is available in worksheet form." The activity is social, not military, and lacks the boyish embellishments that girls find anxiety inducing.

The teachers recognized that they needed to write vastly different programs to motivate students of different genders. While their results appear to rely heavily on gender stereotypes, they do recognize the problem and produce programs that are appropriate to the point of caricature. Based solely on these conditions, one would expect that teachers would be able to find a happy medium when working for "students."

The "student" condition, however, is precisely where the problem arises. Teachers designing without a specified gender did *not* assume that half their users would be male and half would be female. Their answers were resoundingly like the programs that had been written for boys and nothing like the programs that had been written when girls were the focus of attention. A typical description in this condition began, "Here's a fast-paced program for your arcade-game lovers. Just what the teenager spends his quarters on! . . . . Sentences zip across the screen—some correctly punctuated with commas, some not. Correct sentences are "zapped" off the screen by your students as they try to be on the roster of top scorers."

Programs that most teachers wrote for "students" were nearly identical to programs that other teachers wrote for boys. All of the programs were coded and assessed by independent raters and then subjected to a multidimensional scaling analysis.

The results showed that the programs written for students were statistically indistinguishable from the programs written for boys on a dimension that ranged from "learning tool" to "toy." Both were markedly on the "toy" end of the dimension. Programs written when girls were the focus of attention were written as learning tools and were significantly different from both the boy and student programs (Fig. 1.2).

Greater Anxiety and Poorer Performance

Even though it was clear that most educational programs were designed with the boy definition of fun in mind, it still remained to be seen what effect these programs had on girls compared to more gender-neutral—though far less common—alternatives. Cooper, Hall, and Huff (1990) examined this issue using *Demolition Division,* a program intended to teach division in a stereotypically boy manner by employing war-related imagery, competition, and eye-hand coordination. Its manufacturer described it as "an opportunity to practice the division of problems [*sic*] in a war game format. Tanks move across the screen as guns from bulkheads are fired by the student as he answers the problem. Hits and misses (correct and incorrect answers) are recorded at the bottom of the screen." The researchers had middle-school boys and girls learn division with either this program or another one, *Arithmetic Classroom,* a CAI program that lacked all of the features of a stereotypically boy-focused game but taught essentially the same information. The students worked with the program for several minutes in a computer cluster in their school. Following the exercise, the children filled out a questionnaire assessing their liking for the CAI-learning program as well as their level of anxiety and stress. When they returned to the classroom, their ability to perform division problems was assessed.

The data showed that girls liked the *Demolition Division* program less often than boys did, and the girls also were considerably more anxious at the conclusion of the CAI lesson. In addition, the level of anxiety was negatively correlated with performance: the more anxious the student, the less she had learned. In *Arithmetic Classroom,* however, the results were quite different. Remem-

ber that this program had no competitive elements, no eye-hand coordination tasks, and no war-story plot line. The anxiety levels of girls using this program were not any higher than those of boys. Girls felt quite comfortable and experienced slightly less anxiety than did the boys. The results are shown in Fig. 1.3. When the computer software had the formal elements that boys enjoy, girls showed the typical pattern of the digital divide: They experienced stress and anxiety and, consequently, did not perform as well. There was no evidence for the gender divide when the program had the formal features that girls enjoy.

Nearly a decade after Cooper et al., another group of researchers found a similar effect in a more elaborate computer game. In this case, the variable of concern was not flashing lights or explosions, but identification. Littleton, Light, Joiner, Messer, and Barnes (1992) worried that many of the storyline CAI games did not include characters that girls could relate to. These researchers targeted a popular CAI game called *King and Crown,* which taught a series of spatial reasoning skills in an adventure format. The characters in the game, however, were primarily warriors and the game was aggressive. Boys learned the skills necessary for the game and fully succeeded in the adventure approximately 50% of the time. Girls were successful only 8% of the time. Littleton et al. found that the male-oriented world of *King and Crown* caused the girls to disidentify, become anxious, and withdraw. Yet, the problem was with the context, not the content. When performance was examined in another program, one that taught the same skills in a gender-neutral context, girls and boys performed equally well. Now, 50% of both genders completed the game. Taken together, the Cooper et al. (1990) and Littleton et al. studies suggested that girls are not innately inferior at learning through computer programs. The problem is in the design of the game. Programs rich in boy-favorable elements are not efficient learning tools for girls; their anxiety level increases while their interest and performance decrease. Also important is that boys do not do worse on the programs that allow girls to do better.

The problem is not one confined to young children in the lower grades. The problem may start there, but it persists into adulthood. Surveys have revealed negative attitudes and higher

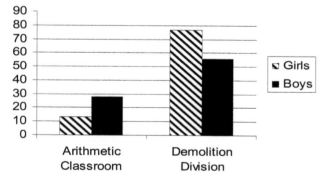

FIGURE 1.2.   Level of computer anxiety after learning from a male oriented (Demolition Division) and control (Arithmetic Classroom) computer assisted learning program. (Source: Huff & Cooper).

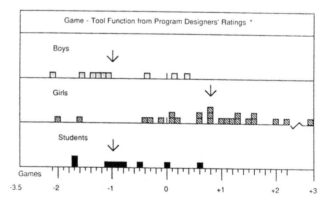

FIGURE 1.3.   Results of multi-dimensional scaling depicting programs written by teachers for instructing boys, girls, and students. The arrows depict the central tendency of each condition. (Source: Huff & Cooper).

computer anxiety among females in college, the workplace, and among retirees (Zhan, 2005). Experimental data collected with college students also support the negative effects of male-favored software on women's levels of computer anxiety. For example, in a study by Robinson-Staveley and Cooper (1990), men and women students at Princeton University played the game of *Zork*, in which players compete to find a buried treasure in an adventure-game format. Women reported a high degree of stress while playing the game and, in turn, performed poorly. Male students, on the other hand, performed considerably better and did not experience computer anxiety.

Another interesting finding in the Robinson-Staveley and Cooper (1990) study is that the social context of computing has a substantial effect on the experience of computer anxiety. In their research, gender differences in performance occurred only in the presence of other people. If the students were asked to solve the *Zork* adventure in complete privacy, the women did well (better than the men) and experienced only a little computer anxiety. Similarly, middle-school girls in the *Demolition Division* study did not experience more computer anxiety than boys if they worked with the CAI program without the presence of others (Cooper et al., 1990).

The social context also matters when the gender composition of learning groups is considered. Girls learning in the presence of boys suffer from increased computer anxiety and learn less. Light, Littleton, Bale, Joiner, and Messer (2000) had boys and girls work with a mildly competitive problem-solving game in which the players' task was to reach a geographical location without being captured by monsters. The children worked in groups of two, either same sex or opposite sex dyads. Light et al. (2000) found that, overall, boys performed better than girls in this game. However, in same-sex dyads, the difference in performance was small. In mixed-sex dyads, the difference was enhanced. Boys' performance was markedly improved relative to their performance in the same-sex group while girls' performance showed significant decrements.

Nicholson, Gelpi, Young, and Sulzby (1998) examined the interactions that occurred when first-grade girls and boys were asked to work together on a computer task. They found that, in mixed-gender groups, girls were likely to have their competence and/or their work criticized or laughed at compared to girls working with other girls. It is unsurprising that taking an already somewhat anxious person and putting her in a position where she may be subject to ridicule or competition worsens performance, but it is still an important consideration. While the centrality of social context in these effects relates more to usage environment than program design, these findings underscore the socially constructed nature of the gender gap. In a more friendly and accepting environment, girls can prosper.

What we have seen through the analysis of patterns in education software is just part of one path through the model we propose. Gender stereotypes have led to the creation of programs that increase the girls' computer anxiety. This anxiety then lowers their computer performance, making them more skeptical about their computer ability and reaffirming the stereotype. However, software and peer environment are not the only means through which society can affect a woman's computer success. What are the effects of an authority figure, a parent, a teacher, a boss, holding such a stereotype?

The Hazards of Low Expectations

Reactions to stereotypes can vary. First, the stereotype can be believed. There is good reason to believe that gender-based stereotypes can have the power of the self-fulfilling prophecy, creating further evidence for the stereotype. The classic self-fulfilling prophecy study was conducted in a classroom setting by Rosenthal and Jacobson (1968). These researchers convinced teachers that a new test of intelligence could predict which of their students were likely to experience sudden improvements in their academic ability during the next school year. Some students were identified as likely to show these "spurts," but for the majority, nothing was said. When they tested the students at the end of the year, they found that students that the teachers had expected to show considerable improvement did perform much better on standardized tests than students not labeled that way, especially among younger students. The interesting part of this study was that the test Rosenthal and Jacobson administered was entirely bogus and the results were intentionally randomized. The teachers responded differently to students whom they expected to perform well. While no data shows exactly the nature of this difference, imagination provides a host of possible explanations; the teachers could have been more encouraging, more attentive, and more supportive of those they "knew" could do well. This study has long been seen as a warning to teachers to be careful in labeling their students either positively or negatively. Students are attuned to the expectations of their superiors and can be persuaded to either try harder or give up based on what is expected.

Word, Zanna, and Cooper (1974) showed that stereotypes about groups of people also impact peoples' performance. Their study made use of implicit racial stereotypes about African Americans. In the context of a job interview, Word et al. showed that the negative racial stereotypes that White job interviewers held about the traits and capabilities of Black candidates subtly and nonconsciously affected the way in which they behaved toward White and Black applicants. Although the interviewers did not consciously realize it, they behaved in subtly more negative ways toward the Black applicants than they did toward the White applicants. They spoke more quickly to African Americans, avoided eye contact, and sat at greater distances. In the end, it was clear that candidates who were treated in these ways performed objectively worse in their job interviews (Word et al., 1974).

These studies, as well as many others, should serve as a caution to parents and teachers. Believing that girls dislike computers or are not competent with them can directly lead to girls being treated differently when interacting with computers. Adult educators and parents who introduce the computer tasks to children may have negatively stereotyped beliefs (like the interviewers in the Word et al. study) and communicate these beliefs in subtle ways to their children. Self-fulfilling prophecies have the added danger of promoting resiliency in the stereotype because those parents and teachers who hold stereotypical beliefs will often see them confirmed. What they do not realize is that their lessons, their examples, and their communications may have contributed to creating the very disparity they believe they are observing (Schofield, 1995).

The self-fulfilling prophecy literature also provides examples of how expectations about *oneself* can affect one's future performance. Zanna, Sheras, Cooper, and Shaw (1975) modified Rosenthal and Jacobson's procedure by telling not only the teachers but also the students of the results of a bogus ability test. As was the case when the teachers believed certain students would excel, students who had been told that they should improve dramatically during the year drastically outperformed their peers in reading and math at year's end. In the context societal messages, a girl who comes to believe the stereotype about her gender can be expected to give up more easily and not become as competent as her peers. This link between computer attitudes and ability makes computer anxiety part of a vicious and mutually reinforcing cycle.

Boys, in believing that technology is in their domain, are likely to benefit from the self-fulfilling prophecy. Their parents, teachers, and other socializing agents act in ways that produce positive feelings about computers, and the boys may respond positively in the way they approach, think about, and perform with computers. This is also a reinforcing cycle. When trying to dismantle the negative cycle that holds back girls, one cannot neglect the positive cycle that enables boys. The message should be that technology is everyone's domain.

Mere Awareness: Stereotype Threat

People can also believe that the stereotype exists but disbelieve in the truth of the stereotype. For example, Devine's (1989) work on racial stereotypes shows that, regardless of whether White participants were prejudiced or not, they could quickly and automatically produce the list of traits that form Whites' stereotypes of Blacks. Even the most self-confident female computer scientist cannot help but be aware that most of the population considers her an anomaly. Unfortunately, even the mere knowledge of a stereotype can have harmful effects. This is another path between the gender stereotype and computer anxiety. It is called "stereotype threat."

Research on stereotype threat has shown that the mere knowledge of a negative stereotype applying to one's group can cause one to perform poorly at a particular task (Spencer, Steele, & Quinn, 1999; Steele, 1997; Steele & Aronson, 1995). Therefore, how does a girl in an introductory programming course get harmed by a stereotype she doesn't believe? She feels that girls are just as good at computers as boys. Or maybe she thinks that, even if the stereotype is true about girls in general, it is not true about her. She likes to use computers and generally does very well with them. Where is the weakness?

She could have two worries that would cause her to feel anxious. First, she may worry that others will judge her based on the stereotype. Even though she is bright and accomplished, she may worry that others still view her as a stereotypical woman. She may still have her work questioned and belittled just because of her sex (Nicholson et al., 1998; Schofield, 1995). It is a worry common to groups subject to discrimination. Such worries can distract her from the task at hand. With cognitive resources devoted to worrying about whether she is being judged according to the gender stereotype, the student may perform less well at the task.

The other worry is linked to the first. She knows there is a stereotype and she wants to disprove it. She doesn't just want to be good; she wants to overcome the deficit that people might believe exists. Task importance has been shown to be a key variable in stereotype threat research. The more something matters, the more it hurts to be reminded of the stereotype.

As in much literature about stereotypes, work on stereotype threat began with studies involving African Americans. Given the stereotype that African-American students are not as academically capable as White students, Steele and Aronson (1995) predicted that making this stereotype relevant would decrease the performance of Black subjects taking a standardized test. Initially, they made the test relevant by saying that the test was a very reliable measure of academic ability. And students who thought the test was relevant performed worse than those who did not. The researchers went on to show that making the stereotype more salient also worked even if they did not tell the test-takers that the exam should accurately reflect their intelligence. Think about the two worries in the context of these experiments. Steele and Aronson's (1995) subjects were so concerned about confirming the stereotype that they did precisely that.

Spencer, Steele, and Quinn (1999) showed a similar effect using gender differences in mathematics. When women were told that the results of a test showed gender differences, they performed more poorly than when gender differences were not mentioned. Interestingly, the experimenters did not actually say that men outperformed women in the "gender-differences" conditions, but all subjects came to the appropriate conclusion.

Other research has shown that stereotype threat can affect a wide array of activities. Stone and colleagues have shown that Whites are likely to experience stereotype threat on activities that are alleged to be indicants of "natural athletic ability" when compared to Blacks (Stone, Lynch, Sjomeling, & Darley, 1999; Stone, Perry, & Darley, 1997). To revisit the math finding, Aronson and his colleagues showed that stereotype threat affected White males' performance on a mathematics test when they thought that the test was diagnostic of their performance relative to that of Asian-American males (Aronson, Lustina, Good, Keough, Steele, & Brown, 1999). Stereotype threat is a pervasive phenomenon that works by increasing anxiety and cognitive load.

The stereotype that links gender to computer performance is as well known as the others. Although there are currently no published studies to link the stereotype threat experienced by women to performance on the computer, the link between the two is clear to see, especially when a student identifies with her gender and gender is a salient aspect of the social situation in which the student finds herself, then the same threat that dealt a blow to women's performance on mathematics tests. White males' performance on an athletic task should occur to women using the computer. Simply through knowledge of the existence of the stereotype, the woman is more likely than the man to succumb to that stereotype, demonstrating greater anxiety and poorer performance.

Compounding the Problem:
Gender Differences in Attribution

One of the more ubiquitous assumptions in the behavioral sciences is that people strive to understand the causes of their behavior in the social world. This axiom has given rise to a field of study known as "attribution" (Heider, 1958; Jones & Davis, 1965;

Kelley 1967; Weiner, 1979). In the wake of witnessing behavior, people are motivated to make attributions for the causes of the act, whether it is their own act or the behavior of another. If an athlete catches a football on the gridiron, was it because he was a good athlete (an internal attribution) or because it was an easy pass to catch (an external attribution) or simply because he was lucky, opening his arms at the right moment with the ball floating into his hands at just the right moment? Similarly, if I answer a question correctly on a standardized test, I wonder if it was because I am smart, because the test question was easy, or because I luckily picked the correct answer from a set of equally obscure alternatives.

In achievement domains that are stereotypically male, a pattern of attributions occurs that is protective for boys but damaging for girls. Boys come to feel that any success they achieve at a stereotypically male task is a function of their ability, whereas any failure is due to lack of trying, bad luck, or an unduly difficult task. This is a protective pattern because success serves to bolster boys' opinions of how good they are. It enables them to complete even more difficult tasks with a strong belief in their own ability to succeed. If they don't succeed, they can rely on the notion that they only need to try harder, pay more attention, or be more judicious in their choice of tasks. Lack of success, in short, does not translate into a belief in lack of ability.

Girls, on the other hand, make a very different pattern of attribution. Success is attributed more so because of external factors such as luck, effort, or an unduly easy task, whereas failure is taken personally as confirmation of a lack of ability. This potentially damaging attributional pattern causes girls to believe that the primary route to success in a stereotypically male domain like math, science, or computer technology is through luck or effort. By continuously working hard, a girl may feel that she can achieve success, but it is not because of her ability or intelligence at the task. Failure, by contrast, can be devastating because it can provide evidence for what she already believed: that she is not a capable performer in the world that has, according to stereotypes, been the province of boys (Diener & Dweck, 1978; Licht & Dweck, 1984; Nicholls, 1975).

Parsons and colleagues examined attributional differences in the field of mathematics (Parsons, Meece, Adler, & Kaczula, 1982). Children were asked to rate their own ability at mathematics, as well the difficulty level of the courses they were taking. The children's parents were also asked to rate their children's math ability and course difficulty. Parents of girls believed that their children had less mathematical ability than did parents of boys. Moreover, the girls' view of their own ability was related to their parents' view, but not to their actual performance in math classes. Objective school records showed neither overall difference in the children's performance nor any difference in the difficulty of the courses. Nonetheless, parents unintentionally socialized their children into thinking that the girls were less gifted in math than their male counterparts.

More recently, Tiedermann (2000) surveyed several hundred elementary-school students and their parents concerning the children's ability in mathematics. As in the Parsons et al. (1982) study, there were no objective differences between boys and girls as measured by their school records. Tiedermann found that both mothers and fathers thought that boys were more skilled in mathematics than girls. And the more strongly parents believed in the gender stereotype about math, the more they attributed greater mathematics ability to their sons, but not to their daughters.

The consensus of studies that have been conducted in the attribution tradition shows that boys and girls do make different attributions for success and failure in stereotypically male domains and that the impact on girls' confidence in their own ability is damaging. Nelson and Cooper (1997) adopted these insights and applied them to the field of information technology. Ten-year-old boys and girls were asked to unscramble anagrams on a computer. After a few trials, half of the boys and half of the girls began to see error messages appear on the screen. The error messages increased in frequency and severity until they finally stated that the computer was shutting down and the drive was about to be destroyed. The other half of the children received no such error messages and, in the end, reported that no computer errors had been detected.

Nelson and Cooper (1997) then asked the children to indicate what they thought their ability level was at computer tasks. Following a successful performance (i.e., without error messages), girls thought that their ability was about average compared with other 10-year-old girls. In the failure message condition, however, they thought their ability was significantly inferior to the average 10-year-old. Boys thought their computer ability was higher than the average 10-year-old, and failure messages did not affect their confidence in their ability. In addition, when the boys and girls in the error-message (failure) condition were asked to describe the reason for their failure to complete the task, girls were three times more likely to attribute the failure to their lack of ability than were boys. In the success condition, boys were much more likely than girls to attribute their smooth and errorless performance to their own ability, while girls were more likely to evenly distribute their attributions to good luck, persistent effort, or an easy task.

Working with college students in Germany, Dickhauser, and Stiensmeyer-Pelster (2002) also asked students to make attributions for success and failure at a computer task. Like the young children in the Nelson and Cooper (1997) study, university males were much more likely to attribute failure to a defective computer, whereas females were more likely to attribute failure to their own ability, causing females to feel greater shame about their performance and lowered expectation about future interactions with the computer.

The conclusion of these studies points to the fact that children are taught by their parents about how to make attributions for success and failure. At least in the academic areas that are stereotypically seen as male domains, parents teach their children that success for boys is due to ability, but success for girls is due to more ephemeral factors that do not rely on girls' internal capabilities. In contrast, girls' failure is an indictment of their capability at the task but boys need only work harder or have better luck in order to achieve in the future. These attributions have consequences. Boys are encouraged to keep trying, because they have the basic ability to succeed. Girls, on the other hand, are not stimulated by success because success is not a reflection of their ability. Failure is. Therefore, it is not surprising that, in the last step of the Nelson and Cooper (1997) study, children were asked if they would like to try another task either on the computer or by traditional paper and pencil. Following failure at the anagrams task, girls were much less likely than boys to want to interact with a computer for the future task.

DISMANTLING THE DIVIDE

The digital divide has made it difficult for women to participate fully in the technological age. As we have seen, women now use computers frequently but continue to feel greater anxiety, more negative attitudes, and lower personal efficacy in their interactions with computer technology. The cycle is continuous. Beginning with the shared knowledge, or stereotype, that computers are the province of men and boys, women and girls either succumb to that stereotype or fight against it. Either way, the mere knowledge that the stereotype exists causes girls and women to experience anxiety. They need not believe the stereotype is true, nor do they need to believe that the stereotype applies to them. This computer anxiety often leads to negative attitudes and lowered performance, which is then interpreted via attributional processes to reflect the accuracy of the stereotype. Parents and educators inadvertently teach girls to attribute any success they may have with computer technology to luck or effort, which limits the ameliorative role that positive performance can have on a girl's estimation of her capability to succeed at computer tasks. The attributional patterns that boys develop allow them to benefit from successful performance and shield them from being dissuaded or discouraged by the occasional occurrence of errors and failures.

Software Design Can Limit the Gender Stereotype

How then can the cycle be disrupted? One place to begin is with the stereotype itself. It cannot be willed out of existence overnight. However, its pervasiveness can be disrupted in a number of ways. Educators and software designers were partially insightful when they saw the opportunity to use technology to make learning fun. With the power of digital technology to accommodate children's varied interests and fantasies, learning can be placed in a context that children find meaningful. What the designers did not see clearly was that their collective decision to model educational software on the image of the video arcade set in motion a series of psychological and sociological factors that helped to reify the image of computers as being the province of boys—i.e., the people who are fascinated by the video arcade.

The design of educational software needs to change. This has already begun to happen, with far more of a variety of educational programs available in the current decade than existed in the past two decades. Nonetheless, the pace of educational software that appeals equally to both genders needs to quicken. The software, along with the various peripheral interfaces, is a significant communicator of the computer stereotype. As we have seen, the educational and entertainment software packages available at the end of the 20th century were communication packages directed at boys. They were written as though educators were speaking directly to boys, encouraging them to learn by having fun, and simultaneously leaving girls out of the conversation.

One means to change the communication pattern of computer technology is to change the metaphor that characterizes the fantasy elements of the game. A war metaphor is a strong communication that that the learning technology is for boys; a

sports metaphor is a less potent, but probably similar, communication. Learning a lesson for the purpose of hitting a home run, making a goal, or scoring a touchdown is less likely to pique the interests of most girls; nor would such metaphors serve to weaken the stereotype that educational computing has been designed for boys.

Learning what girls like, and ultimately designing information technology software that appeals to girls, is an empirical question. We need to study girls' interactions with computer technology and design software that addresses their interests and preferences. Lepper and Malone (1987) led the way by showing that girls prefer educational communication to be in the form of a learning tool. If it is to be preferred over other communications, then it must teach what needs to be learned in a direct and efficient way (Lynn et al., 2003). Moreover, the software attracts the interest of girls to the extent that it is interactive, and involves communication and sharing (Light et al., 2000; Littleton et al., 1992).

Other forms of human–computer interaction (HCI) issues are also important. What kinds of peripheral devices should be used? How will the student ask for help? Will the screen be used to keep score? Will there be a score? Will lights flash and objects explode in order to increase attention, curiosity, and interest? These are not trivial considerations. When girls come to feel that the computer is not a tool intended for them, part of their belief may be through the human factors assumptions made about their interactions with the machine. Passig and Levin (1999) worked with kindergarten children to assess what they liked about computer interfaces. They found that boys, compared to the girls, preferred to use navigational buttons to discover how a game should be played. Girls preferred that writing be part of the game and preferred a way to ask for help directly rather than use computer interfaces such as buttons. Passig and Levin (1999) found that the children's satisfaction with the games, and their time-on-task for learning at the computer, were direct functions of whether the human-factor decisions were consistent with their gender preferences.

The creation of software programs that speak to both genders is a first step. A second step is to encourage parents, school boards, and educators to purchase such programs. This is a business and educational issue with obvious payoffs to society as well as the corporate bottom line. Recent, successful commercial experience with such programs as *Barbie Fashion Designer* (Subrahmanyarn & Greenfield, 1998) and *Purple Moon* attests to the financial viability of such enterprises. Whether it is more appropriate to utilize current gender stereotypes in order to be certain that there are computer programs accessible to each gender, or whether it is best to design programs that have elements that both genders enjoy is a complex issue that goes beyond the scope of the current chapter. However, the conscious and deliberate focus that ascertains that programs educators put in schools and that parents bring to the home is a necessary step to weaken the stereotype that forms the crux of the digital divide.

Decisions about software design can also be informed by examining data collected from adults. Men and women across a broad age range have logged into Massive Online Role Playing Games (MORPGs). Men vastly outnumber women in the game (Yee, 2006a), but it is still informative to illustrate the kinds of activities and parameters within the MMORPGs that interest women. Players can choose a character, or avatar, to represent them in the game. In one data set, 48% of the males chose to be represented by a female character, whereas only 23% of the female players chose a male avatar. This suggests the importance of having female protagonists in computer activities. Those activities, such as learning tools for school-aged children, that require an identification between player and a character on the screen, will most likely cause a female player to feel uncomfortable if she is not represented by a female protagonist. Women also are more likely than men to view their avatar as an idealized representation of themselves and to see their in-game behavior as similar to their real-life behavior (Yee, 2007).

Women are also more likely than men to use the MMORPG environment to build supportive social networks. According to Yee (2006b), women form stronger friendships in the game than do men and shun playing for achievement, dominance, and advancement. On a percentage basis, women are more likely than men to join the game with friends and are much more likely to join with a romantic partner. They are also more likely to share contact information. Men play for power, dominance, and points. Women play to enjoy the communication, social contact, and social interaction. Games that are multifaceted such as the MMORPG environment allow for players to find elements, activities, and goals that appeal to both genders. Careful scrutiny of what appeals to women in these multifaceted games can highlight the features that would make games for young and adult females more interesting and more acceptable.

Disrupting the digital divide at the level of software stereotyping is the most important step in reducing its deleterious impact on women. The self-fulfilling nature of the stereotype and the existence of stereotype threat will cease to be issues that support the divide if the stereotypes are diminished. Software designers need to orient their software to appeal to all groups of users. This is particularly true of programs used in education because, for many children, school represents their first exposure to the computer. It is here that they learn whether computer technology is competitive, like the games in video arcades, or is a medium of communication that is equally accessible to girls and boys.

Adjusting the Social Context

The social context of computing can also help to weaken the digital divide. In line with the research of Robinson-Staveley and Cooper (1990) and Cooper et al. (1990), female college students and middle-school students performed better and experienced less anxiety when computing in private than when computing in public. One suggestion from these studies is that girls should be afforded more private space when working with computer technology. Discomfort about competitive interactions, worry about ridicule (Nicholson et al., 1998), and stress about the stereotype held by others (Steele & Aronson, 1995) can exacerbate the anxiety that perpetuates the digital divide. Where possible, structuring the work and school environments to allow private computer interactions may be quite helpful.

Another strategy for diminishing the gender divide in computing is to consider an educational structure in which girls have an opportunity to engage in computer instruction with other girls rather than boys. This can be accomplished by single-sex schools or by classes within a traditional school that are for girls only. Research with young children (Light et al., 2000) showed that girls performed significantly better in same-sex dyads. Jackson and Smith (2000) applied the single-gender concept to 11–13-year-old girls in their math classes. They found evidence for increased performance and lower stress when the girls took their class in the single-gender format.

Changing the Conclusions

Parents and educators are prone to succumb to the same stereotype as the rest of society: If computer technology is an area that boys do well in, then it must be true that girls' success is a lucky break or a matter of sheer determined effort. In our roles as parents and educators, we must be vigilant about the two elements of that trap. First, it is only a stereotype that supports the first premise: Girls are not intrinsically worse at interacting with computers than boys. Second, girls and boys must be allowed to use the outcomes of their behavior to make judgments about how good they are at computer tasks. If a child does well at computer tasks, it should be taken as evidence for his or her ability. However, research has shown that girls are not given the opportunity to benefit from their success. By unwittingly communicating attributional patterns that differ by gender, parents and educators have systematically deprived girls of benefiting from their success and deprived all children of learning from their mistakes. As Fig. 1.1 depicts, the effects of successful and unsuccessful computer performance pass through an attributional filter that undermines the beneficial effects of success and exacerbates the digital divide.

References

Abdelhamid, I. S. (2002). Attitudes toward computer: A study of gender differences and other variables. *Journal of the Social Sciences, 30*, 285–316.

Alliance of Automobile Manufacturers (2006). Today's automobile: A computer on wheels. Retrieved November 1, 2006, from http://www.autoalliance.org/innovation/cpuWheels.php

American Association of University Women Educational Foundation Commission on Technology, Gender and Teacher Education (2000). *Tech-savvy: Educating girls in the new computer age.* Washington, DC: Author.

Aronson, J., Lustina, M. J., Good, C., Keough, K., Steele, C. M., & Brown J. (1999). When White men can't do math: Necessary and sufficient factors in stereotype threat. *Journal of Experimental Social Psychology, 35*, 29–46.

Atwell, P., & Battle, J. (1999). Home computers and school performance. *The Information Society, 15*, 1–10.

Brosnan, M. J. (1998). The impact of psychology gender, gender-related perceptions, significant others, and the introducer of technology upon computer anxiety in students. *Journal of Educational Computing Research, 18*, 63–78.

Chappel, K. K. (1996). Mathematics computer software characteristics with possible gender-specific impact: A context analysis. *Journal of Educational Computer Research, 15*, 25–35.

Colley, A., & Comber, C. (2003). Age and gender differences in computer use and attitudes among secondary school students: What has changed? *Education Research, 45*, 155–165.

Colley, A. N., Gale, M. T., & Harris, T. A. (1994). Effects of gender role identity and experience on computer attitude components. *Journal of Educational Computing Research, 10*, 129–137.

Cooper, J., & Weaver, K. D. (2003). *Gender and computers: Understanding the digital divide*. Mahwah, NJ: Erlbaum.

Cooper, J., Hall, J., & Huff, C. (1990). Situational stress as a consequence of sex-stereotyped software. *Personality and Social Psychology Bulletin, 16*, 419–429.

Dambrot, F. H., Watkins-Malek, M. A., Silling, S. M., Marshall, R. S., & Garver, J. A. (1985). Correlates of sex differences in attitudes toward and involvement with computers. *Journal of Vocational Behavior, 27*, 71–86.

Devine, P. G. (1989). Stereotypes and prejudice: Their automatic and controlled components. *Journal of Personality and Social Psychology, 56*, 5–18.

Dickhauser, O., & Stiensmeyer-Pelster, J. (2002). Learned helplessness in working with computers? Gender differences in computer-related attributions. *Psychologie in Erziehung und Unterricht, 49*, 44–55.

Diener, C. I., & Dweck, C. S. (1978). An analysis of learned helplessness: Continuous changes in performance, strategy, and achievement cognitions following failure. *Journal of Personality and Social Psychology, 36*, 451–462.

Dundell, A., & Haag, Z. (2002). Computer self-efficacy, computer anxiety, attitudes toward the Internet and reported experience with the Internet, by gender, in an East European sample. *Computers in Human Behavior, 18*, 521–535.

Farina, F., Arce, R., Sobral, J., & Carames, R. (1991). Predictors of anxiety towards computers. *Computers in Human Behavior, 7*, 263–267.

Favio, R-A., & Antonietti, A. (2002). How children and adolescents use computers to learn. *Ricerche-di-Psicologia, 25*, 11–21.

Heider, F. (1958). *The psychology of interpersonal relations*, New York: Wiley

Huff, C., & Cooper, J. (1987). Sex bias in educational software: The effect of designers' stereotypes on the software they design. *Journal of Applied Social Psychology, 17*, 519–532.

Jackson, C., & Smith, I.D. (2000). Poles apart? An exploration of single-sex and mixed-sex educational environments in Australia and England. *Educational Studies, 26*, 409–422.

Jones, E. E., & Davis, K. E. (1965). From acts to dispositions: the attribution process in social psychology, in L. Berkowitz (Ed.), *Advances in experimental social psychology* (Volume 2, pp. 219–266). New York: Academic Press.

Kelley, H. H. (1967). Attribution theory in social psychology. In D. Levine (Ed.), *Nebraska Symposium on Motivation* (Volume 15, pp. 192–238). Lincoln: University of Nebraska Press.

Lepper, M. R., & Malone, T. W. (1987). Intrinsic motivation and instructional effectiveness in computer-based education. In R. E. Snow & M. J. Farr (Eds.), *Aptitude, learning and instruction: Cognitive and affective process analysis*. Hillsdale, NJ: Lawrence Erlbaum Associates.

Licht, B. G., & Dweck, C. S. (1984). Determinants of academic achievement: The interaction of children's achievement orientations with skill area. *Developmental Psychology, 20*, 628–636.

Light, P., Littleton, K., Bale, S., Joiner, R., & Messer, D. (2000). Gender and social comparison effects in computer-based problem solving. *Learning and Instruction, 10*, 483–496.

Littleton, K., Light, P., Joiner, R., Messer, D., & Barnes, P. (1992). Pairing and gender effects on children's computer-based learning. *European Journal of Psychology of Education, 7*, 311–324.

Lynn, K. M., Raphael, C., Olefsky, K., & Bachen, C. M. (2003). Bridging the gender gap in computing: An integrative approach to content design for girls. *Journal of Educational Computing Research, 28*, 143–162.

Merton, R. K. (1948). The self-fulfilling prophecy. *The Antioch Review, 8*, 193–210.

Mitra, A., Lenzmeier, S., Steffensmeier, T., Avon, R., Qu, N., & Hazen, M. (2001). Gender and computer use in an academic institution: Report from a longitudinal study. *Journal of Educational Computer Research, 23*, 67–84.

Mucherah, W.-M. (2003). Dimensions of classroom climate in social studies classrooms where technology is available (middle school students). *Dissertation Abstracts International Section A: Humanities and Social Sciences, 60*, 1451.

Nelson, L. J., & Cooper, J. (1997). Gender differences in children's reactions to success and failure with computers. *Computers in Human Behavior, 13*, 247–267.

Nicholls, J. G. (1975). Causal attributions and other achievement-related cognitions: Effects of task outcome, attainment value, and sex. *Journal of Personality and Social Psychology, 31*, 379–389.

Nicholson, J., Gelpi, A., Young, S., & Sulzby, E. (1998). Influences of gender and open-ended software on first graders' collaborative composing activities on computers. *Journal of Computing in Childhood Education, 9*, 3–42.

Okebukola, P. A., & Woda, A. B. (1993). The gender factor in computer anxiety and interest among some Australian high school students. *Educational Research, 35*, 181–189.

Parsons, J. E., Meece, J. L., Adler, T. F., & Kaczala, C. M. (1982). Sex differences in attributions and learned helplessness. *Sex Roles, 8*, 421–432.

Passig, D., & Levin, H. (1999). Gender interest differences with multimedia learning interfaces. *Computers in Human Behavior, 15*, 173–183.

Pinkard, N. (2005). How the perceived masculinity and/or femininity of software applications influences students' software preferences. *Journal of Educational Computing Research, 32*, 57–78.

Reinen, I. J., & Plomp, T. (1997). Information technology and gender equality: A contradiction in terms? *Computers and Education, 28*, 65–78.

Robinson-Staveley, K., & Cooper, J. (1990). Mere presence, gender, and reactions to computers: Studying human-computer interaction in the social context. *Journal of Experimental Social Psychology, 26*, 168–183.

Rosenthal, R., & Jacobson, L. (1968). *Pygmalion in the classroom: Teacher expectation and pupil's intellectual development*. New York: Holt, Rinehart & Winston.

Schofield, J. W. (1995). *Computers and classroom culture*. Cambridge, MA: Cambridge University Press.

Schumacher, P., & Morahan-Martin, J. (2001). Gender, internet and computer attitudes and experiences. *Computers in Human Behavior, 17*, 95–110.

Snyder, M., & Swann, W. B. (1978). Behavioral confirmation in social interaction: From social perception to social reality. *Journal of Experimental Social Psychology, 14*, 148–162.

Snyder, M., Tanke, E. D., & Berscheid, E. (1977). Social perception and interpersonal behavior: On the self-fulfilling nature of social stereotypes. *Journal of Personality and Social Psychology, 35*, 656–666.

Solvberg, A. M. (2002). Gender differences in computer-related control beliefs and home computer use. *Scandinavian Journal of Educational Research, 46*, 409–426.

Spencer, S. J., Steele, C. M., & Quinn, D. M. (1999). Stereotype threat and women's math performance. *Journal of Experimental Social Psychology, 35*, 4–28.

Steele, C. M. (1997). A threat in the air: How stereotypes shape intellectual identity and performance. *American Psychologist, 52,* 613–629.

Steele, C. M., & Aronson, J. (1995). Stereotype threat and the intellectual test performance of African Americans. *Journal of Personality and Social Psychology, 69,* 797–811.

Stone, J., Lynch, C. I., Sjomeling, M., & Darley, J. (1999). Stereotype threat effects on Black and White athletic performance. *Journal of Personality and Social Psychology, 77,* 1213–1227.

Stone, J., Perry, Z. W., & Darley, J. (1997). "White men can't jump": Evidence for the perceptual confirmation of racial stereotypes following a basketball game. *Basic and Applied Social Psychology, 19,* 291–306.

Subrahmanyarn, K., & Greenfield, P. (1998). Computer games for girls: what makes them play. In J. Cassell & H. Jenkins (Eds.), *From Barbie to Mortal Kombat: Gender and computer games* (pp. 46–71). Cambridge, MA: MIT Press.

Temple, L., & Lips, H. M. (1989). Gender differences and similarities in attitudes toward computers. *Computers in Human Behavior, 5,* 215–226.

Tiedermann, J. (2000). Parent's gender stereotypes and teachers' beliefs as predictors of children's concept of their mathematical ability in elementary school. *Journal of Educational Psychology, 92,* 144–151.

Todman, J., & Dick, G. (1993). Primary children and teacher's attitudes to computers. *Computers and Education, 20,* 199–203.

United States Census Bureau. (2005). Computer and Internet Use in the United States: 2003. October 2005. Washington DC: US Census Bureau. Retrieved March 23, 2007, from http://www.census.gov/prod/2005pubs/p23–208.pdf

Weil, M. M., Rosen, L. D., & Sears, D. C. (1987). The computerphobia reduction program: Year 1. Program development and preliminary results. *Behavior Research Methods, Instruments, and Computers, 19,* 180–184.

Weiner, B. (1979). A theory of motivation for some classroom experiences. *Journal of Educational Psychology, 71,* 3–25.

Whitley, B. E., Jr. (1997). Gender differences in computer-related attitudes and behavior: A metaanalysis. *Computers in Human Behavior, 13,* 1–22.

Wilder, G., Mackie, D., & Cooper, J. (1985). Gender and computers: Two surveys of computer-related attitudes. *Sex Roles, 13,* 215–228.

Wilson, K. R., Wallin, J. S., & Reiser, C. (2003). Social stratification and the digital divide. *Social Science Computer Review, 21,* 133–143.

Word, C. O., Zanna, M. P., & Cooper, J. (1974). The nonverbal mediating of self-fulfilling prophecies in interracial interaction. *Journal of Experimental Social Psychology, 10,* 109–120.

Yee, N. (2006a). The demographics, motivations and derived experiences of users of Massively-Multiuser Online Graphical Environments. *Presence: Teleoperators and Virtual Environments, 15,* 309–329.

Yee, N. (2006b). The psychology of MMORPGs: Emotional investment, motivations, relationship formation and problematic usage. In R. Schroeder and A. Axelsson (Eds.) *Avatars at work and play: Collaboration and interaction in shared virtual environments* (pp. 187–207). London: Springer-Verlag.

Yee, N. (2007). Meta-character: Character creation, gender bending and projection. *The Daedalus Project.* http://www.nickyee.com/daedalus/archissue.php

Yelland, N., & Lloyd, M. (2001). Virtual kids of the 21st century: Understanding the children in schools today. *Information Technology in Childhood Education Annual, 13,* 175–192.

Zanna, M. P., Sheras, P. L., Cooper, J., & Shaw, C. (1975). Pygmalion and Galatea: The interactive effect of teacher and student expectancies. *Journal of Experimental Social Psychology, 11,* 279–287.

•2•

INFORMATION TECHNOLOGY
AND OLDER ADULTS

Sara J. Czaja and Chin Chin Lee
University of Miami Miller School of Medicine

INTRODUCTION

The expanding power of computers and the recent growth of information technologies such as the Internet have made it possible for large numbers of people to have direct access to an increasingly wide array of information sources and services. Network usage is exploding and new interfaces, search engines, and features are becoming available at an unprecedented rate. In 2003, about 61.8% of households in the United States owned a personal computer (PC) and approximately 54.7% of these households had access to the Internet (U.S. Department of Commerce, 2005; Fig. 2.1). In essence, use of technology has become an integral component of work, education, communication, and entertainment. Technology is also being increasingly used within the healthcare arena for service delivery, in-home monitoring, interactive communication (e.g., between patient and physician), transfer of health information and peer support. Use of automatic teller machines, interactive telephone-based menu systems, cellular telephones, and VCRs is also quite common. Furthermore, telephones, television, home security systems, and other communication devices are becoming more integrated with computer network resources providing faster and more powerful interactive services. In essence, in order to function independently and successfully interact with the environment, people of all ages need to interact with some form of technology.

Coupled with the technology explosion is the aging of the population. In 2004, persons 65 years or older represented 12.4% of the U.S. population and it is estimated that people in this age group will represent 20.6% of the population by 2050 (Fig. 2.2). In addition, the older population itself is getting older. Currently, about 44.5 million people are over the age of 75 years, and by the year 2050, almost 50 million people will be 75 years or older (National Center for Health Statistics, 2005; Fig. 2.2).

Given that older people represent an increasingly large proportion of the population and will need to be active users of technology, issues surrounding aging and information technologies are of critical importance within the domain of human–

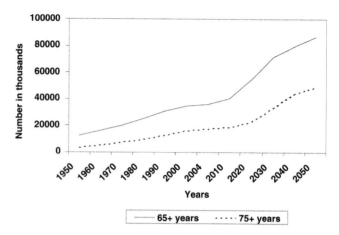

FIGURE 2.2. U.S. population. (Source: National Center for Health Statistics, 2005)

computer interaction (HCI). To ensure that older adults are able to adapt to the new information environment, we need to understand (a) the implications of age-related changes in functional abilities for the design and implementation of technology systems; (b) the needs and preferences of older people with respect to the design of technology interfaces and software applications; and (c) the problems and challenges older adults confront when adopting new technologies. Although this topic has received increased attention within the research community, there are still many unanswered questions.

The intent of this chapter is to summarize the current state of knowledge regarding information technologies and older adults. Topics that will be discussed include adoption and use of information technologies by older adults, training, hardware and software design, and other issues such as privacy and trust in technology. A detailed discussion of the aging process will not be provided. There are many excellent sources of this material (e.g., Fisk, Rogers, Charness, Czaja, & Sharit, 2004; Birren & Schaie, 2001; Fisk & Rogers, 1997). Instead, a brief review of age-related changes in abilities that have relevance to the design of technology systems will be presented. Finally, suggestions for design guidelines and areas of needed research will be summarized. It is hoped that this chapter will serve to motivate researchers and system designers to consider older adults as an important component of the HCI community.

USE OF COMPUTER TECHNOLOGY BY OLDER ADULTS

Usage Patterns and Trends

As noted, there are a number of settings where older people are likely to encounter computers and other forms of communication technologies, such as the Internet, including the workplace, the home, healthcare, and service settings. However, despite a trend toward increased computer use among older people, use of computers and other forms of technology is still

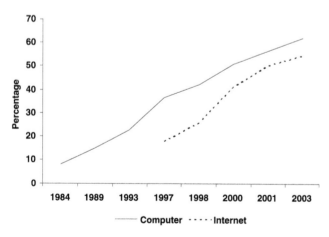

FIGURE 2.1. Trends of computer and Internet access: 1984–2003. (Source: U.S. Department of Commerce, October 2005)

lower among older people as compared to younger people. As shown in Fig. 2.3, more than 32 million older people (55 years and older) have a computer at home as compared to 62 million people between 35 and 54 years. Similarly, although use of the Internet among older people is increasing, it is still lower than that of younger age groups (Fig. 2.4). In 2005, about 26% of people age 65 and older were Internet users as compared to 67% of people age 50 to 64 and 80% of those 30 to 49 years old. Furthermore, people age 65 and older are much less likely than younger people to have a high-speed Internet connection (Pew Internet & American Life Project, 2005). In addition, seniors who do use the Internet tend to be White, highly educated, and living in households with higher incomes (Pew Internet & American Life Project, 2004). Recent data (Czaja et al., in press) also indicated that older adults are less likely than younger adults to use other forms of technology such as automatic teller machines or VCRs and that general use of technology is a predictor of use of computers and use of the Internet.

Lack of technology use puts older adults at a disadvantage in terms of their abilities to live and function independently and to successfully negotiate the built environment. Furthermore, the full benefits of technology may not be realized by older populations. Technology holds great potential for improving the quality of life for older people. For example, computer networks can facilitate linkages between older adults and healthcare providers and communication with family members and friends, especially those who are long distance. It is quite common within the United States for family members to be dispersed among different geographic regions. In fact, nearly 7 million Americans are long-distance caregivers for older relatives (Family Caregiver Alliance, 2005). Clearly, network linkages can make it easier for family members to communicate, especially those who live in different time zones. Computers may also be used to help older people communicate with healthcare providers or other older people and may help older people become involved in continuing education. For example, telemedicine applications allow direct communication between healthcare providers and patients. There are also many opportunities to enroll in distance learning courses online. These opportunities will be more enhanced with future developments in multimedia and video conferencing technologies.

Computers and the Internet can also help older people access information about community services and resources. The Internet may also be used to facilitate the performance of routine tasks such as financial management or shopping. Access to these resources and services may be particularly beneficial for older people who have mobility restrictions or lack of transportation. Finally, many government services, such as Medicare and Social Security, have online information and are moving toward online application processes. Clearly, technology is becoming an integral component of our everyday lives. The following section will present a more detailed discussion of the potential use of computer technology by older adults. This is followed by a discussion of the implications of aging for system design.

Computer and Technology Usage in Everyday Domains

Work Environments

One setting where older people are likely to encounter computer technology is the workplace. The rapid introduction of computers and other forms of automated technology into occupational settings implies that most workers need to interact with computers simply to perform their jobs. Computer-interactive tasks are becoming prevalent within the service sector, office environments, and manufacturing industries. In 2003, 77 million people in the United States used a computer on the job. These workers account for 55% of the labor force and about 2 out of every 5 employed individuals used the Internet on the job (U.S. Bureau of Labor Statistics, 2005). This is in comparison to 45% of workers in 1993 and 25% in 1984 (U.S. Census Bureau, 1999). Currently, more than 2 million adults age 65 and older use a computer at work, and more than 1.3 million use the Internet at work (U.S. Census Bureau, 2003).

The data also suggests that technology will have a major impact on the future structure of the labor force, changing the types of jobs that are available and the way in which jobs are performed. In this regard, most of the growth will come from three occupational groups: computer and mathematical occupations, healthcare practitioners and technical occupations, and education, training, and library occupations. Other occupations that will experience growth include management and financial occupations, sales and related occupations, office and administrative

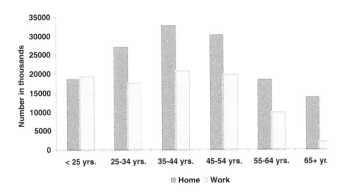

FIGURE 2.3. Percent of U.S. persons using a computer. (Source: U.S. Census Bureau, Current Population Survey, October 2003).

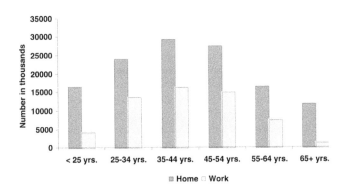

FIGURE 2.4. Percent of U.S. persons using the Internet. (Source: U.S. Census Bureau, Current Population Survey, October 2003)

support operations, and installation, maintenance, and repair occupations especially within the telecommunications industry (U.S. Bureau of Labor Statistics, 2005). Computer and communication technologies will be an integral component of most of these occupations.

The number of people who are telecommuting is also rapidly increasing. Telecommuting encompasses a number of work arrangements including home-based work, satellite office, and neighborhood telework centers. It can be done on a full- or part-time basis. In 2001, about 29 million workers in the United States engaged in some form of telecommuting, and it is estimated that there could be slightly more than 40 million telecommuters by 2010 (Potter, 2003). Telecommuting may be particularly appropriate for older adults, as they are more likely than younger people to be mobility impaired or engaged in some form of caregiving. Telecommuting also allows for more flexible work schedules and autonomy and is more amenable to part-time work. These job characteristics are generally preferred by older people. A recent study by Sharit, Czaja, Hernandez, et al. (2004) found that older people are interested in pursuing this type of work as it provides them with an opportunity to remain productive and engaged in work activities at a more flexible schedule.

However, in today's world telecommuting typically involves the use of computers and the Internet, which may be problematic for older people as they are less likely to have computer skills. The prevalence of telecommuting also raises other interesting issues such as job training. For example, can online training programs be used to train teleworkers, what is the best way to keep these workers updated on changes in job demands, and what strategies can be used to provide teleworkers with technical support? As will be discussed, issues regarding job training are particularly pertinent for older adults.

In summary, to ensure that older adults are able to adapt successfully to new workplace technologies, it is important that older adults be provided with access to retraining programs and incentives to invest in learning new skills and abilities. Greater attention also needs to be given to the design of training and instructional materials for older learners. It is also important to understand how to design technology so that it is useful and usable for older adult populations, especially those with some type of impairment. Adaptive technologies may make continued work more viable for older people, especially those with some type of chronic condition or disability. For example, a number of available technologies can aid people with blindness or low vision, such as portable Braille computers, speech synthesizers, optical character recognition, screen enlargement software, and video magnifiers. Similarly, personal amplifying devices and amplified telephone receivers can aid persons with hearing loss and voice recognition software may be beneficial for persons who have limited ability to use traditional input devices, such as a mouse or keyboard because of hand or finger limitations. Clearly, a number of technologies can improve the ability of older adults to function in work environments; however, the availability of these technologies does not guarantee their success. The degree to which these technologies improve the work life of older persons depends on the usability of these technologies, the availability of these technologies within organizations, the manner in which these technologies are implemented (e.g., training), and the willingness of older people to use these devices.

Home Environments

There are a number of ways older people can use computers at home to enhance their independence and quality of life. Home computers can provide access to information and services and can facilitate the performance of tasks such as banking and grocery shopping. Many older people have problems performing these tasks because of restricted mobility, lack of transportation, inconvenience, and fear of crime (Nair, 1989). For example, about 13% of older people report restrictions in performing activities such as shopping (Federal Interagency Forum on Aging-Related Statistics, 2004). Home computers can also be used to expand educational, recreational, and communication opportunities.

Several studies (e.g., Furlong, 1989; Eilers, 1989; Czaja, Guerrier, Nair, & Landauer, 1993) have shown that older adults are receptive to using e-mail as a form of communication and that e-mail is effective in increasing social interaction among the elderly. Increased social connectivity can also be beneficial for older people, especially those who are isolated or live alone. In 2003, about 31% of all older Americans lived alone, including 41% of older women and 17% of older men (Federal Interagency Forum on Aging-Related Statistics, 2004). A recent study (Cody, Dunn, Hoppin, & Wendt, 1999) found that older adults who learned to use the Internet had more positive attitudes toward aging, higher levels of perceived social support, and higher levels of connectivity with friends and relatives.

The Internet can also be used by older people for continuing education. There are websites and software programs available on a wide variety of topics. In fact, e-learning, which refers to learning delivered or enhanced by electronic technology, is becoming one of the most popular forms of training within industry and in the education industry for the lifelong learner (Willis, 2004). There are also formal online degree programs and opportunities to be linked via videoconferencing and networking facilities to actual classrooms. The American Association for Retired Persons (AARP) has begun to offer several online courses. These learning opportunities can enable older adults to remain intellectually engaged and active, especially those who have difficulty accessing more traditional classroom-based adult education programs. Research (e.g., Baltes & Smith, 1999) clearly shows that cognitive engagement and stimulation is important to successful aging. In fact, lifelong learning is a growing interest among older people. Currently, in the United States, more than 33 million adults age 45 and older are engaged in some form of continuing education (Adler, 2002). The success of SeniorNet also points to the receptivity of older people to use computers for activities such as communication and continuing education. SeniorNet is a nonprofit organization whose mission is to provide people over the age of 50 with access to computer technology. Members learn to use computers for communicating, continuing education, financial management, and other activities, such as desktop publishing. SeniorNet also has online courses on a variety of other topics such as literature and poetry. Currently the organization has over 40,000 members and over 240 Learning Centers throughout the United States (SeniorNet, 2005).

The Internet can also create online learning communities, which bring social interaction to learning and support the learning process. An online community refers to an aggregation of people who have a shared goal, interest, need, or activity and who have repeated interactions and share resources (Preece & Maloney-Krichmar, 2003). The imminent availability of the next generation Internet and interactive multimedia programming will further expand the education experiences that are available to individuals and enable information to be tailored to the specific needs and characteristics of users. This may be particularly beneficial to older adults who often learn at a slower pace than younger people learn and need more instructional support. A recent pilot study (Stoltz-Loike, Morrell, & Loike, 2005) found that e-learning can be an effective tool for teaching older adults technology and business-related skills. The e-learning tool evaluated in the study was customized for older adults. However, as noted by the authors, the results were based on a small sample and the e-learning methods were not compared to other traditional training methods. They also pointed out that many of the e-learning materials available on the market assume a relatively sophisticated knowledge of technology and familiarity with e-learning environments. This may be disadvantageous to older people who have less knowledge of technology and less experience with computers.

In fact, despite the significant market of multimedia user interfaces in training, multimedia design "currently leaves a lot to be desired," and as with other products with which humans interface, it needs to "adopt a usability engineering approach" (Sutcliffe, 2002). Currently, there is little empirical knowledge to guide the development of these applications. In addition, almost no research has been done with older adults. This issue is especially compelling given that multimedia applications place demands on cognitive processes, such as visual search, working memory, and selective attention, which are known to decline with age.

Computers may also be used to augment the memory of older people by providing reminders of appointments, important dates, and medication schedules. Chute and Bliss (1994) showed that computers could be used for cognitive rehabilitation and for the training and retraining of cognitive skills such as memory and attention. Schwartz and Plude (1995) found that compact disc (CD) interactive systems are an effective medium for memory training with the elderly. Computers can also enhance the safety and security of older people living at home. As noted, a large proportion of older women live at home alone. Systems can be programmed to monitor home appliances, electrical and ventilation systems and can be linked to emergency services.

Healthcare

Computer technology also holds the promise of improving healthcare for older people. Electronic links can be established between healthcare professionals and older clients, providing caregivers with easy access to their patients and allowing them to conduct daily status checks or to remind patients of home healthcare regimes. Computers may also be used for healthcare assessment and monitoring. Ellis, Joo, and Gross (1991) demonstrated that older people could successfully use a microcomputer-based health risk assessment. In the future computers may be used to monitor a patient's physical functioning such as measuring blood pressure, pulse rate, temperature, and so forth. The use of computers in healthcare management offers the po-

tential of allowing many people who are at risk for institutionalization to remain at home.

The Internet is also shaping and having a pronounced impact on personal health behavior. Interactive health communication, or e-Health, generally refers to the interaction of an individual with an electronic device or communication technology (such as the Internet) to access or transmit health information or to receive or provide guidance and support on a health-related issue (Robinson, Eng, & Gustafson, 1998). The scope of e-Health applications is broad but mostly encompasses searching for health information, participating in support groups, and consulting with healthcare professionals. Currently, more than 70,000 websites provide health information, and in 2003, 77 million American adults searched the Internet for health information (Pew Internet & American Life Project, 2003). The majority of consumers search for information on a specific disease or medical problem, medical treatments or procedures, medications, alternative treatments or medicine, and information on providers or hospitals. Reasons for the growth of seeking health information online include easier access by a more diverse group of users to more powerful technologies, the development of participative healthcare models, the growth of health information that makes it difficult for any one physician to keep pace, cost containment efforts that reduce physicians, time with patients, and rising concerns about self-care and prevention (Cline & Hayes, 2001).

However, the fact that consumers have access to e-Health applications has significant implications for both patients and providers. On the positive side, access to health information can empower patients to take a more active role in the healthcare process. Patient empowerment can result in more informed decision making, more tailored treatment decisions, stronger patient–provider relationships, increased patient compliance, and better medical outcomes. On the negative side, access to this wide array of health information can overload both patient and physicians, disrupt existing relationships, and lead to poor decision making on the part of consumers. For example, one major concern within the e-Health arena is the lack of quality control mechanisms for health information on the Internet. Currently, consumers can access information from credible scientific and institutional sources (e.g., Medline Plus) and unreviewed sources of unknown quality. Inaccurate health information could result in inappropriate treatment or cause delays in seeking healthcare. In fact, a recent study (El-Attar, Gray, Nair, Ownby, & Czaja, 2005) found that older adults trust health information on the Internet and generally find the Internet to be a valuable source of health information. However, they also indicated that they found health websites somewhat difficult to use. Other concerns relate to the ability of nonspecialists to integrate and interpret the wealth of information that is available and the ability of healthcare providers to keep pace with their patients. Physicians increasingly report that patients come to office visits armed with information on their illness or conditions and treatment options (Ferguson, 1998). Results from an Internet user survey (Pew Internet & American Life Project, 2000) indicated that access to Internet health information has an influence on consumer decisions about seeking care, treatment choices, and their interactions with physicians. Finally, some consumers may find health information difficult to access because of design features that result in usability problems, lack of training, or limited access to technology.

Computer technology may also be beneficial for family caregivers who are providing care for an older person with a chronic illness or disease such as dementia. Generally, the prevalence of chronic conditions or illnesses, such as dementia, diabetes, heart disease, or stroke, increases with age, and consequently, older adults (especially the "oldest old") are more likely to need some form of care or assistance. Approximately 7 million people age 65 years and older have mobility or self-care limitations, and about 4.5 million Americans suffer from Alzheimer's disease (AD). Family members are the primary and preferred source of help for elders. Currently, at least 22 million Americans are providing care for an adult who is ill or disabled (Family Caregiver Alliance, 2005).

Current information technologies offer the potential of providing support and delivering services to caregivers and other family members. Computer networks can link caregivers to each other, healthcare professionals, community services, and educational programs. Information technology can also enhance a caregiver's ability to access health-related information or information regarding community resources. Gallienne, Moore and Brennan (1993) found that access to a computer network, "ComputerLink," increased the amount of psychological support provided by nurses to a group of homebound caregivers of AD patients and enabled caregivers to access a support network that enabled them to share experiences, foster new friendships, and gather information on the symptoms of AD. Technology can also aid caregivers' abilities to manage their own healthcare needs as well as those of their patients by giving them access to information about medical problems, treatments, and prevention strategies. Software is available on several health-related topics such as stress management, caregiving strategies, and nutrition. For example, the Alzheimer's Association has a website that has information for caregivers as does the Family Caregiver Alliance.

The Miami site of the REACH (Resources for Enhancing Alzheimer's Caregiver Health) program evaluated a family-therapy intervention augmented by a computer-telephone system (CTIS) for family caregivers of Alzheimer's patients. The intent of the CTIS system was to enhance the family therapy intervention by facilitating the caregivers' ability to access formal and informal support services. The system enabled the caregivers to communicate with therapists, family, and friends; to participate in online support groups; to send and receive messages; and to access information databases such as the Alzheimer's Association Resource Guide. A respite function was also provided. In addition, the CTIS system provided therapists with enhanced access to both the caregivers and their family members. The experience with the system was very positive with high acceptance of the system by caregivers. The majority of caregivers like the system and find it valuable and easy to use. The most common reason that caregivers used the system was to communicate with other family members, especially those who did not live nearby. The data also indicated that the system facilitated communication with other caregivers. Most caregivers reported that they found the participation in the online support groups to be very valuable (Czaja & Rubert, 2002). Finally, the data showed that caregivers who received the CTIS intervention reported less depressive symptoms, especially those who were Cuban Americans (Eisdorfer et al., 2003).

Clearly, computer technology holds the promise of improving the quality of life for older adults and their families. However, for the full potential of technology to be realized for these populations, the needs and abilities of older adults must be considered in system design. As will be demonstrated in this chapter, older adults generally find technologies such as computers to be valuable and are receptive to using this type of technology. However, available data (e.g., Mead, Spaulding, Sit, Meyer, & Walker, 1997; Czaja, Sharit, Ownby, Roth, & Nair, 2001) also indicated that, although older people are generally willing and able to use computers, they typically have more problems learning to use and operate computer technology than younger adults. They also have less knowledge about potential uses of computers and about how to access computers and other forms of technology. Morrell, Mayhorn, and Bennett (2000) found that the two primary predictors for not using the Internet among people age 60 years and older were lack of access to a computer and lack of knowledge about the Internet. Other barriers to computer and Internet access include cost, lack of technical support, and usability problems. A recent study examining web usability among people age 65 and older found that standard websites are twice as difficult to use for seniors as compared to younger adults (Nielsen Norman Group, 2005). People over the age of 65 also tend to report lower confidence than younger people in their abilities to learn to use computers (American Association of Retired Persons, 2002). Before the full benefits of computer technology can be realized for older people it is important to maximize the usefulness and usability of these technologies for this population. The following section will review characteristics of older adults that have relevance to the design of computer-based systems.

UNDERSTANDING THE OLDER COMPUTER USER

Who Are Today's Elderly?

In general, older Americans today are healthier, more diverse, and more educated than previous generations (Bass, 1995). Between 1970 and 2000, the percentage of adults age 65 and older who had completed high school increased by about 40%, and in 2000, at least 16% of people in this age group had at least a bachelor's degree. As noted, higher levels of education are typically associated with technology adoption, and people who are more educated are more likely to use computers and the Internet.

On some indices, today's older adults are healthier than previous generations. The number of people 65 and older reporting very good health and experiencing good physical functioning, such as ability to walk a mile or climb stairs, has increased in recent years. Disability rates among older people are also declining (Federal Interagency Forum on Aging-Related Statistics,

2004). However, the likelihood of developing a disability increases with age, and many older people have at least one chronic condition such as arthritis or hearing and vision impairments (Fig. 2.5).

Disability among older adults has important implications for system design. People with disabilities, especially disabled elders and minorities with disabilities, are less likely to use technology such as computers both at home and at work (Kaye, 2000; Fig. 2.6). In fact, recent data indicate that Americans of all ages who have a disability have among the lowest rates of Internet access (Pew Internet & American Life Project, 2004).

Consistent with demographic changes in the U.S. population as a whole, the older population is becoming more ethnically diverse. The greatest growth will be seen among Hispanic persons, followed by non-Hispanic Blacks. Currently, individuals from ethnic minority groups are less likely to own or use technologies such as computers. This implies that technology access and training programs need to be targeted for older minority

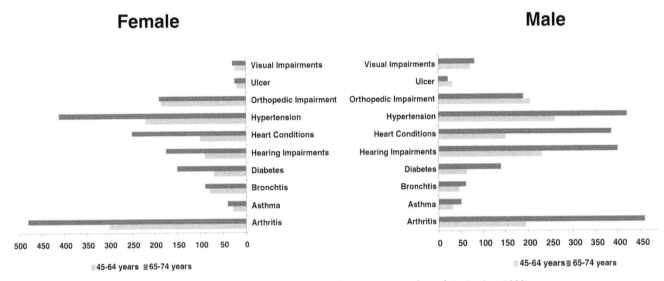

FIGURE 2.5. Chronic illness and older adults. (Source: Clinical Geriatrics, 1999).

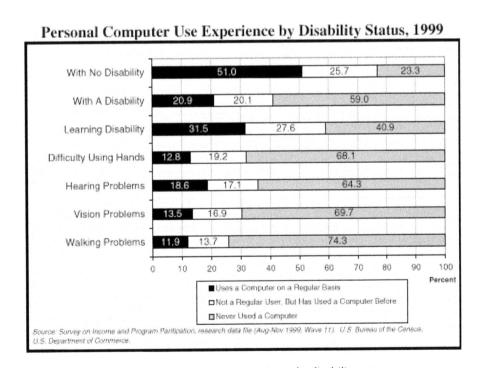

FIGURE 2.6. Computer use experience by disability status.

populations. Finally, there are more elderly women than men, and the proportion of the population that is female increases with age. In sum, health status, race, educational background, cultural traditions, and economic circumstances may all influence the adoption and use of computer-based technologies. Thus, system designers need to understand the heterogeneity of the older adult population and ensure that usability testing is done with truly representative user groups.

Aging and Abilities

Several age-related changes in functional abilities have relevance to the design of computer systems. These include changes in sensory/perceptual processes, motor abilities, response speed, and cognitive processes. In this chapter, a brief review of aging is provided to give a framework for understanding the potential implications of the aging process for system design. It is important to recognize that there are substantial individual differences in rate and degree of functional change. Within any age group, young or old, there is significant variability in range of abilities, and this variability tends to increase with age.

Sensory Processes

A number of changes in visual abilities have relevance to the design of computer systems. Currently, about 14 million people in the United States suffer from some type of visual impairment, and as shown in Fig. 2.5, the incidence of visual impairment increases with age. Generally, with increased age, there is a loss of static and dynamic visual acuity. Older adults also experience a reduction in the range of accommodation, a loss of contrast sensitivity, decreases in dark adaptation, declines in color sensitivity (especially in the blue region), and heightened susceptibility to problems with glare. Visual search skills and the ability to detect targets against a background also decline with age (Kline & Schieber, 1985).

Although most older people will not experience severe visual impairments, they may experience declines in eyesight sufficient to make it more difficult to perceive and comprehend visual information. This has vast implications for the design of computer systems given that computer communications is primarily based on visually presented text. Charness, Schumann, and Boritz (1992) reported that the majority of older participants, in their studies of word processing, experienced some difficulty reading the screen and reported that these difficulties may have contributed to the lower performance of older people. In a more recent study, Charness and Holley (2001) found that older people had more difficulty than younger adults selecting small targets on a computer screen, especially when they were using a mouse as opposed to a light pen. Visual decrements may make it more difficult for older people to perceive small icons on toolbars, read e-mail, or locate information on complex screens or websites. Age-related changes in vision also have implications for the design of written instructions and computer manuals (Fisk et al., 2004).

Aging is also associated with declines in auditory acuity. Many older adults experience some decline in auditory function. Age-associated losses in hearing include a loss of sensitivity for pure tones, especially for high-frequency tones; difficulty understanding speech, especially if the speech is distorted; problems localizing sounds; problems in binaural listening; and increased sensitivity to loudness (Schieber, Fozard, Gordon-Salant, & Weiffenback, 1991). These changes in audition are also relevant to design of computer systems. Older people may also find it difficult to understand synthetic speech as this type of speech is typically characterized by some degree of distortion. Multimedia systems may also be problematic for older adults. High-frequency alerting sounds such as beeps or pings may also be difficult for older adults to detect.

Motor Skills

Aging is also associated with changes in motor skills including slower response times, declines in ability to maintain continuous movements, disruptions in coordination, loss of flexibility, and greater variability in movement (Rogers & Fisk, 2000). The incidence of chronic conditions such as arthritis also increases with age (Fig. 2.5). These changes in motor skills have direct relevance to the ability of older people to use current input devices such as a mouse or keyboard. For example, various aspects of mouse control such as moving, clicking, fine positioning, and dragging are likely to be difficult for older people.

A study by Smith, Sharit, and Czaja (1999) also examined age differences in the performance of basic computer-mouse control techniques (pointing, clicking, double-clicking, and dragging). The data indicated that the older participants had more difficulty performing the tasks than the younger participants, especially the complex tasks such as double-clicking. Furthermore, age-related changes in psychomotor abilities such as manual dexterity were related to performance differences.

Other studies that have examined age differences in mouse performance have found similar results. For example, Riviere and Thakor (1996) found that older adults were less successful in performing tracking tasks with a mouse than were younger adults. Increased task difficulty resulted in greater age differences. Walker, Millians, and Worden (1996) compared older and younger people on a basic target acquisition task and found that older people had more difficulty, especially with smaller and more distant targets. Charness, Bosman, and Elliott (1995) compared people of different ages on a target acquisition task using mice and keyboard control in one study and target acquisition and scroll-bar dragging tasks using mice and light pens in another study. Although the performance of the older people improved with practice, they performed more slowly in all cases, especially with the mouse. Findings from these studies suggest that alternative input devices might be beneficial for older people. In fact, recent data (Charness & Holley, 2001) suggested that for target selection tasks age differences in performance might be minimized with the use of alternative input devices such as a light pen. In this regard, Murata and Iwase (2005) compared target-pointing times among younger, middle-aged, and older adults using a mouse and a touch panel. They found that pointing time was longer for the older age group when using a mouse but that there were no significant age-related differences in pointing time with the touch panel. Based on these results, they recommend that for pointing tasks, a touch panel interface should be used for middle-aged and older adults and provide design guidelines for touch panel interfaces.

Cognitive Abilities

Age-related changes in cognition also have relevance to the performance of computer-based tasks. It is well established that component behaviors comprising cognition decline with age. Processes that decline include attentional processes, working memory, discourse comprehension, problem solving and reasoning, inference formation and interpretation, and encoding and retrieval in memory (Park, 1992). Aging is also associated with declines in information processing speed. Older people tend to take longer to process incoming information and typically require a longer time to respond.

Generally computer tasks are characterized by cognitive demands. For example, component abilities such as psychomotor speed and attention are important predictors of performance of data entry tasks and menu-based tasks (Czaja & Sharit, 1998a; Sharit, Czaja, Nair, & Lee, 2003), and skills such as spatial memory and motor skills are important to word-processing performance. Similarly, searching for information on the Internet is a complex cognitive task and involves cognitive skills such as memory, reasoning, attention, learning, and problem solving. Recent data also indicated that cognitive abilities are also related to technology adoption such that people with higher fluid intelligence are more likely to adopt new technologies such as computers and the Internet (Czaja et al., in press). Given that there are age-related declines in these component abilities, use of current information technologies such as the Internet is likely to be challenging for older adults. Morrell and Echt (1996) posited that age-related declines in cognitive abilities such as working memory, processing speed, and text comprehension are influential in age-related differences in the acquisition of computer skills. They recommended that software, instructional materials, and training protocols be designed to reduce demands on these cognitive mechanisms.

Although the current technologies may present challenges for older adults, the literature generally suggests that older people are receptive to using new technologies. The following section will summarize the literature on aging and attitudes toward computers.

ACCEPTANCE OF INFORMATION TECHNOLOGIES BY OLDER ADULTS

One issue that warrants discussion when considering age and information technology is the degree to which older people are willing to interact with these types of systems and the factors that influence technology adoption. A commonly held belief is that older people are resistant to change and unwilling to interact with high-tech products. However, the available data largely disputes this stereotype. The majority of studies that have examined the attitudes of older people toward computer technology indicate that older people are receptive to using computers. However, older people do report more computer anxiety, less computer self-efficacy, and less comfort using computers than younger adults (Nair, Lee, & Czaja, 2005; Czaja et al., in press). Furthermore, computer anxiety and computer self-efficacy are important predictors of technology adoption (Ellis & Allaire, 1999; Czaja et al., in press). However, the data also indicated that attitudes toward technology and comfort using technology are influenced by experience and the nature of interactions with computer systems as well as system design.

Several studies (e.g., Jay & Willis, 1992; Charness, Schumann & Boritz, 1992; Dyck & Smither, 1994) showed that people who had experience with computers had more positive attitudes and greater computer confidence. Other studies showed that the nature of the experience with technology influences attitudes toward that technology. For example, in a study examining age differences in acquisition of word-processing skills (Czaja, Hammond, Blascovich, & Swede, 1986), the results showed that post training attitudes were related to the training experience such that people who rated the training experience and their own performances positively also had more positive posttraining attitudes toward computers. Furthermore, there were no age effects for either pretraining or posttraining computer attitudes. Danowski and Sacks (1980) also found that a positive experience with computer-mediated communication resulted in positive attitudes toward computers among a sample of older people. A more recent study (Czaja & Sharit, 1998b) examined age differences in computers as a function of computer experience among a community sample of 384 adults ranging in age from 20 to 75 years. The results indicated that, in general, older people perceived less comfort and efficacy using computers than younger people. However, experience with computers resulted in more positive attitudes for all participants irrespective of age. Kalasky, Czaja, Sharit, & Nair (1999) also found that experience with computer systems increased the comfort of older people toward using computers.

Generally, user satisfaction with a system, which is determined by how pleasant the system is to use, is considered an important aspect of usability. User satisfaction is an especially important usability aspect for systems that are used on a discretionary basis such as home computers (Nielsen, 1993). Kelley, Morrell, Park, and Mayhorn (1999) found that the most important predictor of continued use of a bulletin board system, ELDERCOMM, was success at initial training. They also found that the most positive attitudes toward computers were found among persons who used the system most often. Cody et al. (1999), in their study of Internet use among older adults, also found that people who spent the most time online had the least computer anxiety and high computer efficacy. Finally, in our study of e-mail (Czaja, Guerrier, et al., 1993), all participants found it valuable to have a computer in their home. However, the perceived usefulness of the technology was an important factor with respect to use of the technology. When the participants were asked what type of computer applications they would like available, the most common requests included emergency response, continuing education, health information, and banking/shopping.

The available data suggests that older people are not technophobic and are willing to use computers. However, the nature of their experience with technology and available technology applications are important determinants of attitudes, confidence, and comfort using technology. Ultimately confidence and comfort influence technology adoption (Czaja et al., in press).

OLDER ADULTS AND THE ACQUISITION OF COMPUTER SKILLS

A number of studies examined the ability of older adults to learn to use technology such as computers (e.g., Elias, Elias, Robbins, & Gage, 1987; Gist, Rosen, & Schwoerer, 1988; Zandri & Charness, 1989; Czaja, Hammond, Blascovich, et al., 1986; Czaja, Hammond, & Joyce, 1989; Charness et al., 1992; Morrell, Park, Mayhorn, & Echt, 1995; Mead et al., 1997). These studies encompass a variety of computer applications and vary with respect to training strategies such as conceptual versus procedural training (Morrell, Park, et al., 1995) or active versus passive learning approaches (Czaja, Hammond, & Joyce, 1989). The influence of other variables, such as attitude toward computers and computer anxiety, on learning has also been examined.

Overall, the results of these studies indicate that older adults are, in fact, able to use technology such as computers for a variety of tasks. However, they are typically slower to acquire new skills than younger adults and generally require more help and hands-on practice. In addition, when compared to younger adults on performance measures, older adults often achieve lower levels of performance. However, the literature also indicates that training interventions can be successful in terms of improving performance, and it points to the importance of matching training strategies with the characteristics of the learner. For example, Czaja, Hammond, and Joyce (1989) found that older adults benefit from using analogies to familiar concepts and from a more active, hands-on training approach. Similarly, Mead et al. (1997) examined the effects of type of training on efficiency in a World Wide Web search activity. The participants were trained with a hands-on web navigation tutorial or a verbal description of available navigation tools. The hands-on training was found to be superior, especially for older adults. Older adults who received hands-on training increased the use of efficient navigation tools. Mead and Fisk (1998) examined the impact of the type of information presented during training on the initial and retention performance of younger and older adults learning to use ATM technology. Specifically, they compared two types of training: concept and action. The concept training presented factual information whereas the action training was procedural in nature. The action training was found to be superior for older adults. Generally, the literature suggests that these types of strategies are also beneficial for younger people. The literature (e.g., Czaja, Sharit, Ownby, et al., 2001; Charness, Kelley, Bosman, & Mottram, 2001) also suggests that prior experience with technology is an important predictor of ability to learn to use new technology. Finally, it is important to provide older people with training on the potential uses of the technical system (e.g., what the Internet can be used for), as well as training on basic procedural operations (e.g., use of the mouse). As one would expect, the usability of the system from both a hardware and a software perspective is also important (Fisk et al., 2004).

Mayhorn, Stronge, McLaughlin, and Rogers (2004) provided suggestions for development of effective computer training for older adults. They stressed the importance of applying a systems approach to the design of training programs where the goals, abilities, and experience levels of older adults are considered in the design and evaluation of instructional programs and materials. Also, given the important role of anxiety and self-efficacy in technology adoption (Czaja et al., in press), it is important that training environments are relaxed and strategies that reduce anxiety and increase self-efficacy are incorporated into training programs.

AGING AND COMPUTER TASK PERFORMANCE

In terms of actual performance, only a handful of studies have examined the ability of older people to perform computer-based tasks that are common in work settings. For example, Czaja and Sharit (1993, 1998a, 1998b), Sharit and Czaja (1999), and Czaja et al. (1998) conducted a series of studies examining age performance differences on a variety of simulated computer-based tasks (e.g., data entry, inventory management, customer service). Overall, the results of these studies indicate that older adults are willing and able to perform these types of tasks. However, the younger adults generally performed at higher levels than the older people did. Importantly, the data also indicated that there was considerable variability in performance among the older people and that with task experience those in their middle years (40 to 59 years old) performed at roughly the same levels as the young adults. In fact, task experience resulted in performance improvements for people of all ages. The results also indicated that interventions such as redesigning the screen, providing on-screen aids, and reconfiguring the timing of the computer mouse improved the performance of all participants.

Other investigators examined age as a potential factor impacting on ability to use the computer for information search and retrieval. This is an important area of investigation, given that this is one of the most common reasons people use technology such as computers. For example, this type of activity is also central to use of the Internet. Also, in many work settings such as department stores, airlines, hotels, utility and health insurance companies, and educational institutions, workers are required to search through computer databases and access information to respond to customer requests. Generally, the findings from these studies indicated that, while older adults are capable of performing these types of tasks, there are age-related differences in performance. Furthermore, these differences appear to relate to age differences in cognitive abilities.

For example, Westerman, Davies, Glendon, Stammers, and Matthews (1995) examined the relationship between spatial ability, spatial memory, vocabulary skills, and age, and the ability to retrieve information from a computer database which varied according to how the database was structured (e.g., hierarchical vs. linear). In general, they found that the older subjects were slower in retrieving the information than the younger adults; however, there were no age-related differences in accuracy. The learning rates also differed for the two groups such that the older people were slower than the younger people were. They found that the slower response on the part of the older adults was dependent on general processing speed.

Freudenthal (1997) examined the degree to which latencies on an information retrieval task were predicted by movement speed and other cognitive variables in a group of younger and of older adults. The participants were required to search for answers to questions in a hierarchical menu structure. Results indicated that the older subjects were slower than the younger subjects were on overall latencies for information retrieval and that this slowing increased with each consecutive step in the menu. Similar to Westerman et al. (1995), movement speed was a significant predictor of overall latency. Other cognitive abilities, such as reasoning speed, spatial ability, and memory, were also predictive of response latencies. However, memory and spatial abilities predicted only latency on steps further into the menu structure. Freudenthal (1997) suggested that deep menu structures may not be appropriate for older adults as navigation through these types of structures is dependent on spatial skills that tend to decline with age. Vicente, Hayes, and Williges (1987) also found that age, spatial ability, and vocabulary were highly predictive of variance in search latency for a computer-based information retrieval task. They postulated that people with low spatial ability tend to "get lost" in the database. Mead, Sit, Jamieson, Rousseau, and Rogers (1996) examined the ability of younger and older adults to use an online library database. Overall, the younger adults achieved more success than did the older adults in performing the searches. They also used more efficient search strategies. The older adults also made more errors when formulating search queries and had more difficulty recovering from these errors. Czaja, Sharit, Ownby, et al. (2001), in their investigation of this type of task, which simulated a customer service representative for a health insurance company, also found that spatial skills were important predictors of performance as were response speed and prior computer experience. Kubeck, Miller-Albrecht, and Murphy (1999) investigated age differences in finding information using the Internet in a naturalistic setting. They found that older people tended to use less efficient search strategies than younger people and were less likely to find correct answers. However, they also found that, with training, older adults were successful in their searches and had very positive reactions to their Internet experiences.

As discussed, information seeking is a complex process and places demands on cognitive abilities such as working memory, spatial memory, reasoning, and problem solving. Information seeking within electronic environments also requires special skills such as knowledge related to the search system. Given that older adults typically experience declines in cognitive abilities, such as working memory, and that they are less likely than younger people to have knowledge of the structure and organization of search systems, a relevant question is the degree to which they will experience difficulty searching for information in electronic environments. Generally, the available literature suggests that older adults are able to search and retrieve information within electronic environments. However, they appear to have more difficulty than do younger adults and tend to use less efficient navigation strategies. The also appear to have problems remembering where and what they searched. In order to maximize the ability of older people to successful interact with

electronic information systems, such as the Internet, and to have access to the information highway, we need to have an understanding of the source of age-related difficulties. This type of information will allow us to develop interface design and training strategies to accommodate individual differences in performance. Currently, there is very little information on problems experienced by older people when attempting to learn and navigate the Internet, especially in real-world contexts.

DESIGNING COMPUTER SYSTEMS TO ACCOMMODATE OLDER ADULTS

Hardware Considerations

As discussed, age-related changes in functioning have implications for the design of technology systems. For example, careful attention needs to be paid to the design of display screens, placement, size, shape and labeling of controls, and design and layout of instructional materials and manuals. Design features such as character size and contrast are especially important for older computer users. Larger characters and high contrast displays are generally beneficial for older people. This may not be a major problem with most computers used in the home or the workplace, as it is relatively easy to enlarge screen characters. However, it may be an issue for computers in public places such as information kiosks and ATM machines. In addition to character size and contrast, it is also important to minimize the presence of screen glare.

The organization and amount of information on a screen is important, as there are declines in visual search skills and selective attention with age. Only necessary information should be presented on a screen and important information should be highlighted. Further principles of perceptual organization, such as grouping, should be applied. Caution must also be exercised with respect to the use of color coding, as there are declines in color discrimination abilities with age. Finally, as far as possible, information should be in a consistent location and important information should stand out and be in central locations. A flaw with many existing web pages is that advertising information has more prominence than the site information (e.g., pharmaceutical advertisements on health websites).

Design and labeling of controls and input devices need special consideration. In our study of e-mail (Czaja, Guerrier, et al., 1993) people commonly confused the send and the cancel keys. Even though these keys were labeled, they were identical in size, identical in shape, and close in proximity. These findings underscore the importance of clearly differentiating controls so that labeling is easy to read and functions are easily identified.

Although there is a growing body of research examining the relative merits and disadvantages of various input devices, only a few studies have examined age effects. These studies generally suggested that commonly used input devices, such as keyboards and the mouse, might be problematic for older people. More research needs to be directed toward identifying the efficacy of alternative input devices for older people.

For example, Charness and Holley (2001) suggested that older adults may find a light pen may be easy to use because it is a direct addressing device, and it eliminates the need to translate the target selection device onto the display. However, light pens may be difficult to use for people with arthritis or tremors as it may be hard for them to grasp and point the pen. Ellis et al. (1991), in their study of a computerized healthcare system, concluded that a touchscreen version of the appraisal system may have eliminated the interface problems experienced by older users. Casali (1992) evaluated the ability of persons with impaired hand and arm functions (age unknown) and nondisabled persons to perform a target acquisition task with five cursor control devices: a mouse, trackball, cursor keys, joystick, and tablet. She found that even persons with profound disabilities were able to operate each device by using minor modifications and unique operating strategies. The mouse, trackball, and tablet resulted in better performance than the joystick and cursor keys for all participants. Consistent with Charness and Holley (2001, June), Casali found that small targets were problematic for the physically impaired users as was the task of dragging. In a follow-up study (Casali & Chase, 1993) with a sample of persons with arm and hand disabilities, the data indicated that while the mouse, trackball, and tablet tended to result in quicker performance, these devices were more error prone than the keyboard and joystick. In addition, performance improved with practice.

Clearly, much work needs to be done to evaluate which types of input devices are optimal for older people, especially for those who have restrictions in hand function. It appears that input devices, such as mice and trackballs, might not be appropriate for older adults, or they will at least require more practice to effectively use these types of devices. It may be that speech recognition devices will eliminate many of the problems (visual and movement) associated with manual input devices. In this regard, Ogozalek and Praag (1986) found that, although using a simulated listening typewriter as compared to a keyboard editor made no difference in performance of a composition task for both younger and older people, the voice input was strongly preferred by all participants. More recently, Kalasky et al. (1999) found that a commercially available speech-recognition program was robust in accepting the speech input of both younger and older adults. The technology was also acceptable to both user groups, and the data suggested that the older participants found the system more useful than their younger counterparts. These results are encouraging and suggest that speech interfaces may be especially beneficial for older people.

Software Considerations

There has been very little research examining the impact of the design of the software interface on the performance of older computer users. Given that there are age-related changes in cognitive processes, such as working memory and selective attention, it is highly likely that interface style (e.g., function keys vs. menus) will have a significant influence on the performance of older adults. The limited data that are available support this conclusion.

Joyce (1990) evaluated the ability of older people to learn a word-processing program as a function of interface style. The participants interacted with the program using one of three interface styles: on-screen menu, function keys, and pull-down menus. She found that people using the pull-down menus performed better on the word-processing tasks. Specifically, they performed the tasks more quickly and made fewer errors. They also executed a greater number of successful editorial changes. Joyce hypothesized that the pull-down menu reduced the memory demands of the task, as the users did not have to remember editing procedures. Although the on-screen menu provided memory cues, the names of the menu items were not reflective of menu contents, thus requiring the user to remember which items were contained in which menu. The names of the pull-down menus were indicative of menu contents.

Egan and Gomez (1985) found that a display editor was less difficult for older adults that a line editor. The line editor was command based and required the user to remember command language and produce complicated command syntax. Using the display editor, changes were made by positioning a cursor at the location of change and using labeled function keys rather than a command language. Thus, fewer memory demands were associated with the display editor. Further, the display editor was less complex than the line editor, and in accordance with the age-complexity hypothesis, we would anticipate smaller age effects (Cerella, Poon, & Williams, 1980).

In our study of e-mail (Czaja, Hammond, & Joyce, 1989), we found that study participants sometimes had difficulty remembering the address name required to access a particular application. We also had to provide on-screen reminders of basic commands (e.g., "Press enter after entering the address") even though the system was simple and had no complex command procedures. Similarly, in our research on text editing, we found that the older participants had more difficulty remembering editing procedures than younger adults did. Similarly, in our study of interactive telephone menu systems (Sharit, Czaja, Nair, et al., 2003), we found that older adults experienced difficulties using the menu systems and had to use the repeat function more frequently than younger adults. They also reported frustration using the menus.

Charness and Holley (2001) examined the differential effect of a keystroke-based command, menu, and menu + icon interface on word-processing skill acquisition among a sample of younger and older adults. They found that the menu and the menu + icon interface yielded better performance for all participants. The menu conditions provided more environmental support and were associated with fewer memory demands.

Consistent with the cognitive aging literature, these results suggested that systems that place minimal demands on working memory would be suitable for older people. In addition, minimal demands should be placed on spatial abilities. In this regard, interface styles such as windows should prove beneficial. However, windows may be problematic for older adults as windows are spatially organized, and it is sometimes easy to get lost within a windows environment. Also, windows may initially increase cognitive demand as users are required to learn window operations. On-screen aids such as maps and history markers may also prove to be beneficial for older people. We found, for example, that a simple graphical aid that depicted the structure of the menu system helped older adults use interactive telephone menu systems (Sharit, Czaja, Nair, et al., 2003).

Clearly, a number of interface issues need to be investigated. At the present time, we can only offer general guidelines with respect to designing interfaces to accommodate the needs of older users. Table 2.1 presents a summary of these guidelines. An abundance of research needs to be carried out within this area.

Other Important Design Considerations

In addition to general guidelines for the design of training programs, hardware, and software, other issues need to be considered to ensure that the benefits of technology are maximized for older people. As noted, the older adult population is very heterogeneous in terms of culture/ethnicity, health, education, and experience with technology so it is important to take into account the wide range of abilities, needs, and desires of older users to ensure that technology can adapt to their individual differences. For example, it is important to consider different designs (e.g., languages) and approaches (e.g., advertisements) for different cultural/ethnic groups. It is also important to consider how to customize or adjust interfaces and response devices to accommodate those with varying abilities. Privacy issues are also an important concern as the use of technologies becomes more widespread for health and financial and service applications. The issue of trust also needs consideration. Trust and ultimately acceptance of technology may be weakened by unreliability or excessive complexity. There are also potential problems with overtrust, which can be serious if technology fails. Issues of safety and maintainability must also be addressed. Finally, the cost of systems needs to be considered not only in terms of finances but also in terms of effort required on the part of the user with respect to technology access and learning and maintenance requirements.

TABLE 2.1. Summary of Interface Design Guidelines for Older Adults

1. Maximize the contrast between characters and screen background.
2. Minimize screen glare.
3. Avoid small targets and characters that are small (fonts <12).
4. Minimize irrelevant screen information.
5. Present screen information in consistent locations (e.g., error messages).
6. Adhere to principles of perceptual organization (e.g., grouping).
7. Highlight important screen information.
8. Avoid color discriminations among colors of the same hue or in the blue-green range.
9. Clearly label keys.
10. Maximize size of icons.
11. Use icons that are easily discriminated, meaningful and label icons if possible.
12. Provide sufficient practice on the use of input devices such as a mouse or trackball.
13. Provide sufficient practice on window operations.
14. Minimize demands on spatial memory.
15. Provide information on screen location.
16. Minimize demands on working memory.
17. Avoid complex command languages.
18. Use operating procedures that are consistent within and across applications.
19. Provide easy to use online aid and support documentation.

CONCLUSIONS

There are many areas where older people are likely to interact with computer technology including the workplace, the home, service and healthcare settings. Current data indicates that older adults are generally receptive to using this type of technology but often have more difficulty than younger people acquiring computer skills and using current computer systems. They also report more anxiety about using technologies such as computers and express less confidence in their ability to learn to interact with these systems than do younger adults. This presents a challenge for the HCI community.

For many older people, especially those who are frail, isolated, or have some type of mobility restrictions, access to technologies such as computers and the Internet holds promise of enhancing independence and quality of life for older people by providing linkages to goods and resources, facilitating communication, and enhancing opportunities for work and lifelong learning. Technology can also enhance the delivery of health services to older adults and help them access information on health-related topics. However, before the potential of technology is realized for older adults the needs, preferences, skills, and abilities of older people need to be understood by system designers.

Although research in this area has grown, there are many unanswered questions. For example, there are still many issues regarding design of input devices and interface design, such as how to best design speech recognition systems and menus help systems to accommodate older people. We also know little about the efficacy of design aids and support tools for older adults. In addition, we need more information on how to best train older adults to learn to use new technologies, and there are many questions regarding the design of online training programs and multimedia formats. Issues regarding privacy and trust in technology also represent critical areas of needed research. There are also many questions related to the Internet, such as how does access to Internet information impact on healthcare behavior and how do we best train seniors to identify and integrate the enormous amount of information that is available on the Internet that remain unanswered. We also need to examine how technology in the workplace impacts on employment opportunities and the work performance of older people. The issue of telecommuting has received little attention. In addition, we need more information on factors influencing technology adoption especially for minority elderly or those of lower education or economic status. There are also many questions related to quality of life and socialization that need to be addressed.

Many needs of older people would be amenable to technological solutions. However, what is lacking is a systematic effort to understand these needs and to incorporate them into design solutions and the marketplace (National Research Council, 2004). In essence, in order to design information systems so that they are useful and usable for older people, it is important to understand (a) why technology is difficult to use, when it is; (b) how to design technology for easier and effective use; and (c) how to effectively teach people to use and take advantage of technologies that are available. Answers to these questions will not only serve to benefit older adults but all potential users of technology systems.

References

Adler, R. (2002). *The age wave meets the technology wave: Broadband and older Americans.* Retrieved December 9, 2005, from http://www.seniornet.org/downloads/broadband.pdf

American Association of Retired Persons. (2002). *Staying ahead of the curve: The AARP work and career study.* Washington, DC: Author.

Baltes, P. B., & Smith, J. (1999). Multilevel and systemic analyses of old age: theoretical and empirical evidence for a fourth age. In V. L. Bengtson & K. W. Schaie (Eds.), *Handbook of theories of aging* (pp. 153–173). New York: Springer.

Bass, S. A. (1995). *Older and active: How Americans over 55 are contributing to society.* New Haven, CT: Yale University Press.

Birren, J. E., & Schaie, K. W. (2001). *Handbook of the psychology and aging.* San Diego, CA: Academic Press.

Casali, S. P. (1992). Cursor control use by persons with physical disabilities: Implications for hardware and software design. *Proceedings of the 36th Annual Meeting of the Human Factors Society,* Atlanta, GA, 311–315.

Casali, S. P., & Chase, J. (1993). The effects of physical attributes of computer interface design on novice and experienced performance of users with physical disabilities. *Proceedings of the 37th Annual Meeting of Human Factors and Ergonomics Society,* Seattle, WA, 849–853.

Cerella, J., Poon, L. W., & Williams, D. (1980). Age and the complexity hypothesis. In L. W. Poon (Ed.), *Aging in the 1980s* (pp. 332–340). Washington, DC: American Psychological Association.

Charness, N., Bosman, E. A., & Elliot, R. G. (1995). *Senior-friendly input devices: Is the pen mightier than the mouse?* Paper presented at the 103 Annual Convention of the American Psychological Association Meeting, New York, NY.

Charness, N., & Holley, P. (2001, June). *Minimizing computer performance deficits via input devices and training.* Presentation prepared for the Workshop on Aging and Disabilities in the Information Age, John Hopkins University, Baltimore, MD.

Charness, N., Kelley, C. L., Bosman, E. A., & Mottram, M. (2001). Word processing training and retraining: Effects of adult age, experience, and interface. *Psychology and Aging, 16,* 110–127.

Charness, N., Schumann, C. E., & Boritz, G. A. (1992). Training older adults in word processing: Effects of age, training technique and computer anxiety. *International Journal of Aging and Technology, 5,* 79–106.

Chute, D. L., & Bliss, M. E. (1994). ProsthesisWare: Concepts and caveats for microcomputer-based aids to everyday living. *Experimental Aging Research, 20,* 229–238.

Cline, R. J. W., & Hayes, K. H. (2001). Consumer health information seeking on the Internet: The state of the art. *Health Education Research, 16,* 671–692.

Clinical Geriatrics. (1999). Trend watch: Chronic illness and the aging U.S. population. *Clinical Geriatrics, 7*(8), 77.

Cody, M. J., Dunn, D., Hoppin, S., & Wendt, P. (1999). Silver surfers: Training and evaluating Internet use among older adult learners. *Communication Education, 48,* 269–286.

Czaja, S. J., Charness, N., Fisk, A. D., Hertzog, C., Nair, S. N., Rogers, W., et al. (2006). Factors predicting the use of technology: Findings from the Center for Research and Education on Aging and Technology Enhancement (CREATE). *Psychology and Aging.*

Czaja, S. J., & Rubert, M. (2002). Telecommunications technology as an aid to family caregivers of persons with dementia. *Psychosomatic Medicine, 64,* 469–476.

Czaja, S. J., Sharit, J., Ownby, D., Roth, D., & Nair, S. N. (2001). Examining age differences in performance of a complex information search and retrieval task. *Psychology and Aging, 16,* 564–579.

Czaja, S. J., & Sharit, J. (1998a). Ability-performance relationships as a function of age and task experience for a data entry task. *Journal of Experimental Psychology: Applied, 4,* 332–351.

Czaja, S. J., & Sharit, J. (1998b). Age differences in attitudes towards computers: The influence of task characteristics. *The Journals of Gerontology: Psychological Sciences and Social Sciences, 53B,* 329–340.

Czaja, S. J., Sharit, J., Nair, S., & Rubert, M. (1998). Understanding sources of user variability in computer-based data entry performance. *Behavior and Information Technology, 19,* 282–293.

Czaja, S. J., & Sharit, J. (1993). Age differences in the performance of computer based work as a function of pacing and task complexity. *Psychology and Aging, 8,* 59–67.

Czaja, S. J., Guerrier, J. H., Nair, S. N., & Landauer, T. K. (1993). Computer communication as an aid to independence for older adults. *Behavior and Information Technology, 12,* 197–207.

Czaja, S. J., Hammond, K., & Joyce, J. B. (1989). *Word processing training for older adults.* (Final report, Grant # 5 R4 AGO4647-03). National Institute on Aging.

Czaja, S. J., Hammond, K., Blascovich, J., & Swede, H. (1986). Age-related differences in learning to use a text-editing system. *Behavior and Information Technology, 8,* 309–319.

Danowski, J. A., & Sacks, W. (1980). Computer communication and the elderly. *Experimental Aging Research, 6,* 125–135.

Dyck, J. L., & Smither, J. A. (1994). Age differences in computer anxiety: The role of computer experience, gender, and education. *Journal of Education Computing Research, 10,* 239–248.

Egan, D. E., & Gomez, L. M. (1985). Assaying, isolating, and accommodating individual differences in learning a complex skill. *Individual Differences in Cognition, 2,* 174–217.

Eilers, M. L. (1989). Older adults and computer education: "Not to have the world a closed door." *International Journal of Technology and Aging, 2,* 56–76.

Eisdorfer, C., Czaja, S. J., Loewenstein, D. A., Rubert, M. P., Argüelles, S., Mitrani, V. B., et al. (2003). The effect of a family therapy and technology-based intervention on caregiver depression. *The Gerontologist, 43,* 521–531.

El-Attar, T. E., Gray, J., Nair, S., Ownby, R., & Czaja, S. J. (2005). Older adults and Internet health information seeking. *Proceedings of the 49th Annual Meeting of the Human Factors and Ergonomics Society,* Orlando, FL, 163–166.

Elias, P. K., Elias, M. F., Robbins, M. A., & Gage, P. (1987). Acquisition of word-processing skills by younger, middle-aged, and older adults. *Psychology and Aging, 2,* 340–348.

Ellis, E. R., & Allaire, A. J. (1999). Modeling computer interest in older adults: The role of age, education, computer knowledge, and computer anxiety. *Human Factors, 41,* 345–355.

Ellis, L. B. M., Joo, H., & Gross, C. R. (1991). Use of a computer-based health risk appraisal by older adults. *Journal of Family Practice, 33,* 390–394.

Family Caregiver Alliance. (2005). *Selected caregiver statistics.* Retrieved December 10, 2005, from http://www.caregiver.org/caregiver/jsp/content_node.jsp?nodeid=439

Family Caregiver Alliance. (2005). *Women and caregiving: Facts and figures.* Retrieved December 10, 2005, from http://www.caregiver.org/caregiver/jsp/content_node.jsp?nodeid=892

Federal Interagency Forum on Aging-Related Statistics (2004). *Older Americans 2004: Key indicators of well-being.* Washington, DC: U.S. Government Printing Office.

Ferguson, T. (1998). Digital doctoring: Opportunities and challenges in electronic-patient communication. *Journal of the American Medical Association, 280,* 1261–1262.

Fisk, A. D., & Rogers, W. (1997). *The handbook of human factors and the older adult*. San Diego, CA: Academic Press.

Fisk, A. D., Rogers, W., Charness, N., Czaja, S. J., & Sharit, J. (2004). *Designing for older adults: Principles and creative human factors approaches*. London: Taylor & Francis.

Freudenthal, D. (1997). *Learning to use interactive devices; age differences in the reasoning process*. Unpublished master's thesis, Eindhoven University of Technology, Eindhoven, Netherlands.

Furlong, M. S. (1989). An electronic community for older adults: The SeniorNet network. *Journal of Communication, 39,* 145–153.

Gallienne, R. L., Moore, S. M., & Brennan, P. F. (1993). Alzheimer's caregivers: Psychosocial support via computer networks. *Journal of Gerontological Nursing, 12,* 1–22.

Gist, M., Rosen, B., & Schwoerer, C. (1988). The influence of training method and trainee age on the acquisition of computer skills. *Personal Psychology, 41,* 255–265.

Jay, G. M., & Willis, S. L. (1992). Influence of direct computer experience on older adults attitude towards computer. *Journal of Gerontology: Psychological Sciences, 47,* 250–257.

Joyce, B. J. (1990). *Identifying differences in learning to use a text-editor: the role of menu structure and learner characteristics*. Unpublished master's thesis, State University of New York, Buffalo, NY.

Kalasky, M. A., Czaja, S. J., Sharit, J., & Nair, S. N. (1999). Is speech technology robust for older population? *Proceedings of the 43rd Annual Meeting of the Human Factors and Ergonomics Society,* pp. 123–128.

Kaye, H. S. (2000). Computer and Internet use among people with disabilities. *Disability Statistics Reports (13)*. Washington, DC: U.S. Department of Education, National Institute on Disability and Rehabilitation Research. Available online at http://dsc.ucsf.edu/pdf/report13.pdf

Kelly, C. L., Morrell, R. W., Park, D. C., & Mayhorn, C. B. (1999). Predictors of electronic bulletin board system use in older adults. *Educational Gerontology, 25,* 19–35.

Kline, D. W., & Schieber, F. J. (1985). Vision and aging. In J. E. Birren & K. W. Schaie (Eds.), *Handbook of the Psychology of Aging* (pp. 296–331). New York: Van Nostrand Reinhold.

Kubeck, J. W., Miller-Albrecht, S. A., & Murphy, M. D. (1999). Finding information on the World Wide Web: Exploring older adults' exploration. *Educational Gerontology, 25,* 167–183.

Mayhorn, C. B., Stronge, A. J., McLaughlin, A. C., & Rogers, W. A. (2004). Older adults, computer training, and the systems approach: A formula for success. *Educational Gerontology, 30,* 185–204.

Mead, S. E., & Fisk, A. D. (1998). Measuring skill acquisition and retention with an ATM simulator: The need for age-specific training. *Human Factors, 40,* 516–523.

Mead, S. E., Sit, R. A., Jamieson, B. A., Rousseau, G. K., & Rogers, W. A., (1996). *Online library catalog: Age-related differences in performance for novice users*. Paper presented at the Annual Meeting of the American Psychological Association, Toronto, Canada.

Mead, S. E., Spaulding, V. A., Sit, R. A., Meyer, B., & Walker, N. (1997). Effects of age and training on World Wide Web navigation strategies. *Proceedings of the Human Factors and Ergonomics Society 41st Annual Meeting*, Albuquerque, NM, 152–156.

Morrell, R. W., & Echt, C. V. (1996). Instructional design for older computer users: The influence of cognitive factors. In W. A. Rogers, A. D. Fisk, & N. Walker (Eds.), *Aging and skilled performance: Advances in theory and application* (pp. 241–265). Mahwah, NJ: Lawrence Erlbaum Associates.

Morrell, R. W., Mayhorn, C. B., & Bennett, J. (2000). A survey of World Wide Web in middle-aged and older adults. *Human Factors, 42*(2), 175–185.

Morrell, R. W., Park, D. C., Mayhorn, C. B., & Echt, K. V. (1995). *Older adults and electronic communication networks: Learning to use ELDERCOMM*. Paper presented at the 103 Annual Convention of the American Psychological Association. New York, NY.

Murata, A., & Iwase, H. (2005). Usability of touch-panel interfaces for older adults. *Human Factors, 47*(4), 767–776.

Nair, S. N. (1989). *A capability-demand analysis of grocery shopping problems encountered by older adults*. Unpublished master's thesis, State University of New York, Buffalo, NY.

Nair, S. N., Lee, C. C., & Czaja, S. J. (2005). Older adults and attitudes toward computers: Have they changed with recent advances in technology? *Proceedings of the 49th Annual Meeting of the Human Factors and Ergonomics Society,* Orlando, FL, 154–157.

National Center for Health Statistics. (2005). *Health, United States, 2005 with chartbook on trends in the health of Americans*. Washington, DC: U.S. Government Printing Office.

National Research Council. (2004). *Technology for adaptive aging*. Steering Committee for the Workshop on Technology for Adaptive Aging. Richard W. Pew and Susan B. Van Hemel, editors. Board on Behavioral, Cognitive, and Sensory Sciences, Division of Behavioral and Social Sciences and Education. Washington, DC: The National Academics Press.

Nielsen, J. (1993). *Usability engineering*. New York: Academic Press.

Nielsen Norman Group. (2005). *Web usability for senior citizens*. Retrieved January 3, 2006, from http://www.nngroup.com/reports/seniors/

Ogozalek, V. Z., & Praag, J. V. (1986). Comparison of elderly and younger users on keyboard and voice input computer-based composition tasks. *Proceedings of CHI'86 Human Factors in Computing Systems,* New York, NY, 205–211.

Park, D. C. (1992). Applied cognitive aging research. In F. I. M. Crail & T. A. Salthouse (Eds.), *The handbook of aging and cognition* (pp. 449–494). Mahwah, NJ: Lawrence Erlbaum Associates.

Pew Internet & American Life Project. (2000). *The online healthcare evolution: How the web helps Americans take better care of themselves*. Retrieved November 25, 2005, from http://www.pewinternet.org/pdfs/PIP_Health_Report.pdf

Pew Internet & American Life Project (2003). *Internet health resources*. Retrieved November 25, 2005, from http://www.pewinternet.org/pdfs/PIP_Health_Report_July_2003.pdf

Pew Internet & American Life Project (2004). *Older Americans and the Internet*. Retrieved November 25, 2005, from http://www.pewinternet.org/pdfs/PIP_Seniors_Online_2004.pdf

Pew Internet & American Life Project (2005). *Digital divisions*. Retrieved November 25, 2005, from http://www.pewinternet.org/pdfs/PIP_Digital_Divisions_Oct_5_2005.pdf

Potter, E. E. (2003). Telecommuting: The future of work, corporate culture, and American society. *Journal of Labor Research, XXIV,* 73–84.

Preece, J., & Maloney-Krichmar, D. (2003). Online communities: Focusing on sociability and usability. In J. A. Jacko & A. Sears (Eds.), *The human computer-interaction handbook* (pp. 596–620). Mahwah, NJ: Lawrence Erlbaum Associates.

Riviere, C. N., & Thakor, N. V. (1996). Effects of age and disability on tracking tasks with a computer mouse: Accuracy and linearity. *Journal of Rehabilitation Research and Development, 33*(1), 6–15.

Robinson, T. N., Eng, P. K., & Gustafson, D. (1998). An evidence-based approach to interactive health communication: A challenge to medicine in the information age. *Journal of the American Medical Association, 280,* 1264–1269.

Rogers, W., & Fisk, A. (2000). Human factors, applied cognition, and aging. In F. I. M. Craik & T. A. Salthouse (Eds.), *The handbook of aging and cognition*. Mahwah, NJ: Lawrence Erlbaum Associates.

Schieber, F., Fozard, J. L., Gordon-Salant, S., & Weiffenbach, J. W. (1991). Optimizing sensation and perception in older adults. *International Journal of Industrial Ergonomics, 7,* 133–162.

Schwartz, L. K., & Plude, D. (1995). *Compact disk-interactive memory training with the elderly*. Paper presented at the 103 Annual Convention of the American Psychological Association, New York, NY.

SeniorNet. (2005). *Senior net Learning centers*. Retrieved December 5, 2005, from http://www.seniornet.com

Sharit, J., & Czaja, S. J. (1999). Performance of a complex computer-based troubleshooting task in bank industry. *International Journal of Cognitive Ergonomics and Human Factors, 41*, 389–397.

Sharit, J., Czaja, S. J., Hernandez, M., Yang, Perdomo, D., Lewis, et al. (2004). An evaluation of performance by older persons on a simulated telecommuting task. *The Journals of Gerontology: Psychological Sciences, 59B*(6), P305–P316.

Sharit, J., Czaja, S. J., Nair, S. N., & Lee, C. C. (2003). The effects of age and environmental support in using telephone voice menu systems. *Human Factors, 45*, 234–251.

Smith, N. W., Sharit, J., & Czaja, S. J. (1999). Aging, motor control, and performance of computer mouse tasks. *Human Factors, 41*(3), 389–396.

Stolz-Loike, M., Morrell, R.W., & Loike, J.D. (2005). Can e-learning be used as an effective training method for people over age 50? A pilot study. *Gerontechnology, 4*(2), 101–113.

Sutcliffe, A. (2002). Multimedia user interface design. In J. A. Jacko & A. Sears (Eds.), *The human computer-interaction handbook* (pp. 245–262). Mahwah, NJ: Lawrence Erlbaum Associates.

U.S. Bureau of Labor Statistics. (2005). *Computer and Internet use at work in 2003*. Retrieved January 3, 2006, from http://www.bls.gov/news.release/pdf/ciauw.pdf

U.S. Census Bureau. (1999). *Computer use in the United States*. Washington, DC: Author.

U.S. Census Bureau. (2003). *Current population survey*. Washington, DC: Author.

U.S. Department of Commerce (2005). *Computer and Internet use in the United States: 2003*. Retrieved January. 3, 2006, from http://www.census.gov/prod/2005pubs/p23-208.pdf

Vicente, K. J., Hayes, B. C., & Williges, R. C. (1987). Assaying and isolating individual differences in searching a hierarchical file system. *Human Factors, 29*, 349–359.

Walker, N., Millians, J., & Worden, A. (1996). Mouse accelerations and performance of older computer users. *Proceedings of Human Factors and Ergonomics Society 40th Annual Meeting*, Santa Monica, CA, 151–154.

Westerman, S. J., Davies, D. R., Glendon, A. I., Stammers, R. B., & Matthews, G. (1995). Age and cognitive ability as predictors of computerized information retrieval. *Behaviour and Information Technology, 14*, 313–326.

Willis, S. (2004). Technology and learning in current and future older cohorts. In R. W. Pew & S. B. Van Hemel (Eds.), *Technology for adaptive aging* (pp. 209–229). Washington, DC: The National Academies Press.

Zandri, E., & Charness, N. (1989). Training older and younger adults to use software. *Educational Gerontology, 15*, 615–631.

·3·

HCI FOR KIDS

Amy Bruckman, Alisa Bandlow, and Andrea Forte
Georgia Institute of Technology

DESIGNING FOR AND WITH CHILDREN

How is designing computer software and hardware for kids different from designing for adults? Many researchers have addressed questions about the impact of technology on children; less has been said about the impact children can have on the design of technology. Methods for designing with and for children are only recently becoming widespread features of the design literature (see Jensen & Skov, 2005).

In designing for children, people tend to assume that kids are creative, intelligent, and capable of great things if they are given good tools and support. If children cannot or do not care to use technologies we have designed, it is our failure as designers. These assumptions are constructive, because users generally rise to designers' expectations. In fact, the same assumptions are useful in designing for adults. Designers of software for children start at an advantage because they tend to believe in their users. However, they may also be at a disadvantage because they no longer remember the physical and cognitive differences of being a child.

In this chapter, we will

- Describe how children's abilities change with age, as it relates to HCI
- Discuss how children differ from adults cognitively and physically, for those characteristics most relevant for HCI
- Discuss children as participants in the design process
- Review recommendations for usability testing with kids
- Review genres of computer technology for kids, and design recommendations for each genre

HOW ARE CHILDREN DIFFERENT?

As people develop from infants to adults, their physical and cognitive abilities increase over time (Kail, 1991; Miller & Vernon, 1997; Thomas, 1980). The Swiss psychologist Jean Piaget was a leading figure in analyzing how children's cognitions evolve (Piaget, 1970). Piaget showed that children do not just lack knowledge and experience but also fundamentally experience and understand the world differently than adults. He divided children's development into a series of stages:

- Sensorimotor (birth to 2 years)
- Preoperational (ages 2 to 7)
- Concrete Operational (ages 7 to 11)
- Formal Operational (ages 11 and up; Piaget, 1970, pp. 29–33)

Contemporary research recognizes that all children develop differently, and individuals may differ substantially from this typical picture (Schneider, 1996). However, this general characterization remains useful.

In the sensorimotor stage, children's cognitions are heavily dependent on what their senses immediately perceive. Software for children this young is difficult to design. Little interaction can be expected from the child. Obviously, all instructions must be given in audio, video, or animation, since babies cannot read. Furthermore, babies generally cannot be expected to use standard input devices like a mouse effectively, even with large targets.

"Reader Rabbit Toddler" (www.thelearningcompany.com) is targeted at children ages one to three. To eliminate the need for mouse clicking, the cursor is transformed into a big yellow star with room for five small stars inside it. As the mouse is held over a target, the small stars appear one at a time. When the fifth star appears, it counts as clicking on that target. If the child does click, the process simply moves faster. The only downside is the occasional unintended click on the "Go back to the main menu" icon.

In most activities in "Reader Rabbit Toddler," nearly random mouse movement will successfully complete the activity. For example, in the "Bubble Castle" activity, the child needs to rescue animals trapped in soap bubbles that are bouncing around the screen. Random mouse movements will catch the animals relatively quickly. Yet, the parent or teacher watching a child's use of the software over time will typically begin to detect patterns in that mouse movement that become more obviously intentional—the mouse moves more directly toward the bubbles with animals in them. This is a particularly well thought out interface because it mimics how young children learn language. A baby's first attempts at sounds are greeted with great enthusiasm—the child says an unrecognizable phoneme and the parents smile and say, "You said Dada! This is Dada!" Over time, the utterance begins to really sound like the child said "Dada." An initial positive reinforcement for even the most remote attempt at the target behavior puts the child on a good learning trajectory to acquiring that behavior (Holdaway, 1979).

Many examples of software and other cultural artifacts for young children are designed in accordance with adult expectations of what a child should like. There are a few noteworthy exceptions—for example, the television show Teletubbies is out of harmony with those stereotypes. Many adults find the television show bizarre and grating, but it is wildly popular with toddlers. The designers of the original BBC television series, Davenport and Wood (1997), used detailed observations of young children's play and speech in their design. Wood commented:

Our ideas always come from children. If you make something for children, the first question you must ask yourself is, "What does the world look like to children?" Their perception of the world is very different to that of grown-ups. We spend a lot of time watching very young children: how they play; how they react to the world around them; what they say. (Davenport & Wood, 1997)

Focus groups also played an important role (BBC, 1997). Young children are so radically different from adults that innovative design requires careful fieldwork.

While toddlers' interactions with software on a standard desktop computer afford limited possibilities, specialized hardware can expand the richness and complexity of interactions. For example, "Music Blocks" (www.neurosmith.com) is recommended for ages two and up. Five blocks fit in slots in the top of a device rather like a "boom box" portable music player. Each block represents a phrase of music. Each side of the block is a different instrumentation of that musical phrase. Rearranging the blocks changes the music. Interaction of this complexity would be impossible for two-year-olds using a screen-based interface, but is quite easy with specialized hardware. Research on alternative computer interfaces such as tangible technologies (Ishii & Ullmer, 1997; Dourish, 2001) holds great promise for novel children's interface designs (Price, Rogers, Scaife, Stanton, & Neale, 2003; O'Malley & Fraser, 2004).

In the preoperational stage (ages 2 to 7), children's attention spans are brief. They can hold only one thing in memory at a time. They have difficulty with abstractions. They cannot understand situations from other people's points of view. While some children may begin to read at a young age, designs for this age group generally assume the children are still preliterate. It is reasonable to expect children at this age can click on specific mouse targets, but they must be relatively large. Most designers generally still avoid use of the keyboard (except "hit any key" approaches).

In the concrete operational stage (ages 7 to 11), "We see children maturing on the brink of adult cognitive abilities. Though they cannot formulate hypotheses, and though abstract concepts such as ranges of numbers are often still difficult, they are able to group like items and categorize" (Schneider, 1996, p. 69). Concrete operational children are old enough to use relatively sophisticated software, but still young enough to appreciate a playful approach. It is reasonable to expect simple keyboard use. Children's abilities to learn to type grow throughout this age group. It is reasonable to expect relatively fine control of the mouse.

Finally, by the time a child reaches the formal operational stage (ages 11 and up), designers can assume the child's thinking is generally similar to that of adults. Their interests and tastes, of course, remain different. Designing for this age group is much less challenging, because adult designers can at least partially rely on their own intuitions.

Using age as a guide can be useful; however, designers should be aware that designing too young can be just as problematic as designing too old. Children are acutely aware of their own abilities; being asked to interact with technology designed for younger children can be perceived as an affront or boring (Halgren, Fernandes, & Thomas, 1995; Gilutz & Nielsen, 2002).

FIGURE 3.1.   Children playing with Music Blocks.

In addition to considering the implications of cognitive development on children's abilities to use technologies, it is important to remember that different age groups differ culturally too. Understanding what is fun or interesting for a particular age group involves understanding both children's developmental abilities and children's culturally-dependent aesthetic sensibilities. Oosterholt, Kusano, and Vries (1996) suggested that designers should avoid trying to be fashionable—what is cool changes quickly—and should target a limited age range because children's abilities and sensibilities change quickly as well.

In the next sections, we will focus on several characteristics of children that are relevant for HCI research:

- Dexterity
- Speech
- Reading
- Background knowledge
- Interaction style

Dexterity

Young children's fine motor control is not equal to that of adults (Thomas, 1980), and they are physically smaller. Devices designed for adults may be difficult for children to use. Joiner, Messer, Light, and Littleton (1998) noted that "the limited amount of research on children has mainly assessed the performance of children at different ages and with different input devices" (p. 514).

Numerous studies confirm that children's performances with mice and other input devices increase with age (Joiner et al., 1998; Hourcade, 2002). Compared to adults, children have difficulty holding down the mouse button for extended periods and have difficulty performing a dragging motion (Strommen, 1994). This means that many standard desktop interface features pose problems for young users. For example, kids have difficulty with marquee selection. Marquee selection is a technique for selecting several objects at once using a dynamic selection shape. In traditional marquee selection, the first click on the screen is the initial, static corner of the selection shape (typically a rectangle). Dragging the mouse controls the diagonally opposite corner of the shape, allowing you to change the dimensions of the selected area to encapsulate the necessary objects. Dragging the mouse away from the initial static corner increases the size of the selection rectangle, while dragging the mouse toward the initial static corner decreases the size of the selection rectangle. A badly placed initial corner can make it difficult and sometimes impossible to select/encapsulate all of the objects. Berkovitz (1994) experimented with a new encirclement technique: the initial area of selection is specified with an encircling gesture and moving the mouse outside of the area enlarges it.

Kids may have trouble double-clicking, and their small hands may have trouble using a three-button mouse (Bederson, Hollan, Druin, Stewart, Rogers, & Proft, 1996). As with adults, children can use point-and-click interfaces more easily than drag-and-drop (Inkpen, 2001; Joiner et al., 1998). Inkpen (2001) noted, "Despite this knowledge, children's software is often implemented to utilize a drag-and-drop interaction style. Bringing solid research and strong results . . . to the forefront may help make designers of children's software think more about the implications of their design choices" (p. 30).

Strommen (1998) noted that since young children cannot reliably tell their left from their right, interfaces for kids should not rely on that distinction. In his Actimates interactive plush toy designs, the toys' left and right legs, hands, and eyes always perform identical functions. More recent interactive soft toys like "Hug & Learn Baby Tad" by Leapfrog (http://www.leapfrog.com) rely on clear visual markings on left and right paws to indicate their distinct functions.

Speech

Speech recognition has intriguing potential for a wide variety of applications for children. O'Hare and McTear (1999) studied use of a dictation program by 12-year-olds and found that they could generate text more quickly and accurately than by typing. They note that dictation automatically avoids some of the errors children would otherwise make, because the recognizer generates correct spelling and capitalization. This is desirable in applications where generating correct text is the goal. If, instead, the goal is to teach children to write correctly (and, for example, to capitalize their sentences), then dictation software may be counterproductive.

While O'Hare and McTear (1999) were able to use a standard dictation program with 12-year-olds, Nix, Fairweather, and Adams (1998) noted that speech recognition developed for adults will not work with very young children. In their research on reading tutors for children 5 to 7 years old, they first tried a speech recognizer designed for adults. The recognition rate was only 75%, resulting in a frustrating experience for their subjects. Creating a new acoustic model from the speech of children in the target age range, they were able to achieve an error rate of less than 5%. Further gains were possible by explicitly accounting for common mispronunciations and children's tendency to respond to questions with multiple words where adults would typically provide a one-word answer. Even with the improved acoustic model, the recognizer still made mistakes. To avoid frustrating the children with incorrect feedback, they chose to have the system never tell the child they were wrong. When the system detects what it believes to be a wrong answer, it simply gives the child an easier problem to attempt.

Reading

The written word is a central vehicle for communicating information to humans in human-computer interfaces. Consequently, designing computer technology for children with developing reading skills presents a challenge. Words that are at an appropriate reading level for the target population must be chosen. Larger font sizes are generally preferred. Bernard, Mills, Frank, and McKnown (2001) found that kids 9 to 11 years old prefer 14-point fonts over 12-point. Surprisingly, very little empirical work has been done in this area. Most designers follow the rule of thumb that the younger the child, the larger the font should be.

Designing for preliterate children presents a special challenge. Audio, graphics, and animation must substitute for all functions that would otherwise be communicated in writing. The higher production values required can add significantly to development time and cost. Likewise, visually impaired children pose unique challenges for designers. Audio, tactile, and other sensory interfaces can provide opportunities for visually and otherwise impaired children who cannot use traditional interfaces to interact with computers autonomously (McElligott & van Leeuwen, 2004).

Background Knowledge

Many user interfaces are based on metaphors (Erickson, 1990) from the adult world. Jones (1992) noted that children are less likely to be familiar with office concepts like file folders and in-out boxes. In designing an animation system for kids, Halgren et al. (1995) found many kids to be unfamiliar with the metaphor of a frame-based filmstrip and that of a VCR. It is helpful to choose metaphors that are familiar to kids, though kids often have success in learning interfaces based on unfamiliar metaphors if they are clear and consistent (Schneider, 1996).

Interaction Style

Children's patterns of attention and interaction are quite different from those of adults. Traditional task-oriented analyses of activity may fail to capture the playful, spontaneous nature of children's interactions with technology. For example, when adults are the intended users, designers take great pains to create error messages that are informative and understandable based on the assumption that users want to avoid generating the message again. Hanna, Risden, and Alexander (1997) used a funny noise as an error message and found that the children repeatedly generated the error to hear the noise. Similarly, Halgren et al. (1995) found that children would click on any readily visible feature just to see what would happen, and they might click on it repeatedly if it generated sound or motion in feedback. This behavior was causing young users to be trapped in advanced modes they did not understand. The designers chose to hide advanced functionality in drawers—a metaphor that is familiar to children.

By hiding the advanced tools, the novice users would not stumble onto them and get lost in their functionality. Rather, only the advanced users who might want the advanced tools would go looking for more options. This redesign allows the product to be engaging and usable by a wider range of ages and abilities. (Halgren et al., 1995)

Resnick and Silverman (2005) went even further in their design principle: "Make it as simple as possible—and maybe even simpler." They warned against "functionality creep" and suggested removing advanced functionality altogether if it is not clearly required to support children's creative efforts.

Children also bring unique interaction styles to online environments; they respond to information they encounter while browsing the web in markedly different ways than adults. In a study of 55 first through fifth graders, the Nielsen Norman

Group (Gilutz & Nielsen, 2002) found that kids were often unable to distinguish between website content and advertisements. Moreover, they rarely scrolled down to find content; instead, they chose to interact with website elements that were immediately visible. When examining a new website, children were willing to hunt for links in the content by "scrubbing the screen" with the mouse instead of relying solely on visual cues (Gilutz & Nielsen, 2002).

Children are more likely than adults to work with more than one person at a single computer. They enjoy doing so to play games (Inkpen, 1997), and they may be forced to do so because of limited resources in school (Stewart, Raybourn, Bederson, & Druin, 1998). Teachers may also create a shared-computer setup to promote collaborative learning. When multiple children work at one machine simultaneously, they need to negotiate sharing control of input devices. Giving students multiple input devices increases their productivity and their satisfaction (Inkpen, 1997; Inkpen, Gribble, Booth, & Klawe, 1995; Stewart et al., 1998). Inkpen et al. (1995) compared two different protocols for transferring control between multiple input devices: give and take. In a give protocol, the user with control clicks the right mouse button to cede it to the other user; in a take protocol, the idle user clicks to take control. In one study with 12-year-olds and another with 9- to 13-year-olds, they found that girls solve more puzzles with a give protocol, but boys are more productive with a take protocol (Inkpen, 1997; Inkpen et al., 1995).

CHILDREN AND THE DESIGN PROCESS

Users play a variety of roles in HCI design processes. Visionary designers, such as Kay (1972) and Papert (1972), began considering the abilities and sensibilities of children in the design of new technologies as early as the 1970s. Today, ethnographic and participatory (Schuler & Namioka, 1993) methods are becoming increasingly common features of the human-centered design toolkit, as HCI designers attempt to deeply understand the practices and preferences of people who will be using new technologies. When designers enter the world of children, and, conversely, when children enter the laboratory, many of the traditional rules change. As we have seen, children are not just little adults; they engage with the world in fundamentally different ways. Naturally, they bring a host of social, emotional, and cognitive elements to the design process that are unfamiliar to designers who are accustomed to working with adults. In this section, we examine new and traditional methods for working with children in a human-centered design process.

Use of Video With Children

Like adults, children may change their behaviors when a video camera is present. Druin (1999) and a design team found in their early work that children tended to "freeze" or "perform" when they saw a video camera in the room. In subsequent work, Druin's team observed that the problems associated with videotaping had more to do with power relationships than with the video cameras themselves. When the children are in control of the cameras, their discomfort decreases (Alborzi et al., 2000). In addition to considering the social impact of using a camera with children, there are also technical difficulties to deal with. Druin's research team found that, even with smaller cameras, it was difficult to capture data in small bedrooms and large public spaces. The sound and speech captured in public spaces was difficult to understand or even inaudible. Finally, it was difficult to know where to place cameras because they did not know where children would sit, stand, or move in the environment. Druin recommended using multiple data sources to capture "messy" design environments with children, including note takers and participant observers in addition to videotaping. Druin also encouraged the design team to use video cameras (along with journal writing, team discussion, and adult debriefing) as a way to record their brainstorming sessions and other design activities.

Goldman-Segall (1996) explained why video data are an important part of ethnographic interviews and observations. When using video, the researcher does not have to worry about remembering or writing down every detail: "She can concentrate fully on the person and on the subtleties of the conversation." The researcher also has access to "a plethora of visual stimuli which can never be 'translated' into words in text," such as body language, gestures, and facial expressions. It is especially important to be able to review the body language of children as they interact with software. Hanna et al. (1997) stated that children's "behavioral signs are much more reliable than children's responses to questions about whether or not they like something, particularly for younger children. Children are eager to please adults, and may tell you they like your program just to make you happy." MacFarlane, Sim, and Horton (2005) suggested that both signs (behaviors) and symptoms (children's direct responses) should be used together to understand children's enjoyments of and abilities to use new technologies. Video is extremely useful in being able to study behavioral signs as the researchers may miss some important signs and gestures during the actual observation or interview.

Instead of using video in its traditional capacity for ethnographic-style observation, some researchers have attempted to capitalize on children's playful treatment of video cameras to elicit articulation about new technologies. In studies using video probes to capture domestic communication patterns, Hutchinson et al. (2003) observed that images are particularly attractive to young people as an entertaining medium for interacting and communicating. During classroom observations of children using math-learning software, Lamberty and Kolodner (2005) encouraged children to engage in "camera talk" with stationary cameras if they wished. Many of the children regularly talked to the camera. This spontaneous behavior revealed both their preferences for using the software and their developing understanding of fractions. Likewise, Iversen (2002) suggested that, by provoking children to verbalize, video cameras provide a communication link between designers and young informants, thereby enriching both the data collected and the design experience.

Methods for Designing and Testing With Kids

In this section, we review a variety of methods for designing with and for children. These methods differ dramatically in the amount of power they grant to children. Some methods encourage us to view children as codesigners with an equal voice in determining design direction, whereas others place children in a more reactive role as evaluators or subjects in laboratory-based usability tests. In practice, designers use methods from different points on this power spectrum depending on the maturity of the project, and often move back and forth between testing with kids and open-ended exploration (Scaife, Rogers, Aldrich, & Davies, 1997).

Druin (2002) unpacked this spectrum of control by describing four different roles that children can play in the design of new technologies: user, tester, informant, and design partner. The most reactive role she described for children in design is user. As users, children interact with existing technologies and have no direct impact on the design of the technology, except in the form of recommendations for future designs. As testers, children are asked to provide feedback about technology in development so that it can be refined before it is released; however, adult designers determine the goals of the technology much earlier. As informants, children play an earlier, more active role in determining the goals and features of new technologies. When children play the role of informant, they interact directly with designers, but ultimately, the designers decide what the children need or want based on observations, interviews, or other data collection methods. Finally, Druin explains that as design partners, children are equal stakeholders in the design process. Although they may not be able to contribute to the development of the technology in equivalent ways, their expertise is equal in importance to that of other contributors to the design process.

The notion of children as design partners will be explored more fully in the section on cooperative inquiry. Methods for including children as informants and design partners borrow from the tradition of participatory design that emerged in the Scandinavian workplace. Participatory design is an "approach towards computer systems design in which the people destined to *use* the system play a critical role in *designing* it" (Schuler & Namioka, 1993, p. xi). With children, this idea is even more important: Since they are physically and cognitively different from adults, their participations in the design process may offer significant insights. Schuler wrote:

[Participatory Design] assumes that the workers themselves are in the best position to determine how to improve their work and their work life . . . It views the users' perceptions of technology as being at least as important to success as fact, and their feelings about technology as at least as important as what they can do with it. (Schuler & Namioka, 1993, p. xi)

Empowering children in this way and including them in the design process can be difficult due to the traditionally unequal power relationships between kids and adults.

On the other end of the spectrum, methods for including children as users and testers often borrow from the traditional practices of experimental psychology. Usability testing generally takes place in a controlled setting. Sometimes a single design is tested with the goal of improving it; at other times, different design ideas might be compared to establish which ones generate

more positive feedback or enable better task completion. Data collection methods like verbal protocol analysis (Ericsson & Simon, 1993) are commonly used and will be further discussed in the section on adapting traditional usability methods for kids.

Cooperative Inquiry

Druin (1999) developed a systematic approach to developing new technologies for children with children; she created new research methods that included children in various stages of the design process. This approach, called "cooperative inquiry," is a combination of participatory design, contextual inquiry, and technology immersion. Children and adults work together on a team as research and design partners. She reiterated the idea that "each team member has experiences and skills that are unique and important, no matter what the age or discipline" (Alborzi et al., 2000, p. 97).

In this model, the research team frequently observes children interacting with software, prototypes, or other devices to gain insight into how child users will interact with and use these tools. When doing these observations, both adult and child researchers observe, take notes, and interact with the child users. During these observations, there are always at least two note-takers and one interactor, and these roles can be filled by either an adult or a child team member. The interactor is the researcher who initiates discussion with the child user and asks questions concerning the activity. If there is no interactor or if the interactor takes notes, the child being observed may feel uncomfortable, like being "on stage" (Druin, 1999). Other researchers have also found that the role of interactor can be useful for members of the design team. Scaife and Rogers (1999) successfully involved children as informants in the development of ECOi, a program that teaches children about ecology. They wanted the kids to help them codesign some animations in ECOi. Rather than just having the software designer observe the children as they played with and made comments about the ECOi prototypes, the software designer took on the role of interactor to elicit suggestions directly. Through these on the fly, high-tech prototyping sessions, they learned that "it was possible to get the software designer to work more closely with the kids and to take on board some of their more imaginative and kid-appealing ideas" (Scaife & Rogers, 1999).

When working as design partners, children are included from the beginning. The adults do not develop all the initial ideas and then later see how the children react to them. The children participate from the start in brainstorming and developing the initial ideas. The adult team members need to learn to be flexible and learn to break away from carefully following their session plans, which is too much like school. Children can perform well in this more improvisational design setting, but the extent to which the child can participate as a design partner depends on his or her age. Children younger than 7 years may have difficulty in expressing themselves verbally and in being self-reflective. These younger children also have difficulty in working with adults to develop new design ideas. Children older than 10 are typically beginning to become preoccupied with preconceived ideas of the way things are supposed to be. In general, it has been found that children ages 7 to 10 years old are the most effective proto-typing partners. They are "verbal and self-reflective enough to discuss what they are thinking" (Druin, 1999, p. 61), and understand the abstract idea that their low-tech prototypes and designs are going to be turned into technology in the future. They also do not get bogged down with the notion that their designs must be similar to preexisting designs and products.

Through her work with children as design partners, Druin (1999, 2002) has discovered that there are stumbling blocks on the way to integrating children into the design process and to helping adults and children work together as equals. One set of problems deals with the ability of children to express their ideas and thoughts. When the adult and children researchers are doing observations, it is best to allow each group to develop its own style of note taking. Adults tend to take detailed notes, and children tend to prefer to draw cartoons with short, explanatory notes. It is often difficult to create one style of note taking that will suit both groups. Since children may have a difficult time communicating their thoughts to adults, low-tech prototyping is an easy and concrete way for them to create and discuss their ideas. Art supplies such as paper, crayons, clay, and string allow adults and children to work on an equal footing. A problem that arises in practice is that since these tools are childlike, adults may believe that only the child needs to do such prototyping. It is important to encourage adults to participate in these low-tech prototyping sessions.

The second set of problems emerges from the traditionally unequal power relationships between adults and children. In what sense can children be treated as peers? When adults and children are discussing ideas, making decisions, or conducting research, traditional power structures may emerge. In conducting a usability study, the adult researcher might lead the child user through the experiment rather than allowing the child to explore freely on his or her own. In a team discussion, the children may act as if they are in a school setting by raising their hands to speak. Adults may even inadvertently take control of discussions. Is it sensible to set up design teams where children are given equal responsibilities to those of adult designers? Getting adults and children to work together as a team of equals is often the most difficult part of the design process. It is to be expected that it may take a while for a group to become comfortable and efficient when working together. It can take up to six months for an "intergenerational design team to truly develop the ability to build upon each other's ideas" (Druin et al., 2001). To help diffuse such traditional adult–child relationships, adults are encouraged to dress casually, and there always should be more than one adult and more than one child on a team. A single child may feel outnumbered by the adults, and a single adult might create the feeling of a school environment where the adult takes on the role of teacher. Alborzi et al. (2000) started each design session with 15 minutes of snack time, where adults and children can informally discuss anything. This helps both adults and children get to know each other better as "people with lives outside of the lab" (Alborzi et al., 2000) and to improve communication within the group.

Scaife et al. (1997) identified aspects of working with children in the role of informant that require special attention. They found that, when working in pairs, children feel less inhibited about telling strange adults what they were thinking. Other researchers have also found that pairing children, especially with friends, can help ease discomfort (Dindler, Eriksson, Iversen, Ludvigsel, & Lykke-Olesen, 2005; Als, Jensen, & Skov, 2005). Scaife et al. (1997) cautioned that adults also need to become comfortable in the role of facilitator and should take care not to intervene too quickly if children's discussions wander.

In addition to the social challenges associated with mixing adults and children as equal design partners, kids do not always know how to collaborate well with one another in the first place. Because collaborating on a design project is often a novel experience for children, organizing the activities (without imposing too rigid a structure) can help create productive sessions. For example, Guha, Druin, Chipman, Fails, Simms, and Farber (2005) described a technique to support collaboration among kids and adults during cooperative inquiry sessions called "mixing ideas." First, kids generate ideas in a one-on-one session with an adult facilitator, then they work in small groups to integrate these ideas, and finally they work in larger groups until the whole group is finally working together.

Although there have been many successes in having children participate as design and research partners in the development of software, there are still many questions to be answered about the effectiveness of this approach. Scaife and Rogers (1999) attempted to address many of the questions and problems faced when working with children in their work on informant design. The first question deals with the multitude of ideas and suggestions produced by children. Children say outrageous things. How do you decide which ideas are worthwhile? When do you stop listening? The problem of selection is difficult since, in the end, the adult decides which ideas to use and which ideas to ignore. Scaife and Rogers suggested creating a set of criteria to:

Determine what to accept and what not to accept with respect to the goals of the system . . . You need to ask what the trade-offs will be if an idea or set of ideas are implemented in terms of critical 'kid' learning factors: that is, how do fun and motivation interact with better understanding? (Scaife & Rogers, 1999, pp. 46–47)

In addition to deciding which of the children's ideas to use, there is also the problem of understanding the meaning behind what the child is trying to say. Adults tend to assume that they can understand what kids are getting at, but kid talk is not adult talk. It is important to remember that children have "a different conceptual framework and terminology than adults" (Scaife & Rogers, 1999, p. 47).

Another problem with involving children, particularly with the design of educational software, is that "children can't discuss learning goals that they have not yet reached themselves" (Scaife & Rogers, 1999, p. 30). Can children make effective contributions about the content and the way they should be taught, something which adults have always been responsible for? Adults have assumptions about what is an effective way to teach children. Kids tend to focus on the fun aspects of the software rather than the educational agenda. A mismatch of expectations may exist if kids are using components of the software in unanticipated ways. Involving children in the design and evaluation process may help detect where these mismatches occur in the software.

Adapting Usability Evaluation Methods

HCI practices have evolved to address usefulness, enjoyability, and other measures of design success; however, usability remains a fundamental concern for HCI designers. Although efficiency and task completion are often not central to kids' goals in using technology, usability problems can create barriers to achieving other goals. For example, much research done to date has focused on designing educational software, and evaluation is primarily of learning outcomes, not usability. However, usability is a prerequisite for learning. In student projects in Georgia Tech's graduate class "Educational Technology: Design and Evaluation," many student designers are never able to show whether the educational design of their software is successful. What they find instead is that usability problems intervene, and they are unable even to begin to explore pedagogical efficacy. If children cannot use educational technology effectively, they certainly will not learn through the process of using it. MacFarlane et al. (2005) found that measurements of usability and "fun" were significantly correlated in studies of educational software for science. Usability is similarly important for entertainment, communications, and other applications. Many researchers have explored the effectiveness of traditional usability methods with children. In this section, we examine comparative assessments of usability methods and review findings and recommendations.

Traditional usability testing. Several guidelines developed for work with adults become more important when applied to children. For example, when children are asked to work as testers, it is important to emphasize that it is the software that is being tested, not the participant (Rubin, 1994). Children might become anxious at the thought of taking a test, and test taking may conjure up thoughts of school. The researcher can emphasize that even though the child is participating in a test, the child is the tester and not the one being tested (Hanna et al., 1997). Rubin recommends that you show the participant where video cameras are located; let them know what is behind the one-way mirror and whether people will be watching. With children, showing them behind the one-way mirrors and around the lab gives them "a better sense of control and trust in you" (Hanna et al., 1997, p. 12).

Markopoulos and Bekker (2002) described characteristics of kids that can impact the process and outcome of usability testing:

- Children's capacity to verbalize thoughts is still developing.
- Personality may impact both kid's willingness to speak up to adults and their motivation to please authority figures.
- The capacity to concentrate is variable among kids.
- Young children are still developing the capacity for abstract and logical thinking; they may differ in cognitive ability such as remembering several items at once.
- The ability to monitor goal-directed performance develops throughout childhood and adolescence.
- Some ages may have more pronounced gender differences than others may.

- With small children, basic motor skills ability may be a barrier to effective evaluation if kids cannot use prototypes with standard input devices.

Hanna et al. (1997) developed a set of guidelines for laboratory-based usability testing with children:

- The lab should be made a little more child friendly by adding some colorful posters but avoid going overboard as too many extra decorations may become distracting to the child.
- Try to arrange furniture so that children are not directly facing the video camera and one-way mirror, as the children may choose to interact with the camera and mirror rather than doing the task.
- Children should be scheduled for an hour of lab time. Preschoolers will generally only be able to work for 30 minutes but will need extra time to play and explore. Older children will become tired after an hour of concentrated computer use, so if the test will last longer than 45 minutes, children should be asked if they would like to take a short break at some point during the session.
- Hanna et al. (1997) suggested that you "explain confidentiality agreements by telling children that designs are 'top-secret'" (p. 12). Parents should also sign the agreements, since they will inevitably also see and hear about the designs.
- Children up to seven or eight years old will need a tester in the room with them for reassurance and encouragement. They may become agitated from being alone or following directions from a loudspeaker. If a parent will be present in the room with the child, it is important to explain to the parent that he or she should interact with the child as little as possible during the test. Older siblings should stay in the observation area or a separate room during the test, as they may eventually be unable to contain themselves and start to shout out directions.
- Hanna et al. (1997) suggested that you should "not ask children if they want to play the game or do a task—that gives them the option to say no. Instead use phrases such as 'Now I need you to,' 'Let's do this,' or 'It's time to.'"

Think/talk aloud. An important method for collecting usability data with adults is think-aloud protocols. Think-aloud protocols in HCI research are related to verbal protocol analysis methods in psychology, in which subjects are asked to describe what they are thinking about and paying attention to while they complete some set of tasks (Ericsson & Simon, 1993). In usability tests, think-aloud methods are generally used in concert with direct observation (Nielsen, 1993). Researchers who have used this method with children have observed that children may make very few comments during testing (Donker & Reitsma, 2004). In some cases, they seem to have difficulty with concurrent verbalization—verbalizing thoughts while they complete tasks. The cognitive load associated with learning and executing the task itself might interfere with kids' abilities to talk about it (Hoysniemi, Hamalainen, & Turkki, 2003).

Despite potential obstacles, it has been demonstrated that verbal comments from children can play an important role in identifying usability problems. Donker and Reitsma (2004) reported that, although children produced fewer comments, those few comments provided important information about the severity of usability problems that were identified by direct observation. Likewise, other work suggested that, although using think alouds with kids may result in fewer utterances than other approaches, it can be used to generate useful usability data with both older kids aged 8 to 14 (Donker & Markopoulos, 2002; Baauw & Markopoulos, 2004) and younger children aged 6 to 7 (van Kesteren, Bekker, Vermeeren, & Lloyd, 2003).

Active intervention is closely related to think-aloud protocols but involves investigators asking planned questions to encourage testers to reflect aloud on actions at specific points while completing a task. In a small comparative study with kids ages six and seven, van Kesteren et al. (2003) found that active intervention elicited the most comments when compared to think alouds, posttask/retrospection, codiscovery, peer tutoring, and traditional usability testing.

Post-task interviews. In posttask interviews, testers are asked to describe their experiences after they have finished using a new technology to complete a set of tasks. In some cases, video data are reviewed with the participant to evoke comments. This kind of retrospective verbal protocol emerged from the same tradition as think alouds (Ericsson & Simon, 1993). Van Kesteren et al. (2003) raised the question of whether young kids' limited capacities to hold in memory several concepts at once and still-developing abilities to engage in abstract thought limit their abilities to accurately recall and recount past actions. They found that some kids age six and seven were able to recall past actions and describe the ways in which their understanding changed. They note that keeping things interesting is important; children become bored with reviewing videos unless the tasks themselves are engaging to watch. In studies with kids ages 9 to 11, Baauw and Markopoulos (2004) determined that posttask interviews alone revealed fewer usability problems than think alouds; however, when combined with data from observations, which is standard practice, there was no significant difference between the problems that the two methods revealed.

Codiscovery. Codiscovery exploration is a usability method that is used to understand users' experiences and perceptions of new product designs, especially those that may be unfamiliar. In codiscovery sessions, two users who know one another work together to perform a set of tasks using the product. The goal of using two acquainted users is to encourage them to talk about the problems they encounter and their perceptions of the product in the natural course of collaborating on a task instead of relying on a single user's verbal performance for an experimenter (Kemp & van Gelderen, 1996). First, the two users are asked to figure out what a product does and compare it to other products they know about. Next, they are asked to collaborate on a set of specific tasks using the product. Finally, a discussion period allows designers to ask about observed problems and behaviors; in addition, participants can ask questions about the design and the intended purpose of the product. Van Kesteren et al. (2003) found that using co-discovery with kids ages six and seven can be difficult because they often attempt to complete tasks individually. Even when seated next to one another, two children may not interact at all, resulting in very few comments

about the product being tested. When compared with traditional usability tests, think alouds, posttask interviews, peer tutoring and active intervention, codiscovery was found to elicit the fewest comments from kids (van Kesteren et al., 2003).

Peer tutoring. Peer tutoring is a method for usability testing that was developed to capitalize on the ways that children interact with one another in natural, playful settings. When children play together, they regularly teach one another games and invent rules of play. Hoysniemi et al. (2003) explained, "One definition of the usability of a children's software application is that a child is able and willing to teach other children how to use it" (p. 209). Instead of relying on task completion in a lab, peer tutoring is an approach to usability testing that allows kids to engage in exploratory and playful interactions in a naturalistic setting. Peer tutoring involves first helping one or more kids to develop expertise using a piece of software and then asking them to teach other kids how to use it. By observing, recording, and analyzing interactions between tutors and tutees, it is possible to identify usability problems in software as the kids attempt to teach it to one another. Hoysniemi et al. (2003) pointed out that, although it can be useful, the peer-tutoring approach requires time, training, and careful implementation to be effective.

GENRES OF TECHNOLOGY FOR KIDS

Technology for kids falls into two broad categories: education and entertainment. When game companies try to mix these genres, they may use the term *edutainment*. New products for kids increasingly include specialized hardware as well as software.

Entertainment

Designers of games and other entertainment software rarely write about how they accomplish their jobs. Talks are presented each year at the Game Developer's Conference (http://www.gd-conf.com), and some informal reflections are gathered as conference proceedings. Attending the conference is recommended for people who wish to learn more about current issues in game design. *Game Developer* magazine is the leading publication with reflective articles on the game design process.

Game designers are usually gamers themselves and often end up simply designing games that they themselves would like to play. This simple design technique is easy and requires little if any background research with users. Because most game designers have traditionally been male, this approach allowed them to appeal quite effectively to a core gaming audience: young men and teenage boys. However, female designers are becoming more common on design teams and gaming companies are increasingly recognizing that people outside the typical gamer stereotype represent a large potential market for their products. Designing for teenagers is relatively easy. As we have seen, designing for very young children presents substantial challenges. The younger your target audience, the more they should be tightly connected to every stage of the design process.

Brenda Laurel pioneered the use of careful design methods for nontraditional game audiences in her work with the company Purple Moon in the mid-1990s. Laurel aimed to develop games that appeal to preteen girls both to tap this market segment and to give girls an opportunity to become fluent with technology. Many people believe that use of computer games leads to skills that later give kids advantages at school and work. Through extensive interviews with girls in their target age range, Purple Moon was able to create successful characters and game designs. However, the process was so time consuming and expensive that the company failed to achieve profitability fast enough to please its investors. The company closed in 1999, and its characters and games were sold to Mattel. Purple Moon perhaps did more research than was strictly necessary, particularly because their area was so new. The broader lesson is that the game industry typically does not budget for needs analysis and iterative design early in the design process. Playtesting and quality assurance typically take place relatively late in the design cycle. Designers contemplating incorporating research early in their design process must consider the financial costs.

Oosterholt et al. (1996) described several design constraints that are specific to the design of products for children. First, they suggested that trying to be fashionable might result in products that kids quickly perceived as outdated. They also pointed out that fun is just as important to measure as usability, that measurements of fun should be shared with development teams, and moreover, that the product should grow with users over time and continue to be fun long after kids have learned to use it.

Game designer Carolyn Miller (1998) highlighted seven mistakes (or "kisses of death") commonly made by people trying to design games for kids:

"Death kiss #1: Kids love anything sweet"
Miller (1998) wrote, "Sweetness is an adult concept of what kids should enjoy." Only very young children will tolerate it. Humor and good character development are important ingredients. Do not be afraid to use off-color humor or to make something scary.

"Death kiss #2: Give 'em what's good for 'em"
Miller advised, "Don't preach, don't lecture, and don't talk down—nothing turns kids off faster."

"Death kiss #3: You just gotta amuse 'em"
Miller wrote, "Don't assume that just because they are little, they aren't able to consume serious themes."

"Death kiss #4: Always play it safe!"
Adult games often rely on violence to maintain dramatic tension. Since you probably will not want to include this in your game for kids, you will need to find other ways to maintain dramatic tension. Do not let your game become bland.

"Death kiss #5: All kids are created equal"
Target a specific age group, and take into consideration humor, vocabulary, skill level, and interests. If you try to design for everyone, your game may appeal to no one.

"Death kiss #6: Explain everything"
In an eagerness to be clear, some people overexplain things to kids. Kids are good at figuring things out. Use as few words as possible, and make sure to use spoken and visual communication as much as possible.

"Death kiss #7: Be sure your characters are wholesome!"
Miller warned that if every character is wholesome, the results are predictable and boring. Characters need flaws to have depth. Miller identified a number of common pitfalls in assembling groups of characters. It is not a good idea to take a "white bread" approach, in which everyone is White and middle class. On the other end of the spectrum, it is also undesirable to take a "lifesaver approach" with one character for each ethnicity. Finally, you also need to avoid an "off-the-shelf" approach, in which each character represents a stereotype: "You've got your beefy kid with bad teeth; he's the bully. You've got the little kid with glasses; he's the smart one." Create original characters that have depth and have flaws that they can struggle to overcome (Miller, 1998).

Education

To design educational software, we must expand the concept of user-centered design (UCD) to one of learner-centered design (LCD; Soloway, Guzdial, & Hay, 1994). There are several added steps in the process:

- Needs analysis
  - For learners
  - For teachers
- Select pedagogy
- Select media/technology
- Prototype
  - Core application
  - Supporting curricula
  - Assessment strategies
- Formative evaluation
  - Usability
  - Learning outcomes
- Iterative design
- Summative evaluation
  - Usability
  - Learning outcomes

In our initial needs analysis, for software to be used in a school setting, we need to understand not just learners but also teachers. Teachers have heavy demands on their time and are held accountable for their performances in ways that vary between districts and between election years.

Once we understand our learners, and teachers, needs, we need to select an appropriate pedagogy—an approach to teaching and learning. For example, behaviorism views learning as a process of stimulus and reinforcement (Skinner, 1968). Constructivism sees learning as a process of active construction of knowledge through experience. A social-constructivist perspective emphasizes learning as a social process (Newman, Griffin, & Cole, 1989). (A full review of approaches to pedagogy is beyond the scope of this chapter.)

Next, we are ready to select the media we will be working with, matching their affordances to our learning objectives and pedagogical approach. Once the prototyping process has begun, we need to develop not just software or hardware, but (for applications to be used in schools) also supporting curricular materials and assessment strategies.

Assessment should not be confused with evaluation. The goal of assessment is to judge an individual student's performance. The goal of evaluation is to understand to what extent our learning technology design is successful. An approach to assessing student achievement is an essential component of any school-based learning technology. For both school and free time use, we need to design feedback mechanisms so that learners can be aware of their progress. It is also important to note whether learners find the environment motivating. Does it appeal to all learners or to specific gender, learning style, or interest groups?

As in any HCI research, educational technology designers use formative evaluation to understand informally what needs improvement in their learning environments, and guide the process of iterative design. Formative evaluation must pay attention first to usability, and second to learning outcomes. If students cannot use the learning hardware or software, they certainly will not learn through its use. Once it is clear that usability has met a minimum threshold, designers then need to evaluate whether learning outcomes are being met. After formative evaluation and iterative design are complete, a final summative evaluation serves to document the effectiveness of the design and justify its use by learners and teachers. Summative evaluation must similarly pay attention to both usability and learning outcomes.

A variety of quantitative and qualitative techniques are commonly used for evaluation of learning outcomes (Gay & Airasian, 2000). Most researchers use a complementary set of both quantitative and qualitative approaches. Demonstrating educational value is challenging, and research methods are an ongoing subject of research.

This represents an idealized learner-centered design process. Just as many software design projects do not in reality follow a comprehensive UCD process, many educational technology projects do not follow a full LCD process. LCD is generally substantially more time consuming than UCD. While in some cases it may be possible to collect valid usability data in a single session, learning typically takes place over longer periods. To get meaningful data, most classroom trials take place over weeks or months. Furthermore, classroom research needs to fit into the school year at the proper time. If you are using Biologica (Hickey, Kindfield, Horwitz, & Christie, 2000) to teach about genetics, you need to wait until it is time to cover genetics that school year. You may have only one or two chances per year to test your educational technology. It frequently takes many years to complete the LCD process. In the research community, one team may study and evolve one piece of educational technology over many years. In a commercial setting, educational products need to get to market rapidly, and this formal design process is rarely used.

Genres of Educational Technology

Taylor (1980) divided educational technology into three genres:

- Computer as tutor
- Computer as tool
- Computer as tutee

Suppose that we are learning about acid rain. If the computer is serving as tutor, it might present information about acid rain and ask the child questions to verify the material was understood. If the computer is a tool, the child might collect data about local acid rain and input that data into an ecological model to analyze its significance. If the computer is a tutee, the child might program his or her own ecological model of acid rain.

With the advent of the Internet, we must add a fourth genre:

- Computer-supported collaborative learning (CSCL)

In a CSCL study of acid rain, kids from around the country might collect local acid rain data, enter it into a shared database, analyze the aggregate data, and talk online with adult scientists who study acid rain. This is in fact the case in the NGS-TERC Acid Rain Project (Tinker, 1993). See Table 3.1 for an overview of genres of children's software.

Computer as Tutor

In most off the shelf educational products, the computer acts as tutor. Children are presented with information and then quizzed on their knowledge. This approach to education is grounded in behaviorism (Skinner, 1968). It is often referred to as "drill and practice" or "computer-aided instruction" (CAI).

TABLE 3.1. Genres of Children's Software

Genre	Description
Entertainment	Games created solely for fun and pleasure.
Educational	Software created to help children learn about a topic using some type of pedagogy—an approach to teaching and learning.
Computer as Tutor	Often referred to as "drill and practice" or "computer-aided instruction" (CAI), this approach is grounded in behaviorism. Children are presented with information and then quizzed on their knowledge.
Computer as Tool	The learner directs the learning process, rather than being directed by the computer. This approach is grounded in constructivism, which sees learning as an active process of constructing knowledge through experience.
Computer as Tutee	Typically, the learner uses construction kits to help reflect upon what he or she learned through the process of creation. This approach is grounded in constructivism and constructionism.
Computer-supported Collaborative learning (CSCL)	Children use the Internet to learn from and communicate with knowledgeable members of the adult community. Children can also become involved in educational online communities with children from different geographical regions. This approach is grounded in social constructivism.
Edutainment	A mix of the entertainment and educational genres.

The computer tracks student progress and repeats exercises as necessary.

Researchers with a background in artificial intelligence have extended the drill and practice approach to create intelligent tutoring systems. Such systems try to model what the user knows and tailor the problems presented to an individual's needs. Many systems explicitly look for typical mistakes and provide specially prepared corrective feedback. For example, suppose a child adds 17 and 18 and gets an answer of 25 instead of 35. The system might infer that the child needs help learning to carry from the ones to the tens column and present a lesson on that topic. One challenge in the design of intelligent tutors is in accurately modeling what the student knows and what their errors might mean.

Byrne, Anderson, Douglass, and Matessa (1999) experimented with using eye tracking to improve the performance of intelligent tutors. Using an eye tracker, the system can tell whether the student has paid attention to all elements necessary to solve the problem. In early trials with the eye tracker, they found that some of the helpful hints the system was providing to the user were never actually read by most students. This helped guide their design process. They previously focused on how to improve the quality of hints provided; however, that is irrelevant if the hints are not even being read (Byrne et al., 1999).

An interesting variation on the traditional 'computer as tutor' paradigm for very young children is the Actimates line of interactive plush toys. Actimates Barney and other characters lead children in simple games with educational value, like counting exercises. The tutor is animated and anthropomorphized. The embodied form lets young children use the skills they have in interacting with people to learn to interact with the system, enhancing both motivation and ease of use (Strommen, 1998; Strommen & Alexander, 1999).

Computer as Tool

When the computer is used as a tool, agency shifts from the computer to the learner. The learner is directing the process, rather than being directed. This approach is preferred by constructivist pedagogy, which sees learning as an active process of constructing knowledge through experience. The popular drawing program Kid Pix is an excellent example of a tool customized for kids' interests and needs. Winograd (1996) commented that Kid Pix's designer Craig Hickman "made a fundamental shift when he recognized that the essential functionality of the program lay not in the drawings that it produced, but in the experience for the children as they used it" (p. 60). For example, Kid Pix provides several different ways to erase the screen—including having your drawing explode or be sucked down a drain.

Simulation programs let learners try out different possibilities that would be difficult or impossible in real life. For example, Biologica (an early version was called "Genscope") allows students to learn about genetics by experimenting with breeding cartoon dragons with different inherited characteristics like whether they breathe fire or have horns (Hickey et al., 2000). Model-it lets students try out different hypotheses about water pollution and other environmental factors in a simulated ecosystem (Soloway et al., 1996).

The goal of such programs is to engage students in scientific thinking. The challenge in their designs is how to get students to think systematically and not to simply try out options at random. Programs like Model-It provide the student with scaffolding. Initially, students are given lots of support and guidance. As their knowledge evolves, the scaffolding is faded, allowing the learner to work more independently (Guzdial, 1994; Soloway et al., 1994).

Computer as Tutee

Papert (1992) commented that much computer-aided instruction is "using the computer to program the child" (p. 163). Instead, he argued that the child should learn to program the computer and through this process gain access to new ways of thinking and understanding the world. Early research argued that programming would improve children's general cognitive skills, but empirical trials produced mixed results (Clements, 1986; Clements & Gullo, 1984; Pea, 1984). Some researchers argued that the methods of these studies are fundamentally flawed, because the complexity of human experience cannot be reduced to pretests and posttests (Papert, 1987). The counterargument is that researchers arguing that technology has a transformative power need to back up their claims with evidence of some form, whether quantitative or qualitative (Pea, 1987; Walker, 1987). More recently, the debate has shifted to the topic of technological fluency. As technology increasingly surrounds our everyday lives, the ability to use it effectively as a tool becomes important for children's successes in school and later in the workplace (Resnick & Rusk, 1996).

In the late 1960s, Feurzeig (1996) and colleagues at BBN invented Logo, the first programming language for kids. Papert (1980) extended Logo to include turtle graphics, in which kids learn geometric concepts by moving a turtle around the screen. A variety of programming languages for kids have been developed over subsequent years, including Starlogo (Resnick, 1994), Boxer (diSessa & Abelson, 1986), Stagecast (Cypher & Smith, 1995), Agentsheets (Repenning & Fahlen, 1993), MOOSE (Bruckman, 1997), and Squeak (Guzdial & Rose, 2001). Lego Mindstorms (originally "Lego/Logo") is a programmable construction kit with physical as well as software components (Martin & Resnick, 1993). Another programmable tool bridging the gap between physical constructions and representations on the screen is Hypergami, a computer-aided design tool for origami developed at the University of Colorado at Boulder. Students working with Hypergami learn about both geometry and art (Eisenberg, Nishioka, & Schreiner, 1997).

In most design tools, the goal is to facilitate the creation of a product. In educational construction kits, the goal instead is what is learned through the process of creation. So what makes a good construction kit? In an *Interactions* article entitled "Pianos, Not Stereos: Creating Computational Construction Kits," Resnick, Bruckman, and Martin (1996) discussed the art of designing construction kits for learning (constructional design):

The concept of learning-by-doing has been around for a long time. But the literature on the subject tends to describe specific activities and gives little attention to the general principles governing what kinds of

FIGURE 3.2. Penguins created using Hypergami.

"doing" are most conducive to learning. From our experiences, we have developed two general principles to guide the design of new construction kits and activities. These constructional-design principles involve two different types of "connections":

- *Personal connections.* Construction kits and activities should connect to users' interests, passions, and experiences. The point is not simply to make the activities more "motivating" (though that, of course, is important). When activities involve objects and actions that are familiar, users can leverage their previous knowledge, connecting new ideas to their pre-existing intuitions.
- *Epistemological connections.* Construction kits and activities should connect to important domains of knowledge—more significantly, encourage new ways of thinking (and even new ways of thinking about thinking). A well-designed construction kit makes certain ideas and ways of thinking particularly salient, so that users are likely to connect with those ideas in a very natural way, in the process of designing and creating.

Bruckman (2000) added a third design principle:

- *Situated support.* Support for learning should be from a source (either human or computational) with whom the learner has a positive personal relationship, ubiquitously available, richly connected to other sources of support, and richly connected to everyday activities.

Resnick and Silverman (2005) suggested several more design principles for creating construction kits for kids. Some, such as "iterate, iterate—then iterate again," are familiar mantras for HCI designers. Others may be less familiar:

- Low Floor and Wide Walls
  If a technology has a low floor, it means that it is easy for novices to begin using it. Wide Walls suggest a wide range of possible areas of design and exploration. Construction kits define "a place to explore, not a collection of specific activities."
- Make Powerful Ideas Salient—Not Forced
  When designing toward specific learning goals, construction kits should make these ideas visible and useful in design activities rather than imposing the ideas on students as a predetermined solution.
- Support Many Paths, Many Styles
  Kids approach problems in different ways; it is important to support a variety of design approaches in a construction kit.

- Make it as Simple as Possible—and Maybe Even Simpler
  Constraints can be the designer's best friend. Limited functionality sometimes wins out over designs that are more sophisticated because simplicity allows kids to find creative new ways to use a product.
- Choose Black Boxes Carefully
  This principle is related to the previous one; deciding when to reveal complexity and when to conceal it is a difficult question. Resnick and Silverman (2005) suggested that the simplest choice is often the best one.
- A Little Bit of Programming Goes a Long Way
  Because programming is the fundamental mode of construction with computers, designers of construction kits for kids often include some programming functionality. Focusing on powerful, simple commands that kids can do well is often the best way to support a diverse range of activities.
- Give People What They Want—Not What They Ask For
  Observations of kids can often tell designers more than their direct answers to questions. Kids may ask for unrealistic features or may not know themselves why they are having difficulty completing a task.
- Invent Things That You Would Want to Use Yourself
  Although they caution against overgeneralizing one's own personal likes and dislikes, Resnick and Silverman (2005) proposed that the most respectful approach to designing for kids is to create something that the designer herself finds enjoyable.

Computer-Supported Collaborative Learning (CSCL)

Most tools for learning have traditionally been designed for one child working at the computer alone. However, learning is generally recognized to be a social process (Newman et al., 1989). With the advent of the Internet came new opportunities for children to learn from one another and from knowledgeable members of the adult community. This field is called "Computer-Supported Collaborative Learning" (CSCL; Koschmann, 1996).

CSCL research can be divided into four categories:

1. Distance education
   Students attempt to use online environments in ways that emulate a traditional classroom.
2. Information retrieval
   Research projects in which students use the Internet to find information.
3. Information sharing

Students debate issues with one another. One of the first such tools was the Computer-Supported Intentional Learning Environment (CSILE), a networked discussion tool designed to help students engage in thoughtful debate as a community of scientists does (Scardamalia & Bereiter, 1994). They may also collect scientific data and share it with others online. In the "One Sky, Many Voices" project, students learn about extreme weather phenomena by sharing meteorological data they collect with other kids from around the world, and also by talking online with adult meteorologists (Songer, 1996). In the Palaver Tree Online project, kids learn about history by talking online with older adults who lived through that period of history (Ellis & Bruckman, 2001). A key challenge in the design of information sharing environments is how to promote serious reflection on the part of students (Guzdial, 1994; Kolodner & Guzdial, 1996).

Technological Samba Schools

In *Mindstorms*, Papert (1980) had a vision of a "technological samba school." At samba schools in Brazil, a community of people of all ages gather together to prepare a presentation for carnival. "Members of the school range in age from children to grandparents and in ability from novice to professional. But they dance together and as they dance everyone is learning and teaching as well as dancing. Even the stars are there to learn their difficult parts" (Papert, 1980). People go to samba schools not just to work on their presentations, but also to socialize and be with one another. Learning is spontaneous, self-motivated, and richly connected to popular culture. Papert imagined a kind of technological samba school where people of all ages gather together to work on creative projects using computers. The Computer Clubhouse is an example of such a school in a face-to-face setting (Resnick & Rusk, 1996). MOOSE Crossing is an Internet-based example (Bruckman, 1998). A key challenge in the design of such environments is how to grapple with the problem of uneven achievement among participants. When kids are allowed to work or not work in a self-motivated fashion, typically some excel while others do little (Elliott, Bruckman, Edwards, & Jensen, 2000).

Child Safety Online

One challenge in the design of Internet-based environments for kids is the question of safety. The Internet does contain information that is sexually explicit, violent, and racist. Typically, such information does not appear unless one is looking for it; however, it is unusual but possible to stumble across it accidentally. Filtering software blocks access to useful information as well as harmful (Schneider, 1997). Furthermore, companies that make filtering software often fail to adequately describe how they determine what to block, and they may have unacknowledged political agendas that not all parents will agree with. Resolving this issue requires a delicate balance of the rights of parents, teachers, school districts, and children (Electronic Privacy Information Center, 2001). Another danger for kids online is the presence of sexual predators and others who wish to harm children. While

such incidents are rare, it is important to teach kids not to give out personal information online such as their last names, addresses, or phone numbers. Kids who wish to meet online friends face to face should do so by each bringing a parent and meeting in a well-populated public place like a fast-food restaurant. A useful practical guide "Child Safety on the Information Superhighway" is available from the Center for Missing and Exploited Children (http://www.missingkids.com). Educating kids, parents, and teachers about online safety issues is an important part of the design of any online software for kids.

CONCLUSION

To design for kids, we must have a model of what kids are and what we would like them to become. Adults were once kids. Many are parents. Some are teachers. We tend to think that we know kids—who they are, what they are interested in, and what they like. However, we do not have as much access to our former selves as many would like to believe. Furthermore, it is worth noting that our fundamental notions of childhood are in fact culturally constructed and change over time. Calvert (1992) wrote about the changing notion of childhood in America, and the impact it has had on artifacts designed for children and child rearing:

In the two centuries following European settlement, the common perception in America of children changed profoundly, having first held to an exaggerated fear of their inborn deficiencies, then expecting considerable self-sufficiency, and then, after 1830, endowing young people with an almost celestial goodness. In each era, children's artifacts mediated between social expectations concerning the nature of childhood and the realities of child-rearing: before 1730, they pushed children rapidly beyond the perceived perils of infancy, and by the nineteenth century they protected and prolonged the perceived joys and innocence of childhood. (p. 8)

While Calvert (1992) reflected on the design of swaddling clothes and walking stools, the same role is played by new technologies for kids like programmable Legos and drill and practice arithmetic programs: These artifacts mediate between our social expectations of children and the reality of their lives. If you believe that children are unruly and benefit from strong discipline, then you are likely to design CAI. If you believe that children are creative and should not be stifled by adult discipline, then you might design an open-ended construction kit like Logo or Squeak. In designing for kids, it is crucial to become aware of one's own assumptions about the nature of childhood. Designers should be able to articulate their assumptions, and be ready to revise them based on empirical evidence.

References

Alborzi, H., Druin, A., Montemayor, J., Platner, M., Porteous, J., Sherman, L., et al. (2000). Designing StoryRooms: Interactive Storytelling Spaces for Children. *Proceedings of the Symposium on Designing Interactive Systems: Processes, Practices, Methods, and Techniques,* Brooklyn, NY, 95–104.

Als, B. S., Jensen, J. J., & Skov, M. B. (2005). Comparison of think-aloud and constructive interaction in usability testing with children. *Proceedings of the 2005 Conference on Interaction Design and Children,* Boulder, CO, 9–16.

BBC. (1997). *Teletubbies Press Release.*

Baauw, E., & Markopoulos, P. (2004). A comparison of think-aloud and post-task interview for usability testing with children. *Proceedings of the 2004 Conference on Interaction Design and Children*, College Park, MD, 115–116.

Bederson, B., Hollan, J., Druin, A., Stewart, J., Rogers, D., & Proft, D. (1996). Local Tools: An Alternative to Tool Palettes. *Proceedings of the ACM Symposium on User Interface Software and Technology*, Seattle, WA, 169–170.

Berkovitz, J. (1994). Graphical Interfaces for Young Children in a Software-based Mathematics Curriculum. *Proceedings of the ACM Conference on Human Factors in Computing Systems: Celebrating Interdependence*, Boston, MA, 247–248.

Bernard, M.., Mills, M., Frank, T., & McKnown, J. (2001, Winter). Which fonts do children prefer to read online? *Software Usability Research Laboratory (SURL)*. Retrieved March 16, 2007, from psychology .wichita.edu/surl/usabilitynews/w3/fontJR.htm

Bruckman, A. (1997). *MOOSE Crossing: Construction, community, and learning in a networked virtual world for kids*. Unpublished doctoral dissertation, MIT.

Bruckman, A. (1998). Community support for constructionist learning. *Computer Supported Cooperative Work, 7*, 47–86.

Bruckman, A. (2000). Situated support for learning: Storm's weekend with Rachael. *Journal of the Learning Sciences, 9*(3), 329–372.

Byrne, M. D., Anderson, J. R., Douglass, S., & Matessa, M. (1999). Eye tracking the visual search of click-down menus. *Proceedings of the ACM Conference on Human Factors in Computing Systems: the CHI is the limit*, Pittsburgh, PA, 402–409.

Calvert, K. (1992). *Children in the House: The Material Culture of Early Childhood, 1600–1900*. Boston: Northeastern University Press.

Clements, D. H. (1986). Effects of Logo and CAI Environments on Cognition and Creativity. *Journal of Educational Psychology, 78*(4), 309–318.

Clements, D. H., & Gullo, D. F. (1984). Effects of Computer Programming on Young Children's Cognition. *Journal of Educational Psychology, 76*(6), 1051–1058.

Cypher, A., & Smith, D. C. (1995). End user programming of simulations. *Proceedings of the ACM Conference on Human Factors in Computing Systems*, Denver, CO, 27–34.

Davenport, A., & Wood, A. (1997). *TeleTubbies FAQ*. BBC Education. From http://www.bbc.co.uk/cbeebies/teletubbies/grownups/faq.shtml

Dindler, C., Eriksson, E., Iversen, O. S., Ludvigsel, M., & Lykke-Olesen, A. (2005). Mission from Mars: A method for exploring user requirements for children in a narrative space. *Proceedings of the 2005 Conference on Interaction Design and Children*, Boulder, CO, 40–47.

diSessa, A. A., & Abelson, H. (1986). Boxer: A reconstructible computational medium. *Communications of the ACM, 29*(9), 859–868.

Donker, A., & Markopoulos, P. (2002). A comparison of think-aloud, questionnaires and interviews for testing usability with children. *Proceedings of HCI 2002*, 305–316.

Donker, A., & Reitsma, P. (2004). Usability testing with young children. *Proceedings of the Conference on Interaction Design and Children*, College Park, MD, 43–48.

Dourish, P (2001). *Where the action is: The foundations of embodied interaction*. Cambridge, MA: MIT Press.

Druin, A. (1999). Cooperative inquiry: developing new technologies for children with children. *Proceedings of the ACM Conference on Human Factors in Computing Systems: the CHI is the limit*, Pittsburgh, PA, 592–599.

Druin, A. (2002). The role of children in the design of new technology. *Behaviour & Information Technology, 21*(1), 1–25.

Druin, A., Bederson, B., Hourcade, J. P., Sherman, L., Revelle, G., Platner, M., et al. (2001). Designing a digital library for young children: An intergenerational partnership. *Proceedings of the Joint Conference on Digital Libraries*, Roanoke, VA, 398–405.

Eisenberg, M., Nishioka, A., & Schreiner, M. E. (1997). Helping users think in three dimensions: Steps toward incorporating spatial cognition in user modeling. *Proceedings of the International Conference on Intelligent User Interfaces*, Orlando, FL, 113–120.

Electronic Privacy Information Center. (2001). *Filters and Freedom 2.0: Free Speech Perspectives on Internet Content Control*. Washington DC: Author.

Elliott, J., Bruckman, A., Edwards, E., & Jensen, C. (2000). Uneven Achievement in a Constructionist Learning Environment. *Proceedings of the International Conference on the Learning Sciences*, Ann Arbor, MI, 157–163.

Ellis, J. B., & Bruckman, A. S. (2001). Designing palaver tree online: Supporting social roles in a community of oral history. *Proceedings of CHI: Conference on Human Factors in Computing Systems*, Seattle, WA, 474–481.

Erickson, T. (1990). Working with interface metaphors. In B. Laurel (Ed.), *The Art of Human-Computer Interface Design* (pp. 65–73). Reading, MA: Addison Wesley Publishing Company Inc.

Ericsson, K. A., & Simon, H. (1993). *Protocol analysis: Verbal reports as data*. Cambridge, MA: MIT Press.

Gay, L. R., & Airasian, P. (2000). *Education Research: Competencies for Analysis and Application* (6th ed.). Upper Saddle River, New Jersey: Merrill.

Gilutz, S., & Nielsen, J. (2002). Usability of websites for children: 70 design guidelines based on usability studies with kids. Nielsen Norman Group Report.

Goldman-Segall, R. (1996). Looking through layers: Reflecting upon digital video ethnography. *JCT: An Interdisciplinary Journal for Curriculum Studies, 13*(1), 23–29.

Guha, M. L., Druin, A., Chipman, G., Fails, J. A., Simms, S., & Farber, A. (2005). Working with young children as technology design partners. *Communications of the ACM, 48*(1), 39–42.

Guzdial, M. (1994). Software-realized scaffolding to facilitate programming for science learning. *Interactive Learning Environments, 4*(1), 1–44.

Guzdial, M., & Rose, K. (Eds.). (2001). *Squeak: Open Personal Computing and Multimedia*. Englewood, NJ: Prentice Hall.

Halgren, S., Fernandes, T., & Thomas, D. (1995). Amazing Animation™: Movie making for kids design briefing. *Proceedings of the SIGCHI Conference on Human Factors in Computing Systems*, Denver, CO, 519–525.

Hanna, L., Risden, K., & Alexander, K. (1997). Guidelines for usability testing with children. *Interactions, 4*(5), 9–14.

Hickey, D. T., Kindfield, A. C. H., Horwitz, P., & Christie, M. A. (2000). Integrating instruction, assessment, and evaluation in a technology-based genetics environment: The GenScope follow-up study. *Proceedings of the International Conference of the Learning Sciences*, Ann Arbor, MI, 6–13.

Holdaway, D. (1979). *The Foundations of Literacy*. New York: Ashton Scholastic.

Hourcade, J. P. (2002). *It's too small! Implications of children's developing motor skills on graphical user interfaces* (No. CS-TR-4425). University of Maryland Computer Science Department.

Hoysniemi, J., Hamalainen, P., & Turkki, L. (2003). Using peer tutoring in evaluating the usability of a physically interactive computer game with children. *Interacting with Computers, 15*, 203–225.

Hutchinson, H., Mackay, W., Westerlund, B., Bederson, B. B., Druin, A., Plaisant, C., et al. (2003). Technology probes: Inspiring design for and with families. *Proceedings of the SIGCHI Conference on Human Factors in Computing Systems*, Ft. Lauderdale, FL, 17–24.

Inkpen, K. (1997). Three important research agendas for educational multimedia: Learning, children and gender. *Proceedings of Educational Multimedia '97*, Calgary, AB, 521–526.

Inkpen, K. (2001). Drag-and-drop versus point-and-click: Mouse interaction styles for children. *ACM Transactions Computer-Human Interaction, 8*(1), 1–33.

Inkpen, K., Gribble, S., Booth, K. S., & Klawe, M. (1995). Give and take: Children collaborating on one computer. *Proceedings of the ACM*

Conference on Human Factors in Computing Systems, Denver, CO, 258–259.

Ishii, H., & Ullmer, B. (1997). Tangible bits: Towards seamless interfaces between people, bits and atoms. *Proceedings of CHI '97*, 234–241.

Iversen, O. S. (2002). Designing with children: The video camera as an instrument of provocation. *Proceedings of the Conference on Interaction Design and Children*, Eindhoven, Netherlands, CD-ROM.

Jensen, J. J., & Skov, M. B. (2005). A review of research methods in children's technology design. *Proceedings of the Conference on Interaction Design and Children*, Boulder, CO, 80–87.

Joiner, R., Messer, D., Light, P., & Littleton, K. (1998). It is best to point for young children: A comparison of children's pointing and dragging. *Computers in Human Behavior, 14*(3), 513–529.

Jones, T. (1992). Recognition of animated icons by elementary-aged children. *Association for Learning Technology Journal, 1*(1), 40–46.

Kail, R. (1991). Developmental changes in speed of processing during childhood and adolescence. *Psychological Bulletin, 109*, 490–501.

Kay, A. (1972). A personal computer for children of all ages. *Proceedings of the ACM National Conference*, Boston, MA.

Kemp, J. A. M., & van Gelderen, T. (1996). Codiscovery exploration: An informal method for the iterative design of consumer products. In P. W. Jordan, B. Thomas, B. A. Weerdmeester, & I. L. McClelland (Eds.), *Usability evaluation in industry* (pp. 139–146). London: Taylor & Francis Ltd.

Kolodner, J., & Guzdial, M. (1996). Effects with and of CSCL: Tracking learning in a new paradigm. In T. Koschmann (Ed.), *CSCL: Theory and practice*. Mahwah, New Jersey: Lawrence Erlbaum Associates.

Koschmann, T. (Ed.). (1996). *CSCL: Theory and Practice*. Mahwah, New Jersey: Lawrence Erlbaum Associates.

Lamberty, K. K., & Kolodner, J. (2005). Camera talk: Making the camera a partial participant. *Proceedings of the SIGCHI Conference on Human Factors in Computing Systems*, Portland, OR, 839–848.

MacFarlane, S., Sim, G., & Horton, M. (2005). Assessing usability and fun in educational software. *Proceedings of the Conference on Interaction Design and Children*, Boulder, CO, 103–109.

Markopoulos, P., & Bekker, M. (2002). How to compare usability testing methods with children participants. *Proceedings of the Conference on Interaction Design and Children*, Eindhoven, Netherlands, 153–158.

Martin, F., & Resnick, M. (1993). LEGO/Logo and Electronic Bricks: Creating a scienceland for children. In D. L. Ferguson (Ed.), *Advanced Educational Technologies for Mathematics and Science* (pp. 61–90). Berlin Heidelberg: Springer-Verlag.

McElligott, J., & van Leeuwen, L. (2004). Designing sound tools and toys for blind and visually impaired children. *Proceedings of the Conference on Interaction Design and Children*, College Park, MD, 65–72.

Miller, C. (1998). Designing for kids: Infusions of life, kisses of death. Paper presented a the Game Developers Conference, Longbeach, CA.

Miller, L. T., & Vernon, P. A. (1997). Developmental changes in speed of information processing in young children. *Developmental Psychology, 33*(4), 549–554.

Newman, D., Griffin, P., & Cole, M. (1989). *The Construction Zone: Working for Cognitive Change in School*. Cambridge, England: Cambridge University Press.

Nielsen, J. (1993). *Usability Engineering*. London: Academic Press.

Nix, D., Fairweather, P., & Adams, B. (1998). Speech recognition, children, and reading. *Proceedings of the SIGCHI Conference on Human Factors in Computings Systems,* Los Angeles, CA, 245–246.

O'Hare, E. A., & McTear, M. F. (1999). Speech recognition in the secondary school classroom: an exploratory study. *Computers & Education, 3*(8), 27–45.

O'Malley, C., & Fraser, D., (2004). *Literature Review in Learning with Tangible Technologies*. NESTA Futurelab Report.

Oosterholt, R., Kusano, M., & Vries, G. D. (1996). Interaction design and human factors support in the development of a personal communicator for children. *Proceedings of the SIGCHI Conference on Human Factors in Computing Systems*, Vancouver, Canada, 450–457.

Papert, S. (1972). On making a theorum for a child. Paper presented at the ACM National Conference, Boston, MA.

Papert, S. (1980). *Mindstorms: Children, Computers, and Powerful Ideas*. New York: Basic Books.

Papert, S. (1987). Computer criticism vs. technocentric thinking. *Educational Researcher, 16*(1), 22–30.

Papert, S. (1992). *The Children's Machine*. New York: Basic Books.

Pea, R. (1984). On the cognitive effects of learning computer programming. *New Ideas in Psychology, 2*(2), 137–168.

Pea, R. (1987). The aims of software criticism: Reply to Professor Papert. *Educational Researcher, 16*, 4–8.

Piaget, J. (1970). *Science of Education and the Psychology of the Child*. New York: Orion Press.

Price, S., Rogers, Y., Scaife, M., Stanton, D., & Neale, H. (2003). Using 'tangibles' to promote novel forms of playful learning. *Interacting with Computers, 15*(2), 169–185.

Repenning, A., & Fahlen, L. E. (1993). Agentsheets: A tool for building domain-oriented visual programming environments. *Proceedings of the SIGCHI Conference on Human Factors in Computing Systems*, Amsterdam, Netherlands, 142–143.

Resnick, M. (1994). *Turtles, Termites, and Traffic Jams: Explorations in Massively Parallel Microworlds*. Cambridge, MA: MIT Press.

Resnick, M., Bruckman, A., & Martin, F. (1996). Pianos not stereos: Creating computational construction kits. *Interactions, 3*(5), 40–50.

Resnick, M., & Rusk, N. (1996). The Computer Clubhouse: Preparing for life in a digital world. *IBM Systems Journal, 35*(3–4), 431–440.

Resnick, M., & Silverman, B. (2005). Some reflections on designing construction kits for kids. *Proceedings of the Conference on Interaction Design and Children*, Boulder, CO, 117–122.

Rubin, J. (1994). *Handbook of Usability Testing*. New York: John Wiley and Sons, Inc.

Scaife, M., & Rogers, Y. (1999). Kids as informants: Telling us what we didn't know or confirming what we knew already? In A. Druin (Ed.), *The design of children's technology* (pp. 27–50). San Francisco, CA: Morgan Kaufmann Publishers, Inc.

Scaife, M., Rogers, Y., Aldrich, F., & Davies, M. (1997). Designing for or designing with? Informant design for interactive learning environments. *Proceedings of the SIGCHI Conference on Human Factors in Computing Systems*, Atlanta, GA, 343–350.

Scardamalia, M., & Bereiter, C. (1994). Computer support for knowledge-building communities. *The Journal of the Learning Sciences, 3*(3), 265–283.

Schneider, K. G. (1996). Children and information visualization technologies. *Interactions, 3*(5), 68–73.

Schneider, K. G. (1997). *The Internet Filter Assessment Project (TIFAP)*. Retrieved March 16, 2007, from http://www.bluehighways.com/tifap/learn.htm

Schuler, D., & Namioka, A. (Eds.). (1993). *Participatory design: Principles and practices*. Hillsdale, New Jersey: Lawrence Erlbaum Associates, Publishers.

Skinner, B. F. (1968). *The technology of teaching*. New York: Appleton-Century-Crofts.

Soloway, E., Guzdial, M., & Hay, K. E. (1994). Learner-centered design: The challenge for HCI in the 21st century. *Interactions, 1*(1), 36–48.

Soloway, E., Jackson, S. L., Klein, J., Quintana, C., Reed, J., Spitulnik, J., et al. (1996). Learning theory in practice: Case studies of learner-centered design. *Proceedings of the SIGCHI Conference on Human Factors in Computing Systems*, Vancouver, Canada, 189–196.

Songer, N. B. (1996). Exploring learning opportunities in coordinated network-enhanced classrooms: A case of kids as global scientists. *The Journal of the Learning Sciences, 5*(4), 297–327.

Stewart, J., Raybourn, E. M., Bederson, B., & Druin, A. (1998). When two hands are better than one: enhancing collaboration using single display groupware. *Proceedings of the ACM Conference on Human Factors in Computing Systems*, Los Angeles, CA, 287–288.

Strommen, E. (1994). Children's use of mouse-based interfaces to control virtual travel. *Proceedings of the ACM Conference on Human Factors in Computing Systems: Celebrating Interdependence*, Boston, MA, 229.

Strommen, E. (1998). When the Interface is a Talking Dinosaur: Learning Across Media with ActiMates Barney. *Proceedings of the ACM Conference on Human Factors in Computing Systems*, Los Angeles, CA, 288–295.

Strommen, E., & Alexander, K. (1999). Emotional Interfaces for Interactive Aardvarks: Designing Affect into Social Interfaces for Children. *Proceedings of the ACM Conference on Human Factors in Computing Systems: the CHI is the limit*, Pittsburgh, PA, 528–535.

Taylor, R. P. (Ed.). (1980). *The computer in the school, tutor, tool, tutee.* New York: Teachers College Press.

Thomas, J. R. (1980). Acquisition of motor skills: Information processing differences between children and adults. *Research Quarterly for Exercise and Sport, 51*(1), 158–173.

Tinker, R. (1993). *Thinking about science.* Concord, MA: The Concord Consortium.

van Kesteren, I. E., Bekker, M. M., Vermeeren, A. P. O. S., & Lloyd, P. A. (2003). Assessing usability evaluation methods on their effectiveness to elicit verbal comment from children subjects. *Proceedings of the Conference on Interaction Design and Children*, Preston, UK, 41–49.

Walker, D. F. (1987). Logo needs research: A response to Professor Papert's paper. *Educational Researcher*, 9–11.

Winograd, T. (1996). Profile: Kid Pix. In T. Winograd (Ed.), *Bringing design to software* (pp. 58–61). New York: ACM Press.

•4•

INFORMATION TECHNOLOGY
FOR COGNITIVE SUPPORT

A. F. Newell, A. Carmichael, P. Gregor, N. Alm, and A. Waller
University of Dundee

INTRODUCTION

Well-designed communication and information technology systems have great, unrealized potential to enhance the quality of life and independence for those with cognitive dysfunction, including elderly people, by

- Allowing them to retain a high level of independence and control over their lives
- Providing appropriate levels of monitoring and supervision of at-risk people, without violating privacy
- Keeping people intellectually and physically active
- Providing communication methods to reduce social isolation and foster social inclusion

Using technology to augment human cognitive capacity is not a new idea, taking the term *technology* in its widest sense, to include tools and techniques which humans have developed to help them improve upon their limitations. The first cognitive function to be augmented was probably memory. Mnemonic methods help with this and are still in wide use today. One of the most common (e.g., the method of loci) is to link what is to be remembered with a well-established memory structure that is easy to recall, such as the layout of a familiar city, or a narrative that has already been memorized. The introduction of written language extended people's cognitive abilities by allowing memories to be recorded externally. Although at the time some feared this backwards step would allow people's memory powers to wither, it has since become apparent that the overriding effect has been to free cognitive abilities that can further develop on the basis of the external support. The development of information processing technologies has continued this trend of extending the potential of most people's cognitive abilities. Thus, there should be similar potential for computer technology to extend the cognitive abilities of those with some form of impairment or other limitation.

The increasing power and decreasing size of computer technology, along with its capacity to provide communication as well as computation and storage, offers the possibility of quite sophisticated help for cognitive impairments, if we can develop the appropriate software. Computers have the potential to act as a kind of scaffolding for cognitive tasks, taking over functions that have been affected by illness, accident, or aging. They could also provide prompts for daily living, if they were able to track successfully through the user's sequence of tasks and actions.

Particular strengths of computers as assistants for people with cognitive impairments include being consistent, tireless, and not becoming emotionally involved. In addition, multimedia and multimodal systems can provide a very rich interaction, which may be particularly advantageous for users with cognitive dysfunction. For example, such systems have great potential in addressing the problems of memory loss and the related difficulties presented by dementia. In addition, communication systems that use synthetic speech, predictive programs that can facilitate writing, and a range of nonlinguistic methods of communication can be used by those with speech and language dysfunction caused by cognitive impairments. A great deal of work will be required in order to realize this potential both in uncovering new knowledge about how cognition works and in developing assistive systems based on this knowledge.

Because of the wide range of skills and knowledge needed to understand the problems faced by people with cognitive impairment, it is essential that research work in this field should be multidisciplinary, including psychologists, members of the health and therapeutic professions, and engineers. It is also vital to involve potential users of the technology as partners in the research at all stages of development.

In addition, it is essential that interface designers be much more aware of the range of abilities that exist in any population of users. For example, access to the Internet for disabled people is often thought to be synonymous with access for blind people, but in fact, blind people form only a small percentage of the disabled population. Ogozalec (1997) pointed out that, if current trends continue in the United States, by 2030, one fifth of the population will be over 65 years of age and commented, "It is difficult to categorize and draw conclusions about 'the elderly', since they comprise such a diverse and heterogeneous population" (p. 65). This diversity, particularly of cognitive function, ought to be taken into account if we are to make software and the Internet available to as large a percentage of the population as possible.

We will describe the major types of cognitive impairment and include within this discussion the effects of aging. We will then illustrate the development of systems to support people with a variety of cognitive impairment with specific projects in which we have been involved. They will not cover all aspects of cognitive impairment, but they do illustrate a methodology and an approach to developing assistive technology, which may have a wider relevance.

In the concluding part of this chapter, we will address the development of methodologies that we believe will be valuable for designers of systems to support people with cognitive impairment and to assist designers of general systems to take into account the needs of people with cognitive impairments.

COGNITIVE IMPAIRMENT

The use of the term *cognitive impairment* implies that two categories of human cognitive systems exist—impaired and unimpaired. However, this is not the case, although it can reasonably be stated that there are normal or average cognitive systems. The vast majority of experimental cognitive-psychology literature relates to this normal system. In many contexts, this level of explanation is suitable for indicating of what most people are capable. It should always be borne in mind, however, that in real-world situations there is no marked distinction between that which is normal and that which is not. In other words, everyone has some limits to their cognitive abilities. Some have a highly specific impairment, some have more diffuse problems, and some experience interrelated constellations of impairments. In addition, the cognitive abilities of any one human being will change over longer and shorter periods. Aging can have substantial effects on cognitive ability, which is particularly marked in some age-related conditions, such as dementia.

For ease of exposition, the forms of cognitive impairment identified and described below will in the main refer to general categories. It should be noted, however, that all these categories lie somewhere on a continuum and, while they are delineated on the basis of educational and clinical or medical criteria, such cut off points are relatively arbitrary in the context of the wide variability of cognitive ability across the population. It is also worth noting that within the context of normal cognitive systems, there is significant diversity among people in regard to differential preferences for types of material and ways of approaching and processing information. For example, some people may be considered primarily verbal and tend to excel in language-based tasks, relative to those considered visuo/spatial (e.g., Lohman, 2000). Thus, many of the types of impairment addressed below can be construed to an important extent as the extremities of normal diversity.

INTELLIGENCE QUOTIENT

The most widely known dimension of general cognitive ability is probably intelligence. Scientific investigation of this dimension has a controversial past, and many aspects of this are beyond the scope of the present chapter (for a more comprehensive account, see Gould, 1997). One underlying reason for such controversy is that the word *intelligence* has a rather nebulous definition. In day-to-day usage, this is rarely problematic, but the differences between scientific and lay definitions can cause misunderstanding (e.g., Sternberg, 2000). Such misunderstanding can lead to controversy as most definitions of intelligence include connotations that are considered socially important and, thus, can often be highly emotive. Despite these difficulties, the investigation of intelligence has provided many insights into a wide range of more particular cognitive abilities, and has developed methods for quantifying general intellectual ability such as the various forms of IQ (intelligence quotient) tests. Again, controversy has surrounded the use of these tests over the years (see Gould, 1997; Kaufman, 2000), but such tests have been widely used and are accepted as a general benchmark of a person's intellectual capability.

An IQ score of 100 is, by definition, normal with about 50% of the population scoring above and 50% scoring below, but it should be noted that elderly people are not generally included in the standardization of these scores. Approximately 50% of the population is considered within the bounds of normality and deviate either side of 100 by no more than 10 points. The nonnormal 50% are distributed evenly above and below this band. Thus about a quarter of the (nonelderly) population fall below the level of what is considered normal. Although the terminology varies across cultures and over time, around 20% of the population have IQ scores between 75 and 90 and would generally be classified as slow learners. The final 5% will generally have very special needs that overall are best addressed on an individual basis (Kaluger & Kolson, 1987). Further to this and as an example of the emotive connotations associated with the issue of 'intelligence', it is worth noting that the first official classification scheme (see Detterman, Gabriel, & Ruthsatz, 2000) associated with IQ tests further broke down this latter

5%. These classifications were moron (IQ 50–75), imbecile (IQ 25–50), and idiot (IQ < 25), terms which today would be considered wholly unacceptable as a description of anyone with a cognitive impairment.

An IQ score reflects a person's intellectual ability as a whole. A low score may be due to the whole system functioning at a suboptimal level, but a similar result can also be due to one or more component abilities being impaired. There are many tests of IQ, some of which give an indication of this while others do not. Some IQ tests are explicitly broken down into subtests that reflect the relative levels of ability in the component cognitive abilities, such as the verbal and visuo/spatial abilities mentioned above. The more common forms of cognitive impairment are described below in the context of a brief overview of the cognitive system.

For any information in the outside world to enter the cognitive system, it must first be detected and transmitted by the sensory apparatus. In an important sense, this is not simply the start of the process because aspects of attention will influence what is and is not detected/transmitted, and, to a certain extent, how. Basic perceptual processing creates a sensory specific representation of the stimulus event. Streams of such stimulus events are summated into meaningful cognitive entities (e.g., strokes on a page recognized as letters and numerals are summated into a name and telephone number). These will then be either passed immediately to short-term memory (STM), further processed by working memory (WM), or rehearsed for maintenance in STM or for encoding into longer term storage. For example, rehearsal could refer to rote rehearsal of a telephone number between reading and dialing it or to more elaborate processing to associate it with relevant extant memories to improve the chance of subsequent recall (e.g., method of loci, mentioned above).

Output from the cognitive system will generally be initiated in response to some form of external stimulus, or probe, by accessing extant memories relevant to the probe using executive processes to organize them in a task relevant way and then producing a response. Output of this kind has been most commonly studied with the use of memory tests. This minimizes the influence of intellectual processing (problem solving, etc.) per se, and emphasizes the registration, rehearsal, and encoding of information, the effects of decay, interference, and other forms of forgetting, and the effectiveness of different cues (probes) in eliciting specific memories (e.g., recall vs. recognition).

Virtually all aspects of the processing outlined above are shaped by attention, and it is important to note that, regardless of impairment, while we all have some control over attention, it can also be the case that attention can have some control over us. That is, we can utilize attention to focus on searching a list for a particular telephone number while ignoring the chatter of people around us. Having read the number, however, our attention can exert its own control if someone calls our name and asks if we have made that call yet. Despite our best efforts at rehearsal, it is likely that our attention will be grabbed by our name, and the ensuing question and this brief distraction can be enough to lose the information from temporary storage.

In general terms, mild to moderate global cognitive impairment will be associated with decrements in efficiency across most of the processing stages outlined above and in aspects of the utilization of attention. The following will describe some of the main decrements in cognitive ability related to interacting with computer type systems.

INTERFACE DESIGN TO SUPPORT PEOPLE WITH COGNITIVE IMPAIRMENT

It is often important to develop special technology to provide support for people with various types of cognitive impairment, and some such projects are described later. It is also important, however, to address the challenge of providing access to more mainstream technology for people with cognitive impairments. When designing or specifying mainstream technology for such users, it is important to focus on their characteristics and to be fully aware of the range of cognitive diversity, even among those without clinical dysfunction. This is rarely mentioned in human interface design, where the cognitive diversity of the human race has not been the focus of much research. It is also important to consider the effects of age on cognitive function. As Worden, Walker, Bharat, and Hudson (1997) commented:

It is known that, as people age, their cognitive perceptual and motor abilities decline with negative effects on their ability to perform many tasks. Computers can play an increasingly important role in helping older adults function well in society. Despite this little research has focused on computer use of older adults. (p. 266)

A key aspect of any intellectual task, in regard to interactive technology for people with mild or moderate global cognitive impairment, is speed (Salthouse, 1991). Whatever level of performance a person can achieve in any given situation will be made worse if the task must be done under externally imposed time constraints whether actual or simply inferred by the user. Thus, wherever possible, the design of any interaction should allow every step to be carried out at the user's own pace. This issue also raises the first distinction between older and younger people. A relatively greater proportion of the extra time needed by older people is due to age-related declines in their sensory systems, particularly in hearing and vision rather than cognitive impairment per se. For example, given comparable levels of cognitive impairment, an older person will need relatively more time to perceive the relevant stimulus before cognitive processes can be brought to bear on it. Thus, for many older people, the requirement for extra time can be reduced, though rarely removed, if care is taken to present text and other aspects of on-screen layout in a suitably clear way (e.g., Carmichael, 1999; Charness & Bosman, 1994). Attention to text layout can also benefit people with specific learning difficulties, such as those with dyslexia and people with limited literacy levels. Beyond this, clear text and presentation layout will always be worth considering carefully as such aspects have been found to benefit those who do not specifically need it, albeit to a less marked extent (Freudenthal, 1999; Pirkl, 1994).

Another key concept related to interface design for people with cognitive impairment is complexity and its avoidance. Complexity in interfaces can manifest itself in many different ways and at many different levels. A truly comprehensive coverage of this is beyond the scope of the present chapter, but some illustrative examples will be given to elucidate this idea.

The use of language in an interactive system should be given careful consideration, and the syntax and vocabulary should be kept as straightforward and commonplace as the context allows. This is particularly pertinent for any form of instruction. If the requirements of a particular stage of an interaction cannot be captured in a few simple concrete statements, then serious consideration should be given to redesigning the interaction itself. Similarly any on-screen display should be kept as uncluttered as practicable and wherever possible should present the user with only a single issue (menu, subject, decision, etc.) at any particular point in time. Similarly, but at a larger scale, progression through an interaction should be kept, again wherever practicable, as linear as possible. That is, the user should only need to consider one thing at a time. Any requirement to deal with different things in parallel will markedly increase the possibility of errors and general user dissatisfaction (Detterman et al., 2000; Salthouse, 1985).

Unfortunately, as the designers of a system will have a comprehensive understanding of the functions of that system, they are unlikely to assess issues of complexity from the users' point of view, particularly that of a novice user. In addition, prescriptive checklists for avoiding complexity will ultimately be inadequate, as the optimum approach will always depend on the specifics of the task the interactive system is intended to support (Carmichael, 1999). This is one of the main issues that highlights the importance of early and rigorous user involvement in the design of interactive systems. This is particularly important in the case of young designers developing interfaces for older users.

Research into cognitive changes in later life has indicated the heuristic value of the concepts of fluid and crystallized abilities (Horn & Cattell, 1967). In general, fluid abilities (novel problem solving) are those that decline with age, and crystallized abilities (existing world knowledge) are those that do not. Research that has looked at the very old (80 years and older) found that ultimately everything declines but that the crystallized abilities follow a markedly slower trajectory (Bäckman, Small, Wahlin, & Larsson, 2000). This is another distinction between younger and older people, as the former will tend to have much less accumulated knowledge. It is also worth noting that there are advantages and disadvantages on both sides of this distinction. Rabbitt (1993) has shown that, in many circumstances, relevant accumulated knowledge can ameliorate decline in fluid ability (e.g., well-learned strategies). In other circumstances, however, the opposite can be the case, wherein a well-learned, but essentially inappropriate, strategy can put a relatively greater burden on the associated fluid abilities.

The Effects of Attention

Many of the constraints imposed by cognitive impairment can be further shaped by decrements in various aspects of attention. One major aspect of this is generally referred to as "selective attention," which allows us to focus on salient aspects of a task and at the same time helps us actively to ignore irrelevant aspects. The efficiency of selective attention is markedly diminished in most forms of cognitive impairment. This factor further supports the recommendation to present the user with just one thing at a time, which will avoid the user erroneously devoting time and cognitive resources to processing irrelevant information. Similarly, if the nature of the interaction requires the user to attend to some critical information at a particular time/location, appropriately obvious highlighting should be employed to grab the users' attention. These issues become emphasized in situations where selective attention must be maintained over periods of more than just a few minutes.

Another aspect of attention known to be less efficient in cognitive impairment is referred to as "divided attention." In general, this refers to the ability to allocate cognitive resources appropriately when trying to do two or more distinct cognitive tasks or distinct portions of the same task at the same time. Many scenarios, where the user is required to do more than one thing at a time, can simply demand more cognitive resources than are available. However, declines in the efficiency of divided attention can mean that even if the tasks involved demand no more than the resources available, they may not be allocated appropriately. Generally, the interactive system should be designed to relieve the user of this kind of burden. It is difficult to be prescriptive about suitable solutions to this problem, as the appropriate approach will depend on the specifics of the interaction involved, but some general ideas may be of use. For example, the provision of some form of notepad function may be helpful for temporarily recording information for subsequent use, although great care is needed to ensure that the instantiation of such a function and its utilization do not put further cognitive load on the user. Another possibility would be the provision of an overview of the task, which could show or remind the user where they are and what they have and have not done.

Memory Loss and Dementia

Limitations in memory affect people with age- and nonage-related cognitive impairment. Thus, wherever practicable, interactive systems should be designed to take the burden of memory off the user, for example, by judicious use of prompts and reminders. Also, careful consideration of the steps in an interaction and the way they are presented to the user can help mitigate the most common problem of deficient short term and working memory. Even with the best design efforts, however, such problems are likely to make users with cognitive impairment relatively error prone. It is thus very important to ensure that the interactive system allows for error correction in an easy to use form. To ensure that the user spots such errors, the system should provide feedback regarding user actions and where appropriate elicit active confirmation from them.

Various forms of dementia can exaggerate the relatively mild effects of normal aging on the cognitive system. At the age of 60 years, about 1% of the population is diagnosed with dementia. This percentage approximately doubles for every subsequent 5-year age band (e.g., 4% at 70 years and 16% at 80 years; Bäckman et al., 2000). Of the elderly population with dementia, Alzheimer's disease accounts for about 60%, depending on the diagnostic criteria used and a further 25% is vascular dementia (e.g., related to circulatory problems). Most of the vas-cular dementias are referred to as "multi-infarct dementias" and tend to be caused by a series of ministrokes, and thus, they tend to have more diffuse and less predictable effects on ability than a major stroke. The remaining 15% of dementias are made up of various relatively rare conditions (Bäckman et al., 2000).

Regardless of the various causes and effects, all forms of dementia involve damage to the brain, such damage being more or less widespread, affecting cortical and/or subcortical areas. In general, damage to the cortex results in cognitive/perceptual impairment, while damage to the subcortical areas is more related to physical impairment. There are, however, a number of well-known problems related to the diagnosis of dementia. Two of these are relevant here. The first involves the grey area between the worst effects of normal aging and the initial effects of pathological aging at the onset of dementia. The second is that the effects of depression in later life can closely mimic those of dementia. These additional complexities further expand the overall diversity of cognitive impairment in relation to human interface design, both in regard to the general level of ability and in the variation of that level over periods of days, weeks, months, and even years. The convolution of this situation is further added to by the effects of a relatively greater probability of ill health among older people. It is estimated that around 80% of those aged 65 and over have at least one chronic illness and that many will have more than one. In addition to the effects of health per se, there is also potential for cognitive ability to be affected by a variety of medication and by interactions between different medicines.

Despite the above, some systematic changes associated with extreme old age and dementia are relevant to human-interface design. In general, the first ability to deteriorate in dementia, particularly with Alzheimer's disease, is episodic as distinct from semantic memory. Episodic memory is memory for events, usually from the viewpoint of personal experience, rather than for facts. That is, remembering, "X is the capital of Y" is the product of semantic memory, whereas remembering when and where you were while you were reading "X is the capital of Y" is the product of episodic memory. This generalized decrement in episodic memory may be related to findings in normal aging research such as disproportionate decrements in source memory (e.g., specifically remembering where an item was rather than what it was) and to prospective memory (e.g., remembering to do something in the future). These changes have important implications for successful navigation in interactive systems. For example, keeping track of where you have just been is often an important prompt to where you are going now.

Visuo-Spatial, Iconic, and Verbal Abilities

In addition to memory problems, there is, at least at the level of the population, marked deterioration of visuo-spatial and verbal abilities in older people. Decline in visuo-spatial abilities can cause difficulty with decoding layouts and utilizing any inherent organization. Related to this is a deterioration in iconic memory which, given the graphical nature of many interfaces, can be problematic in its own right. Particular difficulties have also been found in the comprehension of abstract and metaphorical phrases, with the tendency being to take them literally.

Such conditions can develop into more global aphasia (e.g., Broca's aphasia, related to the production of speech, and Wernicke's aphasia, related to comprehension). The general difficulty with recall of proper nouns, found in normal aging, can likely develop into more profound anomia. The depth of such problems may not be apparent to an outside observer as the ability to read aloud may be well preserved, although the content may not be properly understood.

Another distinct form of global cognitive impairment is autism, including a set of rarer but related syndromes (Kaluger & Kolson, 1987). The precise causes of autism are not clearly understood. Briefly stated, it is a general neurological disorder that impacts the normal development of the brain particularly in relation to social interaction and communication skills. Its effects will usually become apparent within the first three years of life. People with autism typically have difficulties in verbal and nonverbal communication and social interactions. The disorder makes it hard for them to communicate with others and relate to the outside world, they will also tend to have relatively low IQ scores. Closely related to autism is Asperger's Syndrome. People with Asperger's experience similar social communication difficulties but will generally demonstrate a normal IQ. Further to this, there are several generally similar conditions, some of which have varying physical and behavioral elements associated with them. These come under the collective heading of pervasive developmental disorders and all tend to produce difficulties with communication. An important element of these social communication difficulties in the context of the present chapter is an inability to grasp the implications of metaphorical or idiomatic language. This is similar to that mentioned above for dementia but in autism tends to be more profound. However, some evidence shows that people with autism or Asperger's syndrome are more able to communicate with computers than with people, or with people via computers, rather than face to face, and thus properly designed computer systems may have potential for assisting such user groups. A variety of other impairments could also be ameliorated with suitably designed support for communication.

Augmentative and Alternative Communication

Most interfaces use traditional orthography (the written word). However, cognitive limitations can have a major impact on the ability to encode and decode traditional orthography. People with acquired cognitive impairments (e.g. aphasia resulting from a stroke or cerebral vascular accident; CVA) can experience varying degrees of difficulty with both expressive and receptive language. A person with aphasia may have slight word-finding problems through to more pervasive problems in understanding spoken language. While some individuals may retain some literacy skills, damage to the language processing centers will usually also affect symbolic representations of language.

Individuals with congenital language and/or intellectual disabilities (e.g. congenital aphasia and Down's syndrome) may never become literate. The physical inability to speak (e.g. dysarthria resulting from cerebral palsy) may also impact literacy learning, as basic skills required for reading and writing (e.g. phonemic awareness) may be absent.

Visual images, photographs and drawings, provide augmentative and alternative ways to access communication. Such images can be used to enhance text-based interfaces. In addition, sets of symbols (e.g. the Picture Communication System, PCS, and semantic-based writing systems such as Blissymbols can be used as an alternative to text [Beukelman, & Mirenda, 1998; Wilson, 2003]). The type of picture, symbol, or graphic used will depend on the iconicity (ease of recognition), transparency (guessability), opaqueness (logic organization) and learnability of the image. For instance, a photograph of a house may be transparent—for example, it is easily recognizable—while a Blissymbol representing the emotion of happy has logic (heart = feeling; up arrow = up) and is thus opaque to learning (see Table 4.1). The more concrete a representation, the more recognizable it will be. However, representations of abstract meanings, such as emotions, will involve less transparent images necessitating a longer learning curve (see Table 4.1).

Cognitive Prostheses

In addition to providing better access to everyday software, specially designed computer systems have great potential to offer more specific support for people with a wide variety of cognitive impairments. We will describe some specific examples of such systems so that the reader is able to see how such development can proceed. Within the Applied Computing Department at Dundee University, a number of projects have developed computer-based systems to support people with cognitive dysfunction, and examples from the work of this group can

TABLE 4.1. Examples of Three Symbol Sets which have been Developed for Augmentative and Alternative Communication

	PCS	Rebus	Blissymbols
house			
happy			
sad			
big			
small			
fall			

illustrate the range of areas where this technology has particular potential. These examples will also demonstrate the need for appropriate methodologies for research in this area, including need for research into developing more sensitive and effective methodologies.

Software as a Cognitive Scaffolding and a Prompt for Communication

One sequence of projects took as its starting point the improvement of communication systems for physically impaired nonspeaking people. It became apparent that this could be done very effectively by developing models of the cognitive tasks involved in communication. This research has now spawned a new area of development, which is cognitive support for people with dementia, where communicative impairment is just part of their range of difficulties.

For severely physically impaired nonspeaking people, even with current speech output technology, speaking rates of 2 to 10 words per minute are common, whereas unimpaired speech proceeds at 150 to 200 words per minute. In an attempt to improve this disparity, a certain amount of progress has been made in the area of using computers to replace or augment some of the cognitive aspects of communication. Although the cognitive processes underlying language use are incompletely understood, a number of theories that attempt to explain language use have been used to improve the functionality of communication systems for nonspeaking people. This approach to the problem usually involves taking a sociolinguistic view of language. Instead of focusing on the building blocks, or taking a bottom up approach, the interaction as a whole is analyzed, paying attention to its goals, or taking a top down approach to the communication. This may well be a realistic simulation of the natural process, since the production of speech by an unimpaired speaker occurs at such a rate that conscious processing and controlling of the speech at a microlevel is not possible. In common with other learned skills, speech is produced to some extent automatically, with the speaker being aware of giving high-level instructions to the speech production system, but leaving the details of its implementation to the system.

The nonconscious control of much of speech production has been modeled in the CHAT prototype (Alm, Arnott, & Newell, 1992). This produced quick greeting, farewell, and feedback remarks by giving the user semiautomatic control of exactly what form the remarks would take, within parameters that the user had previously selected. This mimicked the phenomenon of a speaker responding automatically to greetings and other commonly occurring speech routines, without giving the process any detailed thought.

The CHAT-like conversation described illustrates an attempt to achieve a particular communicative goal, achieving social closeness by observing social etiquette. Some recent research efforts have been directed at finding ways to incorporate large chunks of text into an augmented conversation, to help users carry out topic discussion. This has been driven by the observation that a great deal of everyday discourse is reusable in multiple contexts.

Much of this type of discourse takes the form of conversational narratives. Research into the conversational narrative at a sociolinguistic level indicates several interesting characteristics. These include the way in which narratives are told and to whom they are told. For example, a recent event is told repeatedly for a limited time to most people with whom the speaker has contact. As the event recedes in history, the narrative is retold when it is relevant to the topic of conversation. The length of the narrative depends on its age (the older a story is, the more embellished it can become, particularly if it has previously gone down well) and the time available within the conversation. The version of the story (the sequence of events may remain the same while the details or embellishments of a story can differ) depends on factors such as the conversational context and other interlocutors present.

One of the ways to make the retrieval of text chunks easier is to anticipate the chunks that the user may want to use. This has been achieved by modeling the way in which conversational narratives are used using techniques from the fields of artificial intelligence and computational linguistics (Waller, 1992). The prototypes developed constantly adapt to the users' language use, thus mirroring the user's perception of where conversation items are stored. In this way, the system adapts to the way the user thinks instead of having the user learn a new retrieval system.

One of the arguments against using such prediction was raised when word prediction systems were first developed in the early 1980s. Therapists and teachers were concerned that nonspeaking people, especially children, would select what was offered on the screen rather than what they originally wanted to say. Although this may happen, research into predictive systems applied to writing suggests that they may carry over the help they offer and have a wider effect on the users' ability development. Some of this research reports an increase in written output by reluctant writers and people with spelling problems (Newell, Arnott, Booth, & Beattie, 1992). A general improvement in spelling has also been noted. Children with language dysfunction and/or learning disabilities have shown improvement in text composition (Newell, Booth, & Beattie, 1991). This research is in the writing domain, but the results suggest that predictive systems can offer assistance without becoming mere substitutes for creative expression.

Also, it is true of unimpaired conversation that speakers often change direction in their communication depending on chance occurrences, or on the sudden recollection of a point they would like to include. Thus, there is a degree of opportunism in all conversations. Another argument in favor of offering predicted phrases and sequences is that the current situation for most augmented communicators is that their conversations tend to be quite sparse, with control tending to reside with unaided speaking partners. If it is not possible to go boating on the lake, easily going off in any direction you please, is it not preferable to build a boardwalk out over the water than to stay on the shore?

One of the motivations to improve communication systems for nonspeaking people is the fact that they are commonly perceived by people who do not know them as being less intellectually capable than they actually are. It is often reported by nonspeaking people that they are considered unintelligent or immature by strangers. The issue of perceived communicative competence is one that needs increased attention (McKinlay, 1991).

Related to this, an interesting finding emerged from work in which one of the authors was involved. Here, a prototype communication system was used to evaluate listeners' impressions of the content of computer-aided communication based on prestored texts, as compared to naturally occurring dialogues. The nonspeaking user was able to use only prestored texts in order to conduct the conversations. Most of the text was material about one subject (holidays). A number of rapidly accessible comments and quick feedback remarks were also available. The unaided conversations were between pairs of normally speaking volunteers who were asked to converse together on the topic of holidays. Transcripts of randomly sampled sections of the conversations and audio recordings of reenactments of the samples with pauses removed were rated for social competence on a six-item scale (coefficient alpha = 0.83) by 24 judges. The content of the computer-aided conversations was rated significantly higher than that of the unaided samples ($p < .001$). The judges also rated the individual contributions of the computer-aided communicator and the unaided partners on how socially worthwhile and involving these appeared. There was no significant difference between the ratings of their respective contributions ($p > .05$; Todman, Elder, & Alm, 1995).

This finding came as something of a surprise to the researchers, since the original purpose had been to establish whether conversations using prestored material would simply be able to equal naturally occurring conversations in quality of content. Of course, the pauses in actual computer-aided conversation (removed in the above analysis) do have an effect on listeners' impressions of the quality of the communication, but this finding is still of interest, since it suggests that in some ways augmented communication could have an edge over naturally occurring talk. A plausible explanation for this finding is that naturally occurring talk is full of high-speed dysfluencies, mistakes, substitutions, and other messy features that listeners tend to discount with their abilities to infer what the speaker is intending to say. Prestored material is by its nature selected because it may be of particular interest, and it is expressed more carefully than quick flowing talk, and thus may appear more orderly and dense with meaning than natural talk.

In addition to conversational narratives, another common structure in everyday communication is the script, particularly where the speaker is undertaking some sort of transaction. Scripts may be a good basis for organizing prestored utterances to attempt to overcome the problem of memory load when operating a complex communication system based on a large amount of prestored material. Users' memory load can be reduced by making use of their existing long-term memories to help them locate and select appropriate utterances from the communication system. Schank and Abelson (1977) proposed a theory that people remember frequently encountered situations in structures in long-term memory, which they termed *scripts*. A script captures the essence of a stereotypical situation and allows people to make sense of what is happening in a particular situation and to predict what will happen next. Other research (e.g., Vanderheiden, Demasco, McCoy, & Pennington, 1996) has shown the potential that similar script-based techniques offer to this field.

An initial experiment was devised (Alm, Arnott, & Newell, 1999) to investigate the potential of a script-based approach to transactional interactions with a communication system, and a prototype system was developed to facilitate this experiment. The aim was to ascertain whether or not a transactional interaction could be conducted using a script-based communication system. It was decided to simulate a particular transactional interaction that could reasonably be expected to follow a predictable sequence of events, for example, one that would be amenable to the script approach, in order to find out whether a computer-based script could enable a successful interaction.

The transaction chosen for the experiment was that of arranging the repair of a household appliance over the telephone. Although the script interface was a relatively simple one devised for the purpose of this experiment, it was successful in facilitating the interaction, and produced a significant saving in the amount of physical effort required.

To take this work further a large-scale project was undertaken to incorporate scripts into a more widely usable device. The user interface of this system is made up of three main components: scripts, rapidly produced speech acts, and a unique text facility. The scripts component is used in the discussion phase of a conversation and consists of a set of scripts with which the user can interact. The rapid speech act component contains high frequency utterances used in the opening and closing portions of a conversation and in giving feedback and consists of groups of speech-act buttons. This facility is based on previous work with CHAT. The unique text component is used when no appropriate prestored utterances are available and consists of a virtual keyboard, a word prediction mechanism, and a notebook facility.

To provide access to a set of scripts, an interface was devised that involved a pictorial representation of the scenes in the script. The pictorial approach was taken in order to give users easier access to the stored material and to assist users with varying levels of literacy skills. In this interface, scripts are presented to the user as a sequence of cartoon-style scenes. The scenes give the user an indication of the subject matter and purpose of the script and assist the user to assess quickly whether the script is appropriate for current needs. Each scene is populated with realistic objects chosen to represent the conversation tasks that can be performed. Thus, the user receives a pictorial overview of the script, what happens in it, and what options are available. This assists the user to see quickly what the script will be able to do in the context of the current conversation. An example of the interface for the system can be seen in Fig. 4.1 below, which shows a scene within the at the doctor script.

Research into picture recognition and memory structures has demonstrated that groups of objects organized into realistic scenes corresponding to stereotypical situations better assist recognition and memory compared to groups of arbitrarily placed objects (Mandler & Parker, 1976; Mandler, 1984). The scene-based interface using a realistic arrangement of objects within a scene was therefore chosen to facilitate recognition and remembering by the user and thus reduce the cognitive load required to locate suitable objects during a conversation.

As it would be impractical to provide scripts for every conceivable situation, it was decided to provide users with a limited number of scripts together with an authoring package with which they can develop their own custom scripts with help from their therapists.

FIGURE 4.1. The script system user interface showing a scene from the doctor script.

A text preview and display box appears at the top of the user interface. The main interface area (bottom right) contains the scene image. The function buttons on the left side of the interface are, from top to bottom: "I'm listening" rapid speech-act button; button to access the main rapid speech-act interface; scene navigation backtrack to previous scenes button; scene navigation overview button; and tool button to access the notepad and additional system control facilities.

It was initially decided to develop six complete scripts. These were chosen after discussions with a user advisory group about situations in which they found difficulty communicating. The scripts developed were at the doctor, at the restaurant, going shopping, activities of daily living, on the telephone, meeting someone new, and talking about emotions.

The system uses the script to guide the user through a dialogue. There is a prediction mechanism, which predicts the next most probable stage in the dialogue that the user will need (based on the script), so the user can usually follow a predicted path through a conversation. This prediction mechanism monitors the sequences of objects selected and uses this information to modify future predictions.

Help for Aphasia

Communication systems for nonspeaking people have been described which in some way model the cognitive processes underlying communication. In the case of most physically disabled nonspeaking people, this is needed in order to speed up the communication process. However, in the case of speech problems that are caused by a stroke or other trauma (aphasia), the person trying to communicate will also have cognitive problems to deal with. Interestingly, the objections that conversation modeling might provide an active prompt to communication suggested a way of possibly helping people who might need such a prompt in order to initiate communication at all.

In a research project investigating the possibility of prompting people with Broca's aphasia in their communications, a pre-

dictive communication system was developed with a very simple interface (Waller, Dennis, Brodie, & Cairns, 1998). The system held personal sentences and stories, which were entered with the help of a caregiver. The user could then retrieve the prestored conversational items, with the system offering probable items based on previous use of the system. The interface was designed to be as simple as possible and to be usable by people who were unfamiliar with technology, had language difficulties, but had retained the ability to recognize familiar written topic words.

To access the sentences and stories, the user is led through a sequence of choices on the screen. First, they are offered a choice of conversational partners, a list of topics most likely to be appropriate for the chosen partner, and then a list of the four most common sentences for that partner and that topic. The user can choose to speak one of the sentences through a speech synthesizer, or have the system look for more suggestions. The sentences and topic categories are personal to each user, and the order in which topic words or sentences are presented depends on the past use of the system. Thus, the system is specific to users, both in the information content and in how it adapts to individual ways of communicating.

The system was evaluated with five adults with nonfluent aphasia who were able to recognize but not produce familiar written sentences. There was little change in the underlying comprehension and expressive abilities of the participants while not using the system. When making use of the system, the results showed that some adults with nonfluent aphasia were able to initiate and retain control of the conversation to a greater extent when familiar sentences and narratives were predicted. In other words, users' existing/residual abilities (e.g., small vocabulary, pragmatic knowledge of conversation) were to a degree augmented by the computer functioning as a cognitive prosthesis. This project indicated that a communication system based on prompting could be of help to people with cognitive and communication difficulties.

Support for Nonliterate Users

Children with complex disabilities may have a combination of physical and intellectual impairments that impact on the development of speech and language. Some individuals will learn to read and write using this medium in combination with nonverbal means to communicate. Others will use an alternative graphic medium for interaction. One of the drawbacks of providing symbol-based communication is the difficulty in providing access to novel vocabulary. The ongoing BlissWord project (Andreasen, Waller, & Gregor, 1998) is investigating ways in which users with physical and cognitive limitations can explore new vocabulary. Blissymbolics is a semantic-based natural written (pictographic) language, similar to Chinese. Because of its generative characteristics (BlissWords are spelled using a sequence of one or more Bliss characters), predictive algorithms can be applied to Blissymbolics to assist users in the retrieval of words.

BlissWords are sequenced beginning with a classifier (e.g., all emotions begin with a heart). As illustrated in Fig. 4.2, selecting a shape from the Bliss keyboard, the interface produces a list of BlissWords which begin with classifiers using that shape.

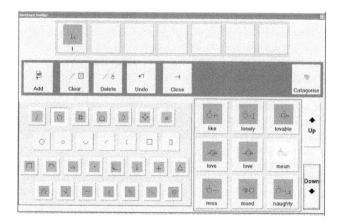

FIGURE 4.2. Screenshot showing Blissymbol keyboard on left and prediction list on right.

Frequency and word lists can be used further to refine the Bliss-Words that are displayed. Users do not need to be literate to explore language and vocabulary. It is envisaged that video clips and spoken explanation could further augment learning through exploration.

Support for Dementia

Dementia, which involves the loss of short-term memory in elderly people, is a very serious problem for the person and for their families and caregivers. It can rule out most social activities and interactions, since these depend on a working short-term memory for effective participation. This includes even the essential ability to communicate.

As well as being valuable with all older people, reminiscence is an important tool used to help elderly people who have dementia (Sheridan, 1992; Feil, 1993). This is because, while their short-term memories may be impaired, their long-term memories are often more or less intact (Rau, 1993). The difficulty is accessing these long-term memories without the capability of keeping a conversation going, which depends on short-term memory. Activities that do not require patients to maintain the structure of a conversation can help, for instance, looking at and commenting on a series of photographs, which can provide a framework for meaningful person-to-person interaction.

The tools used in such reminiscence work can also include videos, sound, music, and written materials. However, traditionally these are all in separate media, and it can be very time consuming searching for a particular photo, sound, or film clip. Bringing all these media together into a digital multimedia scrapbook could mean easier access to content for more lively reminiscence sessions. The intention of this project was to begin to develop a system that could act as a conversation prosthesis, giving the user the support needed to carry out a satisfying conversation about the past.

An investigation was undertaken to determine which aspects of multimedia would be most helpful for such a reminiscence experience, and the best way to present them. A number of prototype interfaces for a multimedia reminiscence experience

were developed. These included text, photographs, videos, and songs from the past life of the city. The materials were collected with the assistance of the University, Dundee City archives, and the Dundee Heritage Project. The prototypes were demonstrated for people with dementia and their care staff at a day center run by Alzheimer Scotland Action on Dementia. The following issues were addressed in these evaluation sessions:

1. *Is it better for the display to use the metaphor of a real-life scrapbook or a standard computer screen?* Six of the staff members preferred the book presentation, three preferred the screen, and two had no preference. Interestingly, this was almost a reversal of the preference shown by the people who had dementia, with the majority preferring the screen presentation. The preference shown for the screen presentation could be due to reduced cognitive ability. The book presentation is a metaphor that may not be interpreted suitably by the person with dementia.
2. *How should the scrapbook material be organized—by subject or by media type?* The majority of the staff evaluators preferred the arrangement by subject saying it was more logical, some were unsure; however, no one showed a preference for the arrangement by media. The clients with dementia reflected these findings. Despite preferring the arrangement being by subject, the majority of evaluators could see benefits from having access to both arrangements. It was concluded that for basic reminiscence sessions the arrangement by subject is preferable. But access to the arrangement by media should be an option, to make the software available for use in other ways.
3. *How do the sounds, pictures, video, and music add to the reminiscence process, and what are their differential effects?* Most of the videos and photographs and all the songs were able to spur conversations. However, it was found that with videos the clients were able only to identify strongly with them when they triggered off specific personal memories, whereas songs and photographs were more generally appreciated. Attention remained focused longest with the songs, which were particularly enjoyed when played repeatedly with everyone singing along. However, the staff felt that some individual clients had enjoyed the videos most.

One general finding was that the multimedia presentation as a whole produced a great deal of interest and motivation from the people with dementia. The staff was also very keen to see the idea developed further. A new project is now beginning which will fully develop the ideas produced by this preliminary work and will produce a fully functional multimedia reminiscence system. The project will be carried out by a multidisciplinary team consisting of a software engineer, a designer, and a psychologist specializing in aging.

In addition to its benefits for the person with dementia, participation in reminiscence activities has been shown to have a positive outcome for the caregivers who take part. Thus, help for reminiscence is not only a tool to stimulate interaction, but also a contributor to improved quality of life, for the person with dementia and his or her family.

Recent work on using videos to present life histories for people with dementia has shown that new technologies, where sensitively and appropriately applied, can bring a substantial added impact to supportive and therapeutic activities for people with cognitive problems (Cohen, 2000). We hope the reminiscence system developed will have the immediate application described above and will also serve as an exploration in developing a range of computer-based entertainment systems for people with dementia.

SYSTEMS TO SUPPORT PEOPLE WITH DYSLEXIA

Dyslexia is a language disorder that is still in the process of being investigated and thoroughly understood. A word-processing environment has been developed to alleviate some of the visual problems encountered by some people with dyslexia when they read and produce text. However, in the absence of a full understanding of the nature of dyslexia, the researchers' approaches with to identify some of the most commonly noted problems with dyslexics. On the basis of these common difficulties, they identified ways in which each individual might be able to minimize the consequences of their own particular problems by manipulating the appearance of their word-processing environment and text presented within it. This stage of the development process involved the start of an iterative cycle of prototype development and evaluation with dyslexic computer users. The work led to a software system which provides a highly (and easily) configurable environment for dyslexic people (who experience a very wide range of differing preferences and problems) to use for reading and producing text. The approach was to examine the parameters of the situation, including cost, cost benefit, existing software, and user demand, with a view to finding an optimal path to the production of a software system which is of real use in practical situations. While users were extensively involved throughout the development process, the researchers did not always rely directly on their individual inputs for further development; rather they ensured that at all times the process was sensitive to user need and opinion, but informed by relevant sources ranging from documented wisdom about the subject to hunches about what might work.

This development of a word processor to assist dyslexics is thus a particularly illuminating example of how the needs of people with a particular type of cognitive impairment can be effectively factored into the design and development process. This research will thus be described in some depth, as an example of a development and to assist the reader in appreciating the generic importance of this approach, which requires knowledge of the underlying syndrome, and a methodology, which encourages an innovative approach to user involvement.

Dyslexia, as has been noted, is still being investigated as a language disorder, and at present, it has a number of definitions. The British Dyslexia Association (2006) offered this description:

Dyslexia is best described as a combination of abilities and difficulties that affect the learning process in one or more of reading, spelling, and writing. Accompanying weaknesses may be identified in areas of speed of processing, short-term memory, sequencing and organisation, auditory and/or visual perception, spoken language and motor skills. It is particularly related to mastering and using written language, which may include alphabetic, numeric and musical notation.

The symptoms of dyslexia vary greatly and the reasons for the existence of these symptoms are also varied. However, two distinct types of dyslexia are recognized: acquired dyslexia and developmental dyslexia.

Acquired dyslexia is associated with those people who have difficulties caused by damage to the brain. Therefore, prior to the occurrence of brain damage, no difficulties would have been identified. This area can be subdivided into disorders in which the visual analysis system is damaged—peripheral dyslexia—and disorders in which processes beyond the visual analysis system are damaged, resulting in difficulties affecting the comprehension and/or pronunciation of written words—central dyslexia.

Peripheral dyslexia is associated with difficulties such as misreading letters within words and migrating letters between words. Central dyslexia on the other hand concerns issues such as bad comprehension due to an impaired semantic system, and the inability to read unfamiliar words, while familiar words are read easily. Examples would include *monkey* being read as *ape* (a semantic error), and *patient* being read as *parent* (a visual error).

Developmental (or congenital) dyslexia is present from birth but will generally become apparent later, and has been described as follows:

Developmental dyslexia is a learning disability which initially shows itself by difficulty in learning to read, and later by erratic spelling and by lack of facility in manipulating written as opposed to spoken words. The condition is cognitive in essence, and usually genetically determined. It is not due to intellectual inadequacy or to lack of sociocultural opportunity, or to emotional factors, or to any known structural brain defect. (Critchley, 1964, p. 5)

One of the main features of dyslexia is the individual nature of the disorder. The condition is not typically characterized by one single difficulty, but by a range of difficulties that will vary in combination and in intensity between individuals, giving rise to an enormous variation between individuals in the problems encountered. Each dyslexic person thus has a range of difficulties that need to be addressed differently from those of others. Dyslexia is an example of the need to design for dynamic diversity.

The wide-ranging characteristics of dyslexia provide a challenge for technological assistance, as a single approach will not be appropriate for the range of problems presented by the population of dyslexic people. Computer technology offers the opportunity to provide reading and writing systems that are highly configurable for each individual user but they need to be based on an understanding of the problems which dyslexics have in reading and writing, and some of the visual problems which can affect them. The approach adopted in this research was to offer dyslexic users a range of appropriate visual settings for the display of a word processor, together with the opportunity to configure easily the way in which text is displayed to them. The user can select, by experimentation, the settings that best suit them. These settings are then saved and later recalled each time that person uses the word processor. It will be seen that this approach affords the potential to make computer-based text significantly easier to read than printed text, as well as improving the usability of computer word-processing systems for a wide range of dyslexics.

Some Common Problems of Dyslexia

It was first necessary to determine the parameters which should be offered for configuration by the user. An initial investigation revealed that some of the most commonly encountered problems are as follows (adapted from Willows, Kruk, & Corcos, 1993; D. Shaw, personal communication, May, 1994):

1. *Number and letter recognition.* One of the fundamental problems faced by dyslexics is the recognition of individual alphanumeric symbols. This is often seen when letters that are similar in shape, such as *n* and *h* and *f* and *t,* are confused. The problem is exacerbated with the introduction of uppercase letters. In addition, many dyslexic adults, who are capable of reading printed letters, have difficulty in reading cursive writing.

2. *Letter reversals.* Many dyslexics are prone to reversing letters, which results in a particular letter being interpreted as another letter. Examples of these characters would be *b, d, p,* and *q.* This problem can result in poor word recognition with words containing reversal characters being substituted for other words such as *bad* for *dad.*

3. *Word recognition.* As well as the substitution effect caused because of letter reversals, words that are similar in their outline shape (word contour) can be substituted by dyslexics. Typical examples of this problem are the words *either* and *enter.* Both words have the same start and finishing characters and this, allied with their similar word contours, make them candidates for being substituted for each other when they occur in the text.

4. *Number, letter, and word recollection.* Even if the ability to recognize numbers and letters is adequate, it can still prove difficult for a dyslexic individual to recall the actual form and shape of a character. Many dyslexics have so much difficulty recalling upper and lower case characters that they continue to print later in life. Similarly, poor visual memory means that dyslexics have little ability to distinguish whether or not a word looks right.

5. *Spelling problems.* Due to the problems discussed above, dyslexics can have great difficulty with spelling, and many dyslexics have very poor spelling. Much of the spelling of dyslexics appears to reflect a phonic strategy with words like *of* and *all* being spelled *ov* and *ohl.*

6. *Punctuation recognition.* As with characters, dyslexics appear to have difficulty recognizing punctuation marks.

7. *Saccadic and fixation problems.* Another problem that is found in many dyslexics is their lack of ability to follow text without losing their places. Many find it difficult to move from the end of one line to the beginning of the next and find themselves getting lost in the text.

8. *Word additions and omissions.* Dyslexics may add or remove words from a passage of text, apparently at random. This is manifested by words being omitted or duplicated, extra words being added, or word order being reversed or otherwise jumbled.

9. *Poor comprehension.* With the variety of errors caused by the factors described above, a dyslexic person may perceive a totally different (or impoverished) passage of text from the one that is actually in front of them. Dyslexics thus display poor comprehension skills due to text which they perceive being significantly different from the actual text.

Computer Aid for the Problems of Dyslexia

In an attempt to alleviate some of the problems discussed above, dyslexics, particularly within the education system, are encouraged to use computers for text manipulation. The use of a computer keyboard has the potential to alleviate the problems of character recollection, but this only really helps with the recollection of characters, not the recognition of them once they are on the screen.

There is, however, strong evidence to suggest that the use of lexical and spelling aids can greatly assist with spelling problems exhibited by dyslexics (e.g., Newell & Booth, 1991). However, merely highlighting an incorrect word and offering a replacement may not be enough, since one of the other problems that some dyslexics face is an inability to tell if a word looks right; thus they will have difficulty selecting the appropriate corrections.

Some dyslexics are sensitive to color and colored acetate screens or tinted glasses as well as lighting conditions can improve their abilities to read text. Many dyslexics also report interference from peripheral vision, indicating that anything that can be done to reduce screen clutter outside the main screen window, such as making the document page fill the whole screen, may be of benefit (D. Shaw, personal communication, May, 1996).

Based on the above difficulties encountered by dyslexics, the researchers considered ways that the screen image of the text could be manipulated. These included foreground and background color, character typeface, font and spacing, making letters or words with similar shapes distinguishable by using different fonts, sizes, and/or colors, and presenting text in narrower columns, or with different spacing between words and/or lines to reduce saccade and fixation problems. The researchers investigated potentially promising ideas by implementing prototypes and evaluating their utility with dyslexic users, with an overall view to developing a configuration system that will enable all dyslexics to set up their own optimized environments.

An Experimental Text Reader for Dyslexics

The first stage of the research was to develop an experimental text reader. This prototype presented the user with an easily configurable interface, which allowed for a number of display variables to be altered. Initially, these were background, foreground, and text colors, font size and style, and the spacing between paragraphs, lines, words, and characters. The interface was designed in such a way that it gave visual feedback on selections before they were confirmed and made minimal use of text instruction.

This was evaluated using 12 computer literate dyslexic students from higher education using think-aloud techniques, as well as questionnaires and interviews. At various development stages, the helpers were asked to try out the system with a view to seeing if it was possible for them to put together a display which improved their abilities to read text from the screen. All the users were able to find a setting that was subjectively superior for them to standard black text on a white background with Times Roman 10- or 12-point text, but the screen layouts that were developed by the test subjects were extremely varied. This highlighted the individual nature of the disorder, and the diverse characteristics of any interface that would be appropriate for this group.

Each appeared to have a favorite color combination, although brown text on a green background was liked by all the testers. Subjects were in greatest agreement about the selection of a typeface: sans serif. Arial was rated the best by almost all the testers. All reported that increasing the spacing between the characters, words, and lines was beneficial. The most interesting point that arose during the testing, however, was the fact that at the beginning of the evaluation period the dyslexic subjects did not appear to be aware that altering these variables might be of any use.

A second prototype was then developed based on Word for Windows (Microsoft, 1994; 1995) macros to provide the required configuration interfaces. This was based on the concept of an evolutionary system, rather than a fixed prototype. It was clear that there would be a substantial advantage in developing a dyslexic configuration, but this design decision raised an interesting deviation from the received wisdom of the desirability of WYSIWYG (What You See Is What You Get). In the case of a dyslexic user, what you see should be whatever you can read best, and print previewing facilities would have to be used to show how the layout will appear when printed.

This prototype provided a facility to enhance characters prone to reversals (e.g., b, d), by using color font type and size. This idea of coloring reversal characters provided very interesting and unanticipated results, which are described below. Fixation problems were tackled by reducing the page width, and a speech synthesizer that could read the text on the screen was included.

There were two distinct parts to the overall solution, a preference program, and a reading/editing program. The first allowed users to experiment with the various parameters and the second made use of these preferences within a reading and editing environment. The preference program menu presents the user with various options and variables, together with a preview facility, to enable the user to experiment with, and finally store his or her data in a preferences file.

The fact that a unique user environment, tailored to the need of each individual, is provided means that the document is (deliberately) not WYSIWYG. A print option thus allows the user to print the document as it appeared with his or her preferred formatting applied to it, or as it would appear without any special formatting.

This second prototype was developed as an add-on module to Microsoft Word, and was evaluated by seven dyslexic users of 15 to 30 years old, in a similar fashion to that above. The users found the system easy and intuitive to use, reporting that each of the options had an effect on their abilities to read. The options that allowed the user to change the color scheme of the document appeared to be the most helpful, but font size and spacing, column width, and indications of reversals were also reported to assist reading by some or all of the users.

The reversals option provided the most interesting results of all. However, the reason for the improvement was not always that the reversal characters were clearly distinguished and easy to read. Instead, it was claimed that the sporadic coloring broke the text up and resulted in the user being less likely to get lost; for example, the system was reducing fixation problems rather than recognition problems.

As the testing progressed, the testers appeared to be surprised at times by the effect some of the changes had on their abilities to read the document. Comments included "I would never have thought of doing that," or "I don't think that will do me much good" before finding that a feature did indeed help.

The prototypes were developed from the perspective that the user population was diverse, and that the design process must accommodate potential changes in preferences over time. The fact that dyslexia is a very idiosyncratic disorder and findings that the users were often unaware of how easy it was to improve their reading potentials by changing visual aspects of the reading environment illustrate how a standard user-centered design methodology is not appropriate for such user groups.

RESEARCH METHODOLOGIES

The research described above gives a flavor of successful approaches to developing human interfaces and software to support people with various types of cognitive impairment. Much of the methodology used in these developments, however, had to be developed *ab initio*. Traditional user-centered design does not have the flexibility for these user groups, and most research and development in the field of communication and information technology to support people with disabilities has, to date, concentrated on the development of special assistive systems and on accessibility features for younger, mainly physically or sensorially disabled people. Similarly, the human interfaces to most computer systems for general use have been designed, either deliberately or by default, for a typical, younger user (Newell & Cairns, 1993; Newell, 1995; Newell & Gregor, 1997). Knowledge from these fields does not necessarily transfer comfortably to the challenges encompassed in universal design (Beirmann, 1997; Hypponen, 1999; Sleeman, 1998; Stephanidis, 2001) and, in particular, the widely varying and often declining abilities associated with the range of cognitive impairments.

This section addresses the particular issues for the design process which accompany cognitive impairment and suggests a paradigm and methodology to support the process of designing software which is as near to the universal accessibility ideal as is possible, derived from the approach to specific projects described above.

Software systems which are aimed at the mainstream (rather than being of a prosthetic nature) need to address the wide variation in the types and severity of cognitive impairment between individuals. This demand is further complicated by the fact that as people grow older their abilities change. This process of change includes a decline over time in the cognitive, physical, and sensory functions, and each of these will decline at different rates relative to one another for each individual. This pattern of capabilities varies widely between individuals, and as people grow older, the variability between people increases. In addition, any given individual's capabilities vary in the short term due, for example, to temporary decrease in, or loss of, function due to a variety of causes, such as the effects of drugs, illness, blood sugar levels, and state of arousal.

This broad range and variability of change presents a fundamental problem for the designers of computing systems, whether they be generic systems for use by all ages, or specific systems to compensate for loss of function. Systems tend to be developed for a typical user and either by design or by default, this user tends to be young, fit, male, and crucially, has abilities that are static over time. These abilities are assumed broadly similar for everybody. Not only is this view wrong, in that it does not take account of the wide diversity of abilities among the wider population of users, but it also ignores the fact that for individuals, these abilities are dynamic over time.

Current software design also typically produces an artifact which is static and which has no, or very limited, means of adapting to the changing needs of users as their abilities change. Even the user-centered paradigm (e.g., ISO 13407, 1999; Nielsen, 1993; Preece, 1994; Shneiderman, 1992) looks typically at issues such as representative user groups, without regard for the fact that the user is not a static entity. Thus, it is important not only to be aware of the diverse characteristics of people with cognitive dysfunction, but also the dynamic aspects of their abilities.

It is clear that people with cognitive impairments, whatever their cause, can have very different characteristics to most human interface and software designers. It is also clear that in these circumstances user-centered design principles need to be employed if appropriate technology is to be developed for this user group (Gregor & Newell, 1999). These methodologies, however, have been developed for user groups with relatively homogonous characteristics. People with dementia, for example, are a diverse group and even small subsets of this group tend to have a greater diversity of functionality than is found in groups of able young people.

An additional complication is that there can be serious ethical issues related to the use of such people as participants in the software development process. Some of these are medically related, but also include, for example, the ability to obtain informed consent. It is thus suggested that the standard methodology of user-centered design is not appropriate for designing for the inclusion of this user group. The importance of research and development taking into account the full diversity of the potential user population, including cognitive diversity, was addressed by Newell in his keynote address to InterCHI '93, where the concept of "Ordinary and Extraordinary Human Computer Interaction" was developed (Newell, 1993; Newell & Cairns, 1993; Newell & Gregor, 1997).

Market share is clearly an important consideration, and this has been given impetus, not only by demographic trends, but also by recent legislation in the United States and other countries, on accessibility of computer systems for people with disabilities. In terms of the workplace, both the Americans with Disabilities Act and the United Kingdom Disability Discrimination Act put significant requirements on employers to ensure that people with disabilities are able to be employed within companies and to provide appropriate technology so that such employees had full access to the equipment and information necessary for their employment. Increasingly, there is political pressure to increase this access, and more requirements for improved access by disabled people are being enshrined in legislation. However, access does not mean only that people with wheelchairs can maneuver around buildings; it also means that there needs to be provision for people with cognitive (and sensory and other physical) impairments to be able to operate computers and other equipment essential to the workplace.

An important additional factor in the value for money equation is that design that takes into account the needs of those with slight or moderate cognitive dysfunction can produce better design for everyone. An example where this has not occurred is illustrated by the problems that the majority of users have had with video tape recorders. If the designers had considered those with cognitive impairments within their user group, they may have been able to design more usable systems. Another example is an e-mail system specifically designed to be simple to use by older people with reduced cognitive functioning, which was found to be preferred by executives to the standard e-mail system that they were used to.

Some people are impaired from birth, but some may become temporarily or permanently disabled by accident or illness (suddenly or more slowly), or even by normal functioning within their employment. This is particularly noticeable in cognitive functioning. Short-term changes in cognitive ability occur with everyone. These can be caused by fatigue, noise levels, blood sugar fluctuations, lapses in concentration, stress, or a combination of such factors, and can produce significant changes over minutes, hours, or days. In addition, alcohol and drugs can also induce serious changes in cognitive functioning, which is recognized in driving legislation, but not in terms of how easy it is to use computer-based systems.

Most people, at one time or another, will exhibit cognitive functional characteristics which are significantly outside the normal range. Although neither they, nor their peers, would consider these people disabled, their abilities to operate standard equipment may well be significantly reduced.

The questions that designers need to consider include:

- Does the equipment that I provide comply with the legislation concerning use by employees who may be cognitively disabled?

- To what extent do I need to take into account the needs of employees who are not considered disabled, but have significant temporary or permanent cognitive dysfunction?

- Should I make specific accommodation for the known reductions in cognitive abilities which occur as employees get older (e.g., less requirement for short-term memory, or the need to learn new operating procedures)?

- What are the specific obligations designers and employers have to provide systems that can be operated by employees whose cognitive ability has been reduced due to the stress, noise, or other characteristics of the workplace?

The argument is that it would be very unusual for anyone to go through their working life without at some stage, or many stages, being significantly cognitively disabled. If equipment designers considered this, it is probable that the effectiveness and efficiency of the work force could be maintained at a higher level than would be the case if the design of the equipment were based on an idealistic model of the characteristics of the user and their work environment.

The Disabling Environment

In addition to the user having characteristics that can be considered disabled, it is also possible for them to be disabled by the environments within which they have to operate. Newell & Cairns (1993) made the point that the human–machine interaction problems of an able bodied (ordinary) person operating in an high workload, high stress, or otherwise extreme (e.g., extraordinary) environment have very close parallels with a disabled (extraordinary) person operating in an ordinary situation (e.g., an office).

High workloads and the stress levels to which this can lead often reduce the cognitive performance of the human operator. For example, a very noisy environment cannot only create a similar situation to hearing or speech impairment, but can also lead to reduced cognitive performance. The stress level in the dealing room of financial houses can be very high and is often accompanied by high noise levels. A significant advance may be made if the software that was to be used in these houses was to be designed on the assumption that the users would be hearing impaired and have a relatively low cognitive performance. It is interesting to speculate as to whether such systems would produce higher productivity, better decision making, and less stress on the operators. Other examples of extreme environments in which people have to operate are the battlefield, under water, or out in space. The stress and fatigue caused by working within such environments means that their performance is similar to that which could be achieved by a very disabled person operating in a more normal environment. It is not always clear that the equipment such people need to operate has been designed with this view of the user.

It is very important to describe the users of technology in terms of their functional abilities related to technology rather than generic definitions of either medical conditions or primarily medical descriptions of their disabilities. Unfortunately, most statistical data is presented as generic and medical categorizations of disability. Gill and Shipley (1999), however, defined disabled user groups in terms of their functional abilities, with specific emphasis on the use of the telephone. They estimated that within the European Union, which has a population of 385 million, there were 9 million people with cognitive impairment, which could lead to problems using the telephone. Theses figures do not take into account multiple impairments, and the authors pointed out that, in the elderly population in particular, there may be a tendency toward cognitive, hearing, vision, and mobility impairments being present to a varying extent and these may interact when considering the use of technological systems. It is this multiple minor reduction in function (often together with a major disability) which means that the challenges to technological support for older people have significantly different characteristics to that of younger disabled people and to the nondisabled, nonelderly population.

There has been some movement in mainstream research and development in technology, both in academia and industry, away from a technology led focus to a more user led approach, and this has led to the development of user-centered design principles and practices in many industries. In addition, a number of initiatives have been launched to promote a consideration of people with disabilities within the user group in mainstream product development teams with titles including "Universal Design," "Design for All," "Accessible Design," and "Inclusive Design." The "Design for All/Universal Design" movement has been very valuable in raising the profile of disabled users of products, and has laid down some important principles. This approach however has tended not to place too much significance on cognitive impairment, and, particularly if this is included as a factor in the design process, then it becomes more difficult to use traditional user-centered design approaches.

Newell and Gregor (2000) suggested that a new design approach should be developed, which would be based on the already accepted user-centered design methodology. There are some important distinctions between traditional user-centered design with able-bodied users and the approach needed when the user group either contains, or is exclusively made up of, people with cognitive dysfunction. These include:

- Much greater variety of user characteristics and functionality
- The difficulty in finding and recruiting representative users
- Situations where design for all is certainly not appropriate (e.g., where the task requires a high level of cognitive ability)
- The need to specify exactly the characteristics and functionality of the user group
- Conflicts of interest between user groups, including temporarily able bodied
- Tailored, personalizable, and adaptive interfaces
- Provision for accessibility using additional components (hardware and software)

The balance in the design process also needs to shift from a focus on user needs to one on the users themselves. There will be additional problems when considering people with cognitive dysfunction, which will include

- The lack of a truly representative user group
- That a different attitude of mind of the designer is required
- Ethical issues (Alm, 1994; Balandin & Raghavendra, 1999)
- It may be difficult to get informed consent from some users
- Difficulties of communication with users
- The users may not be able to (sufficiently) articulate their thoughts, or even may be incompetent in a legal sense

Thus, there can be particularly difficult ethical problems when involving users with cognitive impairments in the design process. In addition, it is often necessary to involve clinicians when such users are involved, so some of the user-centered design actually focuses on professional advice about the user, rather than direct involvement of the user. Even with these problems, however, it is possible to include users with cognitive dysfunction sensitively in the design process.

The Inclusion of Users With Disabilities Within Research Groups

In Dundee users with disabilities have a substantial involvement in the research, and they have made a significant contribution both to the research and to the commercial products that have grown from this research. Users are involved in two major ways:

- As disabled consultants on the research team, where they act essentially as test pilots for prototype systems.
- By the traditional user-centered design methodology of having user panels, formal case studies, and individual users who assess and evaluate the prototypes produced as part of the research.

The contribution made by clinicians is also vital to the research, and these are full members of the research team. Dundee's Applied Computing Department is also one of the few computing departments that has employed speech therapists, nurses, special education teachers, linguists, and psychologists (both clinical and cognitive).

User Sensitive Inclusive Design

Some significant changes must be introduced to the user-centered design paradigm if users with disabilities are to be included, and this is particularly important if the users have cognitive impairment. In order to ensure that these differences are fully recognized by the field, the title "User Sensitive Inclusive Design" has been suggested. The use of the term *inclusive* rather than *universal* reflects the view that inclusiveness is a more achievable, and in many situations, appropriate goal than universal design or design for all. *Sensitive* replaces *centered* to underline the extra levels of difficulty involved when the range of functionality and characteristics of the user groups can be so great that it is impossible in any meaningful way to produce a small representative sample of the user group, nor often to design a product that truly is accessible by all potential users.

Design for Dynamic Diversity

In addition to the aspects of user sensitive inclusive design described above, it is necessary to make designers fully aware of the range of diversity which can be expected with cognitively impaired people, and also the changing nature of the cognitive functioning of people. Thus, Gregor and Newell (2000) suggested that this be drawn particularly to the attention of designers by introducing the concept of "Designing for Dynamic Diversity." This process, described above, entails recognition that people's abilities are diverse at any given age and that as they grow older this diversity grows dynamically; it also involves a recognition that any given individual's abilities will vary according to factors such as mood, fatigue, blood sugar levels, and so on. Only by taking on board the factors associated with Designing for Dynamic Diversity will software design produce artifacts which are not static and which have no, or very limited, means of adapting to the changing needs of users as their abilities change.

As has been seen above, metaphors and processes in use at present are limited in meeting the needs of this design paradigm or addressing the dynamic nature of diversity. New processes and practices are needed to address the design issues; awareness raising among the design, economic, and political communities has to start; and research is needed to find methods to pin down this moving target.

A story-telling metaphor. In addition, researchers need to consider how best to disseminate the concepts behind universal usability and the results of user sensitive inclusive research. User sensitive inclusive design needs to be an attitude of mind rather than simply the mechanistic application of design for all guidelines. This offers a further challenge to the community. The dangers of using such studies to produce more extensive guidelines has been referred to above, but it is important that the results of user sensitive inclusive design are made available to other designers and researchers. However, it is too early to lay down principles and practices that must be followed by designers, and it may even be impossible to do this for some of the contexts and environments in which designers work. Thus, it is suggested that we follow a story-telling approach, in which information about accessibility issues and design methods which focus on accessibility is presented in narrative form, with particular examples to illustrate generic principles. This is, in some sense, an extension of the single case-study methodology. This methodology could provide very useful insights to designers in a form that they will find easy to assimilate and act. Thus, this will assist in their educations and will help them to design more accessible products, and better products for everyone.

The use of theater. As an extension of the story-telling metaphor, the research group in Applied Computing at Dundee University has investigated the use of dramatic techniques and theater as a way of addressing the challenges of user sensitive design. For example, as part of the UTOPIA (Usable Technology for Older People: Inclusive and Appropriate) project, they worked in collaboration with the Foxtrot Theater Company to use theater to encourage interaction between (older) users of technology and designers. The outcome was the "UTOPIA Trilogy," a series of short-video plays addressing problems older people have in using technology (Carmichael, Newell, Morgan, Dickinson, & Mival, 2005). The films were dramatizations of some of the issues the researchers had encountered during the project. These films were based on real events, conversations, and observations, and they were the amalgamation of many and are intended to convey older people's experiences with technology and the situations they encounter. These videos were evaluated with a variety of audiences including academics, practitioners, software engineers, relevant groups of undergraduates, and older people. This established that the videos provided a useful channel for communication between users of technology and designers, and changed the perceptions of both students and more mature designers of IT systems and products about older people's requirements. A similar technique has also been used in the requirements gathering phase for an IT system designed to monitor older people in case of falls at home (Marquis-Faulkes, McKenna, Gregor, & Newell, 2003).

This research showed that the use of theater can be a very powerful method of encouraging dialogue between various professional groups particularly in a clinical environment, for keeping a focus for discussions, and also for providing a channel for communication between users of technology and designers. The researchers view is that the success of this approach was in large part due to the plays being narrative based rather than having a pedagogic style. That is, they illustrated the issues involved within interesting story lines, with all the characteristics of a good narrative—humor, tension, human interest, and antagonists and protagonists. In addition, the quality of the production, having been produced by theater professionals, played a major part in the success of the venture.

Newell, Morgan, Carmichael, and Gregor (2006) discussed the various ways in which actors and theater can play a part in the design process for human–computer interfaces. This could provide a particularly valuable methodology for the design process when the target users have cognitive impairment and thus may not be appropriate for including within standard user-centered design methodologies.

CONCLUSION

Although it is not necessary for human interface designers whose systems may be used by people with cognition impairment to be fully versed in all aspects of cognition, it is important for them to have some background knowledge of the area. They should also be in contact with experts in other disciplines, such as psychology, and have access to appropriate clinical knowledge. In addition, the development of the concept of and a methodology for user sensitive inclusive design, design for dynamic diversity, and development of story-telling methods for communicating results will facilitate researchers in this field and will provide mainstream engineers with an effective and efficient way of including people with disabilities within the potential user groups for their projects. If both of these can be achieved it will go some way towards providing appropriate technological support for people with cognitive impairment.

References

Alm, N. (1994). Ethical issues in AAC research. In J. Brodin & E. B. Ajessibm (Eds.), *Methodological Issues in Research in Augmentative and Alternative Communication: Proceedings of the Third ISAAC Research Symposium* (pp. 98–104). Sweden: University Press.

Alm, N., Arnott, J. L., & Newell, A. F. (1992). Prediction and conversational momentum in an augmentative communication system. *Communications of the ACM, 35*(5), 46–57.

Alm, N., Morrison, A., & Arnott, J. L. (1995). A communication system based on scripts, plans and goals for enabling non-speaking people to conduct telephone conversations. *Proceedings of the IEEE Conference on Systems, Man & Cybernetics*, Vancouver, Canada, 2408–2412.

Andreasen, P. N., Waller, A., & Gregor, P. (1998). BlissWord—full access to Blissymbols for all users. *Proceedings of the 8th Biennial Conference of ISAAC*, Dublin, Ireland, 167–168.

Bäckman, L., Small, B. J., Wahlin, Å., & Larsson, M. (2000). Cognitive functioning in very old age. In F. I. M. Craik & T. A. Salthouse (Eds.), *The handbook of aging and cognition* (pp. 499–558). Mahwah, New Jersey: Lawrence Erlbaum Associates.

Balandin, S., & Raghavendra, P. (1999). Challenging oppression: Augmented communicators' involvement in AAC research. In F. T. Loncke, J. Clibbens, H. H. Arvidson, & L. L. Lloyd (Eds.), *Augmentative and Alternative Communication, New Directions in Research and Practice* (pp. 262–277). London: Whurr.

Beirmann, A. W. (1997). *More than screen deep—Towards an every-citizen interface to the National Information Infrastructure.* Computer Science and Telecommunications Board, National Research Council. Washington DC: National Academy Press.

Beukelman, D. R., & Mirenda, P. (1998). *Augmentative and alternative communication management of severe communication disorders in children and adults* (2nd ed.). Baltimore, MD: Brookes.

British Dyslexia Association (2006, January), http://www.bdaydyslexia.org.uk/facq.html#q1

Carmichael, A. R. (1999). *Style Guide for the Design of Interactive Television Services for Elderly Viewers.* Winchester: Independent Television Commission.

Carmichael, A., Newell, A., Morgan, M., Dickinson, A., & Mival, O. (2005). Using theatre and film to represent user requirements. *Proceedings INCLUDE '05*, Royal College of Art, London. ISBN (CD rom) 1-905000-10-3.

Charness, N., & Bosman, E. A. (1994, January–March). Age-related changes in perceptual and psychomotor performance: Implications for engineering design. *Experimental Aging Research, 20*(1), 45–61.

Cohen, G. (2000). Two new intergenerational interventions for Alzheimer's disease patients and their families. *American Journal of Alzheimer's Disease, 15*(3), 137–142.

Critchley, M. (1964). *Developmental dyslexia.* London: Heinemann.

Detterman, D. K., Gabriel, L. T., & Ruthsatz, J. M. (2000). Intelligence and mental retardation. In R. J. Sternberg (Ed.), *Handbook of Intelligence* (pp. 141–158). Cambridge: Cambridge University Press.

Feil, N. (1993). *The Validation Breakthrough.* Maryland: Health Professions Press.

Freudenthal, A. (1999). *The design of home appliances for young and old consumers.* Delft: Delft University Press.

Gill, J., & Shipley, T. (1999). *Telephones, what features do people need.* London: Royal National Institute for the Deaf.

Gould, S. J. (1997). *The mismeasure of man* (2nd ed.). Harmondsworth: Penguin.

Gregor, P., & Newell, A. F. (1999). The application of computing technology to interpersonal communication at the University of Dundee's Department of Applied Computing. *Technology and Disability, 10*, 107–113.

Horn, J. L., & Cattell, R. B. (1967). Age differences in fluid and crystallised intelligence. *Acta Psychologica, 26*, 107–129.

Hypponen, H. (1999). *The Handbook on Inclusive Design for Telematics Applications.* Helsinki, Finland: European Union INCLUDE Project.

ISO 13407. (1999). *Human-centred design processes for interactive systems.* International Organisation for Standards.

Kaluger, G., & Kolson, C. L. (1987). *Reading and Learning Disabilities* (2nd ed.). Columbus, Ohio: Bell & Howell Company.

Kaufman, A. S. (2000). Tests of intelligence. In R. J. Sternberg (Ed.), *Handbook of Intelligence* (pp. 445–476). Cambridge: Cambridge University Press.

Lohman, D. F. (2000). Complex information processing and intelligence. In R. J. Sternberg (Ed.), *Handbook of Intelligence* (pp. 285–340). Cambridge: Cambridge University Press.

Marquis-Faulkes, F., McKenna, S. J., Gregor, P., & Newell A. F. (2003). Scenario-based drama as a tool for investigating user requirements with application to home monitoring for elderly-people. In D. Harris, V. Duffy, M. Smith, & C. Stephanidis (Eds.), *Human-Centred Computing: Vol. 3. Cognitive, Social and Ergonomic Aspects* (pp. 512–516). Mahwah, NJ: Lawrence Erlbaum.

McKinlay, A. (1991). Using a social approach in the development of a communication aid to achieve perceived communicative competence. In J. Presperin (Ed.), *Proceedings of the 14th Annual Conference of the Rehabilitation Engineers Society of North America* (pp. 204–206). Washington, DC: The RESNA Press.

Mandler, J. M. (1984). *Stories, scripts and scenes: Aspects of schema theory*. Mahwah, NJ: Lawrence Erlbaum Associates.

Mandler, J. M., & Parker, R. E. (1976). Memory for descriptive and spatial information in complex pictures. *Journal of Experimental Psychology: Human Learning and Memory, 2*, 38–48.

Microsoft, (1994). *Word Developers Kit*. Redmond, Washington: Microsoft Press.

Microsoft. (1995). *Word for Windows 95*. Redmond, Washington, Microsoft Corporation.

Newell, A. F. (1995). Extra-ordinary human computer operation. In A. D. N. Edwards (Ed.), *Extra-ordinary human-computer interaction* (pp. 3–18). Cambridge: Cambridge University Press.

Newell, A. F., Arnott, J. L., Booth, L., & Beattie, W. (1992). Effect of the PAL word prediction system on the quality and quantity of text generation. *Augmentative and Alternative Communication, 8*, 304–311.

Newell, A. F., & Booth, L. (1991). The use of lexical and spelling aids with dyslexics. In C. Singleton (Ed.), *Computers & literacy skills* (pp. 35–44). Hull: University of Hull.

Newell, A. F., Booth, L., & Beattie, W. (1991). Predictive text entry with PAL and children with learning difficulties. *British Journal of Educational Technology, 22*, 23–40.

Newell, A. F., & Cairns, A. Y. (1993, October). Designing for extra-ordinary users. *Ergonomics in Design (Human Factors and Ergonomics Society)*, 10–16.

Newell, A. F., & Gregor, P. (1997). Human computer interfaces for people with disabilities. In M. Helander, T. K. Landauer, & P. Prabhu (Eds.), *Handbook of human-computer interaction* (pp. 813–824). Amsterdam: Elsevier.

Newell, A. F., & Gregor, P. (2000). User sensitive inclusive design—in search of a new paradigm. *Proceedings of the ACM Conference on Universal Usability*, Washington, DC, 39–44.

Newell, A. F., Morgan, M. E., Gregor, P., & Carmichael, A. *CHI 2006* (2006). Experience Report in CHI 2006 extended abstracts on Human Factors in Computing Systems Montreal, Quebec, Canada, 22–27 April 2006. pp. 111–117.

Nielsen, J. (1993). *Usability engineering*. London: Academic Press.

Ogozalec, V. Z. (1997, March). A comparison of the use of text and multimedia interfaces to provide information to the elderly. *Proceedings of CHI '97*, Atlanta, Georgia, New York: ACM Press, 65–71.

Pirkl, J. J. (1994). *Transgenerational design, products for an aging population*. New York: Van Nostrand Reinhold.

Preece, J. (1994). *A Guide to Usability—Human Factors in Computing*. London: Addison Wesley & Open University.

Rabbitt, P. M. A. (1993). Does it all go together when it goes? The nineteenth Bartlett Memorial lecture. *The Quarterly Journal of Experimental Psychology, 46A*(3), 385–434.

Rau, M. T. (1993). *Coping with Communication Challenges in Alzheimer's Disease*. California: Singular Publishing Group Inc.

Salthouse, T. (1985). *A Theory of Cognitive Aging*. Amsterdam: North Holland.

Salthouse, T. A. (1991). *Theoretical Perspectives on Cognitive Aging*. Mahwah, NJ: Lawrence Erlbaum Associates.

Schank, R., & Abelson, R. (1977). *Scripts, plans, goals, and understanding*. Mahwah, NJ: Lawrence Erlbaum Associates.

Sheridan, C. (1992). *Failure-free Activities for the Alzheimer's patient*. London: Macmillan Press.

Shneiderman, B. (1992). *Designing the user interface: strategies for effective human-computer interaction*. Reading, MA: Addison-Wesley.

Sleeman, K. D. (1998). Disability's new paradigm, implications for assistive technology and universal design. In I. Placencia Porrero & E. Ballabio (Eds.), *Improving the Quality of Life for the European Citizen: Vol. 4 Assistive Technology Research Series* (pp. xx–xxiv). Amsterdam: IOS Press.

Stephanidis, C. (Ed.). (2001). *User interfaces for all*. Mahwah, NJ: Lawrence Erlbaum Associates.

Sternberg, R. J. (2000). The concept of intelligence. In R. J. Sternberg (Ed.), *Handbook of Intelligence* (pp. 3–15). Cambridge: Cambridge University Press.

Todman, J., Elder, L., & Alm, N. (1995). Evaluation of the content of computer-aided conversations. *Augmentative and Alternative Communication, 11*(4), 229–234.

Vanderheiden, P. B., Demasco, P. W., McCoy, K. F., & Pennington, C. A. (1996). A preliminary study into schema-based access and organization of re-usable text in AAC. *Proceedings of the RESNA '96 Conference*, Salt Lake City, Utah, USA, 59–61. Arlington, VA: RESMA Press.

Waller, A. (1992). *Providing narratives in an augmentative communication system*. Unpublished doctoral dissertation, University of Dundee, Dundee, Scotland, UK.

Waller, A., Dennis, F., Brodie, J., & Cairns, A. (1998). Evaluating the use of TalksBac, a predictive communication device for nonfluent adults with aphasia. *International Journal of Language and Communication Disorders, 33*(1), 45–70.

Willows, D. M., Kruk, R. S., & Corcos, E. (1993). *Visual processes in reading and reading disabilities*. Mahwah, NJ: Lawrence Erlbaum Associates.

Wilson, A. (Ed.). *Communicating with Pictures and Symbols*. CALL Centre publications (Augmentative and Alternative Communication in Practice: Scotland), University of Edinburgh. Edinburgh. ISBN 1-898042-25-X

Worden, A., Walker, N., Bharat, K., & Hudson, S. (1997). Making computers easier for older adults to use. In *Proceedings of CHI '97* (pp. 266–828).

•5•

PHYSICAL DISABILITIES AND COMPUTING TECHNOLOGIES: AN ANALYSIS OF IMPAIRMENTS

Andrew Sears
UMBC

Mark Young
Maryland Rehabilitation Center

Jinjuan Feng
Towson University

INTRODUCTION

Computing devices are becoming smaller, more mobile, more powerful, more ubiquitous, and less expensive, changing the way people work, shop, communicate, and entertain. As a result, an inability to use these technologies can significantly limit employment, educational, and social opportunities. While computing devices can be convenient tools for traditional computer users, they can also create barriers for individuals with physical, cognitive, or perceptual impairments. For example, individuals with certain types of tremors may experience difficulty using small keyboards. However, a design process that considers the impairments of users can turn these potential barriers into powerful tools that increase employment opportunities, provide enhanced communication capabilities, and enable increased independence (Young, Tumanon, & Sokal, 2000).

While cognitive, perceptual, and physical impairments (PI) can all hinder the use of computing technologies, this chapter focuses on specific physical impairments that contribute to disability. More specifically, we focus on PI that may hinder an individual's ability to physically interact with computing technologies. Therefore, we do not address PI that affect the lower body or PI that hinder the production of speech. For additional information regarding cognitive or perceptual impairments, see chapters 41, 43, and 45. Chapters 39 and 40 may also provide useful insights as they discuss the design of technologies for older adults and children, two groups whose physical capabilities may require special attention.

The objectives of this chapter are to:

• Provide an introduction to specific PI that can hinder the use of traditional computing devices;

• Highlight critical characteristics of PI that must be considered when designing computing systems;

• Discuss the relationship between PI that result from health conditions and those that result from environmental or contextual factors;

• Summarize existing HCI research involving individuals with PI;

• Offer observations based upon the literature to guide future research and development efforts.

We believe that understanding PIs will allow for the design of more effective computing technologies that lessen or even eliminate the associated disabilities. We begin by presenting a set of definitions for *impairment*, *disability*, and *handicap* that are offered by the World Health Organization (WHO). By formally defining these terms, and using the terms in a way that is consistent with these definitions, we hope to eliminate potential ambiguity and confusion. Subsequently, we define the subset of PI that is of primary concern when interacting with computing technologies. Through this definition, we refine the scope of this chapter.

Given these definitions, we proceed to discuss the relationship between health conditions and PI. Understanding the underlying health condition is critical since this often provides valuable insights regarding the nature of the resulting impairments. We describe common health conditions (e.g., cerebral palsy, spinal cord injuries) associated with PI that affect interactions with computing devices. We identify the associated PI, important characteristics of these PI, and any additional impairments that may prove critical when designing computing technologies. While health conditions are most often associated with PI, both contextual and environmental factors can also hinder interactions with computing devices. Building on this observation, we briefly discuss the relationship among the environment, context, physical impairments, and disabilities. Most importantly, we highlight similarities and differences between those PI associated with health conditions and difficulties individuals may experience as a result of the environment or the activities in which they are engaged.

We conclude by reviewing recent research, discussing existing technologies, highlighting some challenges associated with adopting these technologies, and offering directions for additional research.

DEFINING IMPAIRMENT

The World Health Organization published the "International Classification of Impairments, Disabilities, and Handicaps" (ICIDH) in 1980. The current version of this classification serves as a foundation for this discussion (WHO, 2006). The ICIDH model acknowledges the complex relationships that exist among health conditions, impairments, disabilities, and handicaps while highlighting the potentially important role of both the context and environment in which activities are taking place. Figure 5.1 highlights the relationships between user capabilities and the influence of both health conditions and context. Both health conditions and context can directly affect the users' capabilities. Often, this influence is gradual rather than abrupt, causing an individual's capabilities to move along

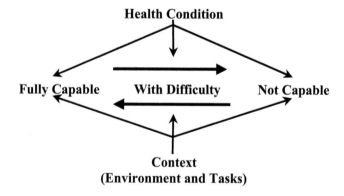

FIGURE 5.1. Model illustrating the relationship between health conditions, context, and an individual's capabilities.

a continuum from fully capable to the point at which they are no longer able to effectively perform a particular task. A specific health condition may, or may not, negatively affect an individual's capabilities, depending on the environment in which the individual is located and the tasks in which they are engaged.

The following definitions are adapted from the ICIDH:

- Health condition: A disease, disorder, injury, or trauma. Examples include spinal cord injuries, arthritis, cerebral palsy, stroke, multiple sclerosis, and amyotrophic lateral sclerosis.
- Impairment: A loss or abnormality of body structure or function. Physical impairments include a loss of muscle power, reduced mobility of a joint, uncontrolled muscle activity, and absence of a limb. Impairments can be caused by health conditions. For example, an individual with arthritis (health condition) may experience weakness, stiffness, and a reduced range of motion (impairments). Impairments can also be caused by context. For example, a cold environment can limit joint mobility.
- Disability (activity limitations): Difficulties an individual may have in executing a task or action. Examples include difficulty communicating, learning, performing tasks, or using computing technologies. Disabilities can be caused by impairments; an individual has a disability; disabilities are activity-specific. For example, a loss of muscle power in both arms (impairment) may interfere with an individual's ability to use a keyboard and mouse (disability). Context can also lead to disabilities. For example, working in a moving vehicle does not necessarily result in an impairment (e.g., the individual's capabilities have not been altered), but can result in a disability (e.g., the individual may not be able to use a mouse due to the movement of the vehicle).
- Handicap (participation restrictions): Problems an individual may experience while involved in life situations. Handicaps can be caused by disabilities (which originate with either a health condition or context); an individual experiences a handicap; handicaps occur at a social level. For example, an individual may have difficulty communicating (disability) which places restrictions on her ability to participate in educational and work activities, participate in the exchange of information, and maintain social relationships (handicaps).

Given these definitions the distinction between a disability and a handicap is, in many ways, dependent upon the individual. If an individual has an impairment (e.g., difficulty bending his fingers) that makes it difficult to use a standard keyboard, then this individual has a disability with respect to the operation of the keyboard. However, this individual only experiences a handicap if the inability to use the keyboard interferes with his normal life activities. Our focus is on designing computing technologies to lessen or eliminate disabilities that may result from either PI or context. As a result, the potentially subtle differences between disabilities and handicaps are not important.

PHYSICAL IMPAIRMENTS AND COMPUTING

Impairments, as described by the WHO, can affect every structure and function of the body. In this chapter, we focus on the subset of PI that are most relevant in the context of physical interactions with computing technologies. Impairments, even PI, that interfere with vision, hearing, and other activities involved in the use of computing devices are not discussed (see chapters 4, 6, and 8 for coverage of some of these issues). Four categories of PI are particularly relevant:

- Structural deviations—Situations in which there is significant deviation or loss with regard to anatomical parts of the body. Examples include the partial or total absence of a body part (e.g., missing finger, hand, arm) as well as situations in which a body part deviates from the norm in either position or dimension.
- Mobility (of bone and joint) functions—Mobility functions address an individual's ability to move a joint or bone. This includes both the range and ease of movement. For example, an individual may have limited range of motion or experience difficulty bending her fingers (e.g., arthritis).
- Muscle power functions—Muscle power functions address an individual's ability to generate force by contracting a muscle or muscle group. Paresis (e.g., weakness) refers to the partial loss of power while paralysis refers to the complete loss of power. Both can be caused by various conditions (e.g., brain injury, spinal cord injuries). Muscle tone and endurance functions may also be important, but tend to be less critical for most interactions with computing technologies.
- Movement functions—While additional movement functions exist, we focus on an individual's ability to control voluntary and involuntary movements. Difficulty controlling movements that involve a rapid change in direction (e.g., dysdiadochokinesia) is an example of a problem with a voluntary movement. Uncontrolled shaking or trembling of the hands (e.g., essential tremors or tremors associated with Parkinson's disease) are examples of involuntary movements.

While there are many other types of PI, including the stability of the joints, motor reflex functions, and gait pattern functions, the four categories identified earlier account for the majority of PI that hinder the use of computing technologies. As we discuss various health conditions, it is useful to identify specific PI that may occur as well as important characteristics of these PI that may influence the design of alternative strategies or technologies. While PI vary in many ways, the following four dimensions are perhaps the most critical:

- PI can be *permanent* or *temporary*;
- PI may be *continuous* or *intermittent*. For example, individuals with Multiple Sclerosis experience relapses and remissions;
- PI may be *progressive* (getting more severe with time), *regressive* (getting less severe with time), or *static* (no change with time);
- The severity of a PI can range from *mild* to *severe*. In other cases, severity is *variable*.

Temporary PI can be challenging because individuals are unlikely to invest significant time or effort in adapting to new interaction techniques. If a PI is *permanent*, individuals will develop accommodation or adaptive strategies and learning new interaction techniques may be acceptable. *Intermittent* PI create difficulties because alternative interaction techniques must be accessible when needed without hindering interactions when the PI is not present. *Progressive* and *regressive* PI are most effectively addressed by accommodations that adapt as the PI changes. This adaptation may be automatic or user directed. Adaptation is also useful when the severity of the PI is *variable*; the solution can be more complex since severity can change at any time. PI that are permanent, continuous, static, and of a fixed severity are perhaps the easiest to address. Under these conditions, the individuals are likely to adapt to their impairment, develop their own accommodation strategies, and be more willing to invest time learning new interaction techniques.

HEALTH CONDITION-INDUCED IMPAIRMENTS

It is tempting to view all individuals with a particular PI as benefiting equally from a given accommodation. Unfortunately, this perspective can result in solutions that fail to adequately address the needs of the individual. Understanding the underlying health condition provides valuable insights into the nature of the associated PI, which in turn provides insights into the characteristics of the accommodations that will prove most beneficial.

We briefly describe some of the more common diseases, disorders, and injuries associated with PI that hinder the use of computing systems. For additional information regarding any of the underlying health conditions, we refer the reader to the Merck Manual (Berkow, 1997) or the references included throughout this chapter. Our descriptions assume that appropriate treatments (e.g., medication, surgery, therapy) have already taken place. While we do not address issues involved in perceptual, speech, or cognitive impairments, we do highlight situations in which these impairments may be present. We do this for two reasons: (a) to provide additional insights into difficulties individuals with specific health conditions may face (e.g., vision, hearing, or cognition difficulties), and (b) to highlight situations in which specific alternative interaction strategies or technologies may not be appropriate (e.g., speech recognition).

ALS (Amyotrophic Lateral Sclerosis)

ALS is also known as Lou Gehrig's disease (Taylor & Lieberman, 1988). ALS is a progressive disease that begins with weakness in the hands (less often the weakness begins in the feet). ALS progresses at a variable rate, but 50% of those affected die within 3 years of the initial symptoms and only 10% live beyond 10 years. As the disease progresses, the weakness spreads to additional muscles and becomes more severe (eventually leading to paralysis throughout the body). With time, spasticity occurs resulting in muscles becoming tight. Spasms and tremors can also occur. ALS only affects those nerves that stimulate muscle action. Cognitive functions and sensation remain intact. ALS is characterized primarily by muscle power PI (weakness progressing to paralysis), but movement PI also occur (stiff muscles, spasms, and tremors). The PI are permanent, continuous,

and progressive. Initially, the PI are mild, but with time they become severe. Eventually, the muscles involved in producing speech are affected.

Arthritis

Arthritis refers to the inflammation of a joint (Hicks, 1988; Schumacher, 1993). The two most common forms of arthritis are osteoarthritis and rheumatoid arthritis. *Osteoarthritis* (OA) is a chronic joint disorder that can cause joint pain and stiffness. OA is characterized by degeneration of joint cartilage and adjacent bone, is the most common joint disorder, and is one of the most common causes of physical impairments (Berkow, 1997). OA is not an inevitable part of aging, but does occur more often in older populations. OA typically progresses slowly after symptoms first appear. OA can result in significant mobility PI (reduced range of motion, weakness, and difficulty with repetitive motions), which tend to be permanent, intermittent or continuous, progressive, and can range from mild to severe.

Rheumatoid arthritis (RA) is an autoimmune disease that can cause swelling, pain, and often the eventual destruction of the joint's interior. RA is characterized by symmetric inflammation of the joints. RA typically appears first in people between 25 and 50 years of age, affecting 1% of the population. Most often, RA starts subtly, progressing at a highly variable rate. In some rare cases, RA spontaneously disappears and treatment is often successful, but RA can result in significant mobility PI (reduced range of motion, weakness, and difficulty with repetitive motions). The PI tend to be permanent, continuous, and progressive. The severity of the PI is typically mild initially, but in 10% of the cases will progress to severe.

Brain Injury

Brain injury (BI) is a term used to describe a collection of injuries (Horn & Zasler, 1996; Rosenthal, Griffith, Bond, & Miller, 1990). Technically, there is a disconnection between the cause of the injury (e.g., penetration of the skull by a foreign object, or the rapid acceleration or deceleration resulting from a forceful blow to the head or vehicular accident) and the injury itself. Common head injuries include skull fractures, cerebral contusions (bruises on the brain), cerebral lacerations (torn brain tissue), concussions (brief loss of consciousness after an injury), and intracranial hematomas (collection of blood in the brain or between the brain and skull). Head injuries are the most common cause of death and disabilities for individuals under 50 years of age. BI is not always the result of trauma. For example, anoxia can also result in BI. The consequences of a brain injury depend upon the area of the brain affected and the severity of the injury, but may include death; language, vision, and motor-control difficulties; and periodic headaches. Other parts of the brain often assume the responsibilities of the damaged portion of the brain, lessening the severity of the resulting impairments. This ability is more prevalent in children, where the brain is more adept at shifting functions to different parts of the brain. The most common PI are movement oriented (e.g., difficulty controlling muscles). After rehabilitation, the PI are considered permanent, continuous, and static. Severity may range from mild to severe.

Cerebral Palsy

Cerebral palsy (CP) is a condition that results from brain injury that typically occurs before, during, or shortly after birth. CP is not a disease and is not progressive (Molnar, 1992). Over 90% of those affected with CP survive to adulthood. There are four main types of CP: spastic, choreoathetoid, ataxic, and mixed. *Spastic CP* occurs in approximately 70% of those individuals with CP and is characterized by both movement and muscle power PI (stiff muscles and weakness). *Choreoathetoid CP* is characterized by movement PI (spontaneous, slow, uncontrolled muscle movements; abrupt and jerky movements) and occurs in 20% of those individuals with CP. *Ataxic CP* occurs in approximately 10% of those individuals with CP and is characterized by movement and muscle power PI (poor coordination, weakness, trembling; rapid or fine movements can be difficult). *Mixed CP* occurs in many individuals and is characterized by a combination of two of the previously mentioned forms. Seizures are also possible (most often with ataxic CP). In all forms of CP, the PI are permanent, continuous, static, and may range from mild to severe. It is important to note that speech and intelligence can be affected in all forms of CP.

Missing Limbs or Digits (Amelia or Amputation)

Technically, missing limbs or digits is an impairment, not a health condition (Banerjee, 1982). However, like tremors, it is possible for limbs or digits to be missing without any associated health condition (e.g., congenital absence of a limb or digit is not considered a health condition). The PI associated with missing limbs or digits is permanent, continuous, static, and can vary from mild to severe depending on the use and effectiveness of prosthetic devices.

Multiple Sclerosis

Multiple sclerosis (MS) is a disorder in which the nerve fibers associated with the eye, brain, and spinal cord lose patches of myelin (a protective, insulating sheath) (Ropper & Brown, 2005). MS is progressive, but no single pattern typifies the progression. MS results in numerous symptoms, including a variety of movement and muscle power PI (clumsiness, tremors, stiff muscles, weakness). MS often results in periods of flare-ups alternating with periods of relatively good health (e.g., relapses and remissions), but in some individuals the symptoms become more severe within weeks or months. As flare-ups become more frequent, the symptoms become more severe and may become continuous. The PI are permanent, intermittent, and progressive. The PI may be mild initially, but often progress to become severe. In addition to multiple PI, MS can result in visual impairments.

Muscular Dystrophy

Muscular dystrophy (MD) is a group of inherited muscle disorders (Pidcock & Christensen, 1997). The two most common forms of MD are Duchenne's MD (affecting 20–30 of every 100,000 boys) and Becker's MD (affecting three of every 100,000 boys). Duchenne's MD is more severe, appearing between the ages of three and seven; most children must use a wheelchair by 10 or 12 and death is common by age 20. Becker's MD is less severe, with symptoms appearing around age 10; very few children are confined to a wheelchair by age 16, and more than 90% live beyond age 20. Both forms begin with muscle power PI (weakness) that spread and become more severe. This is often followed by mobility PI (joints that cannot be fully extended due to contracted muscles). Some PI may not be present early (e.g., inability to fully extend the elbows). In both forms, the PI are permanent, continuous, and progressive. Typically, the PI are mild initially and progressively get worse.

Parkinson's Disease

Parkinson's disease (PD) is a progressive degenerative disorder of the nervous system (Duvaisin, 1991). PD is characterized by degeneration of the nerve cells in the basal ganglia which results in reduced production of dopamine (the main neurotransmitter of the basal ganglia). PD affects approximately one in 250 individuals over the age of 40, and one in 100 individuals over the age of 65. PD typically begins with mild symptoms and progresses gradually. PD often begins with resting tremors (movement PI). As the disease progresses, initiating a movement can become difficult, muscles become stiff, and bending the elbow can become difficult or uncomfortable (movement and mobility PI). The tremors vary dramatically. Some individuals never develop tremors. For others, tremors may be permanent or temporary; continuous or intermittent; progressive, regressive, or static; and can vary from mild to severe. Other PI, including difficulty initiating movements and stiffness of muscles, are permanent, continuous, and progressive. They are mild initially, but progress to become severe. Still other PI, such as difficulty bending the elbow, may never occur. Individuals with PD often speak softly in a monotone and may stutter.

Repetitive Stress Injury

Repetitive stress injuries (RSI) are known by a variety of names including "cumulative trauma disorders" and "repetitive trauma syndromes" (Moore & Garg, 1992). RSI refers to a collection of disorders or injuries that are believed to be associated with repetitive activities, but the precise cause is still a subject of debate. RSI are most commonly associated with the wrists, but can affect other parts of the body. RSI most often result in mobility PI (reduced range of motion, weakness, and difficulty with repetitive motions) that are permanent, continuous or intermittent, progressive, and can vary from mild to severe. A classic example is carpal tunnel syndrome.

Stroke

A stroke is defined as the death of brain tissue resulting from a lack of blood flow and insufficient oxygen to the brain (Wade, Langton Hewer, Skilbeck, & David, 1985). A stroke typically occurs when a blood vessel becomes blocked or bursts. The symptoms depend upon the area of the brain that is affected, but can include vision, speech, cognitive, and physical impairments. Since other parts of the brain can assume the responsibilities of the damaged portion of the brain, it is possible to lessen the severity of the impairments that are apparent immediately following a stroke. Rehabilitation services are critical in this respect. The most common PI are movement oriented (e.g., difficulty controlling muscles). After rehabilitation, the PI can be defined as permanent, continuous, and static. The severity of the PI may range from mild to severe.

Spinal Cord Injury

Spinal cord injuries (SCI) occur when the spinal cord (a collection of nerves extending from the base of the brain through the spinal column) is compressed, cut, damaged, or affected by disease (Stiens, Goldstein, Hammond, & Little, 1997). The spinal cord contains motor nerves (controlling movement) and sensory nerves (providing information about temperature, pain, position, and touch). The consequences of SCI depend upon the level and completeness of the injury.

The level of an injury is based upon the nerves that are affected. The spinal column is divided into four areas: cervical (neck), thoracic (chest), lumbar (lower back), and sacral (tail bone). There are seven cervical vertebrae, numbered C1 through C7 from top to bottom. There are 12 vertebrae in the thoracic region (T1–T12), five in the lumbar region (L1–L5), and five in the sacral region. The level of an SCI refers to the location of the damaged nerves. In the cervical region, injuries are labeled based upon the vertebrae immediately below the damaged nerves (e.g., level C1 refers to damage to the nerves just above vertebrae C1). Damage to nerves just below C7 and above T1 are referred to as "level C8." In all other regions, injuries are labeled based upon the vertebrae immediately above the damaged nerve (level T3 refers to injuries to the nerves between T3 and T4). In general, injuries higher on the spinal cord will result in greater impairment. Table 5.1 summarizes the possible consequences of high-level SCI.

The completeness of an injury is often assessed using a scale defined by the American Spinal Injury Association (ASIA, 2001). While there are additional details, and assessments must be completed by trained professionals, the following summaries provide sufficient detail for our purposes. Table 5.2 provides a brief description of the ASIA scores.

SCI result in muscle power PI with the level of the injury determining which muscles are affected and the completeness of the injury determining how much muscle power is lost. Injuries with ASIA scores of A or B result in a complete loss of muscle power (e.g., paralysis) in the affected muscles. Injuries assigned ASIA scores of C or D are more difficult to describe since the amount of residual muscle power can be highly variable (e.g., weakness or paresis). PI associated with SCI are considered permanent, continuous, and static, with the severity of the PI ranging from mild to severe.

Tremors

A tremor is an involuntary movement (Hallett, 1991). Tremors are produced by involuntary muscle activity and are often described as rhythmic shaking movements. Some tremors occur when muscles are in use (*action tremors*) while others when the muscles are resting (*resting tremors*). *Intention tremors* occur when an individual makes a purposeful motion. Tremors are a common consequence of various health conditions (e.g., MS or stroke). In these situations, the tremors are described under the appropriate health condition (e.g., see MS or stroke).

Essential tremors usually begin in early adulthood and slowly become more obvious (essential tremors that begin in older adults are referred to as *senile tremors*). Essential tremors typically stop when the arms (or legs) are at rest, but become more obvious when the arms are held away from the body or in awkward positions. Essential tremors are relatively fast but result in little movement. The vocal cords can be affected by essential tremors, resulting in inconsistent speech. Essential tremors are permanent, intermittent, and can be progressive. Essential tremors are typically mild initially, but may progress to be severe and can be variable. While tremors are technically a movement PI, essential tremors have no known cause, which is why they are included in this list of health conditions.

TABLE 5.1 Relationship Between the Level of a Spinal Cord Injury and the Resulting Effects on Muscles

Level of Injury	Effect
C1 to C5	Paralysis of muscles used for breathing, controlling the arms, and controlling the legs.
C5 or C6	Paralysis of the legs. Some ability to flex the arms remains.
C6 or C7	Paralysis of the legs. Paralysis of part of the wrists and hands. Much of the ability to move the shoulders and to bend the elbows remains.
C8 or T1	Legs and trunk paralyzed. Hands paralyzed. Arms remain relatively unaffected.

TABLE 5.2 Residual Motor and Sensory Function Associated with Each ASIA Score

ASIA Score/ Completeness	Description of Motor and Sensory Function
A	Complete. No residual motor or sensory function below the level of the injury.
B	Motor complete, sensory incomplete. There is no residual motor function, but some sensory capabilities remain intact.
C	Motor and sensory incomplete. Most muscles below the level of injury are below 60% of normal strength.
D	Motor and sensory incomplete. Most muscles below the level of injury are at or above 60% of normal strength.
E	Normal motor and sensory function.

Locked-In Syndrome

Locked-in syndrome is a rare condition in which a person is conscious and able to think but is so severely paralyzed that communication is possible only by opening and closing the eyes in response to questions (Berkow, 1997). Many diseases or injuries may result in locked-in syndrome, including ALS, subcortical hemorrhage, polyneuropathy, Guillian-Barré syndrome, and brain injuries.

Many cases of locked-in syndrome are the result of various progressive health conditions, such as ALS. In these cases, individuals gradually lose their physical capabilities, usually starting from the lower limbs and progressing to the body and arms. Eventually, speech and movement of the neck are affected. Locked-in syndrome results in muscle power and movement PI including paralysis of most muscles, stiff muscles, and tremors. The PI are permanent, continuous, and severe. The initial severity may be less severe depending on the underlying disease or injury, but by the time an individual is considered locked-in, muscle power PI are always severe.

Summary

While the examples listed are some of the most common health conditions that result in PI, this is by no means a complete list. Table 5.3 provides a summary of the information presented earlier, including both the PI and any additional impairments that are associated with each health condition that may influence the design of alternative strategies or technologies.

Associated PI are listed as well as additional impairments associated with the health condition that could interfere with the use of standard computing technologies.

SITUATIONALLY-INDUCED IMPAIRMENTS AND DISABILITIES

As Fig. 5.1 indicates, both the environment in which an individual is working and the activities in which she is engaged can contribute to the existence of impairments that interfere with the use of computing technologies. Environmental characteristics such as lighting, noise, vibration, and temperature can all affect an individual's ability to interact with computing technologies. Similarly, the user may be engaged in multiple activities, each placing demands on her cognitive, perceptual, or physical capabilities, resulting in an overload such that one or more of the tasks cannot be completed optimally. Stress can also affect an individual's ability to interact with computers.

In some situations, an individual's context creates temporary impairments. In other situations, there is no impairment, but a temporary disability does exist. When the individual's physical capabilities are altered, the result is a temporary PI. For example, extended exposure to cold temperatures can make it difficult for an individual to bend his fingers, creating a temporary mobility impairment (similar to arthritis) that can hinder the use of information technologies. In contrast, when the individual's physical capabilities are not altered (e.g., a vibrating environ-

ment), but are simply rendered inadequate for the task, the result is a temporary disability. An example is an environment that is vibrating (e.g., a moving vehicle). Vibration does not alter any of the body's structures or functions so there is no impairment, but can make it difficult to use a stylus to enter text using gesture recognition, so the individual does experience a disability. Whether the context results in an impairment or disability, our goal is to minimize the negative impact on the user's computing activity.

While extensive research has studied dual-task scenarios, in which participants were required to attend to two tasks at the same time, these studies tend to highlight decreases in cognitive performance rather than in physical performance. At the same time, individuals are finding themselves in situations in which they are engaged in secondary computing tasks with increasing frequency. For example, a paramedic's hands may be busy providing medical care while they engage in a secondary computing activity to take notes and complete required forms. Simply walking can also make the physical activities required to interact with a mobile computing device more complex (e.g., Price, Lin, Feng, Goldman, Sears, & Jacko, 2006).

While many factors can adversely affect an individual's ability to perform the physical actions required to interact with computing technologies, and some studies have been conducted, these relationships have yet to be comprehensively investigated and reported within the HCI literature. At the same time, the fundamental characteristics of these situationally induced impairments and disabilities (SIID) are clear. Most importantly, unlike impairments associated with health conditions which tend to be permanent, SIID are temporary. As a result, individuals experiencing SIID are less likely to develop accommodation strategies and are therefore more likely to benefit from technology-based accommodations. Since SIID are temporary, they are also intermittent. SIID may be progressive, regressive, or static depending on the conditions and can range from mild to severe or could be variable. Effectively addressing the highly variable nature of SIID is an interesting, and challenging, area in need of additional research.

HCI RESEARCH AND PI

Researchers have approached the issue of designing computing technologies for individuals with PI from a variety of perspectives. Many projects were clearly motivated by a desire to address the needs of individuals with disabilities, and a subset of these projects focuses on specific health conditions or PI. For example, researchers have investigated the development of a new interaction alternative for individuals with cerebral palsy (Roy, Panayi, Erenshteyn, Foulds, & Fawcus, 1994), cursor control for individuals with spinal cord injuries (Casali, 1992), and text entry for individuals with spinal cord injuries (Sears, Karat, Oseitutu, Karimullah, & Feng, 2001). Other projects are motivated by specific impairments or disabilities regardless of the underlying cause. For example, researchers have investigated the effect neck range of motion has on the use of head controls (LoPresti, Brienza, Angelo, Gilbertson, & Sakai, 2000), and the use of force feedback for individuals with various movement and

TABLE 5.3 Common Health Conditions that Can Result in PI that Hinder Interactions with Standard Computing Technologies

Health Condition	Common PI and Characteristics	Additional Impairments
ALS	**Muscle power**: Weakness often begins in hands or feet. With time, loss of muscle power spreads and becomes more severe. *Permanent, continuous, progressive, mild progressing to severe (paralysis).* **Movement**: Muscles can become stiff. Spasms and tremors can occur. *Permanent, continuous, progressive, mild to severe.*	In time, muscles throughout the body can be affected, including those involved in producing speech.
Arthritis	Osteoarthritis **Mobility**: Reduced range of motion, weakness, difficulty with flexibility of joints and movement. *Permanent, intermittent or continuous, progressive, mild to severe.*	None.
	Rheumatoid arthritis **Mobility**: Reduced range of motion, weakness, difficulty with flexibility of joints and movement. *Permanent, continuous, progressive, mild to severe.*	None.
Brain injury	**Movement**: A wide range of movement-oriented difficulties are possible depending on the location and severity of the injury. *Permanent, continuous, static, mild to severe.*	Speech, vision, and cognitive impairments are possible.
Cerebral palsy	Spastic **Muscle power**: Weakness. *Permanent, continuous, static, mild to severe. Paralysis does not occur.* **Movement**: Stiff muscles. *Permanent, continuous, static, mild to severe.*	Speech and intelligence can be affected.
	Choreoathetoid **Movement**: Spontaneous, slow, uncontrolled muscle movements as well as jerky, abrupt movements. *Permanent, continuous, static, mild to severe.*	Speech and intelligence can be affected.
	Ataxic **Muscle power**: Weakness. *Permanent, continuous, static, mild to severe. Paralysis does not occur.* **Movement**: Poor coordination, trembling, difficulty with rapid or fine movements. *Permanent, continuous, static, mild to severe.* Mixed Any combination of two forms listed earlier.	Speech and intelligence can be affected.
Locked-in syndrome	**Muscle power**: Locked-in syndrome involves complete loss of power in all muscles except those that control the eyes. *Permanent, static, severe.* **Movement**: Muscles can become stiff. Spasms and tremors can occur. *Permanent, static, severe.*	Speech is affected.
Missing limbs or digits	**Mobility**. *Permanent, continuous, static, mild to severe.*	None.
Multiple sclerosis	**Muscle power**: Weakness. *Permanent, intermittent, progressive, mild to severe. Paralysis does not occur.* **Movement**: Clumsiness, tremors, stiff muscles. *Permanent, intermittent, progressive, mild to severe.*	Vision can be affected.
Muscular dystrophy	**Muscle power**: Weakness that spreads and becomes more severe. *Permanent, continuous, progressive, mild to severe. Paralysis does not occur.* **Mobility**: Joints that cannot be fully extended. *Permanent, continuous, progressive, mild to severe.*	None.
Parkinson's disease	**Movement**: Tremors (pill rolling) of one side are often the first sign. *Permanent or temporary; continuous or intermittent; progressive, regressive, or static; mild to severe.* Difficulty initiating movements and stiff muscles. *Permanent, continuous, progressive, mild to severe.* **Mobility**: Difficulty bending the elbow. *Permanent, continuous, progressive, non-existent to severe.*	Speech can be affected. Cognitive impairments are possible.
Repetitive stress injuries	**Mobility**: Reduced range of motion, weakness, difficulty with repetitive motions. *Permanent, continuous or intermittent, progressive, mild to severe.*	None.
Stroke	**Movement**: A wide range of movement-oriented difficulties are possible depending on the location (in the brain) and severity of the injury. Stroke does not always result in movement disorders. *Permanent, continuous, static, mild to severe.*	Speech, vision, and cognitive impairments are possible.
Spinal cord injuries	**Muscle power**: The specific muscles affected are determined by the level of the injury. How much power is lost is determined by the completeness of the injury. *Permanent, continuous, static, mild to severe.*	None.
Tremors	**Movement**: Tremors often affect the hands. *Permanent, intermittent, progressive or static, mild to severe or variable.*	Speech can be affected.

muscle power PI that affect the hands (Keates, Langdon, Clarkson, & Robinson, 2000). Numerous articles that investigate the needs of individuals with specific health conditions or PIs are discussed later.

Another group of projects addresses issues associated with PI, but does not identify any specific PI or health conditions as the motivation for the research. For example, several articles discuss communication issues for individuals "with severe speech and motor impairments" (McCoy, Demasco, Jones, Pennington, Vanderheyden, & Zickus, 1994), while others explore speech-based interfaces for "motor-control challenged computer users" (Manaris & Harkreader, 1998), new techniques for eye tracking that "could be useful to some people with physical disabilities" (Patmore & Knapp, 1998), and the use of word-prediction for text entry by "people with severe physical disabilities" (Garay-Vitoria & González-Abascal, 1997). Multiple articles that fit into this category will also be discussed.

The goals for this literature review include (a) summarizing the current state of our knowledge with respect to designing computing technologies for individuals with PI, (b) highlighting numerous unsolved problems in need of additional research, and (c) providing pointers to related articles that are not discussed. While there are many ways to organize this review, we group articles by health condition or PI whenever possible. Articles that focus on a specific health condition or PI are discussed, followed by articles that address less-precisely defined PI (e.g., various impairments that affect the hands, regardless of the underlying health condition). Next, we present several articles that have a clear association with PI, but do not discuss any specific PI (e.g., articles that present a technology specifically in the context of individuals with PI, but do not identify the PI or health condition). We conclude by reviewing articles that explore the adoption or abandonment of assistive technologies. Occasionally, articles that focus on a specific health condition are included in other sections due to the PI or technology discussed (e.g., Doherty, Cockton, Bloor, & Benigno, 2000). Throughout this section, we use terminology that is consistent with the definitions presented earlier. As a result, our terminology may differ from that used in the original articles.

Before discussing specific articles, we must acknowledge that several related bodies of literature are not discussed. First, we do not discuss the articles that simply mention individuals with disabilities as one of the possible groups of users who may benefit from the results of the research. For example, many articles discussing multimodal interfaces mention potential benefits for individuals with disabilities at some point, but few are designed for or evaluated using individuals with disabilities. Second, we do not discuss articles that focus on the more general problem of designing accessible computer systems without a significant focus on PI. For articles of this nature, the reader is referred to the proceedings of the ASSETS conferences (ASSETS, 2006), CUU conference (CUU, 2003), and the ERCIM Workshops on User Interfaces for All (ERCIM, 2006). Stephanidis (2001) also discussed these issues. We encourage the reader to review this literature as it often provides valuable insights into the methods that can be used to more effectively integrate accessibility concerns into the development process. Finally, we do not discuss the extensive literature available in the

rehabilitation and assistive-technology communities. We do encourage the reader to review the contributions of these communities. More specifically, we refer the reader to the proceedings of the annual Closing the Gap conferences (CTG, 2006), Rehabilitation Engineering and Assistive Technology Society of North America conferences (RESNA, 2006), and the CSUN conferences (CSUN, 2006). The *IEEE Transactions on Rehabilitation Engineering* often includes articles discussing the technical details of new devices and various books and Internet sites can provide valuable insights (e.g., Edwards, 1995; Gray, Quatrano, & Liberman, 1998; King, 1999; Stephanidis, 2001; Trace Center, 2007).

Spinal Cord Injuries (SCI)

SCI are one of the most frequently studied health conditions in the HCI community. Given the relationship between the level of injury and the resulting impairments, studies typically include individuals with injuries that are no lower than C7. The following articles discuss various topics including cursor control, validating keystroke-level models, a brain-controlled interface, and the use of speech recognition for communication-oriented activities.

Casali (1992) investigated the efficacy of five cursor-control devices when used by individuals with SCI. Twenty individuals with SCI, and ten individuals with no PI, participated in this study. Participants with SCI were divided into two groups using a custom assessment test (see Casali & Chase, 1993; Casali, 1995) that evaluates upper-extremity motor skills. As a result, three groups of participants were discussed: low motor skills, high motor skills, and participants with no PI. Each participant used a trackball, mouse, tablet, and joystick in addition to the cursor keys to complete a series of tasks that required either selecting or dragging a target. Participants in the low-motor-skill group took longer to complete the tasks than the individuals with no PI, but the difference between participants in the high-motor-skill group and the participants with no PI was not significant. For all three groups, the mouse, trackball, and tablet resulted in the shortest task-completion times, the cursor keys took longer, and the joystick was the slowest device. Target size was more important as motor skills decrease. Interestingly, the authors noted that all participants were able to complete the tasks with all five devices with minimal customization. However, individuals in the low- and high-motor-skills groups experienced substantial difficulty when using the mouse, especially when tasks required holding a button down while moving the mouse.

Koester and Levine (1994) investigated the efficacy of keystroke-level models for predicting performance improvements which individuals with SCI would experience when using word prediction software. The authors described a study that provides insights into the benefits word prediction software can provide and the effectiveness of keystroke-level models for predicting performance by individuals with PI. Eight individuals with no PI and six individuals with high-level SCI participated in the study. The individuals with SCI used their normal method of interacting with a keyboard (two used a mouthstick, four used hand splints). Participants completed the tasks with and without word prediction to allow performance improvements to be evaluated.

Two models were explored: (a) a generic model based upon parameters derived from the existing literature combined with new data from individuals with no PI, and (b) a user-driven model based upon parameters derived from the data collected in the current study. The user-driven models resulted in an average error rate of 6% while the generic models resulted in error rates of 53% and 11% for the participants with and without PI, respectively. Perhaps more interesting were the changes in data entry rates observed for the two groups of participants. Overall, the models predicted modest improvements in performance ranging from slight decreases to improvements of over 40%. Results for participants without PI matched the predictions, but participants with PI showed a consistent, and large, decrease in data-entry rates when using word prediction (ranging from approximately 20–50% reductions in data-entry rates).

Mason, Bozorgzadeh, and Birch (2000) and Birch, Bozorgzadeh, and Mason (2002) described a brain-controlled switch that can be activated by imagining movement. Earlier articles described studies in which this technology was used by individuals without PI (Lisogurski & Birch, 1998; Bozorgzadeh, Birch, & Mason, 2000), demonstrating that this technology could effectively identify both real and imagined finger movements. In the current articles, the authors describe a study that involved two individuals with high-level SCI with no residual motor or sensory function in their hands. Their results indicated false-positive rates of less than 1% (e.g., the system detected activity when none was intended) and hit-rates of 35–48% (e.g., the system detected activity when it was intended). While higher hit-rates are desirable, these results were viewed as positive in that they confirmed that individuals with no residual motor or sensory function in their hands could activate the system by imagining finger movements.

Sears, Karat, Oseitutu, Karimullah, and Feng (2001) discussed the results of an experiment designed to explore the effectiveness of speech recognition for communication-oriented activities when used by individuals with SCI as well as individuals with no PI. The individuals with SCI all had injuries at or above C6 with ASIA scores of A or B, resulting in either no use or limited use of their arms and their hands being paralyzed. Overall, there were no significant differences between the two groups of users, with participants producing text at approximately 13 words per minute (wpm) with few errors. An analysis of subjective satisfaction ratings indicated that participants with PI were more positive about their experience. A more detailed analysis revealed significant differences with respect to the processes employed by the two groups of participants. For example, participants with PI spent more of their time dictating and interrupted their dictation more often to correct errors, but they also spent less of their time navigating from one location to another within the document. Oseitutu, Feng, Sears, and Karat (2001) provided additional details that highlighted the different strategies adopted by these two groups of participants.

Sears, Feng, Oseitutu, and Karat (2003) focused on speech-based navigation techniques to help people with spinal cord injuries edit text documents. They found that traditional target-based techniques (e.g., "Select book") and direction-based techniques (e.g., "Move up four lines") were highly ineffective with users spending one third of their time issuing navigation commands while experiencing failure rates of approximately 15%

to 19%, respectively. Further, when navigation commands failed, the consequences were often rather severe. Frequently, failed commands would alter the content of the document, requiring significant time and effort to recover. The authors proposed several changes to existing navigation techniques to address these difficulties. These revisions significantly reduced the time spent on navigation and significantly improved data-entry rates. Feng and Sears (2004) proposed a new speech-based navigation solution, which introduces carefully specified navigation anchors to allow users to select target words efficiently and reliably. A preliminary study confirmed that their new anchor-based navigation technique allowed users to reach targets quickly while experiencing fewer errors.

LoPresti and Brienza (2004) investigated use of head-controlled devices by individuals with SCI. Although head-controlled devices can help people with limited use of their hands, the devices can prove challenging when the users also have limited neck movement. To address this issue, the authors developed software that automatically adjusted the interface sensitivity to the needs of a particular user. The software was evaluated in two stages. The first stage involved 16 novice users with SCI or multiple sclerosis. Icon selection speed increased significantly with the new software. The second stage involved five users with extended use of the original head-controlled device. Only one participant experienced improvements in performance. Unfortunately, calibrating their system required five steps, resulting in a complex and time-consuming process, especially for novice users who have yet to commit to long term use of the system.

Cerebral Palsy (CP)

CP can result in a variety of impairments due to reduced muscle power or a loss of control over voluntary or involuntary movements. CP can also result in both impaired speech and intelligence. As a result, designing technologies for individuals with CP can be particularly challenging. In this section, we discuss the results of a single study that was motivated explicitly by CP. Additional studies discussed elsewhere in this chapter may have been motivated by CP. For example, Alm, Todman, Elder, and Newell (1993) described a system designed for individuals with severe speech and motor impairments, including an evaluation that included one individual with CP.

Roy et al. (1994) discussed the design of a gestural interface designed for individuals with speech and motor impairments due to CP. Fourteen students with CP, ages 5–17, were observed during their regular school activities. Numerous communicative acts were observed that combined facial expressions; eye-gaze; vocalization; dysarthic (slurred, slow, difficult to produce, and difficult to understand) speech; and upper extremity gestures using the head, arms, hands, and upper torso. Subsequently, four students participated in sessions in which data were collected for a predefined set of gestures (using a 3-D magnetic tracker and EMG electrodes). An evaluation of the complete set of gestures, or the subsequent use of such gestures by individuals with CP, was not reported. However, preliminary results that were reported provided encouragement with respect to the use of computer-based gesture recognition as a communications tool for individuals with CP.

Impairments of the Hands and Arms

Multiple studies have investigated the difficulties individuals experience due to movement or muscle power PI that affect their hands and arms. These studies often include participants with a range of underlying health conditions (e.g., CP, SCI, stroke) as well as varying PI. The unifying theme is that each study focused on the ability of the participants to interact with computers using the arms and hands.

Trewin and Pain (1999) discussed the errors that occur when individuals with PI that affect their hands and arms interact with a keyboard and mouse. Twenty individuals with various PI participated, as did six individuals with no PI. Each participant completed tasks using the keyboard, mouse, or both. Several important categories of typing errors were identified: long key presses, additional keys (both local and remote), missing keys, simultaneous keys, and bounces. Each type of error occurred more often for the participants with PI. For additional related work, see Trewin (1996) as well as Trewin and Pain (1998a, 1998b).

Keates et al. (2000) discussed the use of haptic feedback for individuals with PI. More specifically, they investigated the potential of force feedback to facilitate point-and-click activities. They reported on two pilot studies that focused on individuals with movement and muscle power PI but unaffected sensitivity of touch. Six individuals participated in the studies; four participants had choreoathetoid CP, one had Friedrich's ataxia, and one had Kalman-Lamming's syndrome. Participants completed a series of point-and-click activities. In the first study, the efficacy of several alternative forms of feedback was explored. These forms included a pointer trail, changing the color of a target when the cursor is over it, force feedback gravity wells that pulled the cursor toward targets, vibrating the mouse when the cursor is over a target, or a combination of all of these forms of feedback. Their results indicated that gravity wells may reduce target selection times by 10–20%. Similar benefits were also seen for pointer trails—a standard accessibility option in the Windows operating systems. Participants disliked the vibrating cursor, which also increased target selection times. Interestingly, these results are similar to those reported for participants with no PI using the Phantom input device (Oakley, McGee, Brewster, & Gray, 2000).

The researchers' second study explored the relationship between force feedback and target size. Force feedback reduced the time required to select targets by 30–50% and error rates by approximately 80%. Without force feedback, the time required increased as target size decreased, but this pattern was greatly reduced when force feedback was provided. Interestingly, at least one participant was unable to complete the tasks without force feedback, but successfully selected the targets when force feedback was provided. These results suggest that force feedback could prove useful for individuals with PI, but it should be noted that these results may be unique to the specific form of force feedback utilized in the current study. Careful evaluation of other forms of force feedback is still necessary.

Building on the previous study, Hwang, Keates, Langdon, and Clarkson (2003) investigated the use of force feedback gravity wells for point-and-click tasks when an undesired haptic distracter is present on the screen. Ten individuals with PI together with eight individuals without PI participated. Their results indicated that the location of the distracter has an important effect on performance. While selection times and error rates improved when force-feedback gravity wells were present, participants were significantly slower when the distracter was directly between the start point and the target. The greatest improvements were evident for the most impaired users.

Keates, Clarkson, and Robinson (2000) investigated the use of the Model Human Processor (Card, Moran, & Newell, 1983) for describing the behavior of computer users with PI. This model combines perceptual, cognitive, and motor activities to predict the time required to complete a task, but was originally developed with an emphasis on individuals without PI. Two groups of individuals took part in their study. The first group consisted of six individuals with PI caused by various health conditions (e.g., quadraplegia, MD, spastic CP, choreoathetoid CP, and Friedrich's ataxia) while the second group included three individuals with no PI. Their results are summarized in Table 5.4.

Participants with PI appeared to take longer to complete fundamental perceptual and cognitive activities. These results are consistent with those discussed earlier (Koester & Levine, 1994), but should be interpreted carefully given the varied health conditions of the participants. Through observation and an analysis of the different times recorded for motor activities, the authors concluded that additional cognitive or perceptual activities were taking place when individuals with PI completed basic motor tasks (as compared to individuals with no PI). While their results are not definitive, they confirm the need for additional research that focuses on the time required to complete basic perceptual or cognitive activities.

Keates, Hwang, Langdon, and Clarkson (2002) investigated the challenges of the point-and-click tasks for people with PI by studying detailed records of cursor movements. The motivation for their study was the assumption that traditional task measures such as speed and error rate are not sufficient if the goal is to fully understand the interaction process including the reasons users experience difficulties. Five individuals with PI and three without PI completed a series of multidirectional cursor positioning tasks, providing detailed information regarding the path the cursor followed during each activity. Based on those data, six metrics, including the distance traveled relative to cursor displacement, distribution of distance traveled for a range of cursor speeds, submovements, distance traveled away from the target, distribution of distance traveled for a range of curvatures, distribution of distance traveled for a ranged of radii from the target, were proposed which focused on deviations from the optimal path for a number of predefined cursor positioning tasks. Next, a subset of the participants (four with PI and

TABLE 5.4. Average Time Required, By Individuals With and Without PI, to Complete Fundamental Perceptual, Cognitive, and Motor Tasks

	Perceptual	Cognitive	Motor
No PI	80 ms	93 ms	70 ms
PI	100 ms	110 ms	110 ms, 210 ms, 300 ms

three without PI) completed a set of target selection tasks. These tasks differed slightly from those used in the initial study in that a circular starting point was defined near the center of the screen. The results indicated that the new metrics helped highlight differences between users with and without PI, especially those difficulties experienced by the more impaired users.

Building on the results of the previous study, Hwang, Keates, Langdon, and Clarkson (2004) investigated the submovements of the cursor path for individuals with PI. When conducting navigation tasks, users completed a rapid movement that is comprised of a sequence of smaller, discrete submovements. The authors studied the cursor trajectories of six users with PI together with three individuals without PI. The results showed that some individuals with PI paused more often and longer than participants without PI, resulting in up to five times more submovements to complete the same task. There was a correlation between error rate and peak submovement speed for users with PI, which was not observed for users without PI.

Myers, Wobbrock, Yang, Yeung, Nichols, and Miller (2002) focused on individuals with PI that result in a loss of gross motor control, such as the movement of the arm and the wrist, but retention of specific fine motor control, such as the finger. Software was developed allowing the use of a PDA as the mouse and keyboard. Four participants with muscular dystrophy tested the software; all provided positive feedback. The parents of one participant noted that although performance was slower using the PDA as compared to the regular keyboard and mouse, the PDA was much less tiring and could be used much longer without a break.

Impairments in Infants

Infants with severe PI often develop to be passive with limited or nonexistent speech. This is true even when cognitive skills are normal. Fell, Delta, Peterson, Ferrier, Mooraj, and Valleau (1994) discussed the development and evaluation of the Baby-Babble-Blanket as a technology to enable infants with severe PI to control their environment and communicate. One specific goal was to help these infants improve their motor skills. The key component of the system is a pad, containing 12 equally spaced switches, which can be placed on the floor. When a switch is pressed, a computer provides feedback. The feedback could include digitized speech, music, and sounds. The authors report on a single-subject experiment involving a five-month-old infant with poor muscle tone, hydrocephaly, and clubfeet. This infant was able to activate switches, became more active when sounds were played in response to switch activations, associated switches with a desired effect (e.g., hearing his mother's voice), and was able to modify these associations when the feedback was moved from one switch to another.

Significant Speech and Physical Impairments (SSPI)

While the terminology may vary, several studies have explored technologies for individuals with significant speech and physical impairments. The article by Roy, Panayi, Erenshteyn, Foulds and Fawcus (1994) discussed earlier could have been included here, but was listed separately due to its explicit focus on impairments caused by CP.

Alm, Todman, Elder and Newell (1993) described the development of a computer-aided communication technology for individuals with SSPI. This article focused on situations in which the speech impairment results from PI rather than cognitive impairments. The system tracks the conversation, guiding users as they specify the type of utterance they wish to generate. The system was evaluated with two participants. The mean rate of speech was 144.4 words per minute (wpm). One participant with no PI took part in eight conversations with student volunteers, producing speech at a rate of 67.4 wpm while the conversation partners spoke 132.9 wpm. While selecting an utterance could require one to four clicks, less than 0.5% of the selections involved more than two clicks. This suggests that the desired utterances were easily accessed given the current design of the system. The second participant had CP and used a word board containing 400 words as his primary method for communicating. Using the experimental system, this participant was able to express himself more fully, increasing his vocabulary from 143 to 534 words and the number of times he took control of the conversation from 10 to 27. Many important questions must still be addressed for systems such as the one described in this article. One critical issue is the effectiveness of such a system as the number of conversation topics and phrases increases. Another critical issue is how the user can efficiently enter and organize these phrases as they customize the system for use in new situations.

Demasco, Newell, and Arnott (1994) discussed the development of a system to support communication activities based upon principles of visual information seeking. Building on five key principles derived from the literature, a computerized word board was designed. Key features included the separation of navigation and selection which allows these activities to be supported by different devices, and expanding the collection of accessible words using word associations (e.g., is-like, isa, hasa, goes-with). At the time the article was written, the system was under development and a variety of directions for research were presented, but no evaluation had been completed.

McCoy, Demasco, Pennington, and Badman (1997) investigated the application of natural language processing to Augmentative and Alternative Communication (AAC) to improve the communication capability for individuals who have cognitive disabilities that hinder their ability to speak. The objective is to develop intelligent communication aids that provide linguistically "correct" output when speeding the communication rate. The authors proposed a prototype system with a key component called the "Intelligent Parser Generator." Several research issues related to the system and the specific user population were proposed and discussed. However, no user evaluation result has been reported.

Albacete, Chang, Polese, and Baker (1994) and Albacete, Chang, and Polese (1998) discussed the design of iconic languages for use by individuals with SSPI. Their approach is based upon semantic compaction in which ambiguous icons are combined to represent concepts. For example, "APPLE VERB" could mean "eat" while "APPLE NOUN" means "food." An icon algebra was defined, including operators to combine, mark, provide context for, enhance, or invert icons. As a result, this approach can allow a vocabulary of several thousand words to be represented with just 50 to 120 icons. This approach served as the foundation for the Minispeak system.

Input Using Electrophysiological Data

Traditional input techniques involve some kind of physical activity. Most often, the hands and arms are involved, but other technologies allow users to provide input by moving their eyes, heads, tongue, or eyelids. However, a growing number of researchers are investigating input techniques that do not require physical activities. The article by Mason, Bozorgzadeh and Birch (2000) discussed earlier provides one example, in which electrophysiological data were used to generate input in a computer. In this section, several additional articles that discuss the use of electrophysiological data are discussed. This is a new, exciting, and promising form of input for individuals with significant PI.

Patmore and Knapp (1998) discussed an approach to eye tracking that uses the electrooculogram (EOG) and visual evoked potentials (VEP) as input. While a number of EOG-based eye-tracking systems have been discussed, additional research is needed to make such systems reliable (see LaCourse & Hludik, 1990; Patmore & Knapp, 1995). The VEP is a response to a flash of light that is highly sensitive to the stimulus' distance from the center of the field of vision. The authors presented the technical details of a system that combines the EOG, the VEP, and fuzzy logic to improve accuracy. While promising, significant additional research is necessary before such a system will be sufficiently reliable for use under realistic conditions.

Allanson, Rodden, and Mariani (1999) described a toolkit designed to explore the use of electrophysiological data to support interactions with computers. They discussed the variety of electrophysiological data available including the electroencephalograph (EEG), electromyograph (EMG), and galvanic skin resistance (GSR). The EEG is an electrical trace of brain activity, the EMG can detect the state (e.g., complete relaxation, partial contraction, complete contraction) of a muscle, and the GSR measures changes in the resistance of the skin. The authors provided interesting references describing the use of the EEG to turn switches on and off (Kirkup, Searle, Craig, McIsaac, & Moses, 1997) and to differentiate between five tasks (Keirn & Aunon, 1990), as well as references illustrating the use of the EMG to control computer games (Bowman, 1997) and manipulate objects on a computer (Lusted & Knapp, 1996). Their toolkit integrated these signals using predefined widgets that allow the signals to be used to drive applications.

Barreto, Scargle, and Adjouadi (1999) discussed the design of an integrated assistive real-time system as an alternative input device to computers for individuals with significant PI. The system uses EMG biosignals from cranial muscles and EEG from the cerebrum's occipital lobe to control mouse input. The system, combined with an onscreen keyboard, provides full operation functionalities with computers. Six healthy participants tested the system. Participants required an average of 13–21 seconds to select a button. Although the selection speed is quite low, the study demonstrated that the EMG and EEG based system could allow users to complete target selection tasks.

Kübler, Kochoubey, Hinterberger, Ghanayim, Perelmouter, Schauer, et al. (1999) discussed the use of slow cortical potentials in a 2-s rhythm to control cursor movements on a computer screen. Three individuals with advanced ALS participated in extensive practice sessions to learn to control this signal. Ultimately, two of the three reached accuracy rates between 70 and 80%, which allowed these individuals to select letters and words displayed on a computer screen. Building on these results, Birbaumer, Hinterberger, Kübler, and Neumann (2003) developed a Thought-Translation Device (TTD) using slow cortical brain potentials. The system allows the user to directly select letters and words with brain response alone. Eleven locked-in patients tested the device with a training period ranging from six months to six years. Nine participants had end-stage ALS, one had experienced a subcortical hemorrhage, and one had polyneuropathy and Guillian-Barré syndrome. At the beginning of the training session, the seven participants who were not completely locked-in patients were able to communicate "yes" or "no" signals with face or eye muscles. Two of these seven patients lost all motor control during the study but were still able to communicate using TTD. None of the patients who were completely locked-in at the beginning of training were able to gain effective control over the system, but one patient did reach a selection speed of one letter or word per minute. Although only two participants were able to use this system successfully, this study provided additional confirmation that a brain–computer interface has the potential to benefit users with severe PI.

Craig, Tran, McIsaac, Moses, Kirkup, and Searle (2000) developed a system that employs alpha waves, which increase in amplitude when the eyes are closed and decrease in amplitude when the eyes are open, to activate electrical devices. Ten individuals without PI and one individual with severe PI evaluated the system. The participant with PI had childhood-onset poliomyelitis which resulted in the loss of motor function below her neck. Participants without PI completed one 50–60 minute session on each of four consecutive days. In each session, they activated the environmental control system 30 times. The participant with severe PI completed four trials of approximately ten minutes duration in each trial. There was a significant decrease in selection times between day one and all subsequent days, including a 16% reduction from 12.7 seconds on day one to 10.7 seconds on day two. Error rates also decreased significantly from approximately 50% to 23% by the second day.

Doherty et al. (2000) discussed the use of formative experiments and contextual design to assess the potential of applications controlled by eye movements, the EOG signal, and the EEG signal (e.g., the Cyberlink interface). An initial study involving 44 participants with various PI demonstrated some expected limitations (e.g., the mouse was not effective when arm/hand control was limited, eye tracking was not effective when peripheral vision was impaired) as well as unexpected difficulties (e.g., some quadriplegic participants could not produce signals below the neck and therefore could not operate the GSR device). Interestingly, the Cyberlink interface was the only alternative that all 44 participants could use to navigate a maze. Based on those results, Doherty et al. (2001) proposed ten design guidelines for developing interfaces with Cyberlink and brain-injured users. Example guidelines include avoiding strobes when sampling the cursor, using tunnels instead of clicking for selection, avoiding long distances by providing rest stops, and not using timeouts to indicate selection. Overall, their results confirmed that interfaces built upon electrophysiological data can be useful for individuals with PI, but that substantial research is necessary before the full potential of these technologies will be realized.

Gnanayutham, Bloor, and Cockton (2005) improved the CyberLink system described earlier by adding a discrete acceleration function for cursor movement. Two studies were conducted. The first tested the adaptive cursor acceleration algorithm with 30 participants, including 19 with PI. The results suggested that the acceleration function did improve performance, but was not sufficient to overcome the wide variations in capabilities of the study participants. To address this shortcoming, the authors explored personalized tiling for brain interfaces, with the goal of producing a better match between the demands of the device and the capabilities of the users. In the second study, five individuals with brain injuries evaluated the personalized interface. All five participants were able to communicate with the system. Two participants communicated through a two-target, yes/no interface with an average success rate of 70%. Two other participants communicated using a switch, also achieving 70% success rate. Three participants showed interest beyond communication tasks and tried using the interface for television and music systems in their rooms.

Kennedy, Kakay, Moore, Adams, and Goldwaithe (2000) discussed an experimental interface that requires the implantation of special electrodes into specific areas of the brain. For one individual, EMG signals from various muscles were utilized to supplement the neural signals detected by the electrode. With five months of practice, the individual learned to control cursor movements. Results for three tasks were reported. These tasks involved moving the cursor across the screen, placing the cursor on an icon or button, and selecting the icon or button by holding the cursor still for a predefined period (or activating a predefined EMG signal). Through these tasks, this individual was able to invoke synthesized speech for predefined messages, spell names, and answer questions. Moore and Kennedy (2000) identified several directions for additional research including signal understanding, patient training, navigation paradigms, control of neural prosthetics, and communication.

Carroll, Schlag, Kirikci, and Moore (2002) focused on the communications needs of individuals who are locked-in. The authors analyzed log sheets, conducted interviews, and visited users to identify user needs. The style and content of interactions varied according to the role of the person with whom the individual was communicating. Family members engaged in fewer interactions that allowed the individual to participate, while nurses continually questioned the individual about various issues, including their comfort and needs. This finding suggests that the support offered by the system may vary based on the relationship between the individual and her communication partner.

Felzer and Freisleben (2002) developed the "Hands-free" Wheelchair Control System (HaWCoS) that allows individuals to control an electrically powered wheelchair without using their hands. The system uses muscle contractions, a kind of EMG signal associated with any arbitrary muscle, as input signals to communicate with the wheelchair. Three experiments were conducted, comparing one prototype of the HaWCoS system with a traditional joystick control. Although requiring more time than the traditional joystick, the HaWCoS system allowed users to control the wheelchair under a hands-free condition, providing an alternative for people with severe PI.

Borisoff, Mason, Bashashati, and Birch (2004) investigated Low Frequency Asynchronous Switch Design (LF-ASD) as a direct Brain-Computer Interface (BCI) technology for people with severe PI. Earlier research on LF-ASD produced encouraging results, but error rates were too high for practical use. These authors implemented energy normalization and feature space dimensionality reduction features, which normalize the low-frequency signals and increase accuracy. The revised system was evaluated by four individuals with high-level SCI as well as four individuals without PI. Accuracy improved significantly compared to earlier results, with true positive rates increased by approximately 33% and false positive rates by 1% to 2%. Individuals with PI achieved the same accuracy as those without PI, which was not possible when using a traditional keyboard and mouse.

Moore and Dua (2004) developed a biometric input device using galvanic skin response (GSR), which is a change in electric conductivity of the skin caused by increased activity of the sweat glands. Various emotional responses such as fear, excitement, or anxiety can affect the GSR level. A locked-in individual who had lost the ability to use a brain–computer interface tested the system during a one-year period. The results showed that there was a significant difference in accuracy between baseline and controlled states. After several days, an accuracy rate of approximately 62% was achieved. The experiments demonstrated that GSR has the potential to be used by people with severe PI.

Cursor-Control Technologies

Cursor control and text generation are perhaps the two most common activities when interacting with computers. In fact, cursor control can form the foundation for text generation and therefore may be the most fundamental task in which individuals with PI engage. Numerous technologies have been explored for controlling the cursor location and many of the articles summarized earlier focused on cursor control tasks. While new eye-tracking technologies are being explored, commercial systems are readily available and can prove useful for individuals with PI. The issues involved in using these technologies are not discussed in this chapter, but Sibert and Jacob (2000) provided useful pointers to the recent literature on this topic. In the following sections, technologies are discussed that allow individuals to control the cursor using their heads, tongues, eyes and eyelids, or speech.

Head. When an individual has limited use of their hands and arms, other options must be found for interacting with computers. One option is to control the cursor location using head movements. In this section, several articles are discussed that explore the issues involved in interacting with computers via head movements.

Radwin, Vanderheiden, and Lin (1990) discussed the results of a study that compared a standard mouse to a lightweight, ultrasonic, head-controlled pointing device. Ten participants with no PI completed a variety of tasks that required the selection of circular targets. Distance, target size, and device all significantly affected movement times. Interestingly, target size had a greater effect for the head-controlled device, with movement times decreasing more dramatically than when the mouse was used. The direction of the movement also affected movement times, with the horizontal movements resulting in the shortest times for the mouse and vertical movements resulting in the shortest times for the head-controlled device. Results were also reported for two individuals with CP including detailed results for one of these individuals that highlighted the importance of carefully assessing the abilities of individuals with PI when designing computer systems for their use. For example, horizontal movements were efficient for participants with no PI, but the individual with CP experienced more difficulty moving the cursor to the right than any other direction.

Lin, Radwin, and Vanderheiden (1992) reported on a similar study that focused on the relationship between control-display gain and performance with both mouse and head-controlled device. Ten individuals with no PI participated in this study. Movement times and deviation of the cursor from the optimal path were reported. Extensive analyses of the results highlighted the importance of target size, movement distance, and gain. Gain affected movement times, but both target size and movement distance were more influential. A U-shaped curve was observed for both the mouse and head-controlled device. The reader is referred to the original article, as well as Schaab, Radwin, Vanderheiden, and Hansen (1996), for additional information regarding the relationship between control-display gain and performance with head-controlled devices.

Malkewitz (1998) described a system that uses head movements to control cursor location and speech to click the mouse buttons, activate hotkeys, and emulate the keyboard. The goal was to provide access to standard applications by emulating standard input devices (e.g., keyboard and mouse). Speech recognition was restricted to a predefined set of commands and the ability to enter individual letters. Full dictation was not supported for various reasons. While preliminary results were encouraging, the authors note the need for additional research.

Evans, Drew, and Blenkhorn (2000) described the design of a head-operated joystick. The authors briefly summarized the results of an evaluation involving 40 participants including nine individuals with undocumented impairments. It is reported that all 40 individuals used the device successfully, but detailed measures were not provided. The nine individuals with impairments also used a commercially available device that emulated a mouse. Interestingly, all nine preferred the joystick emulation to the mouse emulation provided by the commercial device. It is speculated that this preference is due at least in part to the fact that users can rest their heads in a neutral position once the cursor is placed in the correct location when using a device that emulates a joystick.

LoPresti, Brienza, Angelo, Gilbertson, and Sakai (2000) studied the relationship between neck range of motion and the use of head-controlled pointing devices. Fifteen individuals with no PI as well as ten individuals with various health conditions that resulted in PI that affected neck movements participated in the study. A head-mounted display was used as opposed to a traditional desktop monitor. The results confirmed that the individuals with PI were less accurate and took longer when completing icon selection tasks. They were also less accurate when performing target-tracking tasks. Further, their performance was much more variable than that of the participants with no PI. As with several other studies, the authors found that models developed based upon individuals with no PI do not accurately represent the performance of individuals with PI. A more detailed analysis indicated that vertical movements were faster than horizontal movements and horizontal movements were faster than diagonal movements. LoPresti (2001) reported similar results.

De Silva, Lyons, Kawato, and Tetsutani (2003) developed a vision-based face-tracking system for cursor control. The cursor position was controlled by head movements. The mouse button was activated by the individual opening his mouth. Eight users without impairments tested the system. The face tracking system allowed for an index of performance of 2.0 bits/second, which is greater than a joystick, but slower than a trackball and touchpad. The subjective satisfaction measures showed that the system was rated average for ease of use. The lowest ratings were given for smoothness and accuracy of cursor control. The system has the potential to support hands-free typing.

Tongue. Salem and Zhai (1997) provided a brief description of a system that allows the cursor location to be controlled using a tongue-operated isometric joystick like those available on many portable computers. A Trackpoint joystick was mounted in a mouthpiece, similar to those used by athletes, that was fitted to the individual's upper teeth. Two individuals participated in a pilot study. Neither had experience with an isometric pointing device. Both used the tongue-controlled joystick as well as a standard finger-controlled joystick. Initial performance was substantially better when using their fingers, but with practice the gap between the two devices narrowed. During the last trial, the tongue-based device was 5% slower than the finger-based device for one participant, but it was 57% slower for the other participant.

Eye and eyelid. Shaw, Loomis, and Crisman (1995) presented a system designed to allow individuals to control various devices, including computers, by opening and closing their eyes. Two individuals utilized the system. The first (participant A) had a high-level SCI resulting in no residual use of his arms or hands. The second (participant B) had a stroke and could only control a single eyelid. With six hours of practice, participant A was able to move from interacting with simulation software to controlling a powered wheelchair. He found this device to be less obtrusive than the chin-control he normally used to control his wheelchair. After 12 hours of practice, participant B was able to navigate the maze presented by simulation software and made progress toward controlling a powered wheelchair. Given the minimal motor control this individual exhibited, this was considered a noteworthy accomplishment.

Bates and Istance (2002) studied a zooming interface to enhance the performance of eye-controlled pointing devices. Selecting small targets was challenging due to the way the human eye functions. The authors addressed this issue by introducing automatic zooming to make targets larger and easier to select. Six individuals without PI took part in the study. The result confirmed that target size is an important factor for the effectiveness of eye-based interaction. The zooming interface was significantly faster and more accurate compared to the original solution without zooming. However, the zooming interface was considered less comfortable than the head mouse.

Hornof, Cavender, and Hoselton (2004) developed a system for children with motor impairments to draw pictures and select targets using eye movements. The eye system allowed users to switch between looking and drawing by focusing on a predefined part of the screen. Eight participants without PI evaluated the system. Participants generally found the system easy to learn. The quality of the pictures drawn showed that the system could allow children to draw pictures with their eyes, although it is not an easy task.

Speech. Manaris and Harkreader (1998) and Manaris, Macgyvers, and Lagoudakis (2002) discussed a "listening keyboard" that allowed users with PI to access all of the functionality of a traditional keyboard and mouse. A usability study with three individuals with PI explored the effectiveness of the listening keyboard through a "Wizard-of-Oz" design. Results from a postexperiment questionnaire suggested that participants were comfortable with the concept of a "listening keyboard" and that it would be preferable to mouthstick or handstick systems. The participants were able to complete tasks faster with lower error rates when using the listening keyboard compared to the handstick. However, these results should be interpreted with caution since the number of participants was limited and the "Wizard-of-Oz" methodology results in unrealistically low recognition error rates.

Karimullah and Sears (2002) focused on continuous speech-based navigation techniques and studied a standard speech-based mouse cursor and a predictive cursor. The predictive cursor was proposed to address the delay related to speech processing. As the cursor moves, the predictive cursor appears in front of the actual cursor, indicating the position at which the actual cursor will be located if the command to stop the cursor were issued. Twenty-eight participants without PI evaluated the efficacy of the two solutions. The difficulty users experienced selecting targets was affected by the speed at which the cursor was moving, target size, and the delay between when the command was issued and when it was processed, but no differences were detected between the standard and predictive cursors.

Building on the results of the previous study, Sears, Lin, and Karimullah (2002) conducted an experiment to explore why the predictive cursor did not prove beneficial. This study investigated three sources of delay associated with speech input: the time required for the user to initiate a response, the time required to speak the response, and the time required for the system to produce a response. Fifteen individuals without PI participated in this study. Based on the results of this study, the authors proposed a variable speed cursor to improve the performance as this would allow users to decrease the speed of the cursor when selecting small targets. This is expected to address the problems that users experience and significantly reduce failure rates.

Dai, Goldman, Sears, and Lozier (2004) focused on discrete speech-based navigation techniques, evaluating two grid-based solutions, which allow users to select targets by recursively locating the cursor onto a smaller area of the screen. Both solutions divided the screen into nine regions. One solution provided a single cursor in the center region. The other provided nine cursors, each located in one region. Twenty-four participants without PI tested the two techniques. The nine-cursor solution allowed users to select targets of varying sizes, distance, and direction significantly faster than the one-cursor solution. Through comparisons with earlier studies, the authors suggested that grid-based solutions might prove advantageous when compared to other speech-based navigation solutions such as the continuous cursor.

Text Entry Technologies

Computers are frequently used to generate text. While the keyboard is the most frequently used alternative, other technologies can also be used. Some individuals with PI select letters from on-screen keyboards, others use speech recognition, and still others interact with the standard keyboard. When using an on-screen keyboard, text entry is effectively turned into a cursor-control task. As a result, all of the cursor-control technologies discussed earlier can also be used for text entry. The article by Sears, Karat, Oseitutu, Karimullah, and Feng (2001) discussed the use of speech recognition for text entry, but was included earlier due to its focus on individuals with SCI. Articles in earlier sections could also be included here, but are discussed separately due to their focus on individuals with specific impairments. The articles included later discuss technologies that can be used for text entry without focusing on any health condition or PI.

Keyboard-based text entry. Matias, MacKenzie, and Buxton (1993) described a one-handed keyboard based upon the traditional QWERTY design. This basic design allows users to access all of the keys on one side of the keyboard at a time, switching between sides by pressing and holding the spacebar. A study with ten participants with no PI was reported. Each participant used his or her nondominant hand when typing with one hand. Speed and errors improved significantly over ten sessions. By the end of ten sessions, participants were typing 34.7 wpm with an error rate of 7.44% with the one-handed keyboard as compared to 64.9 wpm and 4.20% errors with the two-handed keyboard. While this system could prove useful for individuals with PI that hinder the use of one hand, it is important to remember that the efficacy of this design for individuals with limited experience using the QWERTY keyboard is uncertain.

Many researchers have investigated techniques that allow users to generate text using fewer keystrokes than are normally required. These techniques typically predict words, given just a couple of letters, using statistics that describe how frequently various words are used. Some systems also use statistics regarding how frequently various words follow words that have already been entered. Other techniques present multilevel interfaces in which users repeatedly select categories until the desired word is available. Once the word is selected, an appropriate ending (e.g., "s," "ing") can be added. Demasco and McCoy (1992) presented a theoretical discussion of a word-based virtual keyboard that uses a level-based interface. McCoy, Demasco, Jones, Pennington, Vanderheyden, and Zickus (1994) discussed a word-based text-entry technique by which users specify uninflected content words and the system adds appropriate endings to the words as well as additional words (e.g., articles, prepositions) to complete the sentence. Garay-Vitoria and González-Abascal (1997) discussed the potential of syntactic analysis of previously entered words to enhance word prediction results. Boissiere and Dours (2000) described a system that predicts word endings without assistance from the user. Unlike many of the articles discussed earlier, all of these articles take either a theoretical or a systems approach to the problem. Consequently, some of the systems have not been implemented and those that have been implemented have not been evaluated by the intended end users.

Trewin (2002) studied overlap errors, one of the common typing errors in which two keys are pressed at once, to improve typing accuracy for people with PI. While some users with PI use keyguards to reduce overlap errors, the author developed a software filter to identify and eliminate overlap errors. The software used keystroke timing and language-based techniques to identify and automatically correct overlap errors. Twenty participants with PI tested the software. The most effective filter was based on keystroke timing characteristics. This filter identified 97% of the errors and misclassified 46% of the deliberate overlaps. The accuracy of the filters was dependent on the typing styles of the users.

Willis, Pain, Trewin, and Clark (2002) developed a system that automatically reconstructs full sentences from abbreviated text, allowing people with PI to type less text. Ten individuals without PI participated in an evaluation study. Four texts were used in the study, each of 500 letters. Each participant completed four abbreviation tasks based on one text and one reconstruction task based on a different text. Both texts were randomly chosen from the four texts. Several abbreviation patterns were identified, such as "double letter" being transformed to single letter and "ch" being transformed to "k." These patterns provided insights that guided the development of an abbreviation expansion system that matches users' natural abbreviation styles.

Trewin, Slobogin, and Power (2002) developed a keyboard optimizer capable of adapting the keyboard to meet the users' needs, with a focus on users with mild to moderate PI that negatively affects their ability to use a keyboard. Given a sample of an individual's typing behaviors, the optimizer generates a customized profile that includes details about how the individual types that can allow the keyboard to be configured to improve typing outcomes for the specific individual. The system was evaluated by eight users with PI that affected their typing. Twenty-one configuration modifications were suggested by the optimizer, with 20 being judged by the clinicians as being appropriate. Participants liked the system and found it easy to use. Trewin (2004) further improved the underlying models and built a dynamic keyboard. Ten individuals with young-onset Parkinson's disease provided input to the system. The automatic adjustment of the key repeat delay feature was acceptable to the users.

Wobbrock, Myers, Aung, and LoPresti (2004) developed two integrated gestural text-entry methods designed for use from power wheelchairs. Both devices can be integrated into existing controls on power wheelchairs and are small, light, inexpensive, and require minimal configuration. One method uses a joystick and the other a touchpad. Both methods use a stylus-based unistroke text-entry method called EdgeWrite designed for people with tremors. EdgeWrite relies on physical edges and corners to provide stability during motion. The users move stylus, finger, or joystick along the physical edges and into the corners of a square bounding the input area. The system recognizes the input based on the order in which corners are hit. Seven power wheelchair users joined the participatory design process and evaluated the two techniques. Although data were too limited for a complete statistical analysis, the results showed potential for the target population. The touchpad application was faster than the joystick application, allowing users to generate an average of one word per minute.

Speech recognition–based text entry. Goette (1998) conducted a field study to identify factors that influence the successful adoption of speech recognition software for both dictation-oriented and environmental control tasks. Individuals with various health conditions (e.g., MS, MD, SP, arthritis) participated. Most individuals who stopped using the software (53%) did so within three months. Interestingly, those individuals who successfully adopted the software had higher expectations for the system and expected greater benefits, but also believed the system would be easy to use as compared to those individuals who stopped using the software. When the software was successfully adopted, it was used for a wide variety of computer-based tasks rather than a few isolated activities. Four guidelines were proposed for successful outcomes: (a) managing expectations by understanding the potential benefits and limitations, (b) selecting the correct system for the tasks to be accomplished, (c) obtaining thorough training, and (d) trying the system for an extended period before purchasing it.

Pieper and Kobsa (1999) developed a speech system to allow severely impaired users who are bedridden to interact with the computer. The authors implemented several modifications to the commercially available system to cater the usage environment. The output of the computer was displayed on the ceiling and the speech recognition system configured to accommodate the use of respiratory devices. One patient who had ALS for five years used the system for seven months to write poetry, diaries, and letters. The result shows that the projected video display was acceptable to the user. Overall, the user was satisfied with the system.

Manasse, Hux, and Rankin-Erickson (2000) studied the use of speech recognition to help a 19-year-old survivor of traumatic brain injury (TBI). This study focused on text generation tasks using Dragon NaturallySpeaking. This individual experienced a recognition error rate of approximately 20%. The participant quickly and easily mastered navigation and error-correction commands. Interestingly, the participant generated more text using the standard keyboard than she did using speech recognition, but she generated more complex sentences when using the speech-recognition system. Overall, minimal qualitative differences were observed in the resulting text.

Bain, Basson, and Wald (2002) focused on using speech recognition to generate lecture notes for students in the classroom. Soft- and hardcopy notes can provide an alternative to traditional note taking for people with PI or perceptual impairments. More recently, Hewitt, Lyon, Britton, and Mellor (2005) developed the "SpeakView" system, which provides speech-to-text displays for meetings and in classrooms. Key issues that were addressed by the "SpeakView" system included automatically adding punctuation, style, and acceptability of the captions, usability, and training. Sixty-nine students evaluated the system by observing two presentations and answering predefined questions. Captions produced by "SpeakView" were provided for one of the presentations. The use of captions did not influence the number of questions answered correctly by the students, but students did prefer presentations with captions. Students indicated that the captions made the talk easier to understand and that they felt the captions helped them answer more questions correctly. This system, if fully implemented with acceptable accuracy, has the potential to benefit individuals with PI or hearing impairments who find taking notes difficult.

Koester (2004) investigated the use of automatic speech-recognition technology by experienced users. Twenty-four users with PI that hindered the use of the keyboard and mouse, with six months of experience with ASR, participated in this study. Each participant completed a series of predefined tasks using two input methods: speech and nonspeech. Participants responded to a survey about their usage patterns. The results showed that speech was significantly slower for command input, but there was no significant difference in text entry rates between speech and nonspeech input. There was a wide variation among text entry rates, ranging from 3–32 words per minute. There was also a wide variation in recognition accuracy, ranging from 72%–94%. The study confirmed that recognition errors are a major limitation for speech-based input, with users spending an average of 56% of their time on error correction. Interestingly, the participants who achieved the best performance were those who tended to employ the prescribed correction strategies.

Eye-gaze-based typing. Lankford (2000) developed an eye-gaze interface that can support both text entry and cursor control tasks. The system used dwell time to provide the user with access to various mouse clicking actions. However, because of the limitations of the human eye, gaze-based clicking was not able to reliably control the Windows environment, particularly for users with PI. To improve the accuracy of gaze-based clicking, a zooming interface was developed to enlarge the objects when needed. Key selection was accomplished by prolonged fixation on the keyboard key. The activation time could be customized to suit individual users. A list of possible words to insert into a document, based on what the user had already typed, was provided to speed up typing and reduce errors. No user study has been reported concerning the effectiveness of this system.

Hansen, Itoh, Tørning, Aoki, and Johanson (2004) investigated the effectiveness of gaze typing compared to hand typing and head tracking. Twenty-seven participants without PI evaluated the three modalities with two interfaces in different languages. Each participant entered predefined text. Participants achieved a typing speed of 6.22 words per minute. The eye-typing system resulted in an error rate of 6% with the Japanese interface and 28% with the Danish interface. Significant learning effects were observed for both interfaces. Overall, gaze typing was the least efficient and least satisfying input modality among the three alternatives. The potential of the gaze-based typing and its advantages compared to other input modalities needs to be investigated by including users with PI, especially those with severe PI for whom gaze typing may be one of the few alternatives that are available.

ADOPTION OR ABANDONMENT OF ASSISTIVE TECHNOLOGIES

It has been estimated that over 1,000 assistive technology products come to the market each year (Scherer & Galvin, 1996). Unfortunately, the adoption rates of assistive technology products are highly disappointing. Many new technologies never reach the majority of the intended users due to a lack of marketing activities or cost. Even when the technologies reach the intended users, many are abandoned within a short period of time. Understanding why users adopt or abandon assistive technologies is critical for the design, implementation, and marketing of these technologies. A number of studies that can provide valuable insights to researchers, designers, as well as healthcare providers are summarized later.

Scherer (1996) studied adoption rates of general assistive technologies and found that many people abandon those technologies. Abandonment rates ranged from 8% for life-support devices to 75% for hearing aids, depending on the type of device. On average, about one third of all assistive technologies were abandoned. However, it was also observed that many people who continue using the technology are unhappy or uncomfortable with the experience, suggesting that the underlying problems are even more significant than the abandonment rates would suggest. It was observed that most abandonment occurs within the first year, especially the first three months, or after five years of use (normally because insurance policies only support the first five years of use). This suggests that users learn relatively quickly whether the specific assistive technology works for them. If not, they abandon it quickly. The author believes that the most significant factor that contributes to the abandonment of assistive technology is that the device does not meet the users' needs or expectations.

King (1999) discussed the factors leading to the failure of assistive technologies, highlighting three major categories of failures. The first category relates to the people surrounding the user. For the people surrounding the user, correctly evaluating and communicating the user's condition is critical since many assistive technologies fail because the users' real needs are not known by the designers. The second category relates to the users themselves. From the perspective of the assistive technology users, design efforts should focus on matching the device's features to the user's needs, instead of the user to the device. The user's age, gender, and literacy can all play important roles in the adoption process. The third category relates to the assistive technology device itself. As to the assistive device, insufficient consideration of human factors during the design process has caused many assistive technologies to fail. Considering ease of use and portability, the designers should focus on lighter and smaller solutions that provide easy access to the power supply. In addition, assistive devices need to be durable and repairable.

Kintsch and DePaula (2002) proposed a framework for facilitating the successful adoption of assistive technologies. The framework proposed four groups of participants in the adoption process: users, caregivers, designers, and assistive technology specialists. All four groups of participants need to be actively involved to support the adoption of assistive technologies. In the development phase, developers need to customize the product to the users' needs and abilities, make products simple and durable, and take into consideration user preferences. In the selection phase, products need to be assessed carefully by assistive technology specialists. The effectiveness of the product for the specific user must be evaluated, currently through a trial and reassessment process. In the learning phase, caregivers need to be trained with the product first so they can teach and support the user. In the integration phase, which is when the device becomes integrated into daily life, special patience is needed since tasks may initially require more time as the user gains experience.

Michaels and McDermott (2003) focused on the adoption of assistive technologies for students in the special education programs. The authors conducted interviews with employees from 356 special education programs and collected 143 surveys from special education teachers. The results show that the major barriers of assistive technology adoption in special education programs include funding, time limitations, lack of faculty knowledge or consistency in assistive technology focus, perceptions that assistive technology is only important for students with certain types of disabilities, and lack of short- and long-term plans.

Goette (2000) studied the use of speech-recognition technology in organizations and the key factors that contributed to successful adoption. A total of 40 individuals with PI or sensory disabilities who were using or had used speech recognition technologies were interviewed. Twenty-three of the participants had successfully adopted the speech-recognition technology, while 17 participants had abandoned it. The results suggest that participants who successfully adopted the technology had realistic expectations about the adoption process. Importantly, they expected to devote more time and effort than stated in the user manual to become proficient with the technology. For those users who had abandoned the technology, unrealistic employer expectations was one of the key reasons for abandonment.

CONCLUSIONS

The most obvious conclusion is that additional research is necessary. PI vary dramatically in terms of severity, temporal variability, and the body parts that are affected. Numerous technologies can be used in a variety of configurations. Given this variability and the limited number of studies reported in the literature, it is clear that only a fraction of the important questions have been investigated. Unfortunately, conducting informative studies involving individuals with PI can be difficult. Unlike computer users with no PI, the pool of potential participants is limited. Even when appropriate participants can be found, additional factors can make such studies difficult.

While additional research is needed, the existing literature does provide insights that can prove useful to both practitioners and researchers. Individually, the articles highlight potential benefits and limitations of various technologies for specific groups of individuals as well as new technologies that may prove useful for specific tasks or individuals in the future. When viewed as a whole, several lessons that extend beyond the technology or specific user groups studied become clear. We conclude by highlighting five such lessons.

PI Does Not Imply Disability

Impairments can, but do not always, result in disabilities (see Fig. 5.1). Several of the studies discussed earlier demonstrate this fact. Casali (1992) included three groups of participants including individuals with: PI resulting in low motor skills, PI resulting in high motor skills, and no PI. The high-motor-skills group was able to complete cursor manipulation tasks just as quickly as the group with no PI. This suggests that the high-motor-skills group did not experience any disability because of their PI. Similarly, Sears, Karat, Oseitutu, Karimullah, and Feng (2001) described a study in which individuals with high-level spinal-cord injuries were able to compose text just as quickly as individuals without PI. Several other studies provide examples where technology can either reduce or eliminate disabilities resulting from specific PI (e.g., Keates, Langdon, Clarkson, & Robinson, 2000). Overall, these results clearly indicate that it is possible, under the right circumstances, to design technology such that specific PI will not interfere with certain computing activities.

PI Affect Cognitive, Perceptual, and Motor Activities

PI are expected to affect an individual's ability to complete basic motor activities, but are not necessarily expected to affect performance on cognitive and perceptual activities. Many of the articles discussed earlier confirm differences in performance for motor activities (e.g., LoPresti, Brienza, Angelo, Gilbertson, & Sakai, 2000; Radwin, Vanderheiden, & Lin, 1990). More importantly, the unexpected affect of PI on cognitive and perceptual activities becomes apparent by analyzing both fundamental interactions and the high-level strategies adopted by individuals with PI.

Sears, Karat, Oseitutu, Karimullah, and Feng (2001) found that individuals with and without PI responded differently when using the same speech recognition system to accomplish a variety of tasks. Individuals with PI interrupted their dictation more often, spent a greater percentage of their time dictating as opposed to issuing commands, and navigated shorter distances. The results reported by Koester and Levine (1994) also highlight differences between individuals with and without PI with regard to both basic cognitive, perceptual, and motor activities as well as higher-level strategies. These studies confirm the importance of integrating individuals with PI into the process when designing new computer systems. Existing models of computer user behavior are based on data from individuals without PI, but it is clear that new models that integrate data from individuals with PI are necessary (Koester & Levine, 1994). These new models must address differences not only for motor activities, but for cognitive and perceptual activities as well.

Basic Actions Can Be Difficult

A number of articles have highlighted the difficulty individuals with PI may experience with even the most basic interaction activities. Trewin and Pain (1999) highlighted a variety of difficulties with activities as fundamental as pressing keys on a keyboard or pointing at and clicking on objects with a mouse. Casali (1992) also illustrated the difficulty individuals with PI could have with activities that require the mouse button to be held down while dragging the mouse. Unfortunately, these difficulties become even more significant as the target of the action gets smaller (Casali, 1992). For example, Radwin, Vanderheiden, and Lin (1990) showed that performance with head-controlled devices was affected by target size even more than the mouse.

Standardized Descriptions of Health Conditions and PI

Perhaps the greatest hindrance for both practitioners and researchers is the lack of accepted standards for describing the capabilities and limitations of the individuals who will make use of the new solutions that are developed. For studies including traditional computer users without PI, we typically provide basic demographics such as age, gender, and computer experience. For these users, we typically assume "normal" perceptual, cognitive, and motor abilities. Occasionally, researchers even may report that participants all had "normal" or "corrected to normal" vision or hearing if these abilities are particularly relevant for a given study. For individuals with PI (or cognitive or perceptual impairments for that matter), we need to describe the individual's perceptual, cognitive, and motor abilities. Unfortunately, existing studies rarely provide sufficient details regarding health conditions or associated PI. Many articles simply list a health condition for each participant (e.g., CP, choreoathetoid CP, incomplete SCI at C6, Friedrich's ataxia). While this provides a general sense as to the PI that may exist, it does not provide a sufficiently detailed understanding of the participants' abilities and limitations.

When a health condition results in a well-defined set of PI, a precise description of the underlying health condition may be sufficient. For example, an SCI at or above C5 with an ASIA score of A or B results in paralysis of the muscles used to control the arms and legs. Given this health condition, we know that an individual would not be able to use her hands or arms when interacting with a computer. In contrast, an SCI at or above C5, with an ASIA score of C or D results in much more ambiguous PI. Similarly, indicating that an individual has CP provides general insights, but the resulting PI can vary dramatically. When a health condition does not map to a precise set of PI that are static, it is critical to describe both the health condition and the resulting PI (and any other impairments that may affect interactions with computers).

Health conditions are best described by physicians. Physical examinations can often provide the necessary information, but access to medical records may be sufficient. Ideally, a standardized scale will be used to describe the health condition, such as the ASIA scale that is used to describe the completeness of a spinal cord injury. However, such assessments focus on the status of the individual's health, not necessarily on their physical abilities. Understanding an individual's health can provide insights into changes that may occur in the future. Unfortunately, most of the time health conditions do not map directly to a well-defined set of PI.

In these situations, an occupational therapist can become involved in assessing each participant's PI. Occupational therapists can evaluate specific skills or abilities, including sensory motor, cognitive, and motor skills as necessary. The primary focus should be on those skills and abilities that may affect the individual's ability to interact with computers. While numerous standardized tests exist, many of these tests focus on activities that are not directly relevant to the use of information technologies. Assessments by occupational therapists focus on the impairments an individual experiences that may affect an individual's ability to perform certain activities without regard to the underlying health condition. These assessments focus on an individual's current status and therefore do not address changes that may occur due to the progression of a disease. New approaches for assessing and documenting the physical capabilities of individuals, that are relevant for interactions with information technologies, could prove useful in a variety of contexts. Effective solutions could be used by rehabilitation specialists as they identify appropriate technologies for patients as well as researchers seeking to provide more comprehensive descriptions of the participants in their research studies.

Successful Adoption of Assistive Technologies Is Critical

Assistive technologies are critical for helping users with PI interact with computer systems, but current adoption rates are highly disappointing. Scherer (1996) indicated that abandonment rates for assistive technologies range from 8% to 75%, with approximately one third of all assistive technologies being abandoned. It is also clear that the majority of new solutions are abandoned relatively soon after the initial interaction. King (1999) suggested that successful adoption of assistive technologies requires sufficient consideration of human factors when designing devices and active involvement of both the end-user and the people surrounding the user, such as the caregivers and healthcare providers. Kintsch and DePaula (2002) also emphasized the importance of active involvement by the end-users, caregivers, designers, and assistive technology specialists. To design and distribute effective assistive technologies that users will readily adopt, active user involvement and sufficient consideration for human factors is required.

FUTURE DIRECTIONS

The existing literature

- Highlighted the potential benefits and limitations of various technologies for specific groups of individuals;
- Provided examples of situations in which PI did not result in disabilities;
- Illustrated the potential impact of PI on cognitive, perceptual, and motor activities;

- Confirmed the difficulties individuals with PI may experience with basic actions such as pressing keys and clicking the mouse button, as well as more complex actions;
- Reinforced the need for standardized descriptions of health conditions and PI;
- Highlighted how infrequently assistive technologies are successfully adopted as well as some of the underlying problems that lead to abandonment.

This same literature confirmed the potential of various new technologies while highlighting the need for additional research. Several studies illustrated the use of body parts that are normally ignored when designing computer interfaces (e.g., Shaw, Loomis, & Crisman, 1995; Salem & Zhai, 1997). Others explored the potential of using electrophysiological data to control computers (e.g., Kennedy, Kakay, Moore, Adams, & Goldwaithe, 2000; Kübler, Kochoubey, Hinterberger, Ghanayim, Perelmouter, Schauer, et al., 1999; Doherty, Cockton, Bloor, & Benigno, 2000). These studies were exploratory, but they highlight the potential of these new technologies, especially for individuals with severe PI.

We must continue to explore technologies that support alternative methods of interacting with computers, but we also need a better understanding of both the potential and limitations of existing technologies. Existing studies provide examples where individuals with PI were just as fast and accurate as individuals with no PI, but important differences were also apparent for satisfaction ratings and the strategies that the individuals with PI employed. Other studies highlighted situations in which PI resulted in longer task completion times, higher error rates, or more subtle differences that only became apparent when performance was analyzed in detail. Future studies should attend to changes in cognitive, perceptual, and motor activities, which may reveal themselves through task completion times, error rates, satisfaction ratings, or through changes in the strategies that individuals adopt when completing a task. To allow both researchers and practitioners to more effectively interpret and generalize the results of such studies, we must carefully document both the health conditions and the PI of study participants. Ideally, new standardized methods for describing the physical capabilities of study participants will be developed and adopted.

While research into the efficacy of various technologies for individuals with PI is important, studies that assess the outcomes when assistive technologies are actually used are also critical (Fuhrer, 2001). Research studies highlight the potential of a technology, but do not guarantee the success of a technology when used in the field. Outcomes research will ultimately determine the success or failure of the technologies developed by the HCI community for individuals with PI. As a result, it is important for HCI researchers and system developers to work closely with those rehabilitation professionals who evaluate the needs of individuals with PI, determine which technologies they should be using, and assess the success or failure of these technologies.

ACKNOWLEDGMENTS

This material is based upon work supported by the National Science Foundation under Grants IIS-9910607, IIS-0121570, IIS-0328391, and IIS-0511954. Any opinions, findings, and conclusions or recommendations expressed in this material are those of the authors and do not necessarily reflect the views of the National Science Foundation (NSF).

References

ACM Conference on Universal Usability (CUU). (2003). Retrieved January 5, 2006, from http://sigchi.org/cuu2003/

ACM SIGACCESS Conference on Computers and Accessibility (ASSETS). (2006). Retrieved March 8, 2007, from http://www.acm.org/sigaccess/assets06/

Albacete, P. L., Chang, S. K., & Polese, G. (1994). Iconic language design for people with significant speech and multiple impairments. In *Proceedings of ASSETS 94* (pp. 23–30). New York: ACM.

Albacete, P. L., Chang, S. K., Polese G., & Baker, B. (1998). Iconic language design for people with significant speech and multiple impairments. In *Lecture notes in computer science: Vol. 1458. Assistive technology and artificial intelligence, applications in robotics, user interfaces and natural language processing* (pp. 12–32). Berlin, Heidelberg: Springer-Verlag.

Allanson, J., Rodden, T., & Mariani, J. (1999). A toolkit for exploring electro-physiological human-computer interaction. In *Proceedings of INTERACT'99* (pp. 231–237). Amsterdam: IOS Press.

Alm, N., Todman, J., Elder, L., & Newell, A. F. (1993). Computer aided conversation for severely physically impaired non-speaking people. In *Proceedings of InterCHI 93* (pp. 236–241). New York: ACM.

American Spinal Injury Association (ASIA). (2001). *ASIA impairment scale*. Retrieved June 27, 2001, from http://www.asia-spinalinjury.org/publications/2001_Classif_worksheet.pdf

Bain, K., Bass, R., & Wald, E. (2002). Speech recognition in university classrooms: liberated learning project. In *Proceedings of ASSETS 2002* (pp. 192–199). Edinburgh: ACM.

Banerjee, S. (1982). *Rehabilitation management of amputees*. Baltimore: Williams & Wilkins.

Barreto, A. B., Scargle, S. D., & Adjouadi, M. (1999). A real-time assistive computer interface for users with motor disabilities. *ACM SIGCAPH Computers and the Physically Handicapped, 64*, 6–16.

Bates, R., & Istance, H. (2002). Zooming interfaces! Enhancing the performance of eye controlled pointing devices. In *Proceedings of ASSETS 2002* (pp. 119–126). Edinburgh.

Berkow, R. (1997). *The Merck manual of medical information home edition*. Whitehouse Station, NJ: Merck Research Laboratories.

Birbaumer, N., Hinterberger, T., Kübler, A., & Neumann, N. (2003, June). The thought-translation device (TTD): Neurobehavioral mechanisms and clinical outcome. *IEEE Trans. on Neural Systems and Rehabilitation Engineering, 11*(2), 120–123.

Birch, G., Bozorgzadeh, Z., & Mason, S. (2002). Initial on-line evaluations of the LF-ASD brain-computer interface with able-bodied and spinal-cord subjects using imagined voluntary motor potentials. *IEEE Trans. on Neural Systems and Rehabilitation Engineering, 10*(4), 219–24.

Boissiere, P., & Dours, D. (2000). VITIPI: A universal writing interface for all. *Proceedings of the 6th ERCIM Workshop: User Interfaces for All.*

Borisoff, J. F., Mason, S., Bashashati, A., & Birch, G. (2004). Brain-computer interface design for asynchronous control applications: Improvements to the LF-ASD asynchronous brain switch. *IEEE Transactions on Biomedical Engineering, 51*(6), 985–992.

Bowman, T. (1997). VR meets physical therapy. *Communications of the ACM, 40*(8), 59–60.

Bozorgzadeh, Z., Birch, G. E., & Mason, S. G. (2000). The LF-ASD BCI: On-line identification of imagined finger movements in spontaneous EEG with able-bodied subjects. *Proceedings of ICASSP 2000.* Vol. 4, pp. 2385–2388.

California State University, Northridge Center (CSUN). (2006). California State University, Northridge Center On Disabilities Annual Conference. Retrieved March 8, 2007, from http://www.csun.edu/cod/conf/2006/proceedings/csun06.htm

Card, S., Moran, T. P., & Newell, A. (1983). *The psychology of human-computer interaction.* Mahwah, NJ: Lawrence Erlbaum Associates.

Carroll, K., Schlag, C., Kirikci, O., & Moore, M. (2002). Communication by neural control. In *Proceedings of CHI 2002* (pp. 590–591). Minneapolis, MN: ACM.

Casali, S. P. (1992). Cursor control device used by persons with physical disabilities: Implications for hardware and software design. In *Proceedings of the Human Factors Society 36th Annual Meeting* (pp. 311–315). Santa Monica: HFES.

Casali, S. P. (1995). A physical skills based strategy for choosing an appropriate interface method. In A. D. N. Edwards (Ed.), *Extra-ordinary human-computer interaction* (pp. 315–341). Cambridge: Cambridge University Press.

Casali, S. P., & Chase, J. D. (1993). The effects of physical attributes of computer interface design on novice and experienced performance of users with physical disabilities. In *Proceedings of the Human Factors and Ergonomics Society 37th Annual Meeting* (pp. 849–853). Santa Monica: HFES.

Closing the Gap: Computer technology in special education and rehabilitation. (2006). Retrieved March 8, 2007, from http://www.closingthegap.com/confTest/index.lasso

Craig, A., Tran, Y., McIsaac, P., Moses, P., Kirkup, L., & Searle, A. (2000). The effectiveness of activating electrical devices using alpha wave synchronization contingent with eye closure. *Applied Ergonomics, 31*, 377–382.

Dai, L., Goldman, R., Sears, A., & Lozier, J. (2004). Speech-based cursor control: A study of grid-based solutions. In *Proceedings of ASSETS 2004* (pp. 94–101). Atlanta: ACM.

De Silva, G., Lyons, M., Kawato, S., & Tetsutani, N. (2003). Human factors evaluation of a vision-based facial gesture interface. *Proceedings, of CVPR-HCI 2003*, Vol. 5. (pp. 52–59).

Demasco, P. W., & McCoy, K. F. (1992). Generating text from compressed input: An intelligent interface for people with severe motor impairments. *Communications of the ACM, 35*(5), 68–78.

Demasco, P., Newell, A. F., & Arnott, J. L. (1994). The application of spatialization and spatial metaphor to augmentative and alternative communication. In *Proceedings of ASSETS 94* (pp. 31–38). New York: ACM.

Doherty, E., Cockton, G., Bloor, C., & Benigno, D. (2000). Mixing oil and water: Transcending method boundaries in assistive technology for traumatic brain injury. In *Proceedings of ASSETS 2000* (pp. 110–117). New York: ACM.

Doherty, E., Cockton, G., Bloor, C., & Benigno, D. (2001). Improving the performance of the Cyberlink mental interface with the 'Yes/No program'. *Proceedings of CHI 2001*, 69–76: ACM.

Duvaisin, R. (1991). *Parkinson's disease: A guide for patients and families.* New York: Raven Press.

Edwards, A. D. N. (1995). *Extra-ordinary human-computer interaction.* Cambridge: Cambridge University Press.

European Research Consortium for Informatics and Mathematics (ERCIM). (2006). 9th ERCIM Working Group: User Interface for All.

Retrieved March 8, 2007, from http://ui4all.ics.forth.gr/workshop2006/

Evans, D. G., Drew, R., & Blenkhorn, P. (2000). Controlling mouse pointer position using an infrared head-operated joystick. *IEEE Transactions on Rehabilitation Engineering, 8*(1), 107–117.

Fell, H. J., Delta, H., Peterson, R., Ferrier, L. J., Mooraj, Z., & Valleau, M. (1994). Using the baby-babble-blanket for infants with motor problems: An empirical study. In *Proceedings of ASSETS 94* (pp. 77–84). New York: ACM.

Felzer, T., & Freisleben, B. (2002). HaWCoS: The hands-free wheel chair control system. In *Proceedings of ASSETS 2002* (pp. 127–134). Edinburgh: ACM.

Feng, J., & Sears, A. (2004). Using confidence scores to improve hands-free speech-based navigation in continuous dictation systems. *ACM Transactions on Computer-Human Interaction, 11*(4), 329–356.

Fuhrer, M. J. (2001). Assistive technology outcomes research: Challenges met and yet unmet. *American Journal of Physical Medicine and Rehabilitation, 80*(7), 528–535.

Garay-Vitoria, N., & González-Abascal, J. (1997). Intelligent word-prediction to enhance text input rate. In *Proceedings of IUI 97* (pp. 241– 244). New York: ACM.

Gnanayutham, P., Bloor, C., & Cockton, G. (2005). Discrete acceleration and personalized tiling as brain-body interface paradigms for neurorehabilitation. In *Proceedings of CHI 2005* (pp. 261–270). Portland: ACM.

Goette, T. (1998). Factors leading to the successful use of voice recognition technology. In *Proceedings of ASSETS 98* (pp. 189–196). New York: ACM.

Goette, T. (2000). Keys to the adoption and use of voice recognition technology in organizations. *Information Technology & People, 13*(1), 67–80.

Gray, D. B., Quatrano, L. A., & Liberman, M. L. (1998). *Designing and using assistive technology.* Baltimore, MD: Paul H. Brooks Publishing Co.

Hallett, M. (1991). Classification and treatment of tremor. *Journal of the American Medical Association, 266*, 1115.

Hansen, J., Itoh, K., Tørning, K., Aoki, H., & Johanson, A. (2004). Gaze typing compared with input by head and hand. *Proceedings of the Eye Tracking Research & Applications Symposium on Eye Tracking Research and Applications*, 131–138.

Hewitt, J., Lyon, C., Britton, C., & Mellor, B. (2005). SpeakView: Live captioning of lectures. *Proceedings of HCII 2005.*

Hicks, J. E. (1988). Approach to diagnosis of rheumatoid disease. *Archives of Physical Medical Rehabilitation, 69*, S–79.

Horn, L. J., & Zasler, N. D. (Eds.). (1996). *Medical rehabilitation of traumatic brain injury.* Philadelphia: Hanley & Belfus.

Hornof, A., Cavender, A., & Hoselton, R. (2004). EyeDraw: A system for drawing pictures with eye movements. *Proceedings of ASSETS 2004*, 86–93. ACM.

Hwang, F., Keates, S., Langdon, P., & Clarkson J. (2003). Multiple haptic targets for motion-impaired computer users. In *Proceedings of CHI 2003* (pp. 41–48). Fort Lauderdale: ACM.

Hwang, F., Keates, S., Langdon, P., & Clarkson, J. (2004). Mouse movements of motion-impaired users: A submovement analysis. In *Proceedings of ASSETS 2004* (pp.102–109). Atlanta: ACM.

Karimullah, A., & Sears, A. (2002). Speech-based cursor control. *Proceedings of ASSETS 2002* (pp. 178–185). Edinburgh: ACM.

Keates, S., Clarkson, J., & Robinson, P. (2000). Investigating the application of user models for motion-impaired users. *Proceedings of ASSETS 2000* (pp. 129–136). New York: ACM.

Keates, S., Hwang, F., Langdon, P., & Clarkson, J. (2002). Cursor measures for motion-impaired computer users. In *Proceedings of ASSETS 2002* (pp. 135–142). Edinburgh: ACM.

Keates, S., Langdon, P., Clarkson, J., & Robinson, P. (2000). Investigating the use of force feedback for motion-impaired users. *Proceedings of the 6th ERCIM Workshop: User Interfaces for All.*

Keirn, Z. A., & Aunon, J. I. (1990). Man-machine communications through brain-wave processing. *IEEE Engineering in Medicine and Biology Magazine, 9*(1), 55–57.

Kennedy, P. R., Kakay, R. A. E., Moore, M. M., Adams, K., & Goldwaithe, J. (2000). Direct control of a computer from the human central nervous system. *IEEE Transactions on Rehabilitation Engineering, 8*(2), 198–202.

King, T. W. (1999). Why AT fails: A human factors perspective. In. T. King (Ed.), *Assistive technology, essential human factors* (pp. 231–255). Boston: Allyn and Bacon.

Kintsch, A., & DePaula, R. (2002). A framework for the adoption of assistive technology. *Proceedings of the SWAAAC 2002: Supporting Learning Through Assistive Technology,* Winter Park, CO, USA, pp. E3 1–10.

Kirkup, L., Searle, A., Craig, A., McIsaac, P., & Moses, P. (1997). EEG-based system for rapid on-off switching without prior learning. *Medical and Biological Engineering and Computing, 35*(5), 504–509.

Koester, H. (2004). Usage, performance, and satisfaction outcomes for experienced users of automatic speech recognition. *Journal of Rehabilitation Research and Development, 41*(5), 739–754.

Koester, H. H., & Levine, S. P. (1994). Validation of a keystroke-level model for a text entry system used by people with disabilities. In *Proceedings of ASSETS 94* (pp. 115–122). New York: ACM.

Kübler, A., Kochoubey, B., Hinterberger, T., Ghanayim, N., Perelmouter, J., Schauer, et al. (1999). The thought translation device: A neurophysiological approach to communication in total motor paralysis. *Experimental Brain Research, 124*(2), 223–232.

LaCourse, J. R., & Hludik, F. C. J. (1990). An eye movement communication-control system for the disabled. *IEEE Transactions on Biomedical Engineering, 37*(12), 1215–1220.

Lankford, C. (2000). Effective eye-gaze input into Windows. *Proceedings of Eye Tracking Research and Applications Symposium 2000,* pp. 23–27. ACM.

Lin, M. L., Radwin, R. G., & Vanderheiden, G. C. (1992). Gain effects on performance using a head-controlled computer input device. *Ergonomics, 35*(2), 159–175.

Lisogurski, D., & Birch, G. E. (1998). Identification of finger flexions from continuous EEG as a brain computer interface. *Proceedings of IEEE Engineering in Medicine and Biology Society 20th Annual International Conference.* Vol. 4 (pp. 2004–2007).

LoPresti, E. F. (2001). Effect of neck range of motion limitations on the use of head controls. In *CHI 2001 Extended Abstracts* (pp. 75–76). New York: ACM.

LoPresti, E., & Brienza, D. (2004) Adaptive software for head-operated computer controls. *IEEE Transactions on Neural Systems and Rehabilitation Engineering, 12*(1), 102–111.

LoPresti, E., Brienza, D. M., Angelo, J., Gilbertson, L., & Sakai, J. (2000). Neck range of motion and use of computer head controls. In *Proceedings of ASSETS 2000* (pp. 121–128). New York: ACM.

Lusted, H. S., & Knapp, B. R. (1996). Controlling computers with neural signals. *Scientific American, 275*(4), 58–63.

Malkewitz, R. (1998). Head pointing and speech control as a hands-free interface to desktop computing. In *Proceedings of ASSETS 98* (pp. 182–188). New York: ACM.

Manaris, B., & Harkreader, A. (1998). SUITEKeys: A speech understanding interface for the motor-control challenged. In *Proceedings of ASSETS 98* (pp. 108–115). New York: ACM.

Manaris, B., Macgyvers, V., & Lagoudakis, M. (2002). A listening keyboard for users with motor impairments—a usability study. *International Journal of Speech Technology, 5,* 371–388.

Manasse, B., Hux, K., & Rankin-Erickson, J. (2000). Speech recognition training for enhancing written language generation by a traumatic brain injury survivor. *Brain Injury, 14*(11), 1015–1034.

Mason, S. G., Bozorgzadeh, Z., & Birch, G. E. (2000). The LG-ASD brain computer interface: On-line identification of imagined finger flex- ions in subjects with spinal cord injuries. In *Proceedings of ASSETS 2000* (pp. 109–113). New York: ACM.

Matias, E., MacKenzie, I. S., & Buxton, W. (1993). Half-QWERTY: A one-handed keyboard facilitating skill transfer from QWERTY. In *Proceedings of InterCHI 93* (pp. 88–94). New York: ACM.

McCoy, K. F., Demasco, P. W., Jones, M. A., Pennington, C. A., Vanderheyden, P. B., & Zickus, W. M. (1994). A communication tool for people with disabilities: Lexical semantics for filling in the pieces. In *Proceedings of ASSETS 94* (pp. 107–114). New York: ACM.

McCoy, K., Demasco, P., Pennington, C., & Badman, A. (1997). Some interface issues in developing intelligent communication aids for people with disabilities. In *Proceedings of the International Conference on Intelligent User Interfaces* (pp. 163–170). Orlando: ACM.

Michaels, C., & McDermott, J. (2003). Assistive technology integration in special education teacher preparation: Program coordinators' perceptions of current attainment and importance. *Journal of Special Education Technology, 18*(3), 29–44.

Molnar, G. E. (1992). Cerebral palsy. In G. E. Molnar (Ed.), *Pediatric rehabilitation* (pp. 481–533). Baltimore: Williams & Wilkins.

Moore, J. S., & Garg, A. (Eds.). (1992). Ergonomics: Low-back pain, carpal tunnel syndrome, and upper extremity disorders in the workplace. *State of the Art Reviews Occupational Medicine, 7*(4), 593–790.

Moore, M., & Dua, U. (2004). A galvanic skin response interface for people with severe motor disabilities. In *Proceedings of ASSETS 2004* (pp. 48–54). Atlanta: ACM.

Moore, M., & Kennedy, P. (2000). Human factors issues in the neural signals direct brain-computer interface. In *Proceedings of ASSETS 2000* (pp. 114–120). Arlington: ACM.

Myers, B., Wobbrock, J., Yang, S., Yeung, B., Nichols, J., & Miller, R. (2002). Using handhelds to help people with motor impairments. In *Proceedings of ASSETS 2002* (pp. 89–96). Edinburgh: ACM.

Oakley, I., McGee, M. R., Brewster, S., & Gray, P. (2000). Putting the feel in 'Look and Feel'. In *Proceedings of CHI 2000* (415–422). New York: ACM.

Oseitutu, K., Feng, J., Sears, A., & Karat, C.-M. (2001). Speech recognition for data entry by individuals with spinal cord injuries. *Proceedings of the 1st International Conference on Universal Access in Human-Computer Interaction.* Vol. 3. (pp. 402–406).

Patmore, D. W., & Knapp, R. B. (1995). A cursor controller using evoked potentials and EOG. In *Proceedings of RESNA 95 Annual Conference* (pp. 702–704). Arlington, VA: RESNA.

Patmore, D. W., & Knapp, R. B. (1998). Towards an EOG-based eye tracker for computer control. In *Proceedings of ASSETS 98* (pp. 197–203). New York: ACM.

Pidcock, F. S., & Christensen, J. R. (1997). General and neuromuscular rehabilitation in children. In B. O'Young, M. Young, & S. Stiens (Eds.), *PM&R Secrets* (pp. 402–406). Philadelphia: Hanley & Belfus.

Pieper, M., & Kobsa, A. (1999). Talking to the ceiling: An interface for bed-ridden manually impaired users. In *CHI 1999, Extended Abstract* (pp. 9–10). Pittsburg: ACM.

Price, K. J., Lin, M., Feng, J., Goldman, R., Sears, A., & Jacko, J. A. (2006). Motion does matter: An examination of speech-based text entry on the move. *Universal access in the information society.* Vol 4 (3), March 2006. 246–257.

Radwin, R. G., Vanderheiden, G. C., & Lin, M. L. (1990). A method for evaluating head-controlled computer input devices using Fitts' Law. *Human Factors, 32*(4), 423–438.

Rehabilitation Engineering and Assistive Technology Society of North America (RESNA). (2006). Retrieved March 8, 2007, from http://www.resna.org/

Ropper, A., & Brown, R. (2005). *Adams and Victor's principles of neurology,* 8th Edition. The McGraw-Hill Companes, Inc.

Rosenthal, M., Griffith, E. R., Bond, M. R., & Miller, J. D. (Eds.). (1990). *Rehabilitation of the adult and child with traumatic brain injury* (2nd ed.). Philadelphia: F. A. Davis.

Roy, D. M., Panayi, M., Erenshteyn, R., Foulds, R., & Fawcus, R. (1994). Gestural human-machine interaction for people with severe speech and motor impairment due to cerebral palsy. In *CHI 94 Conference Companion* (pp. 313–314). New York: ACM.

Salem, C., & Zhai, S. (1997). An isometric tongue pointing device. In *Proceedings of CHI 97* (pp. 538–539). New York: ACM.

Schaab, J. A., Radwin, R. G., Vanderheiden, G. C., & Hansen, P. K. (1996). A comparison of two control-display gain measures for head-controlled computer input devices. *Human Factors, 38*(3), 390–403.

Scherer, M. J. (1996). Dilemmas, challenges, and opportunities. In M. Scherer (Ed.), *Living in the state of struck* (pp. 115–121). Cambridge, IA: Brookline Books.

Scherer, M. J., & Galvin, J. C. (1996). An outcome perspective of quality pathways to most appropriate technology. In J. C. Galvin & M. J. Scherer (Eds.), *Evaluating, selecting and using appropriate assistive technology* (pp. 1–26). Gaithersburg, MD: Aspen Publications.

Schumacher, H. R. Jr. (Ed.). (1993). *Primer on the rheumatic diseases.* Atlanta: Arthritis Foundation.

Sears, A., Feng, J., Oseitutu, K., & Karat, C.-M. (2003). Speech-based navigation during dictation: Difficulties, consequences, and solutions. *Human Computer Interaction, 18*(3), 229–257.

Sears, A., Karat, C.-M., Oseitutu, K., Karimullah, A., & Feng, J. (2001). Productivity, satisfaction, and interaction strategies of individuals with spinal cord injuries and traditional users interacting with speech recognition software. *Universal Access in the Information Society, 1,* 1–12.

Sears, A., Lin, M., & Karimullah, A. (2002). Speech-based cursor control: understanding the effects of target size, cursor speed, and command selection. *Universal Access in the Information Society, 2*(1), 30–43.

Shaw, R., Loomis, A., & Crisman, E. (1995). Input and integration: Enabling technologies for disabled users. In A. D. N. Edwards (Ed.), *Extra-ordinary human-computer interaction* (pp. 263–277). Cambridge: Cambridge University Press.

Sibert, L. E., & Jacob, R. J. K. (2000). Evaluation of eye gaze interaction. In *Proceedings of CHI 2000* (pp. 281–288). New York: ACM.

Stephanidis, C. (2001). *User interfaces for all: Concepts, methods, and tools.* Mahwah, NJ: Lawrence Erlbaum Associates.

Stiens, S., Goldstein, B., Hammond, M., & Little, J. (1997). Spinal cord injuries. In B. O'Young, M. Young, & S. Stiens (Eds.), *PM&R Secrets* (pp. 253–261). Philadelphia: Hanley & Belfus.

Taylor, R. G., & Lieberman, J. S. (1988). Rehabilitation of patients with diseases affecting the motor unit. In J. A. DeLisa (Ed.), *Rehabilitation medicine: Principles and practice* (pp. 811–820). Philadelphia: J. B. Lippincott.

Trace Center (2007). Trace Center, College of Engineering, University of Wisconsin—Madison. Retrieved March 8, 2007, from http://www.trace.wisc.edu

Trewin, S. (1996). A study of input device manipulation difficulties. In *Proceedings of ASSETS 96* (pp. 15–22). New York: ACM.

Trewin, S. (2002) An invisible keyboard. In *Proceedings of ASSETS 2002* (pp. 143–149). Edinburgh: ACM.

Trewin, S. (2004). Automating accessibility: The dynamic keyboard. In *Proceedings of ASSETS 2004* (pp. 71–78). Atlanta: ACM.

Trewin, S., & Pain, H. (1998a). A study of two keyboard aids to accessibility. In *Proceedings of the HCI'98 Conference on People and Computers XIII* (83–97). Heidelberg: Springer-Verlag.

Trewin, S., & Pain, H. (1998b). A model of keyboard configuration requirements. In *Proceedings of ASSETS 98* (pp. 173–181). New York: ACM.

Trewin, S., & Pain, H. (1999). Keyboard and mouse errors due to motor disabilities. *International Journal of Human-Computer Studies, 50,* 109–144.

Trewin, S., Slobogin, P., & Power, M. (2002). Accelerating assessment with self-optimizing devices. *Proceedings of ICCHP 2002,* 273– 275.

Wade, D. T., Langton Hewer, R., Skilbeck, C. E., & David, R. M. (1985). *Stroke: A critical approach to diagnosis, treatment, and management.* Chicago: Year Book.

Wade, T., Langton Hewer, R., Skilbeck, C., & David, R. (1985). *Stroke: A critical approach to diagnosis, treatment, and management.* London: Chapman and Hall.

Willis, T., Pain, H., Trewin, S., & Clark, S. (2002). Informing flexible abbreviation expansion for users with motor disabilities. *Proceedings of International Conference on Computers Helping People with Special Needs (ICCHP) 2002,* 251–258.

Wobbrock, J., Aung, H., Myers, B., & LoPresti, E. (2004). Text entry from power wheelchairs: EdgeWrite for joysticks and touchpads. In *Proceedings of ASSETS 2004* (pp. 110–117). Atlanta: ACM.

World Health Organization. (2006). *International classification of functioning, disability, and health.* Retrieved February 20, 2006, from http://www3.who.int/icf/

Young, M. A., Tumanon, R. C., & Sokal, J. O. (2000, Summer). Independence for people with disabilities: A physician's primer on assistive technology. *Maryland Medicine,* 28–32.

·6·

PERCEPTUAL IMPAIRMENTS: NEW ADVANCEMENTS PROMOTING TECHNOLOGICAL ACCESS

Julie A. Jacko*
Georgia Institute of Technology

V. Kathlene Leonard
Alucid Solution, Inc.

Ingrid U. Scott
Penn State College of Medicine

*Currently at University of Minnesota

INTRODUCTION

The introduction of early computers facilitated new ways for individuals with visual impairment to access information electronically: magnified, in Braille, or aurally via the conversion of digital information. However, the introduction of GUIs that present digital information via visual metaphors and icons contributed to the digital divide, which can hamper the productivity of this population. In many cases, even access to documents and forms can be a difficult to impossible task when available through the direct manipulation paradigm (Fortuin & Omata, 2004). The exclusive reliance of GUIs on the visual interaction paradigm therefore threatens to limit accessibility for anyone whose visual channel is compromised (Dix, Finlay, Abowd, & Beale, 1998). This chapter provides readers with (a) an introduction to the visual sensory channel; (b) a review of research approaches, models, and theories that are relevant to HCI and visual impairment; and (c) a discussion of forms of visual dysfunction in the research, design, and evaluation of human–computer systems.

The interaction strategies and related interaction barriers for individuals with visual impairments in the past 15 years have received growing attention in an attempt to inform judicious, inclusive design for accessible information technologies (e.g., Assistive Technologies for Independent Aging: Opportunities and Challenges, 2004; Arditi, 2002; Brewster, Wright, & Edwards, 1994; Craven, 2003; Fortuin & Omata, 2004; Fraser & Gutwin, 2000; Gaver, 1989; Jacko, 1999; Jacko, Barnard, et al., 2004; Jacko, Barreto, Scott, Chu, et al., 2002; Jacko, Barreto, Scott, Rosa, & Pappas, 2000; Jacko, Moloney, et al., 2005; Jacko, Rosa, Scott, Pappas, & Dixon, 2000; Jacko & Sears, 1998). Visual impairments (that do not lead to blindness) can create barriers to distinguishing fine details of iconic screen targets and to tracking the highly dynamic nature of the pointer used to manipulate these icons (Fraser & Gutwin, 2000). This is largely attributed to difficulty in manipulating objects with the pointer due to reduced visual acuity and visual field.

In terms of the visual sensory channel, it is known that user behavior is strongly influenced by the nature and amount of residual vision the user experiences in combination with computer interface characteristics. As an extreme example, a blind user without any functional vision will use fundamentally different coping skills to navigate an interface as compared to an individual with clouded vision due to cataracts (Jacko & Sears, 1998). Harper, Goble, and Stevens (2001) emphasized that the differences in orientation, navigation, travel, and mobility of visually impaired versus sighted individuals should be considered in the design of technology because there are differences in the mental map and cognitive processes that occur across the spectrum of visual abilities.

The impetus for this chapter has two parts. First, the number of individuals who report low vision is anticipated to rise sharply with the aging baby boomers (who are, on average, living longer) as they experience age-related changes to their functional vision (e.g., reduced visual acuity, presbyopia, contrast sensitivity, color sensitivity, depth perception, and glare sensitivity). They are increasingly predisposed to acquire ocular diseases associated with older age (e.g., macular degeneration, diabetic retinopathy,

glaucoma, and cataracts; for a review, see Orr, 1998; Schieber, 1994). Secondly, the digital divide imposed on the population with visual impairments has been measured in terms of technology access and unemployment (Gerber & Kirchner, 2001). Looking beyond the United States, as the need for information increases globally so does the diversity of the people requiring access. As a result, a potentially large number of users may be disadvantaged with respect to gaining access to a variety of types of information without adequate accommodations.

The framework for the structure of this chapter results from an HCI approach first introduced by Jacko and Vitense (2001), and further clarified by Jacko, Vitense, and Scott (2003). Initial work in this research area included a comprehensive review of the literature to facilitate the development of a categorization scheme to account for categories of impairment. From this literature review, five major categories emerged: (a) hearing impairments, (b) mental impairments, (c) physical impairments, (d) speech impairments, and (e) visual impairments. Fig. 6.1 illustrates that each of the five overarching categories is composed of a collection of clinical diagnoses unique to that category (depicted in Fig. 6.1 by $A_1,...,A_n$, $B_1,...,B_n$, $C_1,...,C_n$, $D_1,..., D_n$, and $E_1,...,E_n$). Each diagnosis, in turn, influences certain functional capabilities that are critical to the access of information technologies (depicted in Fig. 6.1 by $Y_1,...,Y_n$). A subset of these functional capabilities can be directly linked to specific classes of technologies (shown at the bottom of the diagram).

While this framework is applicable to the five identified categories of impairment, its discussion and demonstrated utility in the scope of this chapter will address only visual impairment. Consider, for example, a person who has been diagnosed with a specific type of visual impairment represented as E_1 in Fig. 6.1. This visual impairment results in measurable decrements to certain aspects of this person's functional visual capabilities, Y_7 and Y_8. Observe from Fig. 6.1 that the decrement to the functional capability represented by Y_8 does not impede a person's access to any of the technology classes depicted in Fig. 6.1. In contrast, the functional visual capability represented by Y_7 in Fig. 6.1 impedes this person's ability to successfully access

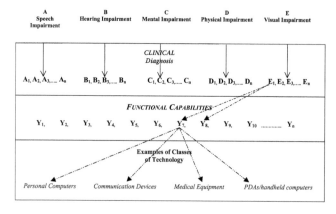

FIGURE 6.1. Framework for the integration of clinical diagnoses, functional capabilities, and access to classes of technologies (adapted from Jacko, Vitense, & Scott, 2003).

information using all four examples of information technology. From this conceptual representation, it is apparent that much more knowledge is needed for researchers to possess an accurate depiction of the empirical relationships that exist between diagnoses, functional capabilities, and access to specific classes of technologies. More specifically, emerging from this conceptual framework are several key research areas in need of investigation:

1. It is critical to establish empirical links between clinical diagnoses and sets of functional capabilities.
2. It is necessary to define the set of functional capabilities required to access information technologies.
3. It is essential to establish empirical bases for the influence of specific functional capabilities on access to specific classes of technologies.

While the framework illustrates five categories of impairment, this chapter will focus on visual impairment and functional visual capability. Auditory/speech, cognitive impairments, and physical (motor) impairments are covered in this handbook in accompanying chapters.

Following this framework, this chapter introduces readers to the clinical definition of visual impairments and diagnoses, in which visual function is first addressed, followed by a discussion of the leading causes of vision loss in the United States and beyond. The section also provides a discussion of specific visual functions. Then, the chapter highlights recent advancements in HCI research, which effectively link the three major areas of the framework: clinical diagnosis, functional impairment, and interaction across a variety of classes of technology and impairments. This research is done in pursuit of HCI solutions for the visually impaired including perceptual interfaces, multimedia interfaces, multimodal interfaces, and adaptive interfaces. Finally, examples of technological advancements for this population are presented. While these technological advancements are discussed briefly in this chapter within the context of visual impairments, it should be noted that additional information on HCI and perceptual-motor interactions, hearing and speech impairments, and motor impairments can be found in their respective chapters.

VISUAL FUNCTION

Definitions

When considering the impact of visual loss on an individual's ability to use a computer effectively, it is first necessary to understand various dimensions of visual performance. Many different terms have been used to refer to abnormalities in visual function, including *disorder, impairment, disability*, and *handicap* (Colenbrander, 1977). Although often used as synonyms, there are distinct differences. For instance, while *disorder* and *impairment* describe aspects of an organ's condition, the term *disability* describes aspects of a patient's condition.

Disorder refers to an abnormality in the anatomy or physiology of an organ and, in the case of a visual disorder, may occur anywhere in the visual system. Examples of visual disorders include corneal scar, cataract, macular degeneration, optic atrophy, or occipital stroke. It is important to recognize that knowing the specific visual disorder provides no information concerning the functional capacity of the eye.

Impairment refers to a functional abnormality in the organ. Thus, varying degrees of visual impairment can be measured in terms of specific visual functions, such as visual acuity, contrast sensitivity, visual field, or color vision. While such impairment measures demonstrate how well the eye functions, they do not reveal the impact of the visual disorder on the patient's ability to perform everyday activities. For example, a physician may state that the patient's visual acuity has dropped by four lines on the eye chart, while the patient reports an inability to see well enough to use a computer.

Disability refers to the ability of a patient (rather than an organ) to perform tasks, such as daily living skills, vocational skills, reading, writing, mobility skills, and so on. Since disability implies a broader perspective (the focus is on the person as a whole rather than on a specific organ), it is no longer entirely vision specific. For instance, while computer skills may be reduced due to vision loss, they may also suffer due to such conditions as arthritis. It is the combination of visual and nonvisual skills that determines the abilities or disabilities of an individual. Vision substitution techniques (such as the use of a white cane and increased reliance on memory and on hearing) may be helpful in improving the ability of an individual to perform specific tasks.

A disorder may cause impairments, and impairments may cause disability. However, these links are not rigid. An analysis of these various dimensions of vision loss permits identification of interventions at each link, which may improve the functional status and quality of life of an individual with visual loss (Fig. 6.1; Fletcher, 1999). For example, if one were interested in improving the ability of an individual with vision loss to use a computer effectively, possible interventions include medical and surgical intervention to impact the visual disorder and impairment, visual aids and adaptive devices to impact visual impairment, and social interventions, training, counseling, and education to impact visual disability. The design of a computer interface that enhances the ability of an individual with vision loss to perceive graphical and textual information would have a beneficial effect on the degree of impairment and resulting level of disability of that individual.

Epidemiology

Low vision has been defined as a permanent visual impairment that is not correctable with glasses, contact lenses, or surgical intervention, and which interferes with normal everyday functioning (Mehr & Freid, 1975). Specifically, low vision is defined as a best-corrected visual acuity worse than 20/40 in the better-seeing eye (The Eye Diseases Prevalence Research Group, 2004a). *Blindness* is defined as a best-corrected visual acuity of 20/200 or worse according to the United States definition and is defined as a best-corrected visual acuity of worse than 20/400 according to the World Health Organization definition (The Eye Diseases Prevalence Research Group, 2004a). Based on demographics from the 2000 U.S. Census, an estimated 937,000 (0.78%) Americans older than 40 years were blind (United States definition); an additional 2.4 million Americans (1.98%) has low vision (The Eye Diseases Prevalence Research Group, 2004a). Thus, blindness or low vision affects approximately 1 in 28 Americans older than 40 years (The Eye Diseases Prevalence Research Group, 2004a). Largely due to the aging of the U.S. population, it is estimated that the number of blind persons older than 40 years in the United States will increase by approximately 70% to 1.6 million (prevalence of 1.1%) by the year 2020 (The Eye Diseases Prevalence Research Group, 2004a). Vision loss has been ranked third, behind arthritis and heart disease, among conditions that cause persons older than 70 years to need assistance in activities of daily living (LaPlante, 1988).

Selected Visual Disorders

In the United States, the most common causes of decreased vision are age-related macular degeneration (AMD), diabetic retinopathy, and glaucoma.

Age-related macular degeneration. AMD is the leading cause of irreversible visual loss in the western world in individuals over 60 years of age. The macula is the part of the retina that is responsible for central vision. AMD affects more than 1.75 million individuals in the United States, and due to the rapid aging of the U.S. population, this number is estimated to increase to almost 3 million by the year 2020 (The Eye Diseases Prevalence Research Group, 2004b). The prevalence of severe visual loss due to AMD increases with age. In the United States, at least 10% of persons between the ages of 65 and 75 years have lost some central vision due to AMD; among individuals over the age of 75 years, 30% have vision loss due to AMD.

Risk factors for this disease and its progression include age, sunlight exposure, smoking, ocular and skin pigmentation, elevated blood pressure, and elevated serum cholesterol levels. The role of nutrition has not been fully identified as a risk factor, but a diet low in antioxidants and lutein may be a contributing factor.

AMD is a bilateral disease in which visual loss in the first eye usually occurs at about 65 years of age; the second eye becomes involved at the rate of approximately 10% per year. The two main types of AMD are atrophic and exudative. The atrophic (dry) form of the disease is generally a slowly progressive disease that accounts for approximately 90% of cases. It is characterized by the deposition of abnormal material beneath the retina (drusen) and by degeneration and atrophy of the central retina (also known as the macula); patients typically note slowly progressive central visual loss. Although much less common, the exudative (wet) form of the disease is responsible for about 88% of legal blindness attributed to AMD. This form of the disease, which often occurs in association with atrophic AMD, is characterized by the growth of abnormal blood vessels beneath the central retina (macula); these abnormal blood vessels elevate and distort the retina and may leak fluid and blood beneath or into the retina. Vision loss may be sudden onset and rapidly progressive (in contrast to the atrophic form of the disease, where vision loss generally occurs progressively over several

months or years). AMD can cause profound loss of central vision, but the disease generally does not affect peripheral vision, and therefore, patients typically retain their abilities to ambulate independently.

Currently, no proven treatment reverses the retinal damage that has already occurred due to AMD. In order to try to prevent further vision loss, recommendations made to patients include eye protection against ultraviolet light exposure (sunglasses with ultraviolet light protection), no smoking, optimal control of blood pressure and serum cholesterol level, and a diet rich in dark green leafy and orange vegetables (antioxidants are believed to reduce the damaging effects of light on the retina through their reducing and free-radical scavenging actions; lutein is a macular pigment). The Age-Related Eye Disease Study (AREDS) was a randomized, controlled clinical trial that demonstrated a statistically significant reduction in rates of at least moderate visual acuity loss in persons with moderate AMD who received supplementation with vitamins C and E, beta carotene, and zinc compared to persons with moderate AMD assigned to placebo (Age-Related Eye Disease Study Research Group, 2001). Laser photocoagulation and photodynamic treatment of the abnormal blood vessels found in patients with exudative macular degeneration may help to prevent severe vision loss in some cases. Intravitreal injections of pegaptanib, a pegylated modified oligonucleotide that binds to extracellular vascular endothelial growth factor (VEGF) isoform 165 (the isoform widely considered to be the primary pathologic form of VEGF), are associated with a higher likelihood of visual preservation and a slowing of visual loss among patients with exudative AMD (Gragoudas, Adamis, Cunningham, Feinsod, & Guyer, 2004). Surgical rotation of the retina away from the area of abnormal blood vessels has also been effective in some cases. Other treatment modalities currently under investigation include such drugs as corticosteroids, other anti-VEGF agents, and combination treatments.

Diabetic retinopathy. Approximately 16 million Americans suffer from diabetes mellitus, most of whom will develop diabetic retinopathy within 20 years of their diagnoses. In fact, after 20 years of diabetes, nearly 99% of those with insulin-dependent diabetes mellitus and 60% with non-insulin-dependent diabetes mellitus have some degree of diabetic retinopathy. Diabetic retinopathy is the leading cause of legal blindness in Americans aged 20 to 65 years, with 10,000 new cases of blindness annually. One million Americans have proliferative diabetic retinopathy, and 500,000 have macular edema. Among an estimated 10.2 million American adults aged 40 years and older known to have diabetes mellitus, the estimated crude prevalence rates for retinopathy and vision-threatening retinopathy are 40.3% and 8.2%, respectively (The Eye Diseases Prevalence Research Group, 2004b). The estimated U.S. general population prevalence rates for retinopathy and vision-threatening retinopathy are 3.4% (4.1 million persons) and 0.75% (899,000 persons), respectively (The Eye Diseases Prevalence Research Group, 2004c).

The major risk factor for diabetic retinopathy is duration of diabetes; it is estimated that at 15 years, 80% of diabetics will have background retinopathy and that, of these, 5% to 10% will progress to proliferative changes. Other risk factors include long-term diabetic control (as reflected in serum levels of glycosylated hemoglobin), hypertension, smoking, and elevated serum cholesterol.

There are two main types of diabetic retinopathy: nonproliferative and proliferative. Nonproliferative diabetic retinopathy refers to retinal microvascular changes that are limited to the confines of the retina and include such findings as microaneurysms, dot and blot intraretinal hemorrhages, retinal edema, hard exudates, dilation and bleeding of retinal veins, intraretinal microvascular abnormalities, nerve fiber layer infarcts, arteriolar abnormalities, and focal areas of capillary nonperfusion. Nonproliferative diabetic retinopathy can affect visual function through two mechanisms: intraretinal capillary closure resulting in macular ischemia, and increased retinal vascular permeability resulting in macular edema. Clinically significant macular edema is defined as any one of the following: (a) retinal edema located at or within 500 μm of the center of the macula; (b) hard exudates at or within 500 μm of the center if associated with thickening of adjacent retina; and (c) a zone of retinal thickening larger than one optic disc area if located within one disc diameter of the center of the macula.

Proliferative diabetic retinopathy is characterized by extraretinal fibrovascular proliferation; that is, fibrovascular changes that extend beyond the confines of the retina and into the vitreous cavity. Fibrovascular proliferation in proliferative diabetic retinopathy may lead to tractional retinal detachment and vitreous hemorrhage. High-risk proliferative diabetic retinopathy is defined by any combination of three of the four following retinopathy risk factors: (a) presence of vitreous or preretinal hemorrhage; (b) presence of new vessels; (c) location of new vessels on or near the optic disc; and (d) moderate to severe extent of new vessels (Diabetic Retinopathy Study Research Group, 1979).

Management of diabetic retinopathy includes referring the patient to an internist for optimal glucose and blood pressure control. In the Early Treatment Diabetic Retinopathy Study, focal or grid laser photocoagulation treatment for clinically significant macular edema reduced the risk of moderate visual loss, increased the chance of visual improvement, and was associated with only mild loss of visual field (Early Treatment Diabetic Retinopathy Study Research Group, 1995). Intravitreal triamcinolone acetonide and other intravitreal anti-VEGF agents are currently under investigation for the treatment of diabetic macular edema. Panretinal laser photocoagulation treatment of eyes with high-risk proliferative diabetic retinopathy reduced the risk of severe visual loss by 50% compared to untreated control eyes (Diabetic Retinopathy Study Research Group, 1981). Surgery is often indicated for nonclearing vitreous hemorrhage and for tractional retinal detachment involving or threatening the macula.

Glaucoma. Primary open angle glaucoma (POAG) is the most prevalent type of glaucoma, affecting 1.3% to 2.1% of the general population over the age of 40 years in the United States. In the United States, the disease is the leading cause of irreversible blindness among Blacks and the third leading cause among Whites (following AMD and diabetic retinopathy), and is responsible for 12% of legal blindness. Risk factors for the disease include increasing age (especially greater than 40 years),

African ethnicity, positive family history of glaucoma, diabetes mellitus, and myopia (nearsightedness).

POAG is a chronic, slowly progressive optic neuropathy characterized by atrophy of the optic nerve and loss of peripheral vision. Central vision is typically not affected until late in the disease. Because central vision is relatively unaffected until late in the disease, visual loss generally progresses without symptoms and may remain undiagnosed for quite some time. While usually bilateral, the disease may be quite asymmetrical. POAG is associated with increased intraocular pressure, but normal-tension glaucoma may cause glaucomatous vision loss in patients with normal intraocular pressure. Thus, normal eye pressure does not rule out the presence of glaucoma.

Treatment of POAG includes topical or systemic medications, laser, or surgery to lower the intraocular pressure to a level at which optic nerve damage no longer occurs. Visual field testing is performed regularly in order to evaluate for progressive loss of peripheral vision, and the optic nerve is examined regularly to evaluate for evidence of progressive optic atrophy (clinical signs of glaucoma in the optic disc include asymmetry of the neuroretinal rim, focal thinning of the neuroretinal rim, optic disc hemorrhage, and any acquired change in the disc rim appearance or the surrounding retinal nerve fiber layer).

Specific Visual Functions

Visual acuity. Visual acuity is the most common measure of central visual function and refers to the smallest object resolvable by the eye at a given distance. It is defined as the reciprocal of the smallest object size that can be recognized. Visual acuity is expressed as a fraction in which the numerator is the distance at which the patient recognizes the object and the denominator is the distance at which a standard eye recognizes the object. For instance, a visual acuity of 20/60 means that the patient needs an object three times larger or three times closer than a standard eye requires. The traditional visual acuity chart presents symbols of decreasing size with fixed high contrast. The visual acuity chart used most often in the clinical setting is the Snellen acuity chart, which is comprised of certain letters of the alphabet; the size of the letters is constant on a given line of the eye chart, and decrease in size the lower the line on the chart. In accurate Snellen notation, the numerator indicates the test distance and the denominator indicates the letter size seen by the patient.

Contrast sensitivity. Contrast sensitivity refers to the ability of the patient to detect differences in contrast and is defined as the reciprocal of the lowest contrast that can be detected. This may be measured with the Pelli-Robson chart, in which letters decrease in contrast rather than size, or the Bailey-Lovie chart, in which letters of a fixed low contrast are varied in size. Contrast sensitivity is considered a more sensitive indicator of visual function than Snellen acuity and may provide earlier detection of such pathology as retinal and optic nerve disease.

Visual field. Visual field is classically defined as a three-dimensional graphic representation of differential light sensitivity at different positions in space. *Perimetry* refers to the clinical assessment of the visual field. Typically, visual field is assessed with kinetic or static perimetry. During kinetic perimetry, a test object of fixed intensity is moved along several meridians toward fixation and points where the object is first perceived are plotted in a circle. During static perimetry, a stationary test object is increased in intensity from below threshold until perceived by the patient, and threshold values yield a graphic profile section. While peripheral visual field loss often produces difficulty for patients in orientation and mobility functions, macular field loss (either centrally or paracentrally) often causes difficulties with reading. For instance, the presence of central or paracentral visual field loss is a more powerful predictor of reading speed than is visual acuity (Fletcher, Schuchard, Livingstone, Crane, & Hu, 1994).

Color vision. Evaluation of color vision may be performed using pseudoisochromatic color plates, which are quick and commonly available; they consist of circles in various colors such that a person with normal color vision function can distinguish a number from the background pattern of circles. Ishihara or Hardy-Rand-Rittler pseudoisochromatic color plates are designed to screen for congenital red/green color deficiencies, while Lanthony tritan plates may be used to detect blue/yellow defects, which are frequently present in acquired disease. With the Farnsworth-Munsell 100-hue test, the patient must order 84 colored disks; the time-consuming nature of this test limits its clinical use. The Farnsworth Panel D-15 is a shorter and more practical version (using 15 disks), but is less sensitive. Most color-vision defects are nonspecific.

Visual Function and Age

This section provides an overview of how levels of visual function vary with age, and to what degree. Aging is synonymous with natural declines in a person's sensory abilities. As such, the process of aging is accompanied by changes to the eye, including the retina and visual nervous system that can impact functional vision (Schieber, 1994). Additionally, older adults are more likely to acquire ocular conditions that can compromise visual functioning beyond normally anticipated changes, such as macular degeneration, diabetic retinopathy, and cataracts. Age-related vision loss commonly impinges on the ability to complete near vision tasks such as reading and using the computer (Arditi, 2004). An understanding of these functional declines provides direction for strategies aimed to mitigate the negative impact of these changes. HCI designers, developers, and usability specialists should be fully aware of these needs, as the needs of this growing user population will become an increased priority with the shift in demographics of population segment.

Aspects of visual function that are known to normally decline as part of the aging process include

- Visual acuity
- Visual field
- Contrast sensitivity
- Color perception
- Floaters

- Dry eyes
- Increased need for light
- Difficulty with glare
- Dark/light adaptation
- Reduced depth perception (Orr, 1998)

Beyond these factors, eye movement efficiency and accuracy are observed to decline with old age. Older adults are known to be less accurate and/or slower in locating a target in the peripheral vision (see also Kline & Scialfa, 1997; Lee, Legge, & Oritz, 2003). Age-related differences have also been observed with the effectiveness with which older adults visually track targets with higher velocities. In both cases, these trends are typically aggravated by the presence of distracting stimuli (visual, auditory, tactile, etc.) in the background or foreground that contribute to the complexity. The perception of moving stimuli, for older adults, is both less effective and less efficient in tasks aimed at the detection of small target movement/change such as those found on dials and controls (Kline & Scialfa, 1997). Furthermore, deficits in central, para-central, and peripheral visual field can pose different demands on vision, resulting in different search strategies related to eye movements (Coeckelbergh, Cornelissen, Brouwer, & Kooijam, 2002).

Older adults tend to exhibit a greater degree of difficulty with visual search tasks, especially when the number of items to be searched increases (Kline & Scialfa, 1997). Older adults have a propensity for longer visual reaction times, especially in cases where attention is divided. Furthermore, this population segment experiences difficulties ignoring extraneous information, or background noise (Schieber, 1994). Research also suggests that visual search is slower and less effective for older adults due to a shrinking of the useful field of view to which attention can be simultaneously allocated. The size of the useful field of view, for older adults, is especially susceptible to context related factors, such as complexity and cognitive task load (Schieber, 1994).

Summary

Studies have demonstrated that ophthalmic patients are at high risk for decreased functional status and quality of life (Parrish et al., 1997; Scott, Schein, Feuer, Folstein, & Bandeen-Roche, 2001; Scott, Schein, West, Bandeen-Roche, Enger, & Folstein, 1994). Patients' functional statuses and qualities of life may be improved by interventions that increase visual function, such as surgery to repair retinal detachment or remove epiretinal membrane (Scott, Smiddy, Feuer, & Merikansky, 1998) and surgery to remove cataract (Applegate et al., 1987; Brenner, Curbow, Javitt, Legro, & Sommer, 1993; Donderi & Murphy, 1983; Javitt, Brenner, Curbow, Legro, & Street, 1993; Steinberg et al., 1994). In addition, functional status and quality of life may be improved by interventions, such as low vision devices and services, which permit patients to use their remaining vision more effectively (Scott, Smiddy, Schiffman, Feuer, & Pappas, 1999). In addition, as the proportion of older adults multiplies, the number of individuals experiencing some degree of vision dysfunction that normally occurs with age has created an increased demand for information technology that affords use despite the dysfunction. Prior studies have demonstrated the effect of low vision interventions on objective task-specific measures of functional abilities such as reading speed, reading duration, and ability to read a certain print size (Nilsson, 1990; Nilsson, & Nilsson, 1986; Rosenberg, Faye, Fischer, & Budicks, 1989; Sloan, 1968).

However, historically there has been little data available concerning the abilities of people with visual impairments to use computers. It has been even more challenging to find data concerning how modifications of graphical user interface features may increase accessibility of computers to patients with visual impairments. An exception to this is the research agenda established by Jacko and colleagues (see Jacko, Moloney, et al., 2005; Jacko, Scott, et al., 2003; Scott et al., 2002a,b; Scott, Jacko, et al., 2006; and Table 6.1 for examples that have yielded systematically derived HCI performance thresholds for users according to their ocular profile (diagnoses and functional ability). Table 6.1 summarizes the research products resulting from this effort, chronologically organized by the ocular pathology investigated, the age of the users involved, the GUI interaction investigated and the corresponding specific interface feature in question. It goes without saying that the rapid proliferation of visual displays beyond the desktop to handheld and wearable computers and other mobile devices, such as cellular telephones, increases the importance of graphical user interface innovations that facilitate efficient and rewarding usage by people with visual impairments.

HIGHLIGHTING TECHNOLOGICAL ADVANCEMENTS IN HCI RESEARCH

With nearly every aspect of today's society involving some type of computer technology, there is an ever-growing need to understand how individuals with visual impairments can access technology. Currently, without special modifications, the typical PC poses several challenges to users who experience limited bandwidth to their visual sensory channels, as well as other perceptual impairments. As a result, the HCI research community is placing great emphasis on the design of universally acceptable technologies. According to Stephanidis, Salvendy, Akoumianakis, Bevan, et al. (1998), "Universal access in the Information Society signifies the right of all citizens to obtain equitable access to, and maintain effective interaction with, a community-wide pool of information resources and artifacts" (p. 6). *Accessibility* has been a term traditionally associated with elderly individuals, individuals with disabilities and others who possess special needs (Stephanidis, Salvendy, Akoumianakis, Arnold, et al., 1999). However, because of the current influx of new technologies into the market, the population of users who may possess special needs is growing. As a result, accessibility has taken on a more comprehensive connotation. This connotation implies that all individuals with varying levels of abilities, skills, requirements, and preferences be able to access information technologies (Stephanidis, Salvendy, Akoumianakis, Arnold, et al., 1999). Universal access also implies more than just adding features to existing technologies. Rather,

the concept of universal access emphasizes that accessibility be incorporated directly into the design (Stephanidis, Salvendy, Akoumianakis, Bevan, et al., 1998). Perceptual and adaptive interfaces are two ideal examples of how universal accessibility can be achieved.

Perceptual Interfaces

The concept of perceptual design describes a perspective of design that defines interactions in terms of human perceptual capabilities. In a sense, it strives to humanize interaction. The design of perceptual interfaces adheres to the idea that lessons learned from psychological research about perception can be applied to interface design (Reeves & Nass, 2000). In adopting the concept of perceptual design, several opportunities surface for the creation of innovative perceptual user interfaces. Interactions with these interfaces can be described in terms of three particular human perceptual capabilities: chemical senses (e.g., taste and olfaction), cutaneous senses (e.g., skin and receptors), and vision and hearing (Reeves & Nass, 2000). Although commonly used computer technology limits the effectiveness of chemical senses, the technology can be extended to incorporate the cutaneous, visual, and hearing senses.

In terms of vision, research has focused on topics such as visual mechanics, color, brightness and contrast, objects and forms, depth, size, and movement. Hearing research includes psychophysical factors such as loudness, pitch, timbre, and sound localization; physiological mechanisms such as the auditory components of the ear, and the neural activity associated with hearing; and the perception of speech such as units of speech and the mechanics of word recognition (Reeves & Nass, 2000). With respect to the cutaneous senses, augmented GUIs with haptic feedback have been around since the early 1990s. Akamatsu and Sate conducted the first research with a haptic mouse that produced haptic feedback via fingertips and force feedback via controlled friction (as cited in Oakley, McGee, Brewster, & Gray, 2000). Engle, Goossens, and Haakma found that directional two degrees of freedom force feedback improved speed and error rates in a targeting task (as cited in Oakley, McGee, Brewster, & Gray, 2000).

The strength of perceptual user interfaces comes from the ability of designers to combine an understanding of natural human capabilities with computer input/output devices, and machine perception and reasoning (Turk & Robertson, 2000). General examples of how capabilities can be combined with technology include speech and sound recognition and generation, computer vision, graphical animation, touch-based sensing and feedback, and user modeling (Turk & Robertson, 2000).

From an applied research standpoint, the concepts of perceptual interfaces are housed within multimedia and multimodal interfaces. Both multimedia and multimodal interfaces offer increased accessibility to technologies for individuals with perceptual impairments. Distinctions can be drawn between perceptual, multimedia, and multimodal interfaces, shown in Fig. 6.2. Perceptual interfaces prescribe human-like perceptual capabilities to the computer. Multimedia and multimodal interfaces can be considered applied extensions of

TABLE 6.1. Visualization of the Breadth and Depth of the Investigations, To-Date, for the Derivation of HCI Performance Thresholds for Users According to Their Ocular Profile (Diagnoses and Functional Ability). A Shaded Cell Indicates that a Given Topic is Addressed by the Corresponding Investigation, Listed in the Leftmost Column

Investigation		Ocular Pathology				Age		GUI Interaction						GUI Interface Feature								
Year	Authors	None	AMD	Diabetic Retinopathy	Other	18-55 years	55+ years	Visual Search	Icon Selection	Cursor Movement	Drag & Drop	Menu Selection	Distract-ers	Icon Size	Set Size	Back-ground	Visual Profiles	Tactile Cues	Auditory Cues	Text Size	Desk-top PC	Hand-held PC
1998	Jacko & Sears																					
1999	Jacko																					
1999	Jacko, Dixon et al.																					
2000	Jacko,Barreto et al.																					
2000	Jacko, Rosa et al.																					
2001	Jacko et al.																					
2002[a]	Scott, Feuer & Jacko																					
2002	Scott, Feuer & Jacko																					
2002[b]	Jacko,Barreto et al.																					
2002	Vitense,Jacko et al.																					
2003	Jacko, Vitense & Scott																					
2003	Vitense, Jacko & Emery																					
2003	Emery,Edwards et																					
2003	Jacko, Scott et al.																					
2004	Jacko,Barnard et al.																					
2004	Jacko, Emery et al.																					
2004	Edwards,Barnard et al.																					
2005	Edwards,Barnard et al.																					
2005	Jacko,Moloney et al.																					
2005	Moloney,Leonard et al.																					
2005	Leonard, Edwards & Jacko																					
2005	Leonard, Jacko & Pizzimenti																					
2006	Scott et al.																					
2006	Moloney, Shi et al.																					

100

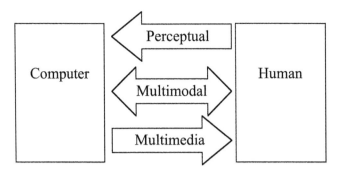

FIGURE 6.2. Perceptual, multimodal, and multimedia interfaces (flow of information).

this concept. Multimedia interfaces elicit perceptual and cognitive skills to interpret information presented to the user, whereas multimodal interfaces use multiple modalities for the HCI. Multimedia interfaces focus on the media while multimodal interfaces focus on the human perceptual channels (Turk & Robertson, 2000). The strength and capabilities of multimedia and multimodal interfaces with respect to individuals with perceptual impairments are described in more depth in the next two sections.

Multimedia interfaces. Multimedia interfaces have grown from the need to display diverse forms of information in a flexible and interactive way. Multimedia can be simply defined as computer-controlled interactive presentations (Chignell & Waterworth, 1997). The broadness of this definition directly corresponds to the broadness of the field of multimedia research. For the purpose of this chapter, the discussion of multimedia is limited to a brief overview of types and potential strengths of multimedia as they relate to enhancing accessibility to information technologies for individuals with perceptual impairments.

There are three approaches to multimedia: performance, presentation, and document (Chignell & Waterworth, 1997). In the performance approach, multimedia is a kind of theatrical play that is conveyed through "actors." The timing of the actors' performances is orchestrated in an effort to entertain and educate (Waterworth & Chignell, 1997). Presentation multimedia is a modern version of slide shows, where video clips and animation enhance a sequence of slides. The goal of the presentation approach is to convey ideas to the user (Chignell & Waterworth, 1997). Lastly, the document approach focuses on text and ideas. It can be thought of as an enhanced document that elaborates ideas in the text. All of these approaches provide additional opportunities to convey perceptual information to the user.

Multimedia allows communication between users and computers in a sensory manner. As such, the essential aspect of designing multimedia interfaces is selecting the media and modalities. Information can be taken from one modality and presented in another modality (Chignell & Waterworth, 1997). For instance, data presented visually could be converted and displayed audibly. For instance, multimedia could be used

to enhance a GUI for a user with a hearing impairment. Information that would be commonly conveyed via auditory feedback could be provided visually. An example of this would be to have an icon display when an error has been committed, rather than the traditional beep sounding. Other examples of how multiple modalities can enhance accessibility to information technologies are discussed in the multimodal section.

The potential strength of multimedia interfaces comes from its ability to use images, text, and animation to connect with users. However, in order to facilitate that type of exchange, much consideration must be paid to users' sensations and perceptions via the auditory, speech, and visual channels. For instance, when designing graphical images, basic knowledge of color vision is needed to ensure that colors can be discriminated, foreground can be separated from background, highlights will attract attention, and grouping of objects is apparent (Gillan, 1998). An understanding of human sensation and perception is necessary if perpetual interfaces are to reach their full potentials. When perceptual interfaces are used, however, they provide individuals with perpetual impairments, such as those related to the visual channel, alternative modalities to access technology. Multimodal interfaces are a second type of perceptual interface that provide this same benefit.

Multimodal interfaces. Multimodal interfaces, as they are discussed in this chapter, are interfaces that support a wide range of perceptual capabilities (e.g., auditory, speech, and visual) as a means to facilitate human interaction with computers.

With the growing complexity of technology and applications, a single modality no longer permits users to interact effectively across all tasks and environments (Oviatt, Cohen, Suhm, et al., 2000). The strength of a multimodal design is its ability to allow users the freedom to use a combination of modalities or the best modality for their needs. These interfaces make the most effective use of the variety of human sensory channels, alone and in combination. Ultimately, multimodal interfaces offer expanded accessibility of computing and promote new forms of computing not previously available to individuals with perceptual impairments.

The development and application of multimodal interfaces for making technologies accessible to users with visual impairments is growing. Some forms of unimodal, bimodal, and trimodal feedback within multimodal interfaces have been investigated and their advantages have been documented for users with visual impairments (Jacko, Emery, et al., 2004; Jacko, Barnard, et al., 2004; Jacko, Moloney, et al., 2005; Jacko, Scott, Sainfort, et al., 2003). Since most information presented on a GUI is visual, there is great interest in the research community to find alternative ways of displaying this information. One of the common approaches to conveying visual information in a nonvisual way is through the use of the tactile modality. Specific research has been conducted in the realm of tactile displays with respect to visual impairments. Not only can tactile displays provide information regarding a graphic's identity but also the depth, location, and perception of its purpose. The use of tactile systems can also provide navigational information. Research has

shown that tactile output of directional information offers support to the blind as they explore images (Kurze, 1998). Research
has also been conducted in the area of movable dynamic tactile
displays that present information to one or several fingertips in
a Braille type manner. The Braille dots move in a wave of lifted
and lowered series of pins (Fricke & Baehring, 1994). Along the
same line of research, the use of a bidimensional single cell
Braille display combined with a standard Braille cell has been
evaluated. Although initial research found this new combined
device is not an improvement over a standard stand-alone
Braille display, continued research is yielding improvement
(Ramstein, 1996). Tactile output via force feedback has been
looked to as a means of conveying numerical information. Yu,
Ramloll, and Brewster (2000) worked on a system that converts
data typically displayed visually into haptic and auditory output. Since data visualization techniques are not appropriate for
blind people or people with visual impairments, this system
translated graphs into friction and textured surfaces along with
auditory feedback.

Another common approach to converting visual information
in a nonvisual way is through the use of the speech modality.
Speech recognition systems serve as an alternative modality for
users and computers to interact. These systems recognize human speech and translate it into commands or words understood by the computer. Chapter 10 in this handbook offers an
introduction to the technology of speech recognition systems
and design issues associated with incorporating speech recognition into applications. However, this chapter's discussion of
speech recognition systems concentrates on its use related to
individuals with visual impairments. The integration of speech
input and output into applications offers an alternative to purely
graphical environments (Yankelovich, Levow, & Marx, 1995).
This type of technology-driven design allows applications to be
suited to a wide variety of individuals with disabilities (Danis &
Karat, 1995). For instance, individuals with visual impairments
can use a computer solely by voice activation.

Speech recognition systems are traditionally associated with
the concept of dictation. Products such as Dragon Systems, Inc.
Dragon Naturally Speaking offer a line of speech recognition
products for dictation. Some specific packages are geared toward particular professions, such as medical and legal (Cunningham & Coombs, 1997). Other common dictation systems
are IBM ViaVoice, Lernout & Hauspie Voice Xpress, and Philips
Speech Processing FreeSpeech98. Along the lines of recognition
engines, Verbex Voice Systems and SRI Corp.'s DECIPHER offer
continuous speech recognition technology. Microsoft's Whisper
provides speaker-independent speech recognition with online
adaptation, noise robustness, and dynamic vocabularies and
grammars (Huang, Acero, Alleva, et al., 1995). Speech driven
menu navigation systems, for instance Command Corp. Inc., IN
CUBE Voice Command for window navigation, have also been
developed (Karshmer, Brawner, & Reiswig, 1994; Karshmer, Ogden, Brawner, Kaugars, & Reiswig, 1994).

Often, circumstances may affect more than one sensory
channel, and individuals may experience multiple impairments.
In addition to providing heightened challenges to interaction,
this impacts the efficacy of perceptual interface in meeting
users' needs. For example, many disabilities are associated to
some degree with speech degeneration, and therefore speech

recognition systems may not fully accommodate a user's needs
(Rampp, 1979). Older adults commonly experience multiple impairments, as all the sensory channels normally decline with
age, and are more susceptible to conditions, such as stroke,
which can cause reduced speech, vision, and motor skills. For
cases in which speech is degraded, speech recognition systems
must be sensitive enough to adapt to impaired speech. Regardless of what caused the speech impairment, devices and techniques can be applied to augment the communicative abilities
of individuals who experience difficulty speaking in an understandable manner.

Augmentative and alternative communication (AAC) is a field
of study concerned with providing such devices and techniques.
McCoy, Demasco, Pennington, and Badman (1997) described a
prototype system aimed at users with cognitive impairments.
This prototype is designed to aid communication and provide
language intervention benefits across several user populations.
An iconic language approach has been applied to aid individuals
with significant speech and multiple impairments (SSMI). Research has been conducted on the use of icon language design
based on the theory of icon algebra (Chang, 1990) and the theory of conceptual dependency (Schank, 1972). These methodologies are then used in interactive interface design (Albacete,
Chang, & Polese, 1994). Individuals with SSMI also commonly
have difficulty with word processing. Using an animated graphical display, phoneme probabilities of speech can be more easily
isolated and recognized. This allows users the opportunity to
interpret their speech rather than forcing the computer to do it
automatically (Roy & Pentland, 1998). The more feedback and
modalities users are provided, the more efficiently they can
interact with speech recognition by assisting the computer to interpret their speech correctly. Speech recognition systems must
also take into account the possible extent of cognitive burden
placed on the user. If voice is used to navigate through a GUI,
certain prosodic features (e.g., pauses) of the user's speech,
resulting from a high cognitive load, may affect performance
(Baca, 1998). Additional challenges associated with the implementation of speech recognition systems are discussed in this
handbook as well. Overall, these systems need to account for
situations when users with visual impairment experience varying levels of hearing, speech, and cognitive abilities.

The auditory channel serves not only to convey information,
but also to receive information. The use of auditory feedback is
extremely useful to many computer users. Ongoing research has
looked at the use of bidirectional sound as a standard element of
an interface environment. A prototype named the Voice Enabled
Reading Assistant (VERA), written using an Aural-Oral User Interface (A-OUI) model, was developed to provide bidirectional
sound. The A-OUI model captures qualities and functions of plain
text files needed for user interfaces to present information to the
auditory, visual, and tactile senses. The use of the VERA prototype
can be applied to many types of office and Internet text and data.
Similar speech-enabled products are Emacspeak and IBM Voice-
Type Simply Speaking Gold (Ryder & Ghose, 1999).

Much work with respect to audio feedback has focused on
conveying information from a GUI. For instance, research conducted by Darvishi and colleagues (Darvishi, Guggiana, et al.,
1994; Darvishi, Munteanu, et al., 1994) looked at mapping
GUIs into auditory domains through impact sounds based on

physical modeling. Ultimately, these sounds are used to convey information to the user regarding the objects in a GUI environment. Audio feedback has also been used to provide visually impaired users with a sense of depth perception, by varying the location of the sound sources in a three-dimensional environment. Because depth perception is a function of vision, any cues that can be conveyed though other modalities are vital. This is particularly crucial when working with three-dimensional computer applications, such as a CAD package (Mereu & Kazman, 1996). The use of auditory feedback has also been studied with regard to enhancing synthesis speech output. Through the development of spatial audio processing systems, a greater benefit of synthesis speech was achieved (Crispien, Würz, & Weber, 1994). IBM's Screen Reader/2 also uses synthesis speech to make GUIs accessible to those who are visually impaired or blind, by converting screen information to speech or Braille. The users are continuously kept informed of screen activity and cursor movement. Reading typed characters, words, and sentences are features that can be made automatic. This software aids in the use of windows, menus, dialog boxes, and other controls (Thatcher, 1994). Microsoft Corporation's Whistler, a trainable, text-to-speech system that produces synthetic speech, sounds very natural by reproducing the characteristics of original speech (Huang, Acero, Adcock, et al., 1996). WRITE:OUTLOUD and OutSPOKEN are two additional text-to-speech products commonly used (Friedlander, 1997). Regardless of what product is used, text-to-speech software enables information to be collected and utilized in a more rapid manner.

In addition to the benefits already discussed that enable users with visual impairments to access technology, multimodal interfaces also provide superior support of error handling. More specifically multimodal interfaces have been shown to have the ability to avoid errors and recover more efficiently from errors when they do occur. This is a very important element of universal accessibility. Enabling users to use a technology includes not only being able to access the technology but also the ability to use it in an efficient manner. Reducing or avoiding errors is a major key to improving efficiency. A multimodal interface provides better error handling than a unimodal interface for several reasons. The following are factors which represent reasons why a multimodal interface may be better at error avoidance.

- Users will select the input mode considered less error prone for a particular context, which is assumed to lead to error avoidance.
- The ability to use several modalities permits users the flexibility to leverage their strengths by using the most appropriate modality. It is possible that the more comfortable the users, the less likely they are to make errors.
- Users tend to switch modes after systems errors. For instance, users are likely to switch input modes when they encounter a system recognition error.
- Users are less frustrated with error when interacting within a multimodal interface, even when errors are as frequent as in a unimodal interface. The reduction in frustration may be the result of the user feeling as though they have more control over the system because they can switch modes (Oviatt, 1999).

Ultimately, multimodal systems aid in reducing errors (error handling) because they offer parallel or duplicated functionality that allows users to accomplish the task using one of several modalities (Oviatt & Cohen, 2000). These benefits stem from the user-centered and system-centered design perspectives from which multimodal interfaces are built (Oviatt & Cohen, 2000; Oviatt, Cohen, Suhm, et al., 2000). Like multimodal interfaces, adaptive interfaces also benefit from a user-centered perspective.

Adaptive Interfaces

Adaptive interfaces have great potential for accommodating a wide range of users in a variety of work contexts. As a result, much research has been conducted on the design and implementation of adaptive interfaces. An example of early research in the field of adaptive GUI is illustrated by the work of Mynatt and Weber (1994) with the Mercator project, and the GUIB (Textual and Graphical User Interfaces for Blind People) project (Petrie, Morley, & Weber, 1995). Both of these projects focused on making environment-level adaptations to GUIs in order to make them more accessible (Stephanidis, 2001a). Mercator interfaces model the graphical objects and their hierarchical relationships. The model serves to predict a user's interaction (Edwards, W. K., & Mynatt, 1994; Edwards, Mynatt, & Stockton, 1994). The goal is to provide visually impaired users with an interface that is more accessible. By better understanding how low vision users interact with a computer, interfaces can be designed more effectively. Outputs such as synthetic and digitized speech, refreshable Braille, and nonspeech sounds also make an interface easier to use. Auditory icons and earcons are two predominate areas of nonspeech sound research. Auditory icons, developed by Gaver (1989), are everyday sounds that occur in the world mapped to the computer world. Gaver first applied the concept of auditory icons to SonicFinder. Earcons are a second method for employing nonspeech audio in GUIs. Earcons, developed by Blattner, Sumikawa, and Greenberg (1989), are audio messages that provide the user with information about objects or operations of the computer. The first version of GUIB adapted the GUI by combining Braille speech and nonspeech audio together to construct a nonvisual interface so that users who are blind can access the GUI (Emiliani, 2001).

Configurable interface designs based on user models have been heavily researched. The user models have been defined with respect to visual, cognitive, motor, and other abilities. With these models, custom computer systems, including both hardware and software, can be created (McMillan & Wisniewski, 1994). Semantic abstraction of user interaction, named abstract widgets, is another modeling approach that provides great flexibility. Abstract widgets separate the user interface from the application functionality. This allows users to interact with interfaces as they choose, independently from their environment (Kawai, Aida, & Saito, 1996). The use of adaptation determinates, constituents, goals, and rules is yet another approach of an adaptation strategy. This strategy is based on the fact that these important attributes, which categorize adaptation, can be used to formulate adaptation rules. These adaptation rules, in turn, assist the development of intelligent user interfaces. This approach can be customized to the requirements of different application domains and user groups (Stephanidis, Karagiannidis, & Koumpis, 1997). Approaches such as described by the Pervasive Accessible Technology (PAT) allow individuals with disabilities to use standard interface devices that adapt to the user in order to communicate with information technology infrastructures (Paciello, 1996). Based on an individual's disability, the implementation of a User Interface Management System (UIMS) model provides the versatility needed to adapt interfaces to individuals. The selection of input devices, presentation of information on the screen, and choice in selection/activation method can all be adapted to fit specific user needs (Bühler, Heck, & Wallbruch, 1994). Current research focuses on operationally reliable software infrastructures that support alternative physical realizations through the abstractions of objects. More specifically, systems such as Active X (by Microsoft and JavaBeans (by SunSoft, represent component-ware technology (Stephanidis, 2001b). These systems represent a mainstream effort to provide technological structures that provide more adequate support for accessibility. They are also two examples of currently available tools that can be used to generate code for the adaptation of various interface components.

Some hypermedia systems research has focused on adaptive hypermedia applications. These adaptive hypermedia systems keep track of evolving aspects of the user, such as preferences and domain knowledge. This information is stored to create a user model, which in turn is the basis for user interface adaptation (De Bra, Houben, & Wu, 1999). Adaptive hypermedia can also be designed based on task models. In the latter case, task models are used as the basis from which hypermedia systems are developed. These task models support the design and development of hypermedia. Different task models are associated with different types of users (Paternó & Mancini, 1998). Task models reflect the user's view of the activities to be performed. In addition to personalizing the content of hypermedia systems, adaptive navigation support has also been researched. Prototype systems have been developed to demonstrate how different navigational possibilities can be made available based on a user model (Pilar da Silva, Van Durm, Duval, & Oliviè, 1998).

Another example of adapting interfaces to users is through the use of the EZ Access protocols developed by the Trace Center at the University of Wisconsin-Madison. EZ Access protocols are a set of techniques that modify an interface to fit a specific user's need. These protocols work across disabilities and with a range of products. The most common use of EZ Access protocols is in touchscreen kiosks (Vanderheiden, 1998).

While research and development in the areas of perceptual and adaptive interfaces, like those examples discussed, have gained substantial attention in generating potential solutions for individuals with visual sensory impairments, the majority of work and realized products are positioned in two areas. The most readily available products and the majority of the research have been somewhat narrowly focused on (a) the magnification of screen elements (Fraser & Gutwin, 2000) and (b) the accessibility of text (Craven, 2003). Comparatively less has been accomplished in terms of critical aspects of the graphical user interface as has been done with magnification and vocalization of text. While the solutions generated have afforded access to both those individuals who are blind and those who retain a range of visual impairment, they tend to be one size fits all solutions, which entirely abandon the visual sensory channel (as in the case of Braille interfaces or Screen Readers), or focus only on the augmentation of the visual channel (e.g., magnification, increased contrast, increased text size, etc.), and very few are multimodal, or perceptual. Table 6.2 summarizes a selection of popular assistive devices. Following this table, a discussion of research that has evaluated the efficacy of such approaches to accessibility is presented.

TABLE 6.2. Popular Assistive Devices

Competitor	Product Overview	Key Limitation(s)
Screen Reader	Software program that reads to the user elements that appear on an interface via synthesized voice. The program reads left to right, starting at the very top of the screen. When an image is encountered the program reads the associated ALT text	Solution abandons any remaining vision the user has, using only their auditory ability Efficacy depends on the organization of the interface (e.g., anything not modeled in a left to right organization is not compatible)
Braille Display 	Similar to the screen readers, but gives the reader the information via tactile cues (Braille characters)	Same limitations as a screen reader, plus the user has to learn Braille, which is not likely if they have residual vision, and are losing vision later in life
Screen Magnifiers	Physical device or software program that enlarges the entire screen image Software programs, such as Zoomtext, developed by Ai squared, allow for "adaptable" magnification of the interface, and some versions incorporate a screen reader	A 'one-size-fits all' solution that is not adaptable between users For people with visual impairments, magnification is not always the most effective strategy (especially with obstructed visual fields)

Kline and Glinert (1995) presented UnWindows V1, a set of interface tools to support selective magnification of window area, and tracking the location of the mouse pointer on the display screen. The authors noted, "Magnification is one method commonly employed to help low vision users deal with the small type fonts, illustrations, and icons present in much of today's printed media and computer displays" (Kline & Glinert, 1995, p. 2). Key components of the UnWindows system included (a) a dynamic magnifier to compensate for the loss of global context imposed by static magnification and changing display content and (b) visual and aural feedback to aid the users in locating the mouse pointer. Kline and Glinert placed emphasis on the problematic nature of visual tracking in the presence of a screen densely populated with icons and windows. Interestingly, they received mixed reaction to their interface by users with and without visual impairment, especially in terms of the auditory feedback provided whenever the mouse pointer entered a new window (users found this annoying). While no formal empirical testing was performed in relation to UnWindows, questions surface as to the effectiveness of nonvisual, multimodal feedback in a complex display.

A usability review of currently available technologies for the conversion of GUI technology for use by individuals who were blind or possessed low vision, the ability of magnification, synthetic speech, and Braille were reviewed for their ability provide the respective users 100% access to GUIs on nine test areas (Becker & Lundman, 1998). These tests included

- Installing and configuring the device/software
- Uninstalling the device/software
- Performance reliability/stability
- Program manager to read and manipulate windows, menus, and icons
- Word processing based tasks such as opening and saving files, reading and editing text, text attributes, and toolbars
- Spreadsheet tasks such as reading cells, tables figures, and editing data and formulas
- Internet use, including dialing up, accessing World Wide Web pages, navigating with link buttons, sending e-mail, and reading graphics
- Screen searching, such as searching for characters, strings, formats, and icons
- Operating start menu, exploring and controlling settings (Becker & Lundman, 1998)

For the assessment of seven magnification programs (synthetic speech and Braille displays are not relevant to individuals with visual impairments who have residual vision), the evaluators comprised a system engineer, ergonomic engineer, computer science expert, and three individuals with visual impairments.

The use of magnification as a strategy to afford access to GUIs proved somewhat successful, providing 89% or higher access to GUIs, except for Internet use (84% access) and screen searching (0%; Fortuin & Omata, 2004). It was concluded that the essential problem for the design of interactive systems for users with visual impairments is to (a) to determine what the users need and (b) how to represent the requested information based on key psychological and physical attributes of the user. The result of ineffective assistive technologies is a lack of usable contextual cues for the users to provide feedback in the case of errors, and this translates to large amounts of imposed workload on the user and frustration (Fortuin & Omata, 2004).

In a case study of an English teacher who was having difficulty reading student papers, typing, and proofreading, Whittaker (1998) discovered that magnification was not affording optimal performance. Typically, the authors found that users with visual acuity of 20/40 or better would respond well to simple optical magnification. The author investigated other visual functioning to find that the individual had severely diminished contrast sensitivity (13% contrast threshold, with 2% representing normal sensitivity). Furthermore, this individual's visual field was 20 degrees horizontally (180 degrees is normal). Magnification was likely reducing the number of letters viewable simultaneously in the presence of scotoma within the visual field. The author warned that magnifiers and large monitors are not always the most effective solution for users with impaired vision.

Arditi (2004) addressed the reading difficulties of individuals with low vision. According to the author, successfully overcoming this difficulty is accomplished through the exploitation of remaining vision. The easiest way to do this is through magnification, but as shown in this study, it is not a one size fits all solution. Several parameters of the font, including height, stroke, spacing, and serif size, must be selected in a combination that best suits a given user. Arditi presented the prototype and initial user testing of computer-based software that lets a user customize fonts for maximized legibility. Those users studied were able to adjust fonts to a usable, legible level, to positively impact reading times and the reading acuity.

An in-depth review of accessibility tools aimed at improving interactions of computer users with low vision informed the design of *MouseLupe* (Silva, Regina, & Bellon, 2002). *MouseLupe* simulates a magnifying glass, enabling users to magnify select portions of text or display graphics, inspired by the problematic nature of screen magnification software. The authors suggested that magnification improves the readability of smaller text, but occludes the visible area of the document. Furthermore, graphics that contain text (like most icons), a critical element of the graphical user interface, when enlarged, are difficult to read (for a comprehensive review of magnification tools, see Silva, Regina, & Bellon, 2002).

Several researchers have considered the effect of visual impairments on web browsing (Arditi, 2003; Craven, 2003; Harper, Goble, & Stevens, 2001; Silva, Regina, & Bellon, 2002). Harper, Goble, and Stevens (2001) addressed this problem in terms of "Web Mobility." These authors provided guidelines for movement through and around complex hypermedia environments, such as the web, for users with visual impairments. The problem, according to these authors, is that visual impairment inhibits individuals' ability to efficiently assimilate page structure and visual cues that lead to the following problems:

- Failure to get a feel for the content on the website
- Failure to have a sense for the magnitude of the display or where in a website the interaction takes place

- Disorientation
- Obstacles and distracters such as spacer images, tables, and large images
- Too much complex detail that cannot be resolved
- Frustration

Arditi (2003) observed the problems of web browsing in terms of the allocation of screen space resources. According to the author, conflicts arise in the implementation of web browsing solutions for individuals with low vision including: (a) high magnification requirements; (b) variable typography color, size, and contrasts of the content presented; (c) embedded text messages to augment web images; and (d) accessible web browsing controls (icons, buttons, menus). The author presented a novel approach for effectively using screen resources, providing evidence that the strategic layout of a display is a critical factor to successful interaction. The layout of screen elements was interpreted as more critical than magnification of the screen elements.

Craven (2003) questioned the accessibility of electronic library resources on the World Wide Web for individuals with visual impairments. The results of her study with 20 sighted and 20 visually impaired users revealed the browsing times of those individuals with visual impairments were significantly greater, depending on the design of the website (layout complexity and distracters). Navigation time for the group of users with visual impairments was significantly longer due to visual functioning, but also due to artifacts of assistive technology use in navigation (magnification and screen readers).

Summary

Through technological advances in HCI research, the concepts of perceptual and adaptive interfaces have emerged. These two categories of technologies provide vast opportunities for individuals with perceptual impairments to fully access electronic information. More specifically, multimedia and multimodal systems have been shown to accommodate users of varying abilities. However, while these are the most promising approaches for this population, there still exists a chasm between research and practice. Unfortunately, it is rare to see the knowledge generated through HCI research actually implemented into commercially-available products. Much work still needs to be done, both in terms of advancing knowledge—through empirical HCI research—and in terms of increasing awareness. Translational efforts need to be made to help disseminate this new knowledge into the design of technologies and devices, so that individuals with perceptual (and other forms of) impairment can better leverage the increasingly ubiquitous technological tools used in everyday life. An evidence-based need for improved design of technology—especially for individuals with sensory or perceptual impairments—is needed to help ensure that the new knowledge continually generated through research is ultimately used to improve human interaction with technology.

CONCLUSIONS

To provide a context for the topic of perceptual impairments and computing technologies, Fig. 6.1 demonstrated that, in order to achieve universal access across classes of computing technologies, researchers must be prepared to address several very challenging issues:

1. Establish empirical links between clinical diagnoses and sets of functional capabilities
2. Define the set of functional capabilities required to access information technologies
3. Establish empirical bases for the influence of specific functional capabilities on access to specific classes of technologies

This chapter aimed to establish a basis for addressing such issues by examining, specifically, visual impairment, several specific diagnoses, and the resulting functional abilities. Finally, groundbreaking advancements in perceptual interfaces, multimodal interfaces, multimedia interfaces, and adaptive interfaces were discussed, which can be applied across a variety of classes of technology in order to enhance the perceptual experience of people who possess such impairments.

ACKNOWLEDGMENTS

This chapter was made possible through funding provided to the first author by the National Science Foundation, as well as the Intel Corporation.

References

Age-Related Eye Disease Study Research Group. (2001). A randomized, placebo-controlled, clinical trial of high-dose supplementation with vitamins C and E, beta carotene, and zinc for age-related macular degeneration and vision loss (AREDS Report No. 8). *Archives of Ophthalmology, 119,* 1417–1436.

Albacete, P. L., Chang, S. K., & Polese, G. (1994). *Iconic Language Design for People with Significant Speech and Multiple Impairments.* Paper presented at the First Annual International ACM/SIG-CAPH Conference on Assistive Technologies (ASSETS '94), Marina del Rey, CA.

Applegate, W. B., Miller, S. T., Elam, J. T., Freeman, J. M., Wood, T. O., & Gettlefinger, T. C. (1987). Impact of cataract surgery with lens implantation on vision and physical function in elderly patients. *JAMA, 257,* 1064–1066.

Arditi, A. (2002). Web accessibility and low vision. *Aging & Vision, 14*(2), 2–3.

Arditi, A. (2003). Low vision web browsing and allocation of screen space resources. *Investigative Ophthalmology and Visual Science, 44*(5), 2767.

Arditi, A. (2004). Adjustable typography: an approach to enhancing low vision text accessibility. *Ergonomics, 47*(5), 469–482.

Assistive Technologies for Independent Aging: Opportunities and Challenges: Hearing Before the Special Committee on Aging, 108th Cong., 2nd Sess., 145 (2004) (statement of the American Foundation for the Blind).

Baca, J. (1998). *Comparing effects of navigational interface modalities on speaker prosodics.* Paper presented at the Third Annual International ACM/SIGCAPH Conference on Assistive Technologies (ASSETS '98), Marina del Rey, CA.

Becker, S., & Lundman, D. (1998). *Improving access to computers for blind and visually impaired persons: The development of a test method for usability.* Paper presented at the Telemeatics for the Integration of Disabled and Elderly People (TIDE) 1998 Conference, Helsinki, Finland.

Blattner, M., Sumikawa, D., & Greenberg, R. (1989). Earcons and icons: Their structure and common design principles. *Human Computer Interaction, 4*(1), 11–44.

Brenner, M. H., Curbow, B., Javitt, J. C., Legro, M. W., & Sommer, A. (1993). Vision change and quality of life in the elderly: Response to cataract surgery and treatment of other chronic ocular conditions. *Archives of Ophthalmology, 111*(5), 680–685.

Brewster, S. A., Wright, P. C., & Edwards, A. D. N. (1994). *The design and evaluation of an auditory-enhanced scrollbar.* Paper presented at the ACM Conference on Human Factors in Computing Systems (CHI), Boston, MA.

Bühler, C., Heck, H., & Wallbruch, R. (1994). *A Uniform Control Interface for Various Electronic Aids.* Paper presented at the 4th international conference on computers for handicapped persons, Vienna, Austria.

Chang, S.-K. (Ed.). (1990). *Principles of visual programming systems.* Upper Saddle River, NJ: Prentice-Hall, Inc.

Chignell, M., & Waterworth, J. (1997). Multimedia. In G. Salvendy (Ed.), *Handbook of Human Factors and Ergonomic* (pp. 1808–1861). New York, NY: John Wiley & Sons.

Coeckelbergh, T. R. M., Cornelissen, F. W., Brouwer, W. H., & Kooijman, A. C. (2002). The effect of visual field defects on eye movement and practical fitness to drive. *Vision Research, 42,* 669–677.

Colenbrander, A. (1977). Dimensions of visual performance. *Transactions—American Academy of Ophthalmology and Otolaryngology, 83*(2), 332–337.

Craven, J. (2003). Access to electronic resources by visually impaired people. *Information Research, 8*(4), paper no. 156 [Available at: http://informationr.net/ir/8-4/paper156.html].

Crispien, K., Würz, W., & Weber, G. (1994). *Using spatial audio for the enhanced presentation of synthesised speech within screenreaders for blind computer users.* Paper presented at the 4th international conference on computers for handicapped persons, Vienna, Austria.

Cunningham, C., & Coombs, N. (1997). *Information access and adaptive technology.* Phoenix, AZ: American Council on Education and The Oryx Press.

Danis, C., & Karat, J. (1995). *Technology-driven design of speech recognition systems.* Paper presented at the conference on Designing interactive systems: processes, practices, methods, & techniques, Ann Arbor, MI.

Darvishi, A., Guggiana, V., Munteanu, E., Schauer, H., Motavalli, M., & Rauterberg, M. (1994). *Synthesizing nonspeech sound to support blind and visually impaired computer users.* Paper presented at the ICCHP, Vienna, Austria.

Darvishi, A., Munteanu, E., Guggiana, V., Schauer, H., Motavalli, M., & Rauterberg, M. (1994). *Automatic impact sound generation for using in nonvisual interfaces.* Paper presented at the First Annual International ACM/SIGCAPH Conference on Assistive Technologies (ASSETS '94), Marina Rey, CA.

De Bra, P., Houben, G.-J., & Wu, H. (1999). *AHAM: A Dexter-based reference model for adaptive hypermedia.* Paper presented at the tenth ACM conference on hypertext and hypermedia: returning to our diverse roots: Returning to our diverse roots, Darmstadt, Germany.

Diabetic Retinopathy Study Research Group. (1979). Four risk factors for severe visual loss in diabetic retinopathy: DRS Report 3. *Archives of Ophthalmology, 97,* 654–655.

Diabetic Retinopathy Study Research Group. (1981). Photocoagulation treatment of proliferative diabetic retinopathy: clinical application of diabetic retinopathy study (DRS) findings (DRS Report 8). *Archives of Ophthalmology, 88,* 583–600.

Dix, A., Finlay, J., Abowd, G., & Beale, R. (1998). *Human-computer interaction* (2nd ed.). New York, NY: Prentice Hall.

Donderi, D. C., & Murphy, S. B. (1983). Predicting activity and satisfaction following cataract surgery. *Journal of Behavioral Medicine, 6,* 313–328.

Early Treatment Diabetic Retinopathy Study Research Group. (1995). Focal photocoagulation treatment of diabetic macular edema (ETDRS Report 19). *Archives of Ophthalmology, 113,* 1144–1155.

Edwards, P. J., Barnard, L., Emery, V. K., Yi, J., Moloney, K. P., Kongnakorn, T., et al. (2005). Understanding users with diabetic retinopathy: Factors that affect performance in a menu selection task. *Behavior and Information Technology, 24*(3), 175–186.

Edwards, P. J., Barnard, L., Emery, V. K., Yi, J., Moloney, K. P., Kongnakorn, T., et al. (2004, October). Strategic design for users with diabetic retinopathy: Factors influencing performance in a menu-selection task. *Proceedings of the Sixth International ACM SIGACCESS Conference on Computers and Accessibility (ASSETS'04),* Atlanta, GA, 118–125.

Edwards, W. K., & Mynatt, E. D. (1994). *An architecture for transforming graphical interfaces.* Paper presented at the ACM Conference on User Interface Software and Technology (UIST '94), Marina del Rey, CA.

Edwards, W. K., Mynatt, E. D., & Stockton, K. (1994). *Providing access to graphical interfaces, not graphical screens.* Paper presented at the ACM Conference on Assistive and Enabling Technology (ASSETS '94), Marina del Rey, CA.

Emery, V. K., Edwards, P. J., Jacko, J. A., Moloney, K. P., Barnard, L., Kongnakorn, T., et al. (2003, November). Toward achieving universal usability for older adults through multimodal feedback. *Proceedings of the 2003 ACM Conference on Universal Usability,* Vancouver, British Columbia, 46–53.

Emiliani, P. L. (2001). Special needs and enabling technologies: An evolving approach to accessibility. In C. Stephanidis (Ed.), *User interfaces for all—concepts, methods and tools* (pp. 97–113). Mahwah, NJ: Lawrence Erlbaum Associates.

The Eye Diseases Prevalence Research Group. (2004a). Causes and prevalence of visual impairment among adults in the United States. *Archives of Ophthalmology, 122,* 477–485.

The Eye Diseases Prevalence Research Group. (2004b). Prevalence of age-related macular degeneration in the United States. *Archives of Ophthalmology, 122,* 564–572.

The Eye Diseases Prevalence Research Group. (2004c). The prevalence of diabetic retinopathy among adults in the United States. *Archives of Ophthalmology, 122,* 552–563.

Fletcher, D. C. (Ed.). (1999). *Low vision rehabilitation: Caring for the whole person.* Hong Kong: Oxford University Press.

Fletcher, D. C., Schuchard, R., Livingstone, C., Crane, W., & Hu, S. (1994). Scanning laser ophthalmoscope macular perimetry and applications for low vision rehabilitation clinicians. *Ophthalmology Clinics of North America, 7,* 257–265.

Fortuin, F. T., & Omata, S. (2004). Designing universal user interfaces—The application of universal design rules to eliminate information barriers for the visually impaired and the elderly. Retrieved May 19, 2004, from http://www.visionconnection.org

Fraser, J., & Gutwin, C. (2000). *A framework of assistive pointers for low vision users.* Paper presented at the ACM Conference on Assistive Technologies, Arlington, VA.

Fricke, J., & Baehring, H. (1994). *Displaying laterally moving tactile information.* Paper presented at the 4th international conference on computers for handicapped persons (ICCHP'94), Vienna, Austria.

Friedlander, C. (1997). Speech facilities for the reading disabled. *Communications of the ACM, 40*(3), 24–25.

Gaver, W. (1989). The SonicFinder: An interface that uses auditory icons. *Human Computer Interaction, 4*(1), 67–94.

Gerber, E., & Kirchner, C. (2001, March). Who's surfing? Internet access and computer use by visually impaired youths and adults. *Journal of Visual Impairment & Blindness*, 176–181.

Gillan, D. J. (1998). *The psychology of multimedia: Principles of perception and cognition.* Paper presented at the 1998 ACM Conference on Human Factors in Computing Systems (CHI'98), Los Angeles, CA.

Gragoudas, E. S., Adamis, A. P., Cunningham, E. T., Jr., Feinsod, M., & Guyer, D. R. (2004). Pegaptanib for neovascular age-related macular degeneration. *New England Journal of Medicine, 351,* 2805–2816.

Harper, S., Goble, C., & Stevens, R. (2001). Web mobility guidelines for visually impaired surfers. *Journal of Research and Practice in Information Technology, 33*(1), 30–41.

Huang, X., Acero, A., Adcock, J., Hon, H.-W., Goldsmith, J., Liu, J., et al. (1996). *Whistler: A trainable text-to-speech system.* Paper presented at the Fourth International Conference of Spoken Language Processing (ICSLP '96), Philadelphia, PA.

Huang, X., Acero, A., Alleva, F., Hwang, M.-Y., Jiang, L., & Mahajan, M. (1995). *Microsoft Windows highly intelligent speech recognizer: Whisper.* Paper presented at the International Conference on Acoustics, Speech, and Signal Processing, Detroit, MI.

Jacko, J. A. (1999, August) The importance of clinical diagnoses in the prediction of performance on computer-based tasks for low vision users. *Proceedings of the 9th International Conference on Human-Computer Interaction,* Munich, 787–791.

Jacko, J. A., Dixon, M. A., Rosa, R. H., Jr., Scott, I. U., & Pappas, C. J. (1999). Visual profiles: A critical component of universal access. *ACM Conference on Human Factors in Computing Systems (CHI 99),* Pittsburg, PA, 330–337.

Jacko, J. A., Barnard, L., Kongnakorn, T., Moloney, K. P., Edwards, P. J., Emery, V. K., et al. (2004). Isolating the effects of visual impairment: Exporing the effect of AMD on the utility of multimodal feedback. *CHI Letters, 6*(1), 311–318.

Jacko, J. A., Barreto, A. B., Scott, I. U., Chu, J. Y. M., Vitense, H. S., Conway, F. T., et al. (2002). Macular degeneration and visual icon use: Deriving guidelines for improved access. *Univeral Access in the Information Society, 1,* 197–296.

Jacko, J. A., Barreto, A. B., Scott, I. U., Rosa, R. H., Jr., & Pappas, C. J. (2000). Using electroencephalogram to investigate stages of visual search in visually impaired computer users: Preattention and focal attention. *International Journal of Human-Computer Interaction, 12*(1), 135–150.

Jacko, J. A., Emery, V. K., Edwards, P. J., Ashok, M., Barnard, L., Kongnakorn, T., et al. (2004). The effects of multimodal feedback on older adults' task performance given varying levels of computer experience. *Behavior and Information Technology, 23*(4), 247–264.

Jacko, J. A., Moloney, K. P., Kongnakorn, T., Barnard, L., Edwards, P. J., Emery, V. K., et al. (2005). Multimodal feedback as a solution to ocular disease-based user performance decrements in the absence of functional visual loss. *International Journal of Human-Computer Interaction, 18*(2), 183–218.

Jacko, J. A., Dixon, M. A., Rosa, R. H., Scott, I. U., & Pappas, C. J. (1999). Linking visual capabilities of partially sighted computer users to psychomotor task performance. *Proceedings of the 9th International Conference on Human-Computer Interaction,* Munich, Germany, August 22–27, 975–979.

Jacko, J. A., Rosa, R. H., Scott, I. U., Pappas, C. J., & Dixon, M. A. (2000). Visual impairment: The use of visual profiles in evaluations of icon use in computer-based tasks. *International Journal of Human-Computer Interaction, 12*(1), 151–165.

Jacko, J. A., Scott, I. U., Barreto, A. B., Bautsch, H. S., Chu, J. Y. M., & Fain, W. B. (2001, August). *Iconic visual search strategies: A comparison of computer users with AMD versus computer users with normal vision.* Paper presented at the 9th International Conference on Human-Computer Interaction, New Orleans, LA.

Jacko, J. A., Scott, I. U., Sainfort, F., Moloney, K. P., Kongnakorn, T., Zorich, B. S., et al. (2003). Effects of multimodal feedback on the performance of older adults with normal and impaired vision. *Lecture Notes in Computer Science (LNCS), 2615,* 3–22.

Jacko, J. A., & Sears, A. (1998). *Designing interfaces for an overlooked user group: Considering the visual profiles of partially sighted users.* Paper presented at the ACM Conference on Assistive Technologies, Marina del Rey, CA.

Jacko, J. A., & Vitense, H. S. (2001). A review and reappraisal of information technologies within a conceptual framework for individuals with disabilities. *Universal Access in the Information Society (UAIS), 1,* 56–76.

Jacko, J. A., Vitense, H. S., & Scott, I. U. (2003). Perceptual impairments and computing technologies. In J. A. Jacko & A. Sears (Eds.), *Human-computer interaction handbook* (pp. 504–522). Mahwah, NJ: Lawrence Erlbaum Associates.

Javitt, J. C., Brenner, M. H., Curbow, B., Legro, M. W., & Street, D. A. (1993). Outcomes of cataract surgery. Improvement in visual acuity and subjective visual function after surgery in the first, second, and both eyes. *Archives of Ophthalmology, 111*(5), 686–691.

Karshmer, A. I., Brawner, P., & Reiswig, G. (1994). An experimental sound-based hierarchical menu navigation system for visually handicapped use of graphical user interfaces. *Proceedings of the first annual ACM conference on assistive technologies,* Marina del Rey, CA, 123–128.

Karshmer, A. I., Ogden, B., Brawner, P., Kaugars, K., & Reiswig, G. (1994). *Adapting graphical user interfaces for use by visually handicapped computer users: Current results and continuing research.* Paper presented at the 4th international conference on computers for handicapped persons (ICCHP'94), Vienna, Austria.

Kawai, S., Aida, H., & Saito, T. (1996). *Designing interface toolkit with dynamic selectable modality.* Paper presented at the second annual ACM conference on assistive technologies, Vancouver, Canada.

Kline, D. W., & Scialfa, C. T. (1997). Sensory and perceptual functioning: Basic research and human factor implications. In A. D. Fisk & W. A. Rogers (Eds.), *Handbook of human factors and the older adult* (pp. 27–54). San Diego, CA: Academic Press.

Kline, R. L., & Glinert, E. P. (1995). *Improving GUI accessibility for people with low vision.* Paper presented at the SIGCHI Conference on Human Factors in Computing Systems, Denver, CO.

Kurze, M. (1998). *TGuide: A guidance system for tactile image exploration.* Paper presented at the Third Annual International ACM/SIGCAPH Conference on Assistive Technologies (ASSETS '98), Marina del Rey, CA.

LaPlante, M. P. (1988). Prevalence of conditions causing need for assistance in activities of daily living. In M. P. LaPlante (Ed.), *Data on disability from the National Health Interview Survey, 1983–1985* (p. 3). Washington, DC: National Institute on Disability and Rehabilitation Research.

Lee, H.-W., Legge, G. E., & Oritz, A. (2003). Is word recognition different in central and peripheral vision? *Vision Research, 43,* 2837–2846.

Leonard, V. K., Edwards, P. J., & Jacko, J. A. (2005, September). Informing accessible design through self-reported quality of visual health. *Proceedings of the Human Factors and Ergonomics Society 49th Annual Meeting,* Orlando, FL, 994–998.

Leonard, V. K., Jacko, J. A., & Pizzimenti, J. J. (2005, October). An exploratory investigation of handheld computer interaction for older adults with visual impairments. *Proceedings of the 7th International ACM SIGACCESS Conference on Computers and Accessibility (ASSETS 2005),* Baltimore, MD, 12–19.

McCoy, K. F., Demasco, P., Pennington, C. A., & Badman, A. L. (1997). *Some interface issues in developing intelligent communication aids for people with disabilities.* Paper presented at the 1997 international conference on intelligent user interfaces (IUI'97), Orlando, FL.

McMillan, W. W., & Wisniewski, L. (1994). *A rule-based system that suggests computer adaptations for users with special needs.* Paper presented at the First Annual International ACM/SIGCAPH Conference on Assistive Technologies (ASSETS '94), Marina Del Rey, CA.

Mehr, E. B., & Freid, A. N. (1975). *Low vision care.* Chicago: Professional Press.

Mereu, S. W., & Kazman, R. (1996). *Audio enhanced 3D interfaces for visually impaired users.* Paper presented at the 1996 ACM Conference on Human Factors in Computing Systems (CHI'96), Vancouver, Canada.

Moloney, K. P., Shi, B., Leonard, V. K., Jacko, J. A., Vidakovic, B., & Sainfort, F. (in press). Leveraging data complexity: Pupillary behavior of older adults with visual impairment during HCI. *Transactions on Computer-Human Interaction (TOCHI).*

Mynatt, E. D., & Weber, G. (1994). *Nonvisual presentation of graphical user interfaces: contrasting two approaches.* Paper presented at the 1994 ACM Conference on Human Factors in Computing Systems (CHI'94), Boston, MA.

Nilsson, U. L. (1990). Visual rehabilitation of patients with and without educational training in the use of optical aids and residual vision: a prospective study of patients with advanced age-related macular degeneration. *Clinical Vision Sciences, 6,* 3–10.

Nilsson, U. L., & Nilsson, S. E. G. (1986). Rehabilitation of the visually handicapped with advanced macular degeneration. *Documenta Ophthalmologica, 62*(4), 345–367.

Oakley, I., McGee, M. R., Brewster, S. A., & Gray, P. (2000). Putting the feel in 'look and feel'. *Proceedings of the SIGCHI conference on human factors in computing systems,* Hague, Netherlands, 415–422.

Orr, A. L. (1998). *Issues in aging and vision.* New York: American Foundation for the Blind.

Oviatt, S. (1999). *Mutual disambiguation of recognition errors in a multimodel architecture.* Paper presented at the SIGCHI conference on human factors in computing systems: The CHI is the limit (CHI '99), Pittsburgh, PA.

Oviatt, S., & Cohen, P. (2000). Multimodal interfaces that process what comes naturally. *Communications of the ACM, 43*(3), 45–53.

Oviatt, S. L., Cohen, P. R., Wu, L., Vergo, J., Duncan, L., Suhm, B., Bers, J., Holzman, T., Winograd, T., Landay, J., Larson, J., & Ferro, D. (2000). Designing the user interface for multimodal speech and gesture applications: State-of-the-art systems and research directions, *Human Computer Interaction, 15*(4), 263–332.

Paciello, M. G. (1996). Designing for people with disabilities. *Interactions, 3*(1), 15–16.

Parrish, R. K. I., Gedde, S. J., Scott, I. U., Feuer, W. J., Schiffman, J. C., Mangione, C. M., et al. (1997). Visual function and quality of life among patients with glaucoma. *Archives of Ophthalmology, 115*(11), 1447–1455.

Paternó, F., & Mancini, C. (1998). *Developing adaptable hypermedia.* Paper presented at the 4th international conference on intelligent user interfaces (IUI '99), Los Angeles, CA.

Petrie, H., Morley, S., & Weber, G. (1995). *Tactile-based direct manipulation in GUIs for blind users.* Paper presented at the conference companion on human factors in computing systems (CHI '95), Denver, CO.

Pilar da Silva, D., Van Durm, R., Duval, E., & Olivié, H. (1998). *Adaptive navigational facilities in educational hypermedia.* Paper presented at the ninth ACM conference on hypertext and hypermedia: Links, objects, time and space—Structure in hypermedia systems (HyperText '98), Pittsburgh, PA.

Rampp, D. L. (1979). Hearing and learning disabilities. In L. J. Bradford & W. G. Hardy (Eds.), *Hearing and Hearing Impairment* (pp. 381–389). New York: Grune & Stratton.

Ramstein, C. (1996). *Combining hepatic and Braille technologies: Design issues and pilot study.* Paper presented at the Second Annual International ACM/SIGCAPH Conference on Assistive Technologies (ASSETS '96), Vancouver, Canada.

Reeves, B., & Nass, C. (2000). Perceptual user interfaces: Perceptual bandwidth. *Communications of the ACM, 43*(3), 65–70.

Rosenberg, R., Faye, E., Fischer, M., & Budicks, D. (1989). Role of prism relocation in improving visual performance of patients with macular dysfunction. *Optometry and Vision Science, 66*(11), 747–750.

Roy, D., & Pentland, A. (1998). *A phoneme probability display for individuals with hearing disabilities.* Paper presented at the Third Annual International ACM/SIGCAPH Conference on Assistive Technologies (ASSETS '98), Marina del Rey, CA.

Ryder, J. W., & Ghose, K. (1999). *Multi-sensory browser and editor model.* Paper presented at the 1999 ACM symposium on applied computing, San Antonio, TX.

Schank, R. C. (1972). Conceptual dependency: A theory of natural language understanding. *Cognitive Psychology, 3*(4), 532–631.

Schieber, F. (1994). *Recent developments in vision, aging and driving: 1988–1994.* Report No. UMTRI-94-26. Ann Arbor, MI: University of Michigan, Transportaion Research Institute.

Scott, I. U., Feuer, W. J., & Jacko, J. A. (2002a). Impact of graphic user interface screen features on computer task accuracy and speed in a cohort of patients with age-related macular degeneration. *American Journal of Ophthalmology, 134*(6), 857–862.

Scott, I. U., Feuer, W. J., & Jacko, J. A. (2002b). Impact of visual function on computer task accuracy and reaction time in a cohort of patients with age-related macular degeneration. *American Journal of Ophthalmology, 133*(3), 350–357.

Scott, I. U., Jacko, J. A., Sainfort, F., Leonard, V. K., Kongnakorn, T., & Moloney, K. P. (2006). The impact of auditory and haptic feedback on computer task performance in patients with age-related macular degeneration and control subjects with no known ocular disease. *Retina, 26*(7), 803–810.

Scott, I. U., Schein, O. D., Feuer, W. J., Folstein, M. F., & Bandeen-Roche, K. (2001). Emotional distress in patients with retinal disease. *American Journal of Ophthalmology, 131*(5), 584–589.

Scott, I. U., Schein, O. D., West, S., Bandeen-Roche, K., Enger, C., & Folstein, M. F. (1994). Functional status and quality of life measurement among ophthalmic patients. *Archives of Ophthalmology, 112*(3), 329–335.

Scott, I. U., Smiddy, W. E., Feuer, W., & Merikansky, A. (1998). Vitreoretinal surgery outcomes: Results of a patient satisfaction/functional status survey. *Ophthalmology, 105*(5), 795–803.

Scott, I. U., Smiddy, W. E., Schiffman, J., Feuer, W. J., & Pappas, C. J. (1999). Quality of life of low-vision patients and the impact of low-vision services. *American Journal of Ophthalmology, 128*(1), 54–62.

Silva, L., & Pereira Bellon, O. R. (2002). A novel application to aid low vision computer users. *Lecture Notes in Computer Science, 2398,* 455–462.

Sloan, L. L. (1968). Reading aids for the partially sighted: Factors which determine success or failure. *Archives of Ophthalmology, 80*(1), 35–38.

Steinberg, E. P., Tielsch, J. M., Schein, O. D., Javitt, J. C., Sharkey, P., Cassard, S. D., et al. (1994). National study of cataract surgery outcomes: Variation in 4-month postoperative outcomes as reflected in multiple outcome measures. *Ophthalmology, 101*(6), 1131–1140.

Stephanidis, C. (2001a). The concept of unified user interfaces. In C. Stephanidis (Ed.), *User interfaces for all: Concepts, methods, and tools* (pp. 371–388). Mahwah, NJ: Lawrence Erlbaum Associates.

Stephanidis, C. (Ed.). (2001b). *User interfaces for all: Concepts, methods, and tools.* Mahwah, NJ: Lawrence Erlbaum Associates.

Stephanidis, C., Karagiannidis, C., & Koumpis, A. (1997). *Decision making in intelligent user interfaces*. Paper presented at the 2nd international conference on intelligent user interfaces (IUI '97), Orlando, FL.

Stephanidis, C., Salvendy, G., Akoumianakis, D., Arnold, A., Bevan, N., Dardailler, D., et al. (1999). Toward an information society for all: HCI challenges and R&D recommendations. *International Journal of Human-Computer Interaction, 11*(1), 1–28.

Stephanidis, C., Salvendy, G., Akoumianakis, D., Bevan, N., Brewer, J., Emiliani, P. L., et al. (1998). Toward an information society for all: An international research and development agenda. *International Journal of Human-Computer Interaction, 10*(2), 107–134.

Thatcher, J. (1994). *Screen reader/2: access to OS/2 and the graphical user interface*. Paper presented at the first annual ACM conference on assistive technologies (ASSETS '94), Marina Del Rey, CA.

Turk, M., & Robertson, G. (2000). Perceptual user interfaces. *Communications of the ACM, 43*(3), 32–34.

Vanderheiden, G. C. (1998). Universal design and assistive technology in communication and information technologies: alternatives or complements? *Assistive Technology, 10*(1), 29–36.

Vitense, H. S., Jacko, J. A., & Emery, V. K. (2002). Foundations for improved interaction by individuals with visual impairments through multimodal feedback. *Universal Access in the Information Society (UAIS), 2*(1), 76–87.

Vitense, H. S., Jacko, J. A., & Emery, V. K. (2003). Multimodal feedback: An assessment of performance and mental workload. *Ergonomics, 46*(1–3), 58–87.

Waterworth, J. A., & Chignell, M. H. (1997). Multimedia interaction. In M. Helander, T. K. Landauer, & P. Prabhu (Eds.), *Handbook of human-computer interaction* (pp. 915–946). New York, NY: Elsevier Science.

Whittaker, S. G. (1998). *Choosing assistive devices when computer users have impaired vision*. Paper presented at the Center on Disabilities, Technology, and Persons with Disabilities Conference, Northridge, CA.

Yankelovich, N., Levow, G.-A., & Marx, M. (1995). *Designing SpeechActs: Issues in speech user interfaces*. Paper presented at the SIGCHI conference on human factors in computing systems, Denver, CO.

Yu, W., Ramloll, R., & Brewster, S. (2000). *Haptic graphs for blind computer users*. Paper presented at the First Workshop on Haptic Human-Computer Interaction, Glasgow, Scotland.

·7·

UNIVERSAL ACCESSIBILITY AND FUNCTIONALLY ILLITERATE POPULATIONS: IMPLICATIONS FOR HCI, DESIGN, AND TESTING

William M. Gribbons
Bentley College

INTRODUCTION

Over the past decade, the topic of universal accessibility has received close attention in the field of system design. While considerable progress has been made in addressing the needs of the physically disabled and aging, the cognitively disabled have largely been overlooked (Gribbons, 1992; Newell, Carmichael, Gregor, & Alm, 2003). With nearly 45 million adult Americans suffering from the debilitating effects of illiteracy, the Universal Accessibility movement has clearly not been universal. While many have called for action (Dickinson, Eisma, & Gregor, 2003; Shneiderman, 2000), an extensive review of the leading human factors and HCI journals revealed a small number of studies focused on this critical issue. The topic of the less literate and learning disabled is also glaringly omitted from the leading HCI and human factors textbooks.

The research that does exist in literacy and learning disability studies has been confined, for the most part, to the fields of educational psychology, instructional design, and health-care informatics. Here, a rich research tradition dates back many decades. However, most of these findings are limited to the design of the classroom experience, standalone computer-based training systems, and paper-based communication products. Research in noneducational settings for general interface and interaction design is severely limited. Further, most of this work focuses on children and not the adult population.

The ubiquitous nature of the Internet has forced some progress on accommodations for the cognitively disabled. The Web Content Accessibility Guidelines 1.0 (1999) released by the W3C urges designers to accommodate this population. Unfortunately, many emerging guidelines tend to be vague and nonspecific.

This chapter will address the challenge of accommodating functionally illiterate users and/or users with learning disabilities in system design. It will explore the nature of this population and the magnitude of the problem, the relationship between literacy and learning disability, characteristics of the disability, and recommendations for design practice and usability testing. Ultimately, it is hoped that this effort will encourage further research in this critical area and the integration of that work in our profession's educational programs, guidelines, and best practices.

Defining the Population

Part of the challenge of addressing this population is building consensus on the most effective definition of this group. Common labels include "functionally illiterate," "cognitively disabled," and "the learning disabled" (a subset of the cognitively disabled group). An underlying assumption in this work is that the three categories are not mutually exclusive. For the purposes of this chapter, discussion will be limited to functionally illiterate and learning-disabled populations. Functional illiteracy, a lack of document and quantitative literacy needed to function in modern society, was selected since it provides the most comprehensive picture of the total population. This perspective brings to the discussion a very detailed set of demographics and statistics. Unfortunately, this perspective is weak on underlying causes and effective accommodations. Fortunately, the learn-

ing-disabled perspective, while focused most often on the school-age population, provides an extensive research base that defines the characteristics of this population, the underlying source of the disability, and possible accommodations. Learning disability is an umbrella term used to describe a wide range of disorders in information processing and it is generally believed that learning disabilities are linked to a dysfunction in the central nervous system. The larger cognitively disabled category was not chosen since it includes a much broader class of disabilities including Down syndrome, autism, emotional disabilities, and Alzheimer's disease. Naturally, these more severe disabilities require more extreme accommodations.

The causes of functional illiteracy are complex and often inextricably intertwined; these causes include a lack of educational attainment, dyslexia, other learning disabilities, and social deprivation. Further complicating matters, many adults with low literacy skills were never diagnosed with a learning disability as children and entered adulthood suffering the effects of the disability without knowing the cause. Consequently, it is all but impossible to characterize whether the cause of adult illiteracy is a product of low educational attainment, an underlying disability, or whether an undiagnosed disability was the cause of an individual dropping out of the educational system. While it may never be possible to determine how many in the lower-literacy population have learning disabilities, it is a safe assumption that the contribution to the population is significant.

Population Size

The current state of functional illiteracy in the United States presents a shameful picture. According to the 1992 National Adult Literacy Survey, some 23%, or nearly 45 million of approximately 200 million adult Americans, function at the lowest level of literacy, "Level 1." Those at Level 1 literacy are, for the most part, able only to read a simple form or understand rudimentary information in a short news article; others are not capable of even this (Kirsch, Jungeblut, Jenkins, & Kolstad, 1993). Within the context of these figures, it is surprising that greater attention has not been directed to this population, since its numbers are larger than all other disability groups combined.

The other literacy levels, two through five, represent progressively higher levels of literacy skills. Not surprisingly, one contributor to functional illiteracy is low educational attainment. The National Adult Literacy Survey states that "adults with relatively few years of education were more likely to perform in the lower literacy levels than those who completed high school or received some postsecondary education (Kirsch, Jungeblut, Jenkins, & Kolstad, 1993). The National Center for Educational Statistics (2001) reported the following:

"Between 1972 and 1985, high school completion rates climbed by 2.6 percentage points (from 82.8 percent in 1972 to 85.4 percent in 1985); since 1985, the rate has shown no consistent trend and has fluctuated between 85 and 87 percent. This net increase of about 3 percentage points over 29 years represents slow progress toward improving the national high school completion rates" (High School Completion Rates, Executive Summary; Kirsch, Jungeblut, Jenkins, & Kolstad, 1993).

Given this slow progress, it is highly probable that educational attainment will continue to be a key contributor to lower literacy skills among adults for the foreseeable future.

The consensus of most literacy experts is that people at Levels 1 and 2, the lowest levels, lack the essential skills to function successfully in society. In 2003, this survey was readministered and the results showed no significant improvement in prose and document literacy, with some improvement in quantitative literacy. Cleary the data from these two studies show the problem is large and gains in literacy—like increases in high school completion rates—will come slowly.

Unlike the measure provided by the National Adult Literacy Survey, there is no large-scale study measuring the prevalence of learning disabilities in the general population. There is, however, a range of estimates from a variety of sources. The National Dyslexia Association (n.d.) estimates that 70–80% of the population of students receiving special education services has deficits in reading and that 15–20% of the general population has language-based learning disabilities (how common are language-based learning disabilities simply to document the scope of the problem). The Interagency Committee on Learning Disabilities believes that 5–10% of the population is affected by learning disabilities. The President's Committee on Employment for People with Disabilities found that 10–14% of the adults in the workforce have learning disabilities. Finally, the National Institutes of Health estimated that 15% of the population in the United States has some type of learning disability (National Institute for Literacy, n.d.). Despite the probable link between those with learning disabilities and the larger lower-literacy population, no attempt has been made to validate this connection.

Not surprisingly, functional illiteracy and illiteracy are not problems confined to the United States. There are an estimated 876 million illiterate adults in the world, which represents nearly a quarter of the world's population. In developed nations, it is estimated that there are approximately 100 million functionally illiterate people (Kickbusch, 2001). In India alone, nearly 45% of the adult population is illiterate according to the 2001 Indian Census (Huenerfauth, 2002). These statistics and others clearly convey the global scale of this problem.

Making the Case for Including Functional Illiteracy

The lack of research addressing literacy and learning disability in the HCI discipline was understandable 10 or 15 years ago when this population was highly unlikely to interact with technology outside of an educational setting. With the ubiquitous nature of information technology in modern society (for example, the Internet, computers in all aspects of work, public kiosks, ATMs, electronic voting machines, e-health, and consumer electronics) this research gap is now indefensible. Accessible technology is no longer a luxury or an option; rather it is a prerequisite to gaining access to information that affects one's quality of life, participating in government, and contributing to the economy. A report from the U.S. Department of Commerce (2002) stated that "[b]etween December 1998 and September 2001, Internet use by individuals in the lowest-income households (lower socio-economic groups represent a disproportionately high percentage of the lower literacy pop-

ulation) increased at a 25 percent annual growth rate" (p. 56). The study also reported that "as of September 2001, about 65 million of the 115 million adults who were employed and 25 or over use a computer at work" (p. 57). While the percentage of workers using computers was significantly higher for "professional" positions, 20% of those working as operators, fabricators, laborers, farmers, and fishermen use a computer as part of their work (p. 58). Lastly, shifts in the economy from "hands-on work" to "information-and-technology work" will seriously disadvantage lower-literacy groups unless there is a corresponding increase in functional literacy skills.

If the size of this population alone were not enough to mobilize action, the cognitively disabled are a protected group under the Americans with Disabilities Act. Under the act, employers are required to provide workplace accommodations for employees who disclose their disability. That last stipulation of the act could very well be one of the reasons this population has been marginalized. Low literacy carries a stigma for most people and creates a reluctance to disclose—no disclosure, no requirement for accommodation.

Functional illiteracy and healthcare. Perhaps the most poignant case of a digital divide between the literate haves and the functionally illiterate have-nots is in the healthcare community. After a long delay, this community has made tremendous progress in implementing information technologies to enhance the quality of patient care, particularly in disseminating health-related information, managing patient care, accessing patient information, and promoting health services. As a result, the trend for seeking health-related information online is increasing dramatically. An estimated 70 million Americans have sought health information online (Cain, Mittman, Sarasohn-Kahn, & Wayne, 2000) from the nearly 10,000 or more health-related websites (Benton Foundation, 1999). Birru et al. (2004) estimated that between 40% and 54% of patients use the Internet to search for information on ailments and treatments. This development would be exciting, given the documented contributions of information to improved health, if it were not for the fact that equal access to this information does not exist for all Americans. Similar to the progression that played out in the larger development community, initial design efforts in the healthcare sector have focused first on the "typical" fully functioning user and then on the aging and physically disabled. Most healthcare materials—web-based or paper—are written well above the literacy level of the average American, approximately a 10th-grade level or greater (Doak, Doak, & Root, 1996; Birru et al., 2004; Weiner et al., 2004).

The healthcare sector has both a moral and economic responsibility to address the needs of lower literacy populations. Morally, it is only right that all citizens benefit equally from the value information technology brings to the quality of healthcare services. Cashen, Dykes, and Gerber (2004) provided support for this position by reporting that literacy is a better predictor of health status than age, income, employment status, education, or race. There is clearly an economic motivation as well. A conservative estimate places excess healthcare costs tied to low literacy at $73 billion a year (Rudd, Moeykens, & Colton, 1999). It is not surprising, then, that low-literacy adults are twice as likely to be hospitalized as their functionally literate counterparts (Birru et al., 2004).

In recent years the definition of the digital divide has shifted from economics to functional literacy. The availability of low-cost equipment and easy access to the Internet in public facilities has lessened the economic barriers. According to Shneiderman (2000), poor interface and interaction design remain as one of the defining variables between the technology haves and have-nots. And while we might hold out hope that we may one day eradicate illiteracy at the source, the fact that learning disabilities are lifelong and that fewer than one in eight low-literacy workers receives literacy training in the workplace (Sum, 1999, p. 156) strongly suggests that the burden falls on better design and appropriate accommodations.

CHARACTERISTICS OF THE POPULATION

Functional illiteracy is often referred to as the "invisible" disability. In many cases, the individual is unaware of the disability or the individual "hides" the disability. Findings from The National Literacy Survey (1993) suggest that as many as "66 to 75 percent of the adults in the lowest [literacy] level and 93 to 97 percent in the second lowest level described themselves as being able to read or write English 'well' or 'very well'" (The Literacy Skills of American Adults). In another study by Moon, Cheng, Patel, and Scheidt (1998), 70% of the participants reported they read "really well," while in actuality their reading scores reflected a 7th- to 8th-grade ability. Also contributing to the "invisibility" of this disability is the stigma attached to functional illiteracy and learning disabilities that increases an individual's reluctance to disclose the problem.

The learning disability label encompasses a wide range of information processing disorders, such as dyslexia (language), dyscalculia (mathematics), and dysgraphia (handwriting). They are thought to be neurobiological in origin and present themselves in various combinations and levels of severity. In addition, an individual with a deficit in one area may have strengths in other areas. Understanding the heterogeneous nature of the learning-disabled population is critical to formulating appropriate accommodations. Learning disabilities generally persist over a lifetime, but the manner in which the disability presents itself will change with life stages and accommodation strategies adopted by an individual (Gerber, 1998). The lifelong persistence of the disability, combined with the aforementioned lack of adult-literacy training, suggests this problem will not go away.

Overall, learning disabilities affect an individual's ability to develop and use reading, writing, reasoning, and mathematical skills (Karande, Sawant, Kulkarni, Galvankar, & Sholapurwala, 2005). In the absence of these fully developed skills, the individual's ability to access, process, and retain (learn) information is severely constrained. In addition, the many challenges and failures experienced by the learning disabled over a lifetime often result in a poor concept of self-worth and low self-esteem (Gerber, 1998).

One point at which some in the learning-disabled population break with the larger functionally illiterate community is at the level of underlying intelligence. Many individuals with one or more learning disabilities have normal or above normal intelligence (Rowland, 2004; Doak et al., 1996; Gerber, 1998). Conse-

quently, once appropriate accommodations are provided to mitigate the effects of the disability, adequate intelligence exists to accomplish most cognitive tasks.

Functional illiteracy and aging. One remaining characteristic of this population is the disproportionately high rates of lower literacy levels in the aging population. Kirsch et al. (1993) reported, "Older adults were more likely than middle-aged and younger adults to demonstrate limited literacy skills. For example, adults over the age of 65 have average literacy scores that range from 56 to 61 points (or more than one level) below those of adults 40–54 years of age" (p. 5) (The Literacy Skills of America's Adults). In the most recent national literacy survey, 14% of the adult population is below basic prose literacy skills, with 26% of this group 65 years of age or older (NAAL, 2003). The cause for the disproportionate representation of older adults is likely a combination of a number of factors including higher school dropout rates for the older population, lack of special education services when they were in school, and the general decline in cognitive abilities associated with aging. Regardless of the cause, design accommodations that benefit lower-literacy populations are likely to have the added benefit of supporting the aging population as well.

The next section will move from this general overview to a detailed discussion of the functional characteristics of the learning disabled. A sizable portion of the research that fuels this discussion is from the study of dyslexia. The rationale for this focus is twofold: (a) dyslexia is the most common of the learning disabilities, affecting nearly 80% of the learning-disabled population (Karande et al., 2005); and (b) deficits associated with dyslexia are most likely to impact information processing as it relates to HCI.

Functional Characteristics of the Learning Disabled

Because of the tremendous diversity in the learning disability community and the high degree of overlap among discrete disabilities, it is easy to become overwhelmed by the complexity of the area. A more manageable strategy is to identify the functional characteristics shared across disabilities, with close attention to those that affect a particular interaction environment (Brown & Lawton, 2001; Bohman & Anderson, 2005). Figure 7.1 summarizes a number of proposed groupings.

After carefully examining these characteristics, I simplified this list into four dynamically interconnected categories:

1. Reading
2. Memory
3. Metacognitive
4. Search and navigation

These categories were selected because of their likely effect on the computer-interaction environment. In addition, many of the characteristics in Fig. 7.1 are directly linked to deficits in a common underlying process, such as metacognition or working memory. These four categories, as highlighted in the discussion that follows, are not mutually exclusive. For example,

Source	Deficiency
Rowland (2005)	• Perception and processing • Problem solving • Low resilience to frustration—not persistent • Memory • Attention
Jiwnani (2001)	• Difficulty with sequential operations • Difficulty with complex, cluttered displays and controls • Difficulty choosing from large sets • Timed responses— particularly when they involve text
Bohman and Anderson (2005)	• Memory • Problem solving • Attention • Reading, linguistic, and verbal comprehension • Math comprehension • Visual comprehension
Cromley (2005)	• Poor decoding • Limited background knowledge • Low vocabulary • Dysfunctional beliefs about reading • Low-strategy use • Motivational Barriers • Working memory issues
Kolatch (2000)	• Trouble with abstract reasoning and organization • Short- and long-term memory loss • Word finding and syntax development difficulties • Diminished capacity to organize, assimilate, and retain information (web section) (Introduction)
Web Aim	• Difficulty interpreting what is heard and read • Difficulty connecting information
The International Dyslexia Association	• Difficulties with accurate and/or fluent word recognition • Poor spelling and decoding abilities • Deficit in the phonological component of language • Problems in reading comprehension and reduced reading experience that can impede growth of vocabulary and background knowledge
The TRACE Center (2003)	• Memory difficulty recognizing and retrieving information • Perception difficulty taking in, attending to, and discriminating sensory information • Difficulties in problem solving • Difficulty evaluating outcomes • Difficulty generalizing previously learned information

FIGURE 7.1. Functional characteristics of the learning disabled.

skilled reading draws on memory and requires the deployment of metacognitive strategies. Similarly, search and navigation place a load on working memory, draw on long-term memory, and require metacognitive monitoring. The following section will review each of these categories more closely, and the final section will offer recommendations for accommodating each of these categories in system design.

Reading

Given the prominent role of language in everyday communication as well as the more demanding software and web environments, reading is one of the first barriers encountered by the functionally illiterate. The alarming rate of poor prose and document-literacy skill was noted earlier. Also noted was the fact that poor reading is often tied to an underlying learning disability, inadequate schooling, or lack of exposure to reading. As in so many other cases in this area, an individual can suffer from all three conditions, two of them, or just one.

In dyslexic children, the learning disability affects reading at the most basic levels of phonological processing (Snowling, Deftry and Goulandris, 1996). Deficiencies at this basic level lead to further difficulties acquiring the complex skill of reading, building a rich vocabulary, and decreasing word-retrieval times. Most significantly, disabled readers allocate a disproportionately high amount of their attention to decoding letters and words, a process that is quickly automated by the nondisabled reader (van Gelderen et al., 2004). For the skilled reader, the process of automation frees the attention resource to focus on comprehension, the most critical component of the reading act. Skilled reading requires the management of a complex series of parallel actions, such as recognizing words, connecting those words to what came before, using those words to anticipate what will come next, and relating this combined experience to what one already knows. For the learning disabled, focusing most of the available attention on recognizing letters and decoding words severely compromises comprehension. Birru et al. (2004) reported that even when poor readers were able to read a passage, their inability to express answers related to the passage in their own words suggested minimal comprehension. Naturally, because reading is such a frustrating experience for the learning disabled, they are also less likely to read on a frequent basis, lowering the likelihood they will increase their reading skill through practice (Stanovich & West, 1989).

Given the difficulty experienced by this population at the most basic level of decoding, it becomes clear that the problem is greatly exacerbated by materials written at a grade level beyond their ability. When information is presented at a level beyond the capability of the reader, mental workload is increased significantly and comprehension is greatly diminished. From an accessibility perspective, grade-level readability is a persistent problem throughout the web development community. A recent study by West (2003) found that:

"89% of government websites are not easily accessible to the citizenry because the site read at higher than an eighth grade level of literacy. Fully two-thirds of all sites have language consistent with a 12th grade reading level, which is higher than the average American [half of all Americans read no higher than the eighth grade level]." (p. 3)

In another review conducted by Croft and Peterson (2002) the investigators found the mean readability score of 145 asthma-related websites was above the 10th grade, with 27 of the websites at the maximum 12th-grade level. Finally, Graber, D'Alessandro, and Johnson-West (2002) conducted a study of the readability level of privacy policies on Internet health websites. The investigators found that the average readability of the statement was at a level equal to that of a second-year college student, far outside the reach of the average American.

Despite the presence of a reading disability, most of this population possesses average or above average intelligence. As a result, if materials are presented at the appropriate grade level, the learning-disabled reader is capable of understanding. In separate studies conducted by Nielsen (2005) and Birru et al. (2004), poor readers performed well when interacting with easier-to-read materials.

While there remains considerable disagreement on how best to measure grade-level reading, certain measures and tests have gained wide acceptance. One such test, the Flesch Kincaid Grade Level Readability test, measures readability by dividing the sentence length (number words divided by the number of sentences) by the average number of syllables per word (number of syllables divided by the number of words). The resulting calculation is then correlated with an appropriate grade level. While this is a far from perfect measure, it does provide a useful benchmark of readability when evaluating the suitability of text for a given population. Unfortunately, this test along with others, such as the SMOG Readability Formula and the FRY Graph Reading Index, apply only to prose and not to lists, labels, and headings. With this type of text, the use of familiar words improves accessibility. Designers can consult reference works, such as the *Word Frequency Book,* (Carroll, 1972), to identify the most commonly used words in the English language. A more complex form of language support can be found on the CAST website (http://www.cast.org/) where the sponsors provide a language tool to support low-literacy users visiting the CAST site.

Because the functionally illiterate commonly experience difficulty with reading, illustrations, audio output, and video are often offered as effective accommodations. While illustrations can assist the less-able reader (Doak et al., 1996; Weiner et al., 2004), positive effects vary with the reader and nature of the information. According to Beveridge and Griffiths (1987), illustrations contributed positively to reading performance in easier-to-read passages. In contrast, they also noted that illustrations degraded reading performance in the most difficult passages. In a study related to the supplemental use of animation, Larsen (1995) reported learning-disabled children had trouble due to the distraction of the animation. Similarly, Jiwnani (2001) reported system designers must be careful using auditory output since it can confuse the learning disabled. The pace of the output must be slow, free of background noise, and repeatable under the control of the user.

Clearly, interface and interaction designers must choose word, sentence length, and sentence structure with due consideration for the readability level of the audience. Readability can be seen as the first barrier in software or web design encountered by the functionally illiterate population. If this population is unable to process the underlying language used in the display, all other accommodations are meaningless.

Memory

Long-term memory. The long-term and working memory systems each have implications for the functionally illiterate and learning-disabled population. As noted in the previous section, this population is less likely to read as often and, consequently, fails to benefit from a major source of acquiring domain knowledge. Without well-established domain models, interacting with web-based information or a software system is considerably more demanding. Cognitive science has long recognized that long-term memories supplement and extend the limited capacity of working memory. This topic will be explored further in the navigation-and-search section that follows.

Working memory. Cognitive science has also recognized the limited capacity of working memory and the major bottleneck it represents in information processing and human–computer interaction. While learning disabilities are very diverse in their causes and in how they manifest themselves, one consistent variable in all learning disabilities is working memory capacity. The dynamic interaction between memory capacity and the time information can be actively maintained also imposes severe constraints on the learning disabled. Debate has raged for years whether the learning disabled suffer from diminished working memory capacity or whether the complex dynamics of the underlying cognitive deficiencies place an excessive burden on a "normal" capacity (Daneman & Carpenter, 1980; Swanson, 1993; Vellutino et al., 1996; Ransby & Swanson, 2003). From the perspective of design accommodations, the precise source of the problem is insignificant.

Clearly, the attention devoted to low-level decoding places a heavy demand on working memory capacity. In addition, the increased likelihood of weaker mental models minimizes the opportunity to "offload" processing burden from the working memory. Further, the load imposed by metacognition, search, and navigation—as discussed in the sections that follow—places additional demands on this precious resource. Finally, the high level of anxiety and frustration experienced by this population also operates in working memory and will negatively affect performance on an information-processing task (Lee, 1999). Consequently, this issue, more than any other, guides many of the design and support recommendations in the final section. For example, limited working memory resource accounts for the problems the learning disabled experience with sequenced operations in software or in retracing their paths of travel navigating a website. Interface and interaction designs must compensate for the limitations of working memory or suffer the consequences of lower performance, increased errors, or abandonment of the task.

While the underlying learning disability can make the acquisition of expertise more difficult, it would be wrong to think this population does not bring some level of learning and a variety of conceptual models to an interaction experience. As with so many other variables in accessibility design, we must recognize that the goals and models of the lower-literacy population may not align with those of the general population. Dickinson, Eisma, and Gregor (2003) reported that "users made remarks which indicated that they were not trying to understand the system or find generic rules that could help them to use it better" (p. 63). In other words, they were not attempting to understand the model underlying the system; they were simply trying to get something accomplished. System design should focus on that goal and avoid excessive functionality and extraneous information. From a development perspective, it is always easy to imagine occasions when one user or another might need this feature or that piece of information. Unfortunately, the lower-literacy population comes to the system with a greatly limited need, for both functions and information. Accessible design must consider these needs. Interface clutter increases mental workload on users as they attempt to locate information and features that align with their needs, a difficult challenge given their attentional impairments.

Metacognitive

Another critical deficiency in most dyslexics is in metacognition, the process of thinking about thinking. Metacognition can take many forms, including strategic planning, monitoring, self-appraisal, document-processing strategies, and reading strategies. Metacognition operates actively in working memory, parallel to the main information-processing task. Consequently, this activity places yet another demand on the memory resource. Also, because these strategies are learned behaviors, it suggests an interaction with long-term memory. Research has shown that the learning disabled are generally deficient in the metacognitive area. While nondisabled learners acquire most of these strategies without formal instruction, studies have shown these strategies can effectively be taught to the learning disabled (Collins, Dickson, Simmons, & Kameenui, 1998). Again, we must be reminded that skilled readers become skilled through years of practice with different reading materials that were processed to support diverse purposes. Because proficient readers and learners use these strategies on a frequent basis, they migrate, over time, to the level of an automatic skill. In contrast, the absence of persistent exposure to reading places the learning disabled at a further disadvantage.

Cromley (2005) indicated that the learning disabled are less likely to know when they do not understand and are less likely to reread, synthesize, generate questions, or make predictions. They are also more likely to "satisfice" and accept partial or incorrect information rather than persevering to gain a more complete understanding. A distinguishing attribute of good readers is their ability to monitor the information-processing activity; in other words, they ask themselves: "Do I understand what I am reading? Do I see the relevance of the information to the task? What do I expect to follow in the next passage? How does this relate to what I already know?" Corley and Taymans (2002) organized this activity in three parts: setting goals before the task and establishing a plan; monitoring comprehension and understanding while engaging in the task; and evaluating one's learning after the task and making adaptations when faced with similar tasks. Danielson (2002), referencing users navigating the web, suggested that once goals are made, strategies are selected for achieving that goal. Further, strategies are constantly assessed and adjusted many times in a successful interaction experience. Given the mental demand of these activities, monitoring and adjusting activities pose a challenge for the disabled. Finally, Britt and Gabrys (2002) indicated that document literacy requires four metacognitive skills: sourcing (evaluating the credibility of source), corroboration (seeking independent confirmation), integration (creating mental representation of information), and search (a skill woven throughout the other three areas). The inability of the learning disabled to engage in self-evaluating activ-

ities is evident in the work of Birru et al. (2004). Although most of the participants in their study failed to answer the questions correctly, seven out of eight reported feeling "very comfortable" or "comfortable" with their Internet search experience. On the sourcing and corroboration skills, five out of eight used information provided on sponsored sites, yet still reported it was "very easy" to find trustworthy information on the Internet.

In short, metacognition requires a tremendous amount of cognitive activity, activity that occurs in parallel with the primary decoding task. All of this activity occurs in the executive control component of working memory. As noted in the preceding section, this is a limited resource for the learning-disabled reader who devotes much of this resource to low-level decoding.

We take for granted that nondisabled users engage in all of these activities effectively and effortlessly. Design support for this variable amounts to providing constant feedback to the disabled user, explicitly connecting the information to the task, establishing checkpoints where the system asks questions of the reader to monitor understanding, and engineering performance support in the form of cognitive scaffolding. On the matter of feedback, for example, Dickinson et al. (2003) noted that longer tasks are tolerated by the learning disabled so long as they receive frequent feedback on what is happening and confirmation they are proceeding correctly. In the absence of such feedback, the poor reader becomes anxious, a state that further degrades performance since anxiety occupies the working memory space.

Navigation and Search

The learning disabled have traditionally experienced tremendous difficulty reading paper-based materials. They experience this difficulty despite the fact the linear format found in most paper-based materials offers considerable support for the reader. In these materials, the author assumes the burden for communicating organization through meaningful structures, logical sequences, helpful transitions, and explicit connections. In contrast, the nonlinear web requires the reader to infer organization and build a coherent model of the subject matter as the information unfolds in less predictable ways (Britt & Gabrys, 2002). McDonald and Stevenson (1998) identified disorientation in the nonlinear environment as one of the most challenging elements of the web. While disorientation is an obstacle that can be overcome by able users, this barrier is often insurmountable for the learning disabled. This population, at best, experiences difficulty building a mental model of the information, monitoring where they are in the experience, or retracing their path. Each of these activities draws on their overtaxed working memory.

As one examines the factors affecting readers' ability to navigate, a number of variables emerge:

- Basic navigation skills related to the interface
- Topology formats: nonlinear, hierarchical, and mixed topologies
- Depth versus breadth of the structure
- Navigational aids

Interaction skills. While there is limited research examining the effects of learning disabilities on navigation, the work of Zarcadoolas, Blanco, Boyer, and Pleasant (2002) suggested learning-disabled users can quickly learn and retain basic web-interaction skills. These skills include the behavior of links and the act of scrolling. Participants in this study did require training and prompting to use graphic links. Overall, the positive reaction to the linking convention is encouraging since Zarcadoolas et al. also found the participants, when given a choice, preferred following links rather than using search. Worth noting, Birru et al. (2004) found that users in their study, although comfortable with the linking convention, would seldom click on more than one or two links to answer questions. Finally, although basic scrolling behavior was understood, none of the participants in the Zarcadoolas et al. study scrolled to view additional information. Without additional corroborating evidence it is difficult to generalize from these limited studies. However, the behaviors exhibited in these studies warrant close scrutiny as we continue to study and observe this class of users.

In contrast to the relative ease with which this population learned basic interaction behaviors, a more significant challenge is faced navigating the larger system. Danielson (2002) suggested users begin the navigation task with a decision about whether they are looking for a specific item, a group of items, or general information about the contents of a domain area. As noted in the metacognitive area, the learning disabled are generally weak at planning tasks, which immediately places them at a disadvantage in the navigation area. Unfortunately, there is little or no research that directly examines the learning-disabled population and the structural issues of nonlinear versus hierarchical, or depth versus breadth. There is, however, considerable speculation on these issues since performance differences for general-population users are typically discussed within the context of working memory capacity.

Topology. The work of McDonald and Stevenson (1998) shed some light on the issue of the efficacy of hierarchical, nonlinear, and mixed (hierarchical with referential links) topologies. Most significant perhaps for the purposes of this discussion, they found novices benefited most from the mixed structure since it offered a balance of freedom and control. The referential links supported exploration of a site without the support of a well-formed mental model, while the hierarchical framework served to constrain movements and minimize disorientation. While the strict hierarchical structure provided the greatest control and guidance for participants, it proved inefficient when participants wanted to make distal movements in the structure. Finally, the nonlinear structure simply provided too many options and placed the heaviest demands on expertise and working memory. In an earlier study, McDonald and Stevenson (1996) found users stopped reading too soon when faced with too many decisions related to "what" and "how much" to read. In each of these situations, it is easy to speculate that the problems faced by the participants in these studies would only be exacerbated for the learning disabled, given their underlying deficiencies in working memory and metacognition.

Depth vs. breadth. The efficacy of breadth versus depth in the hierarchy is also open to speculation. As with topology, the depth issue is framed by limitations of working memory and the users' ability to distribute the navigation workload to their own conceptual models of the domain. In this discussion, we see greater diversity of opinion. In the world of web design, best practice favors breadth over depth, resulting in a greater number of choices at the highest level. A study by Larson and Czerwinski (1998) suggested that a medium depth produced the best search performance over either the broader or the deeper options. Benard (n.d.) suggested depth versus breadth is not the best framing mechanism. Instead, he suggests the shape of the hierarchy is the best predictor of search performance, finding a broad first level, a narrow middle level, and a broad base to be most efficient. While not based on a controlled study, Kolatch (2000) proposed a simpler top-level interface, offering fewer choices and a deeper structure for cognitively disabled users. Kolatch was careful to point out that this contradicts most research on the topic (web section Analyses). Until new research suggests otherwise, the balanced structure proposed by Larson and Czerwinski (1998) provides guidance consistent with our understanding of the adverse effects of too many choices and the mental load imposed by deep structures.

Navigation aids. Danielson (2002) suggested that a variety of visual navigational aids can mitigate the aforementioned problems encountered in navigation. Navigational aids fall into two categories: index and table of contents lists, and site maps. Danielson noted that graphical site maps are superior to lists and table of contents at conveying the relationships between distal nodes on the site. The study also showed that the site maps benefited nonknowledgeable users more than the knowledgeable. Finally, when compared to a control group without site maps, Danielson reported that subjects with maps were less likely to abandon the task, reported information-seeking confidence, moved deeper into the site, and made more movements outside the hierarchy. Given the previously defined tendencies of the learning-disabled population, it's highly probable that they would benefit from the assistance offered by site maps. Supporting this theory, Mirchandani (2003) found detailed site maps helped the disabled user find information within a click or two. See the ADA Insights website (http://www.adainsights.org) for an example of a site map designed to support disabled users. As a visualization, the benefit of the site map is its ability to shift the cognitive load from working memory to the visual and thereby transform the load-intensive cognitive task to a less labor-intensive perceptual task. A more extensive navigational support tool can be found on the CAST website (http://www.cast .org/) The tool displays a taxonomical view of the site, the user's current location, the user's navigational path, quick links to frequently visited areas, and a notepad that allows the creation and storage of notes. Combined, this support helps compensate for deficiencies in monitoring skills.

Search. In contrast to navigation, research is slowly emerging related to the learning disabled and the search task. Not surprisingly, this research reports the learning disabled experience tremendous difficulty with search tasks. At the most basic level, the poor spelling displayed by many in the lower-literacy population creates a major barrier to successful searches. Search engines, such as Google, that offer alternative spellings are best. Weak self-monitoring skills also decrease the likelihood that the user will recognize that a mistake was made. Even when search terms were spelled correctly, Birru et al. (2004) found that subjects rarely retyped search terms to locate more relevant items—a serious problem since they also found the low-literacy adults in this study rarely used optimal search terms as they attempted to retrieve health-related information. Birru et al. (2004) concluded by highlighting the benefits of a categorizing search engine as one means of minimizing the adverse effects of poor search and long lists of results. This engine would sort results by category and minimize the need to evaluate long, unsorted lists. This is a good example of a technology assuming the burden for a weak metacognitive skill in the disabled population.

DESIGN ACCOMMODATIONS

There are three approaches to accommodating the needs of the learning-disabled population. Each of these recommendations is based on the previously discussed characteristics of this population.

1. Assistive technologies: Building technologies customized to compensate for the disability;
2. Layered design: Designating special sections of the website or software for a disabled population;
3. Universal design: Adopting design practices that enable the learning-disabled population while benefiting the larger population as well.

The following sections will briefly review the first two categories and provide more detailed recommendations for the third category, universal design. This emphasis is based on three considerations. First, the initial two categories require greater effort and increase development costs. Second, they benefit a smaller portion of the population, typically the most severely disabled. Third, universal design principles are less likely to increase development time or costs and will likely benefit a larger number of the disabled and fully functioning populations.

Assistive technologies. A number of studies have demonstrated the benefits of assistive technologies for the disabled (Brown, 1992; Cole & Dehdashti, 1998; Newell & Gregor, 2000; McGrenere et al., 2003). A typical application in this genre is a stand-alone computer-based educational system, designed to assist severely disabled children. Other, less intrusive accommodations, including software agents and embedded scaffolding, have been explored by Shaw, Johnson, and Ganeshan (1999); Quintana, Krajcit, and Soloway (2002); and Shneiderman (2003).

Scaffolding and agents offer performance support to the disabled user in the early stages of interaction with a new product. Shneiderman (2000) described this approach as "evolutionary learning." As the user becomes more proficient with the requirements of the system and task, the scaffolding is slowly torn down, either under the control of the user or through intelligent software agents. Most typically, the agent replaces or supports a deficient metacognitive process by assisting with planning, monitoring, self-appraisal, strategy setting, and feedback. Quite simply, this burden is shifted from the users' working memory to the system. Shneiderman (2000) and Quintana, Krajcik, and

Soloway (2002) highlighted the efficacy of this approach for supporting less-knowledgeable and disabled populations.

Layered design. The layered design strategy requires the design of two versions of the system: one for the functionally literate users and one for the functionally illiterate. While it is generally preferred not to separate disabled and fully functioning populations, layered design is simply an extension of a practice that has been used for years for separating domain experts from nonexperts or native speakers from non-native speakers. In this case, one area of the site would be written at the population average 8th-grade level, while another section of the site would be written at the 5th-grade level or lower. Lobach, Arbanas, Mishra, Campbell, and Wildemuth, (2004) described the benefits of a two-tier literacy system in their study. Sites such as this may include other assistance, such as agents, site maps, illustrations, and the like.

Universal design. In contrast to the first two approaches, universal design accommodations are seamlessly embedded in the system interface and interaction design. This class of accommodations is less likely to affect development costs or create an obstacle for the nondisabled. Consequently, these accommodations are more likely to be embraced by the wider development and user communities. Further, it is widely accepted in the accessibility community that these accommodations improve the usability of the product for all users (Shneiderman, 2000; Dickinson, Eisma, & Gregor, 2003). The following recommendations are organized in the previously defined categories, although support for a given recommendation is often found in multiple categories. Finally, recommendations designated with an asterisk identify recommendations that require additional research and warrant close monitoring. All others enjoy wide support in the literature.

Reading

When designing the language component of any system, support the following:

- Maintain an appropriate reading level (use readability formulas to establish a benchmark; measure word difficulty and sentence length);
- Use active voice;
- Use the *Word Frequency Book* to identify common words;
- Place information and instruction in context;
- Employ lists;
- Chunk information;
- Present content in sequence;
- Repeat information from screen to screen (do not assume the user will carry over);
- Maintain consistency in language and procedures;
- Communicate directly and concretely;
- Emphasize actions users must complete;
- Highlight critical information, information structure, or new information;
- Use familiar terms, and avoid acronyms and jargon;

- Use visual and auditory prompts;
- Provide definitions of critical terms through direct linking to glossary;
- Use illustrations to complement text, communicate structure, and emphasize connections;
- Highlight (using circles, arrows, and the like) critical areas of an illustration and explicitly connect to the text;
- Avoid the gratuitous use of animations and other movement;*
- Use 12-point type;
- Use familiar typefaces (there is conflicting research regarding the efficacy of serif versus sans serif);*
- Avoid tight letter-spacing (makes low-level decoding more difficult);
- Pace auditory output slowly, and allow user control to repeat output;
- Avoid background noise with auditory output;
- Provide "looser" versus tighter line spacing;
- Support easy, user controlled, style changes;
- Make it "look easy" by lowering information density;
- Limit use of italics and uppercase (effects are much more severe than for the general population);
- Maintain higher contrast;*
- Avoid light text on a dark background;
- Use ragged right formats to preserve consistent word space.

Memory

In an effort to decrease mental workload, support the following:

- Maintain consistency in all aspects of the design;
- Leverage existing knowledge, behaviors, and tasks;
- Avoid splitting attention between two tasks;
- Focus on the user goals;
- Limit information and features to what is really needed;
- Limit chunking complexity for audio output (particularly IVR);
- Focus on behaviors and tasks rather than facts;
- Partition tasks in reasonably sized groups;
- Avoid the need to retain information over long tasks or across multiple screens;
- Support mental calculations, decisions, and comparisons;
- Complete mathematical calculations;
- Limit choices;
- Provide a list of options for entry fields;
- Complete information automatically in forms and fields whenever possible;
- Employ advanced organizers;
- Use mnemonics;
- Minimize screen clutter;
- Provide extra time for tasks;
- Eliminate the anxiety of timeouts.

Metacognitive

In an effort to strengthen or replace deficient metacognitive processes, support the following:

- Maintain a consistent design;
- Convey associations between new information or process and that which is known;
- Communicate goal or purpose of the site immediately;
- Communicate prerequisite knowledge for the task and provide convenient links;
- Communicate required sequences or organizational structures;
- Allow the user to interact with the information;
- Use checklists to support self-monitoring;
- Minimize choices that require fine discriminations or close monitoring;
- Design for immediate/early success (which lowers anxiety and builds confidence);
- Align with user goals, and provide reward or convey value proposition (motivation);
- Support cooperative work activities;
- Provide reminders;
- Provide source information for material presented;
- Minimize embedded links;
- Avoid taking user to other pages for ancillary information;
- Employ error prevention and recovery support;
- Query when choices or decisions are required;
- Review information entered or validate the successful completion of a process;
- Use auditory and visual cues to mark stages of a work cycle, helping the user self-monitor.

Navigation and Search

In an effort to improve the effectiveness of search and navigation, support the following:

- Make information and features supporting goals readily accessible to minimize navigation and search;
- Provide persistent presentation of path history;
- Offer persistent opportunity to exit, backup, or return to start;
- Use redundancy: repetition, signal design, and channel;
- Provide status indicators;
- Minimize scrolling;
- Label all links;
- Provide linked paths for probable scenarios;
- Maintain alerts onscreen until dismissed by user;
- Place information or process in context;
- Partition information into categories defined by clear rules;
- Use site maps, tables of contents, and indexes;
- Use a topology that is primarily hierarchical with referential links to areas aligned with goals;*

- Maintain a medium breadth and depth—not too broad, not too deep;*
- Use clear, thematic labels at each level;
- Provide productive terms for search;
- Offer suggested spellings;
- Use categorizing search engines;
- Provide performance support for evaluating search results (corroboration and sourcing).

Guidance for Usability Testing

To produce the best data and protect the well-being of participants, support the following:

- Avoid embarrassment or humiliation (anxiety lowers performance);
- Consider ethical obligations when screening for lower-literacy samples (full-disclosure issues, informed consent, and IRB);
- Minimize anxiety in testing situation;
- Test in context (actual work environment);
- Avoid using the term "testing" since this population equates the term with failure;
- Break tasks in the test script into small, yet logical, units;
- Shorten test times to minimize effects of fatigue on performance and to compensate for attention deficit;
- Use direct interaction protocol since users may not freely think aloud (inadequate working memory to support both think-aloud and to acclimate to the task);*
- Accommodate slower processing speeds when setting performance levels (e.g., task times);

- Expect participants to report exaggerated levels of performance or system quality in self-reporting as a means of covering poor literacy skills.

CONCLUSION

Ten years ago, our concern for the effects of lower literacy was confined exclusively to the processing of print media. With the explosion of information technology in the workplace and the role of the Internet as the repository for information of all types, literacy has become a major barrier for millions of Americans as they compete for work and attempt to improve the quality of their lives. While the problems encountered by lower-literacy populations with print media are significant, the problems become greatly exacerbated with the increased mental workload imposed by computer systems.

As advocated in this chapter, many of the barriers to accessibility for this population require simple modifications in the interface and interaction design. Accessible design is an art, in which the needs of one group are carefully balanced against those of another, and the designer recognizes that "accessible for most" is more achievable than "accessible for all." As one reviews the introductory list of design accommodations required by the lower-literacy population, HCI professionals will recognize most, if not all, of these variables as ones considered in each and every interface design. The difference—and a significant one—is that fully capable users have flexible learning, problem solving, and interaction skills that allow them to adjust to less-than-perfect design specifications. The learning disabled, in contrast, lack cognitive flexibility and suffer varying degrees of performance degradation in nonoptimum design environments. By increasing our understanding of this sizable population and embracing the design practices outlined here, we may finally ensure that universal accessibility is truly universal.

References

Benard, M. (n.d.). *Criteria for Optimal Web Design.* Retrieved January 13, 2006, from http://psychology.wichita.edu/optimalweb/structure.html

Benton Foundation. (1999). *Networking for Better Care: Health Care in the Information Age.* Retrieved December 28, 2003, from htpp://www.Benton.org/library/health/

Beveridge, M., & Griffiths, V. (1987). The effects of pictures on the reading processes of less able readers: A mis-cue analysis approach. *Journal of Research in Reading, 10,* 29–42.

Birru, M., Monaco, V., Charles, L., Drew, H., Njie, V., Bierria, T., Detlefsen, E., & Steinman, R. (2004). Internet Usage by low-literacy adults Seeking Health Information: An Observational Analysis. *Journal of Medical Internet Research, 6*(3), Retrieved December 12, 2005, from http://www.jmir.org/2004/3/e25/

Bohman, P., & Anderson, S. (2005). A Conceptual Framework for Accessibilty Tools to Benefit Users with Cognitive Disabilities. *Proceedings of the W4A at WWW2005,* ACM, 85–89.

Britt, M., & Gabrys, G. (2002). Implications of Document-Level Literacy Skills for Website Design. *Behavior Research Methods, Instruments, & Computers, 34*(2), 170–176.

Brown, C. (1992). Assistive Technology Computers and Persons with Disabilities. *Communications of the ACM, 35*(5), 36–45.

Brown, D., &, Lawton, J. (2001). Design guidelines and issues for website design and use by people with a learning disability. Centre for Educational Technology Interoperability Standards. Retrieved March 28, 2007 from http://www.cetis.ac.uk/members/accessibility/links/disabilities/cogdis.

Cain, M. M., Mittman, R., Sarasohn-Kahn, J., & Wayne, J. (2000). *Health E-People: The Online Consumer Experience.* Oakland, CA: Institute for the Future, California Health Care Foundation.

Carroll, J. (1972). *The American Heritage Word Frequency Book.* New York: Houghton Mifflin.

Cashen, M., Dykes, P., & Gerber, B. (2004). eHealth Technology and Internet Resources: Barriers for Vulnerable Populations. *The Journal of Cardiovascular Nursing, 19*(3), 209–214.

Cole, E., & Dehdashti, P. (1998). Computer-Based Cognitive Prosthetics: Assistive Technology for the Treatment of Cognitive Disabilities. *Proceedings of the Third International ACM Conference on Assistive Technologies,* ACM: Arrington, VA, 11–18.

Collins, V., Dickson, S., Simmons, D., & Kameenui, E. (1998). *Metacognition and its Relations to Reading Comprehension: A Synthesis of the Research*. National Center to Improve the Tools of Educators, University Washington.

Corley, M., & Taymans, J. (2002). *Adults with Learning Disabilities: A Review of the Literature*. Retrieved December 30, 2005, from http://www.ncsall.net/?id+575

Croft D. R., & Peterson M. W. (2002). An Evaluation of the Quality and Contents of Asthma Education on the World Wide Web. *Chest, 121*(4), 1301–1307.

Cromley, J. (2005). Metacognition. *Cognitive Strategy Instruction, and Reading in Adult Literacy*. Retrieved January 17, 2006, from http://www.ncsall.net/index.php?id=773

Daneman, M., & Carpenter, P. (1980). Individual Differences in Working Memory and Reading. *Journal of Verbal Learning and Verbal Behavior, 19*, 450–466.

Danielson, D. (2002). Web Navigation and the Behavioral Effects of Constantly Visible Site Maps. *Interacting With Computers, 14*(5), 601–618.

Dickinson, A., Eisma, R., & Gregor, P. (2003). Challenging Interfaces/ Redesigning Users. *CUU 2003*. British Columbia, Canada, 61–68.

Doak, C., Doak, L., & Root, J. (1996). *Teaching Patients with Low Literacy Skills*. Philadelphia: J.B. Lippincott Company.

Dropout Rates in the United States. National Center for Educational Statistics. Retrieved December 10, 2003, from http://nces.ed.gov/pubs 2002/dropoutpub_2001/11.

"Fact Sheet:" Workforce Literacy. National Institute for Literacy, Washington, DC, Retrieved January 4, 2006, from http://www.nifl.gov/nifl/facts/workforce.html

Gerber, P. (1998). Characteristics of Adults with Specific Learning Disabilities. Retrieved December 28, 2003, from http://www.idonline.org/ld_indepth/adult/characteristics.html

Graber, M. A., D'Alessandro, D. M., & Johnson-West (2002). Reading Level of Privacy Policies on Internet Websites. *The Journal of Family Practice, 51*(7), 642–645.

Gribbons, W. (1992). The Functionally Illiterate: Handicapped by Design. *Proceedings of the International Professional Communications Conference*. IEEE, 302–307.

Huenerfauth, M. (2002). Design Approaches for Developing User Interfaces Accessible to Illiterate Users. *American Association of Artificial Intelligence Conference (AAAI 2002), Intelligent and Situation-Aware Media and Presentations Workshop*. Edmonton, Alberta, Canada.

Jiwnani, K. (2001). *Designing for Users with Learning Disabilities*. Retrieved December 28, 2005, from http://www.otal.umd.edu/uupractice/cognition/

Karande, S., Sawant, S., Kulkarni, M., Galvankar, P., & Sholapurwala, R. (2005). Comparison of Cognitive Abilities Between Groups of Children with Specific learning Disability Having Average, Bright Normal and Superior Nonverbal Intelligence. *Indian Journal of Medical Sciences, 59*(3), 95–103.

Kickbusch, I. (2001). Health Literacy: Addressing the Health and Education Divide. *Health Promotion International, 16*(3), 289–297.

Kirsch, I. Jungeblut, A., Jenkins, L., & Kolstad, A. (1993). *Adult Illiteracy in America: a first look at the results of the National Literacy Survey*. Washington: Department of Education, Center for Educational Statistics. Washington, DC.

Kirsch, I, Jungeblut, A., Jenkins, L., & Kolstad, A. (1993). *Executive Summary of Adult Literacy in America: A first Look at the results of the National Literacy Survey*. Retrieved October 27, 2003, from http://www.nces.ed.govnaal/resources/execsumm.asp

Kolatch, E. (2000). *Designing for Users with Cognitive Disabilities*. Retrieved December 28, 2005, from http://www.otal.umd.edu/uuguide/erica/

Larsen, S. (1995). What is "Quality" in the Use of Technology for Children with Learning Disabilities? *Learning Disability Quarterly, 18*(2), 118–130.

Larson, K. & Czerwinski, M. (1998). Web Page Design: Implications of Memory, Structure and Scent for Information Retrieval. *Proceedings of CHI 98*, USA, ACM, 25–32.

Lee, J. (1999, Spring). Test Anxiety and Working Memory. *The Journal of Experimental Education, 67*, 218–240. Retrieved December 4, 2005, from ProQuest database.

Lobach, D., Arbanas, J., Mishra, D., Campbell, M., & Wildemuth, B. (2004). Adapting the Human-Computer Interface for Reading Literacy and Computer Skill to Facilitate Collection of Information Directly from Patients. *MEDINFO 2004*, Amsterdam, IOS Press, 1143–1146.

McDonald, S., & Stevenson, R. (1996). Disorientation in Hypertext: The Effects of Three Text Structures on Navigation Performance. *Applied Ergonomics, 27*, 61–68.

McDonald, S. & Stevenson, R. (1998). Effects of Text Structure and Prior Knowledge of the Learner on Navigation in Hypertext. *Human Factors, 40*(1), 18–27.

McGrenere, J., Davies, D., Findlater, L., Graf, P., Klawe, M., Moffatt, K., Purves, B., & Yang, S. (2003). Insights from the Aphasia Project: Designing Technology For and With People who Have Aphasia. *CUU 2003*. ACM: Arlington, VA, 112–118.

Mirchandani, N. (2003). *Web Accessibility for People with Cognitive Disabilities: Universal Design Principles at Work*. Retrieved December 28, 2005, from http://www.ncddr.org/du/researchexchange/v08n03/8_access.html

Moon, R., Cheng, T., Patel, K., & Scheidt, P. (1998). Parental Literacy Level and Understanding of Medical Information. *Pediatrics, 102*(2). Retrieved from Medline.

National Assessment of Adult Literacy (NAAL). (2003). *A First Look at the Literacy of Americas Adults in the 21st Century*. National Center for Educational Statistics, U.S. Department of Education, Institute of Educational Sciences, NCES 2006-470, Washington, DC.

National Dyslexia Association (n.d.). *What is Dyslexia?* Retrieved November 10, 2003, from http://www.interdys.org

Newell, A., Carmichael, A., Gregor, P., & Alm, N. (2003). Information Technology for Cognitive Support. In *The Human Computer Interaction Handbook*. Eds. Julie Jacko and Andrew Sears. Mahwah, NJ: Lawrence Erlbaum Associates, 464–481.

Newell, A. & Gregor, P. (2000). User Sensitive Inclusive Design: In Search of a New Paradigm. *CUU 2000*. Arlington, VA: ACM, 39–44.

Nielsen, J. (2005). *Lower-Literacy Users*. Retrieved December 9, 2005, from http://www.useit.com/alertbox/20050314.html

Quintana, C., Krajcik, J., & Soloway, E. (2002). A Case Study to Distill Structural Scaffolding Guidelines for Scaffolded Software Environments. *CHI 2002*. Minneapolis, Minnesota: ACM, 81–88.

Ransby, M., & Swanson, H. L. (2003). Reading Comprehension Skills of Young Adults with Childhood Diagnoses of Dyslexia. *Journal of Learning Disabilities, 36*(6), 538–555.

Rowland, C. (2004). *Cognitive Disabilities Part 2: Conceptualizing Design Considerations*. Retrieved December 28, 2005, from http://www.webaim.org/techniques/articles/conceptualize/?templatetype=3

Rudd, R. E., Moeykens, B. A., & Colton, T. C. (1999). *Health and Literacy: A Review of Medical and Public Health Literature*. New York: Josey-Bass.

Shaw, E., Johnson, W., & Ganeshan, R. (1999). Pedagogical Agents on the Web. *Autonomous Agents 1999*. Seattle, WA, 283–290.

Shneiderman, B. (2000). Universal Usability. *Communications of the ACM, 43*(5), 84–91.

Shneiderman, B. (2003). Promoting Universal Usability with Multi-Layer Interface Design. *CUU 2003*. Vancouver, British Columbia, Canada, 1–8.

Snowling, M., Defty, N., & Goulandris, N. (1996). A Longitudinal Study of Reading Development in Dyslexic Children. *Journal of Educational Psychology, 88*(4), 653–669.

Stanovich, K., & West, R. (1989). Exposure to Print and Orthographic Processing. *Reading Research Quarterly, 24*, 402–433.

Sum, A. (1999). *Literacy in the Labor Force.* National Center for Education Statistics, US Department of Education, Washington, DC.

Swanson, H. L. (1993). Working Memory in Learning Disability Subgroups. *Journal of Experimental Child Psychology, 56*, 87–114.

TRACE Center, University of Wisconsin. *A Brief Introduction to Disabilities.* Retrieved November 23, 2003, from htpp://trace.wisc.edu/docs/population/populat.htm.

U.S. Department of Commerce. (2002). *A Nation Online: How Americans are Expanding Their Use of the Internet.* Economics and Statistics Administration, National Telecommunications and Information Administration. Washington, DC.

van Gelderen, A., Schoonen, R., de Glopper, K., Hulstijn, J., Simis, A., Snellings, P., & Stevenson, M. (2004). Linguistic Knowledge, Processing Speed, and Metacognitive Knowledge in First and Second Language Reading Comprehension: A Component Analysis. *Journal of Educational Psychology, 96*(1), 19–30.

Vellutino, F., Scanlon, D., Sipay, E., Small, S., Chen, R., Pratt, A., & Denckla, M. (1996). Cognitive Profiles of Difficult to Remediate and Readily Remediated Poor Readers: Early Intervention as a Vehicle for Distinguishing Between Cognitive and Experimental Deficits as Basic Causes of Specific Reading Disability. *Journal of Educational Psychology, 88*(4), 601–638.

W3C (1999). Web Content Accessibility Guidelines 1.0 [Electronic edition] Retrieved April 3, 2007, from http://www.w3.org/TR/WAI-WebContent/

Weiner, J., Aquirre, A., Ravenell, K., Kovath, K., McDevit, L., Murphy, J., Asch, D., & Shea, J. (2004). Designing an Illustrated Patient Satisfaction Instrument for Low Literacy Populations. *The American Journal of Managed Care, 10*(11), 853–859.

West, D. (2003). *State and Federal E-Government in the United States.* Center for Public Policy, Brown University, Providence, RI. 1–29.

Zarcadoolas, C., Blanco, M., Boyer, J., & Pleasant, A. (2002). Unweaving the Web: An Exploratory Study of Low-Literate Adults' Navigation Skills on the World Wide Web. *Journal of Health Communication, 17*, 309–324.

·8·

COMPUTING TECHNOLOGIES FOR DEAF AND HARD OF HEARING USERS

Vicki L. Hanson
IBM T. J. Watson Research Center

INTRODUCTION

Traditionally, human interaction with computers has relied most heavily on visual perception and motor ability. Graphical user interfaces (GUIs) have dominated computer displays, both large and small, while mouse and keyboard devices have dominated computer input. Increasingly, however, newer interfaces are using sound and speech. The use of attention-grabbing multimedia computer presentations plus the increasing use of conversational speech interfaces are examples.

As exciting as these new technologies may be, they have the potential to disenfranchise deaf and hard of hearing users. This chapter will discuss interface technologies as they relate to the needs of users unable to hear auditory information. In addition, language considerations associated with limited hearing will be discussed.

This chapter begins with a discussion of hearing loss, followed by issues of language acquisition as they relate to hearing loss. Auditory user interfaces that present difficulties will be discussed next, along with information about interface alternatives that enable access for deaf and hard of hearing users. The chapter will conclude with a brief overview of technologies that have been developed to assist with communication.

HEARING LOSS

According to the National Institute on Deafness and Other Communication Disorders (NIDCD), about 28 million people in the United States have some degree of hearing loss (National Institutes of Health [NIH], 2004; see also, Mitchell, 2005, 2006). This is a sizable population to be considered in the design of computer interfaces. Numbers alone, however, obscure some significant differences among the individuals who experience hearing loss. Degree and type of loss, age of onset of loss, as well as family, educational, and societal influences will all contribute to the experience and abilities of an individual who has a functional hearing loss.

Some individuals will have relatively little hearing loss, while others will experience a profound loss. Individuals who are hard of hearing will generally have some hearing. The ability that an individual user will have to make use of their residual hearing, however, is not a straightforward calculation of dB loss. People with hearing loss will experience difficulty with pitch, timbre, and loudness, but, critically, will also experience difficulty with speech perception. Factors such as type of loss (e.g., conductive, sensorineural, mixed, or central) will have a major effect on the user experience. People with sensorineural hearing loss (such as resulting from lengthy exposure to loud noises or as a result of aging) generally will have more difficulty perceiving speech than people with conductive hearing losses that result

from difficulties in middle ear functioning. The extent to which an individual makes use of this hearing for communication and whether the individual can hear computer sounds, however, varies greatly.

In addition, the way in which an individual having a hearing loss interacts with their environment may also be influenced by societal factors. Whether individuals identify themselves as Deaf (with an uppercase "D" as a member of the Deaf Community), deaf, or hard of hearing is indicative of cultural identity (Lane, 1992; Padden & Humphries, 1988, 2005).[1] Membership in the Deaf Community is determined more by shared language and worldviews rather than by results of audiometric tests. For example, people may lose their hearing with age. These people, while deaf by the audiometric definition, would not share the culture of the Deaf Community.

In the United States, the language of the Deaf Community is American Sign Language (ASL). Other countries and locales have their own native sign languages shared by members of Deaf Communities in those areas. Interestingly, these signed languages are not based on the spoken languages of the region. People are often surprised to learn that ASL is more similar to French Sign Language, from which it originated (Lane, 1984), than it is to British Sign Language. Beginning with the seminal work of Ursula Bellugi and colleagues in the 1970s, linguists, cognitive psychologists, and brain researchers have studied native sign languages and their users for clues as to the origin of language and the biological nature of language (Emmorey & Lane, 2000; Klima & Bellugi, 1979; Erard, 2005).

Cochlear implants are a medical intervention that has received much attention in the last couple of decades. The decision as to whether or not to have a cochlear implant is often a complex one, as was explored in the movie *Sound and Fury* (Aronson, 2000; see also Hyde & Power, 2006). From the standpoint of a user, an implant is not the same as perfect hearing, but does allow the user to hear sounds and, with training, may greatly aid in the perception of speech (Chorost, 2005).

HEARING AND LANGUAGE

Some, but not all, deaf and hard of hearing individuals use sign language. The type of signed language depends on the user's life experiences. Deaf children born to deaf parents, regardless of severity of hearing loss, will generally acquire a sign language, such as ASL, natively as hearing children of hearing parents acquire spoken language. People who lose their hearing late in life generally will not master a sign language. In between, there are many variations. Deaf signers may be exposed to a native sign language as adults and as a result, may only acquire partial mastery (Newport, 1990). Other deaf signers will be exposed primarily to manual forms of English (or other spoken languages), rather than natural sign languages. Many schools that use sign

---

[1]Terms such as hearing-impaired and deaf-mute are generally considered to have negative connotations. For a discussion of this, see "What Is Wrong with the Use of These Terms: 'Deaf-mute,' 'Deaf and dumb,' or 'Hearing-impaired'?" by the National Association of the Deaf (available at http://www.nad.org/site/pp.asp?c=foINKQMBF&b=103786). The style manual of the American Psychological Association (APA) recommends use of nondiscriminatory language in all publications (see http://www.apastyle.org/disabilities.html).

language in the classroom do not use a natural sign language, but rather a representation of the spoken language that is signed. Forms of signed English borrow signs from ASL, but these signs are produced in English word order rather than using ASL sentence structures (for an extended discussion, see Lane, Hoffmeister, & Bahan, 1996). Children attending schools that use some form of signed English may not be exposed to ASL. Moreover, schools that educate students using an oral approach, in which speech is the primary means of communication in the classroom, also may not be exposed to ASL.

A person who is profoundly deaf from birth may have difficulty acquiring mastery of the spoken language, be it presented auditorily or in print. It is not surprising that someone who has never heard speech will have difficulty perceiving or producing it. In a large-scale study of deaf and hard of hearing children attending schools using an oral approach to education, Conrad (1979) reported that profoundly deaf children rarely acquired sufficient lipreading skills to allow easy participation in conversations. He found, on average, that these children (with hearing loss greater than 85 dB) could only comprehend about 25% to 28% of the words through lipreading that they could comprehend through reading. Even among these orally educated students, fewer than 20% had speech that was rated even fairly easy to understand. While the statistics are better for hard of hearing children, Conrad found that even these students (with hearing loss less than 65 dB) could only comprehend about 36% of the words through lipreading that they could comprehend through reading. Nearly 85% of these hard of hearing students, however, had speech that was rated at least fairly easy to understand.

Perhaps more surprising may be the fact that many deaf and hard of hearing individuals have difficulty with reading (Conrad, 1979; Gallaudet Research Institute, 2003). To understand this, it is necessary to realize that reading is based on the underlying spoken language. Spoken languages are composed of sounds, linguistically defined as "phonemes." These phonemes correspond, albeit not always in a one-to-one relationship in English, to letters or letter combinations. This is true of all alphabetic languages. Learning to read is generally considered to be learning to map the print onto the spoken language the person already knows (Brady & Shankweiler, 1991). In addition, speech plays a critical role in the short-term memory processes that serve understanding of grammar and text comprehension (Lichtenstein, 1998). In the case of deaf readers, however, it cannot be taken for granted that reading will build on a firm understanding of the structure of the spoken language, but there are no absolutes. Some prelingually, profoundly deaf children become excellent readers, while some with lesser degrees of hearing loss do not (Conrad, 1979; Hanson, 1989).

Interestingly, research has recently turned to an examination of signed languages as an influencer in the development of skilled reading. It has long been known that deaf children of deaf parents, on average, acquire greater mastery of reading and writing than deaf children of hearing parents. Is this due to early exposure to sign language? While a number of both intellectual and societal factors have been considered as contributors to this disparity, interest has focused on the issue of language. Evidence is now emerging as to the important role that early mas-

tery of a sign language can have on second language learning for deaf students as they acquire reading and writing skill (Padden & Hanson, 2000; Padden & Ramsey, 2000).

In short, for any deaf or hard of hearing individual, language experience cannot be assumed. That individual may or may not sign, speak clearly or lipread well, or have reading skills consistent with those of the hearing population. This knowledge has implications for designers who seek to address the needs of deaf and hard of hearing users.

DISPLAY TECHNOLOGIES

In many ways, computers and other technologies have proven to be of great benefit to deaf and hard of hearing users. The largely visual nature of information on the Internet makes this information accessible to deaf and hard of hearing users. Instant messaging and e-mail facilitate communication with deaf and hearing family, friends, and coworkers, while network-connected wireless devices such as PDAs and pagers are seen as lifelines that can be used in place of cell phones. As would be suspected from the previous discussion, however, effective interfaces for these and other technologies for use by deaf and hard of hearing individuals must take into account both sensory and language considerations. In particular, the increasing reliance on sound and speech interfaces to convey information can have serious consequences for individuals who have a hearing loss.

Audible Signals

Sounds have become increasingly popular in computer interfaces. They have long been used to convey information about new messages and have become popular as problem alerts, such as when an error has been committed. These sound events are considered attention-grabbing events for users whose visual attention may otherwise be engaged. For any user who has a hearing loss, however, sounds will be a problem.

A number of considerations can help provide the necessary visual support for a user who is deaf or hard of hearing (Vanderheiden, 1994). It is necessary to provide visual forms for all auditory information. Critically, these visual cues should be sufficiently noticeable so that they catch the attention of a person who may not be looking directly at the computer screen. Operating systems have features that can provide such visual alerts. For example, Windows® has accessibility features that allow users to set up their system to have captions or visual warnings displayed for sound events. These are helpful, although they may not give the full range of information carried by a sound event. For example, the meaning of a sound event may differ based on the tone of the signal or when the sound is produced. While visual captions and warnings may alert a deaf user that a sound event has occurred, they will be unable to convey these more subtle distinctions. It is important for designers to give careful consideration to sound events to ensure that crucial information is available by a nonauditory means for deaf and hard of hearing users.

Multimodal interfaces, as the name implies, are designed to support a range of perceptual capabilities. In theory, this would seem ideal for users who are deaf or hard of hearing as visual alternatives to auditory materials should be available. Multi-modal interfaces, however, do to always present all information on both modalities. The emphasis in many multimodal interfaces is representing information via speech that would otherwise be conveyed by print or some other visual means. To the detriment of deaf and hard of hearing users, often, less attention is given to ensuring that all auditory material be visually conveyed as well.

Multimedia Interfaces

Multimedia uses a combination of text, sound, pictures, animation, and video to present information. Traditionally, games and educational software have exploited the richness of multimedia, but the advent of high-speed Internet communications has enabled the use of multimedia for a number of engaging applications on the web. For the present discussion, consideration will be given to multimedia presentations as they may impact access for deaf and hard of hearing users.

Multimedia material is inherently sensory. The technology offers eye-catching visual displays and attention-grabbing sound effects. To the extent that the information conveyed visually and auditorally is the same, information can be reinforced for users who have both channels available to them. To the extent that different information is presented in the two channels, however, users who have a functional loss of one of the channels will not have full access to that information. As it relates to the present discussion, this means that any information that is carried solely by sound or speech will be unavailable to deaf and hard of hearing users. Accessibility guidelines for multimedia products and web pages require equivalent visual presentations (e.g., see Brewer & Dardailler, 1999; U.S. General Services Administration, n.d.).

Consider the growing popularity of video on the web. Video material is an extremely effective way to convey information and younger generations of computer users have come to expect video to be part of their computer experience. For deaf and hard of hearing users, the voice-over that is common in video will be inaccessible. Words spoken by persons in view of the camera also are largely inaccessible. Even people skilled at lipreading cannot lipread video conversations that have poor lighting or poor resolution, or speakers who turn away from the camera. Additionally, the video may present music or sound events (e.g., doorbells or animals noises) that contribute significantly to events on the video. These, too, are unavailable to deaf and hard of hearing users. Technologies such as captioning and sign-language translation exist to provide alternative presentations of sound events and speech. Designers wishing to make their applications universally accessible should consider these alternatives and incorporate them into their applications.

Captioning. Captioning provides a print alternative to speech and sound events. It is much like subtitling, except that it is specifically designed for deaf and hard of hearing users and, thus, will include comments in the captioning about sounds (e.g., "<music playing>" or "<sounds of child crying>") that

may not be included in subtitles of foreign language videos. Captioning of certain television programming is mandated in the United States by the Federal Communications Commission and is considered to be beneficial to a large number of users, not only those with hearing loss. It has been shown, for example, to improve reading abilities of children and to benefit second language learners. Designers who use multimedia materials have an obligation to their full audience to provide captioning of audio and video materials. The listing of a number of resources for captioning software is available at the Closed Captioning website (n.d.).

As might be anticipated from the previous review of reading levels of deaf and hard of hearing users, there has been some controversy about what language level should be used for captioning. Simply put, the issue revolves around the question of whether captions should be verbatim transcripts or simplified captioning should be provided. Verbatim captioning is generally preferred by users themselves (NIH, 2002) and is often cited as having the potential to improve reading skills (Steinfeld, 2001).

Signing. For sign language users, there are sign language alternatives to captioning. These sign interfaces have been defined as ways of representing signed languages on a computer such that signing can be stored, displayed, and manipulated to facilitate computer interaction (Frishberg, Corazza, Day, Wilcox, & Schulmeister, 1993). It should be noted that these interfaces have often been employed not only for making audio and speech materials accessible to deaf and hard of hearing users, but they have also been used for language learning by both deaf and hearing people. A number of software applications have been developed that use sign language as a means of teaching reading skills and writing to deaf signers or teaching sign language skills to individuals wishing to learn to sign.

Importantly, software applications that purport to use "sign language" or ASL differ in significant ways in the language that is being used. Notably, many of these applications use fingerspelling. Fingerspelling is not a natural signed language, but rather is a derivative of print. Specifically, in fingerspelling, there is one handshape for each letter of the alphabet and words are spelled out letter by letter on the hands. Thus, the word "language" would be spelled out by eight distinct handshapes spelling L-A-N-G-U-A-G-E. Other sign language interfaces use a form of signed English, rather than the native sign language. For young children and others not fluent in English, interfaces that use fingerspelling or signed English transliterations of text may not meet their needs.

In many cases, the goal when using sign interfaces is to provide ASL translations. Given the present state of the art, automatic translation from English to ASL is not possible; however, many current efforts are directed at facilitating this translation, acknowledging the need for translation into ASL. In what follows, a few of the options for sign language presentation of audio and multimedia will be discussed.

Ideal in many respects would be to have a live person signing an ASL version of print and multimedia materials. Hanson and Padden (1989, 1990) used videodisc technology in the earliest computer-based attempt at a bilingual ASL/English approach to reading and writing instruction for signing children. The work combined ASL video and the translated English text on one screen. In this and other language learning situations where the users are young children still developing both ASL and English skills, this use of live signing has been particularly effective (Frishberg et al., 1993).

The advent of high-speed networks and Internet video has created opportunities for web applications that use live signing interfaces. For example, classroom applications have been developed (e.g., see King, 2000; Laurent Clerc National Deaf Education Center, n.d.; for a demo, see "Sample Web Page," Gallaudet University, n.d.). Additionally, video blogs have created the opportunity for blogs to be signed by the blogger, rather than written (Lamberton, 2005), and video e-mail allows signers to communicate through signing by creating video recordings to be transmitted as e-mail messages (e.g., see "Road Runner Video Mail," n.d.; also, "Vibe Video Mail," n.d.).

Although ideal from the language perspective, other constraints may argue against the use of live signing in an interface. Lack of access to high-speed networks for video transmission may be an issue, but often the problem is the desire for automatic translation of software and web content that live signing does not provide. Live signing requires the prior recording of the signed material. Once recorded, changes to the video require a new recording. Because of this, live signing is not practical in applications that require that sign versions of audio or multimedia be created in real time. Short of having an interpreter doing the translations, live signing is not possible in these situations.

One approach to automatic sign presentation is what might be called "concatenated signing" (e.g., see "iCommunicator," n.d.; "Signtel," n.d.). With this, a software program is used to create word strings or sentences by concatenating signs produced by a live signer. The program starts with a vocabulary of stored individual signs. These signs can be strung together to form phrases and sentences. To prevent a jerky appearance that would occur through the simple production of a list of signs, algorithms are used to smooth the transitions between these signs. While these transitions are not as natural as live signing, the signing is legible and suggests an interesting approach to sign language interfaces.

A technique that has generated much interest in recent years is signing avatars (e.g., see Cox et al., 2002; Karpouzis, Caridakis, Fotinea, & Efthimiou, in press; Kennaway, 2002; Sims, 2004). These avatars use virtual-reality techniques to produce animated signing. The specific techniques differ for the various avatars, but central to all is that they display computer generated signing. Some systems are able to display not just hands, but full signers, so that facial expressions as well as hand movements are shown. See, for example, the SigningAvatar® shown in Fig. 8.1 that illustrates the sophistication of these animations.

Signing avatars have been used in education and have potential for applications such as translation of web pages, television programs, and conversational dialog. Chief among the virtues is that, unlike natural (live) sign language applications that are limited to prerecorded materials, avatars can generate signed versions of English words "on the fly." They also have the advantage of not requiring large downloads for web usage.

While the avatars can sign ASL when preprogrammed, the language translation work needed for automatic translation into

FIGURE 8.1. An ASL sign avatar. The full animation of this avatar can be viewed on http://www.vcom3d.com/ASL.htm. © Vcom3D, Inc., 2004. All rights reserved.

ASL (or other signed language) is not ready to support this rendering on the fly. A current research focus in avatar work is on natural language translation and exploring techniques to display native sign languages by avatars (e.g., see Huenerfauth, 2005).

INPUT TECHNOLOGIES

Standard keyboard and mouse input technologies offer no barriers to users who have a hearing loss. The emergence of new interfaces, however, presents alternatives that will impact the way in which we are able to interact with computers. We consider here both speech and gestural interfaces as they may influence interactions for deaf and hard of hearing users.

Speech Interfaces

Conversational speech interfaces have appeal as a natural means of interacting with computers. These interactions can range from simple, even one-word commands to full dictation of documents and user collaborations. For deaf and hard of hearing individuals, such interactions require consideration of user needs. First, these interactions can involve speech output which, as already discussed, requires a visual display alternative. Second are problems with speech input. Some deaf and hard of hearing individuals utilize speech and would be interested in taking advantage of speech input. Speech recognition, however, may be more problematic for these speakers than for hearing speakers.

As mentioned briefly in a previous paragraph, the speech of deaf and hard of hearing individuals is not always highly intelligible to hearing listeners. Since speech recognition engines can be trained to the voice and pronunciations of an individual speaker, however, couldn't a recognizer be trained to understand the speech of an individual deaf speaker, even if the speech is not completely intelligible to listeners? The difficulty is that recognizers require consistent speech.

Research has shown that the speech of deaf and hard of hearing speakers often is more variable than the speech of hearing speakers (McGarr, 1987; McGarr & Lofqvist, 1988). For example, there is more acoustic variation in the pronunciation of a single phoneme by a deaf or hard of hearing speaker than there is in the pronunciation of that phoneme by a hearing speaker. Hearing listeners are very tolerant of this variability; speech recognizers are less so. Thus, deaf and hard of hearing users who many wish to use their speech for input may well find recognition less accurate than it is for hearing speakers.

Speech interfaces are often seen as useful alternatives to visual interfaces in situations when computers or keyboards are not available. For hearing users, this alternative of a speech interface may be highly desirable. Many deaf and hard of hearing users, however, will not be interested in speech interfaces; others may experience difficulties in using them. Although situational demands may sometimes dictate the use of conversational interfaces, care must be taken that outside of the situational context there exists a means for deaf and hard of hearing users to access the same information that hearing users access.

Sign Interfaces

Designed specifically for deaf and hard of hearing users, sign recognition is a specific and complex subset of gesture recognition technologies having the goal of automatically converting signed language to text or speech. These technologies have been investigated for a number of years, primarily to address the need to facilitate communication between signers and nonsigners. They also have the potential to provide an alternative means of natural language input to computers.

The task of recognizing full ASL or other natural sign languages is a difficult problem. The first reason is that an individual sign varies depending upon its context. Take, for example, the sign for the word GIVE. This sign has a specific shape combined with a variable movement. The movement reflects who is giving and to whom. I GIVE-TO YOU, YOU GIVE-TO ME, I GIVE-TO HIM, HE GIVES-TO ME, and I GIVE-TO ALL-OF-YOU each has a different movement that indicates subject and object. A second reason for difficulty in sign-language recognition is that several pieces of linguistic information are produced in parallel. For example, facial expression carries critical grammatical information. Thus, a full language recognizer needs to recognize not only the hand gestures of ASL, but must also recognize certain facial information relevant to the grammar. Such facial elements include eyebrow position, eye gaze, and mouth movements.

As with interfaces that produce signs, many systems that purport to perform sign-language recognition deal with recognition of fingerspelling handshapes rather than recognition of ASL sentences or even ASL signs. Using fingerspelling handshapes rather than ASL signs certainly constrains the size of the problem. Because there are only 26 handshapes in the English alphabet, this represents a much more manageable problem space than the recognition of full ASL signing. While it presents

an alternative to keyboard typing of words, it doesn't provide the type of natural language interaction with computers that is afforded by speech interfaces.

Technologies for recognizing signs have tended to use either instruments worn by the signer or computer vision techniques. The first of these approaches has the signer wear a specially designed glove or sensors placed on their joints that allow a computer to track movement (e.g., see Braffort, 1996; Fang, Gao, & Zhao, 2003; Hernandez-Rebollar, Lindeman, & Kyriakopoulos, 2002; Kadous, 1996; Wang, Gao, & Ma, 2002). In contrast, computer vision techniques use cameras to provide input to a computer about a signer's movements and facial gestures (e.g., see Brashear, Starner, Lukowicz, & Junker, 2003; Kadous, 1996; Lee, et al., 2005; Vogler & Metaxas, 2001). These inputs are then analyzed using a variety of techniques such as neural networks or Hidden Markov Models that then recognize the sign.

Critically, however, this recognition does not do language translation. Thus, the output will be a one-to-one mapping of a sign into print or speech. As mentioned with the avatar work presented earlier, the ASL/English rules that would be needed for such translation are not developed to a state where automatic translation can occur; however, advances are being made. Recently, for example, Hernandez-Rebollar (2005) presented work designed to translate signed input into English phrases. In that work, the translation was enabled by having a limited number of phrases that the system could recognize, thus constraining the problem space.

The current state of the art for sign recognition is not as advanced as speech recognition. As researchers continue to work on the problem, however, advances can be expected. For signers, it might be the case that conversational sign interactions will one day be possible, much as speech recognition now allows speakers to benefit from conversational speech interactions.

TECHNOLOGY AND COMMUNICATION

Deaf and hard of hearing individuals have difficulties not only with computer interfaces, but also experience significant difficulties in certain communication situations. Telephone conversations, as well as one-on-one conversations, group discussions, and presentations or classroom lectures, are all problematic. No discussion of HCI for deaf and hard of hearing individuals would be complete without at least a brief mention of technologies that have been developed to aid in these situations.

Telephones have long been a source of difficulty for deaf and hard of hearing people. Alexander Graham Bell was a teacher of deaf students and was married to a woman who was deaf. His interest in finding improved ways to communicate with deaf speakers led to his invention of the telephone. Ironically, however, over the years, telephones created a number of barriers for deaf and hard of hearing individuals in the workplace and other situations. To overcome these barriers, a number of assistive devices have been developed (Lazzaro, 1993). Amplification and adapters exist for many phones that will allow hard of hearing individuals or people with cochlear implants to hear phone conversations. Teletype devices (TTYs and TDDs) as well as some computer applications allow deaf users to type conversations that are carried over phone lines.

Operator-assisted relay services have been established to enable conversations between deaf and hearing individuals. The relay personnel serve as a bridge between the two conversation participants, translating typed information into speech for the hearing participant and translating speech into written text for the deaf participant. Hard of hearing speakers can speak directly with others using a captioning service that provides, nearly in real time, a printed transcript of the conversation to support hard of hearing users (CapTel, 2005). Various means of enabling signed conversations over the telephone have been explored over the years, but the advent of video phones has now made feasible the option of signed phone conversations.

The current prevalence of conference calls in the workplace has created a new set of problems for deaf and hard of hearing workers. Similarly, classroom situations or lecture presentations also create significant difficulties for deaf and hard of hearing attendees. Even individuals skilled in lipreading have difficulty in these situations because the speaker's face is rarely visible with sufficient resolution for lipreading. Sign language interpreting and captioning are two means of providing accommodation for participants unable to hear or understand the speech in these situations.

Sign language interpreters allow signers to participate in meetings by providing real-time translation of the spoken conversation into sign. This is a two-way translation service, such that the signing participant is also able to participate by having the interpreter speak their signed utterances. Remote interpreting is a technology that addresses the problem of a shortage of skilled interpreters, particularly in some locations. The interesting aspect of this is that the interpreter need not be present at the location of any of the conference participants. The deaf or hard of hearing participant views the remotely located interpreter on a computer display or TV screen. The interpreter listens to hearing participants by telephone and provides sign-language interpreting for the deaf participant. The interpreter also voices what the deaf person signs for hearing participants. This remote interpreting can be used similarly to the telephone relay service, with participants in different locations (Video Relay Service, or VRS), or in situations where the participants are located in the same room (Video Relay Interpreting, or VRI).

Communication Access Real-time Translation (CART) provides real time captioning, enabling discussions and presentations to be transcribed into text for deaf and hard of hearing participants. Typically, this is done by a person who creates the captions. As with sign language interpreting, the captioner can be physically present or remotely located, listening to the conversation by phone and transmitting the text via a network to a computer screen or projected display. The deaf participant types their questions or comments. That input is then read aloud for the other participants, either by a captioner or using computer text-to-speech technology (e.g., see Caption First, n.d.; Viable Technologies, n.d.)

The ability to automatically transcribe speech was envisioned more than 100 years ago as a technology that held great promise for deaf and hard of hearing people, even though such technology was hardly imaginable at the time (Fay, 1883). As speech-recognition systems have matured over the last quarter century, a number of applications designed for deaf users have been explored (Bain, Basson, Faisman, & Kanevsky, 2005; Stinson & Stuckless, 1998).

The Liberated Learning Project is one example of using automatic speech recognition in the classroom (Bain, Basson, & Wald, 2002). In this effort, the classroom teacher speaks into a microphone that transmits his or her speech to a computer to perform the recognition. The transcript of the lecture is displayed in close to real time on a screen at the front of the classroom. Shown in Fig. 8.2 is a classroom situation for the Liberated Learning Project.

FIGURE 8.2. An example classroom lecture transcribed in real time. From Bain et al. (2005); republished with permission.

While recognition technologies have improved over the years, they still do not attain 100% correct performance for dictation. To provide students with accurate transcripts for lecture notes, the professors who participate in the Liberated Learning Project review the transcript for inaccuracies and make corrections after a lecture is completed. The corrected transcripts are then made available to students (Bain et al., 2002). Other efforts that use automatic speech recognition for captioning use different methods for correcting errors in transcripts. These other methods include having a trained speaker "shadow" the speech to produce more reliable recognition, and/or having a person correcting errors in real time (e.g., see Bain et al., 2005; Robson, 2001; Viable Technologies, n.d.).

CONCLUSIONS

Computer technologies are very important in the lives of many deaf and hard of hearing people. These technologies can ease communication between coworkers, friends, family members, neighbors, and a variety of services. These technologies have also provided access for deaf students, employees, and individuals to have full access to rich multimedia services, shopping, news, and general information. In short, barriers long encountered by deaf and hard of hearing people are now being overcome through the use technology.

Deaf and heard of hearing users represent a population that, in itself, is diverse in both hearing and language experiences and skills. Applications that offer flexibility of language (e.g., captioning, signed English, or ASL) will be most accessible to users who have a hearing loss. At a minimum, however, developers and designers need to ensure that information is not carried by the auditory channel alone. Paramount is the need to ensure that any audible information, be it a sound alert, speech prompt, or other auditory event, has a visual counterpart. Such considerations are not only good design, but are also mandated by a growing number of regulations worldwide. The ability of everyone to participate in our increasingly technological society is crucial.

ACKNOWLEDGMENTS

I'd like to thank Al Noll, Debra Noll, and John Richards for valuable comments on early drafts of this chapter.

References

Aronson, J. (2000). *Sound and Fury* [Motion picture]. United States: New Video Group.

Bain, K., Basson, S., Faisman, A., & Kanevsky, D. (2005). Accessibility, transcription, and access everywhere. *IBM Systems Journal, 44*(3), 589–605.

Bain, K., Basson, S., & Wald, M. (2002). Speech recognition in university classrooms: Liberated learning project. In *Proceedings of the 5th International ACM SIGCAPH Conference on Assistive Technologies (ASSETS'02)* (pp. 192–196). New York: ACM.

Brady, S., & Shankweiler, D. (1991). *Phonological processes in literacy: A tribute to Isabelle Y. Liberman.* Hillsdale, NJ: Lawrence Erlbaum Associates.

Braffort, A. (1996). A gesture recognition architecture for sign language. In *Proceedings of the 2nd International ACM SIGCAPH Conference on Assistive Technologies* (ASSETS'96), 102–109.

Brashear, H., Starner, T., Lukowicz, P., & Junker, H. (2003). Using multiple sensors for mobile sign language recognition. In *Proceedings of the 7th IEEE International Symposium on Wearable Computers*, 45–52.

Brewer, J., & Dardailler, D. (1999). *Web Content Accessibility Guidelines 1.0.* Retrieved December 7, 2005, from University of Wisconsin-Madison, Trace Research and Development Center website: http://www.w3.org/TR/WAI-WEBCONTENT/

CapTel. (n.d.). *Introducing the Captioned Telephone.* Retrieved December 7, 2005, from http://www.captionedtelephone.com/about captel.phtml

Caption First. (n.d.). *Overview of our services.* Retrieved January 20, 2006, from http://www.captionfirst.com/overview.htm

Chorost, M. (2005). *Rebuilt: How becoming part computer made me more human.* New York, NY: Houghton Mifflin Company.

Closed Captioning Web. (n.d.) *Closed Captioning Web: Software Links.* Retrieved December 7, 2005, from http://www.captions.org/soft-links.cfm

Conrad, R. (1979). *The deaf schoolchild.* London: Harper & Row.

Cox, S., Lincoln, M., Tryggvason, J., Nakisa, M., Wells, M., Tutt, M., et al. (2002). TESSA, a system to aid communication with deaf people. In *Proceedings of the 5th International ACM SIGCAPH Conference on Assistive Technologies (ASSETS'02)* (pp. 205–212). New York: ACM.

Emmorey, K., & Lane, H. (2000). *The signs of language revisited: An anthology to honor Ursula Bellugi and Edward Klima.* Mahwah, NJ: Lawrence Erlbaum Associates.

Erard, M. (2005). The birth of a language. *New Scientist, 188*(2522), 46–49.

Fang, G., Gao, W., & Zhao, D. (2003). Large vocabulary sign language recognition based on hierarchical decision trees. In *Proceedings of the 5th International Conference on Multimodal Interfaces* (pp. 125–131). New York: ACM.

Fay, E. A. (1883). The glossograph. *American Annals of the Deaf, 28,* 67–69.

Frishberg, N., Corazza, S., Day, L., Wilcox, S., & Schulmeister, R. (1993). Sign language interfaces. In *Proceedings of the ACM SIGCHI Conference on Human Factors in Computing Systems* (pp. 194–197). New York: ACM.

Gallaudet Research Institute. (2003). *Literacy and deaf students.* Retrieved December 7, 2005, from http://gri.gallaudet.edu/Literacy/

Gallaudet University. (n.d.). *Sample web page with embedded real media video.* Retrieved January 20, 2006, from http://academic.gallaudet.edu/pages/iced2000/real/stellaluna_smil.html

Hanson, V. L. (1989). Phonology and reading: Evidence from profoundly deaf readers. In D. Shankweiler & I. Y. Liberman (Eds.), *Phonology and reading disability: Solving the reading puzzle* (pp. 69–89). Ann Arbor: University of Michigan Press.

Hanson, V. L., & Padden, C. A. (1989). The use of interactive video for bilingual ASL/English instruction of deaf children. *American Annals of the Deaf, 134,* 209–213.

Hanson, V. L., & Padden C. A. (1990). Bilingual ASL/English instruction of deaf children. In D. Nix & R. Spiro (Eds.), *Cognition, education, and multimedia: Exploring ideas in high-technology* (pp. 49–63). Hillsdale, NJ: Lawrence Erlbaum Associates.

Hernandez-Rebollar, J. L. (2005). Gesture-drive American Sign Language phraselator. In *Proceedings of the 7th International ACM Conference on Multimodal Interfaces (ICMI'05)* (pp. 288–292). New York: ACM Press.

Hernandez-Rebollar, J.-L., Lindeman, R. W., & Kyriakopoulos, N. (2002). A multi class pattern recognition system for practical fingerspelling translation. In *Proceedings of the 4th IEEE International Conference on Multimodal Interfaces* (ICMI'05) (pp. 185–190). Washington, DC: IEEE Computer Society.

Huenerfauth, M. (2005). Representing coordination and non-coordination in an American Sign Language animation. In *Proceedings of the 7th International ACM SIGACCESS Conference on Assistive Technologies (ASSETS'05)* (pp. 44–51). New York: ACM Press.

Hyde, M., & Power, D. (2006). Some ethical dimensions of cochlear implantation for deaf children and their families. *Journal of Deaf Studies and Deaf Education, 11*(1), 102–111.

iCommunicator. (n.d.). *iCommunicator.* Retrieved January 4, 2006, from http://www.myicommunicator.com/

Kadous, M. W. (1996). Machine recognition of Auslan signs using Powergloves: Towards large-lexicon recognition of sign languages. In L. Messing (Ed.), *Proceedings of WIGLS, The Workshop on the Integration of Gestures in Language and Speech, Wilmington Delaware*, 165–174.

Karpouzis, K., Caridakis, G., Fotinea, S.-E., & Efthimiou, E. (in press). Educational resources and implementation of a Greek sign language synthesis architecture. *Computers in Education.*

Kennaway, R. (2002). *Synthetic animation of deaf signing gestures.* Gesture and Sign Language in Human-Computer Interaction: International Gesture Workshop (GW 2001, LNAI 2298), 146–157.

King, C. (2000). *Online learning at Gallaudet University.* Retrieved January 20, 2006 from Gallaudet University website: http://academic.gallaudet.edu/pages/iced2000/iced2000_GUonlinelearning.PDF

Klima, E. S., & Bellugi, U. (1979). *The signs of language.* Cambridge, MA: Harvard University Press.

Lamberton, J. (2005). *Jason Lamberton's video blog.* Retrieved January 20, 2006, from http://video.google.com/videoplay?docid=6012463606293405795&q=gallaudet+university

Lane, H. (1984). *When the mind hears: A history of the deaf.* New York: Random House.

Lane, H. (1992). *The mask of benevolence: Disabling the deaf community.* New York: Alfred A. Knopf.

Lane, H., Hoffmeister, R., & Bahan, B. (1996). *A journey into the deaf-world.* San Diego, CA: Dawn Sign Press.

Laurent Clerc National Deaf Education Center, Gallaudet University. (n.d.). *Shared reading project: Chapter by chapter—The Thinking Reader.* Retrieved January 20, 2006 from http://clerccenter.gallaudet.edu/Literacy/programs/chapter.html

Lazzaro, J. J. (1993). *Adaptive technologies for learning and work environments.* Chicago, IL: American Library Association.

Lee, S., Henderson, V., Hamilton, H., Starner, T., Brashear, H., & Hamilton, S. (2005). A gesture-based American Sign Language game for deaf children. In *Proceedings of the ACM Conference on Human Factors in Computing Systems (CHI '05)* (pp. 1589–1592). New York: ACM Press.

Lichtenstein, E. H. (1998). The relationships between reading processes and English skills of deaf college students. *Journal of Deaf Studies and Deaf Education, 3*(2), 80–134.

McGarr, N. S. (1987). Communication skills of hearing-impaired children in schools for the deaf. *ASHA Monographs, Oct*(26), 91–107.

McGarr, N. S., & Lofqvist, A. (1988). Laryngeal kinematics in voiceless obstruents produced by hearing-impaired speakers. *Journal of Speech and Hearing Research, 31*(2), 234–239.

Mitchell, R. E. (2005, February 15). *Can you tell me how many deaf people are there in the United States?* Retrieved December 7, 2005, from Galludet Research Institute, Graduate School and Professional Programs website: http://gri.gallaudet.edu/Demographics/deaf-US.php

Mitchell, R. E. (2006). How many deaf people are there in the United States? Estimates from the survey of income and program participation. *Journal of Deaf Studies and Deaf Education, 11*(1), 112–119.

National Institutes of Health, National Institute on Deafness and Other Communication Disorders. (2002). *Captions for deaf and hard of hearing viewers.* NIH Publication No. 00-4834. Retrieved December 7, 2005, from http://www.nidcd.nih.gov/health/hearing/caption.asp#edit

National Institutes of Health, National Institute on Deafness and Other Communication Disorders. (2004). *Statistics about hearing disorders, ear infections, and deafness.* Retrieved December 7, 2005 from http://www.nidcd.nih.gov/health/statistics/hearing.asp

Newport, E. (1990). Maturational constraints on language learning. *Cognitive Science, 14*, 11–28.

Padden, C. A., & Hanson, V. L. (2000). Search for the missing link: The development of skilled reading in deaf children. In K. Emmorey & H. Lane (Eds.), *The signs of language revisited: An anthology to honor Ursula Bellugi and Edward Klima* (pp. 435–447). Mahwah, NJ: Lawrence Erlbaum Associates.

Padden, C., & Humphries, T. (1988). *Deaf in America: Voices from a culture.* Cambridge, MA: Harvard University Press.

Padden, C., & Humphries, T. (2005). *Inside deaf culture.* Cambridge, MA: Harvard University Press.

Padden, C., & Ramsey, C. (2000). American Sign Language and reading ability in deaf children. In C. Chamberlain, J. Morford, & R. Mayberry (Eds.), *Language acquisition by eye* (pp. 165–189). Mahwah, NJ: Lawrence Erlbaum Associates.

Road runner video mail. (n.d.). Retrieved January 20, 2006, from http://vmail.vibephone.com/vm/vm_player?vmfile=4dbc51d033b4cdbac583bf8ec3ef9cd3&vmpid=1007

Robson, G. (2001). *Can realtime captioning be done using realtime voice recognition systems?* Retrieved December 7, 2005, from http://www.robson.org/capfaq/online.html#VoiceRecognition

Signtel, Inc. (n.d.). *Signtel.* Retrieved December 7, 2005, from http://www.signtelinc.com/main.htm

Sims, E. (2004). *Using emerging visualization technologies to provide sign language access to the Web.* Retrieved January 20, 2006 from http://www.w3.org/WAI/RD/2003/12/Visualization/VCom3D.html

Steinfeld, A. (2001). *The case for real time captioning in classrooms.* Retrieved December 7, 2005, from http://www.cartinfo.org/steinfeld.html

Stinson, M., & Stuckless, R. (1998). Recent developments in speech-to-print transcription systems for deaf students. In A. Weisel (Ed.), *Issues unresolved: New perspectives on language and deaf education* (pp. 126–132). Washington, DC: Gallaudet University Press.

U.S. General Services Administration, Office of Governmentwide Policy, IT Accessibility & Workforce Division. (n.d.). *Section 508.* Retrieved January 20, 2006 from http://section508.gov/

Vanderheiden, G. C. (1994). *Application software design guidelines: Increasing the accessibility of application software for people with disabilities and older users.* Retrieved January 20, 2006, from University of Wisconsin-Madison, Trace Research and Development Center website: http://trace.wisc.edu/docs/software_guidelines/software.htm

Viable Technologies. (n.d.). *Frequently asked Questions.* Retrieved December 7, 2005, from http://www.viabletechnologies.com/faq.php

Vibe video mail. (n.d.). Retrieved January 20, 2006, from http://www.vibephone.com/vsg/htdocs/products/video-mail/index.jsp

Vogler, C., & Metaxas, D. (2001). Framework for recognizing the simultaneous aspects of American Sign Language. *Computer Vision and Image Understanding* (81), 358–384.

Wang, C., Gao, W., & Ma, J. (2002). A real-time large vocabulary recognition system for Chinese Sign Language. In I. Wachsmuth and T. Sowa (Eds.), *International Gesture Workshop, Springer Lecture Notes in Artificial Intelligence* (Vol. 2298, pp. 86–95). New York, NY: Springer-Verlag.

APPLICATION/DOMAIN SPECIFIC DESIGN

HUMAN–COMPUTER INTERACTION IN HEALTHCARE

François Sainfort and Julie A. Jacko*
Georgia Institute of Technology

Paula J. Edwards
Georgia Institute of Technology & Children's Health Care of Atlanta

Bridget C. Booske
University of Wisconsin-Madison

*Currently at University of Minnesota

INTRODUCTION

U.S. healthcare expenditures were nearly $1.7 trillion in 2003 and were expected to grow to $2 trillion by 2005 (BlueCross BlueShield Association 2006). Despite such large spending, many Americans remain uninsured and do not have access to healthcare services. Furthermore, while our country has the most formidable medical workforce in the world and develops and uses the most modern medical technologies, the World Health Organization (2000) recently rated the quality and performance of the U.S. healthcare system as being worse than most of its counterparts in the western world. Chassin, Galvin, and the National Roundtable on Health Care Quality (1998) documented three types of quality problems: (a) overuse, (b) underuse, and (c) misuse. The results of an extensive review of over 70 publications covering years 1993 through 2000 provide "abundant evidence that serious and extensive quality problems exist throughout American medicine resulting in harm to many Americans" (Institute of Medicine, 2001, p. 24). In its first report, *To Err Is Human*, the Institute of Medicine (2000) reported serious and widespread errors in healthcare delivery that resulted in frequent avoidable injuries to patients. The Institute of Medicine (2001) suggested four key underlying reasons for inadequate quality of care in the U.S. healthcare system: (a) the growing complexity of science and technology, (b) the increase in chronic conditions, (c) a poorly organized delivery system, and (d) constraints on exploiting the revolution in information technology.

Reengineering the delivery of healthcare services through innovative development, application, and use of information and medical technologies can result in tremendous cost savings, improved access to healthcare services, as well as improved quality of life for all citizens. Healthcare professionals in the United States have recognized that both the information revolution and the biological revolution will offer tremendous opportunities—and challenges—for (re)designing the healthcare system of the future. They are aware of the need to better utilize new information and communication technologies and incorporate computing power into care delivery and clinical practice. They are also aware that the widely publicized biological revolution (which includes both advances in genetics and advances in biomedical engineering) will soon bring a large number of screening and diagnostic tests as well as new treatment strategies and disease-management tools. It is clear that the combination of biotechnology, computing power, information and communication technologies, distance technology, and sensor technology will make future delivery of healthcare in the United States unrecognizable from the care we deliver today.

However, unlike other parts of the American economy, the healthcare system has been slow to embrace modern information and communication technology, which is often viewed as too expensive, very unusable, and quite divergent from current practice (Schoen, Davis, Osborn, & Blendon, 2000). In addition, technologies do not typically come "ready-to-use" off the shelf and it is extremely difficult, if not impossible, for healthcare provider organizations to have the range of expertise, in house, that is needed to design, adapt, and implement technologies to meet an organization's needs. In fact, many healthcare organizations are often not even fully aware of their own needs, do not know which technologies are available for what, and do not know how modern information and communication technologies can be effectively used to improve and simplify care delivery. In many cases, recent information technology adoptions among healthcare providers have been driven by Federal and state regulations and requirements rather than well-recognized internal needs and growth opportunities. For example, public and private groups (i.e., the Federal government and the Leap Frog group, a collaboration of businesses) are strongly encouraging implementation of Computerized Physician Order Entry (CPOE) and other information technologies in order to achieve improvements in the quality and efficiency of providing care (Kaushal, Shojania, et al., 2003).

Healthcare informatics is a field that can be widely defined as the generation, development, application, and testing of information and communication principles, techniques, theories, and technologies to improve healthcare delivery. It includes the understanding of data, information, and knowledge used in the delivery of healthcare and an understanding of how these data are captured, stored, accessed, retrieved, displayed, interpreted, used, and made more efficient. While healthcare informatics intersects with the field of medical or bioinformatics, it is different in the sense that it focuses on healthcare delivery, and hence is centered on the patient (and/or consumer), the clinician (or healthcare professional, or "provider"), and, more importantly, the patient–provider interaction. Human–computer interaction (HCI), from the perspective of both the patient/consumer as well as the provider, is essential to the success of "healthcare informatics." In this chapter, we first review the characteristics of the healthcare industry in the United States. We then review information systems used by consumers, patients, and providers and raise HCI issues and challenges associated with both perspectives. Then we propose a framework for evaluating healthcare applications and conclude with a discussion of future opportunities and challenges for HCI in healthcare.

CHARACTERISTICS OF THE HEALTHCARE INDUSTRY

The Healthcare Industry

As previously discussed, the Institute of Medicine (2001) puts forth four key underlying reasons for inadequate quality of care in the U.S. healthcare system today: (a) the growing complexity of science and technology, (b) the increase in chronic conditions, (c) a poorly organized delivery system, and (d) constraints on exploiting the revolution in information technology. In addition, a growing trend toward consumerism is becoming a major force in shaping the future organization of the healthcare industry. These five trends, detailed further in the following discussion, are shaping the future of healthcare in the United States.

Complexity of Science and Technology

The sheer volume of new healthcare science and technologies—the knowledge, skills, interventions, treatments, drugs, and devices—is very large today and has advanced much more rapidly that our ability to use and deliver them in a safe, effective, and efficient way. Government as well as private investment in pharmaceutical, medical, and biomedical research and development has increased steadily. The healthcare delivery system has not kept up with phenomenal advancement in science and technology and proliferation of knowledge, treatments, drugs, and devices. With current advances in genomics (offering promise in diagnosis as well as, possibly, treatment), sensor technologies (offering promise in automated detection, measurement, and monitoring), nanotechnologies (offering promise in diagnosis, treatment, and control), and information and communication technologies (enabling remote delivery, telemedicine, e-health, and patient empowerment), the complexity of science and technology in healthcare is only going to increase.

Chronic Conditions

As noted by the Institute of Medicine (2001), "because of changing mortality patterns, those age 65 and over constitute an increasingly large number and proportion of the U.S. population" (p. 26). Therefore, there is an increase in both the incidence and prevalence of chronic conditions (defined as conditions lasting more than three months and not self-limiting). Hoffman, Rice, and Sung (1996) estimated that patients with chronic conditions make up 80% of all hospital bed days, 83% of prescription drug use, and 55% of emergency-room visits. Compared to acute illnesses, effectively treating chronic conditions requires disease management and control over long periods of time, collaborative processes between providers and patient, as well as patient involvement, self-management, and empowerment.

Organization of the Delivery System

The healthcare delivery system in the United States is a highly complex system that is nonlinear, dynamic, and uncertain. The system is further complicated by a large number of agents who are multiple stakeholders, each with multiple, sometimes conflicting, goals, aspirations, and objectives. As a result, the entire system leads to a lack of accountability; it is has frequently misaligned reward as well as incentive structures, and it suffers from inefficiencies embedded in multiple layers of processes. The healthcare "product" or "service" is often ill defined or difficult to define and evaluate. The processes involved in delivering healthcare services are complex, ill specified, and difficult to measure, monitor, and control. Health outcomes are difficult to measure, manage, and analyze. The system experiences growing cost pressures, faces potential insurance premium increases, and is extremely fragmented. Wagner, Austin, and Von Korff (1996) identified five elements needed to improve patients' outcomes in a population increasingly afflicted by chronic conditions:

- evidence-based, planned care
- reorganization of practices to meet the needs of patients who require more time and/or resources, and closer follow-up

- systematic attention to patients' need for information and behavioral change
- ready access to necessary clinical knowledge and expertise
- supportive information systems

Regarding this last point, the Institute of Medicine (2001) pointed to the fact that:

healthcare organizations are only beginning to apply information technology to manage and improve patient care. A great deal of medical information is stored on paper. Communication among clinicians and with patients does not generally make use of the Internet or other contemporary information technology. Hospitals and physician groups operate independently of one another, often providing care without the benefit of complete information on the patient's condition or medical history, services provided in other settings, or medications prescribed by other providers. (p. 30)

Information Technology

The revolution in information technology holds great promise in a number of areas for consumers, patients, clinicians, and all organizations involved in the delivery of healthcare services. A recent report by the National Research Council of The National Academies (2000) identified six major information technology application domains in healthcare: (a) consumer health, (b) clinical care, (c) administrative and financial transactions, (d) public health, (e) professional education, and (g) research. While many applications are currently in use (such as online search for medical information by patients), others, such as remote and virtual surgery and simulation of surgical procedures, are in early stages of development (Institute of Medicine, 2001). While the Internet (and intranets) has been a driving force for changes in the information technology landscape in the past decade, many healthcare applications are not web based. Applications such as administrative billing systems, electronic medical records (EMR), and computerized physician order entry systems frequently remain on legacy systems, which are often built around older mainframe systems and lack integration with other applications.

With respect to consumers/patients and providers, the Committee on the Quality of Health Care in America identified five key areas in which information technology could contribute to an improved delivery system (Institute of Medicine, 2001):

- access to medical knowledge base
- computer-aided decision support systems
- collection and sharing of clinical information
- reduction in medical errors
- enhanced patient and clinician communication

Consumerism

The Internet and other developments in information and communication technologies are contributing to greater consumerism with stronger demands from individuals for information and convenience. People are more demanding; they want timely and easy access to medical information, the latest in technology, and the latest in customer service. Patients are starting to have access to tools that can lead to empowerment and shared decision making regarding their own healthcare. In addition, because of increased fragmentation and specialization in medical care, patients need to take a more active role in managing their health and healthcare to ensure that the various providers involved in providing their care (i.e., primary care physician, specialist, pharmacist, etc.) have the information they need to provide appropriate, quality care.

There are, however, many technical, organizational, behavioral, and social challenges and barriers to increased use of information technology by individuals in managing their health/healthcare. These technological challenges not only include the design of optimal, effective, flexible human–computer interfaces (HCIs), but also issues of privacy and security of information. Those are discussed in the following section.

The Healthcare Regulatory Environment

As Kumar and Chandra (2001) mentioned, the healthcare industry has unique legislative challenges. Among them, two in particular have implications on the field of healthcare informatics: (a) the Health Insurance Portability and Accountability Act (HIPAA); and (b) health information and other business data security. The Health Insurance Portability and Accountability Act was passed and signed into law in 1996 and is designed to improve the portability of health-insurance coverage in the group and individual markets, limit healthcare fraud and abuse, and simplify the administration of health insurance. The act has serious, impending implications for healthcare providers and information managers. Of all its mandates, administrative simplification is perhaps the most critical for healthcare information managers, who are faced with everything from establishing standardized financial and clinical electronic data interchange (EDI) code sets to adopting, assigning, and using unique numerical identifiers for each healthcare provider, payer, patient, and employer. Both HIPAA and the growing role of the Internet-based technologies in delivering healthcare create an even greater and critical concern for data security and privacy.

HIPAA contains five key Titles that apply to every provider, payer (including self-insured employers), and healthcare clearinghouses. Title I pertains to healthcare access, portability, and renewability. Title II relates to healthcare fraud and abuse. Title III addresses tax-related provisions and medical savings accounts. Title IV discusses the application and enforcement of group health-plan requirements. Lastly, Title V relates to revenue offsets.

Currently, the healthcare industry is focusing its efforts on Title II of the Act, also known as Administrative Simplification. Unlike past legislation that has imposed regulations on only Medicare and Medicaid, all healthcare providers, payers, and clearinghouses must meet HIPAA Administrative Simplification regulations within 26 months of each subsection's finalization. There are four subsections to Title II. They are:

- transaction standards
- coding sets
- patient privacy
- security

Transaction Standards and Coding Sets

The Department of Health and Human Services has adopted national standards for electronic administrative and financial healthcare transactions. This is one of the most positive attributes of Title II as it eliminates the conflicting transaction standards, coding sets, and identifiers used by the various players in the industry. Initial transaction and code sets were implemented in October 2002, with revised code set standards defined and implemented by October 2003. The standards relate to enrollment, referrals, claims, payments, eligibility for a health plan, payment and remittance advice, premium payments, health-claim status, and referral certifications and authorizations. By developing national standards, it is anticipated that Electronic Data Interchange (EDI) of healthcare data will significantly reduce administrative costs. The Workgroup on Electronic Data Interchange (2000) estimated that EDI can reduce administrative costs by $26 billion per year by streamlining pre-certification, enrollment status, and reimbursement processes. Code sets have also been defined in which variations are not permitted. Code sets are based on the following standards:

ICD-9 Volumes 1 & 2—Diagnosis Coding
ICD-9 Volume 3—Inpatient Hospital Service Coding
CPT 4—Physician Service Coding
CDT-2—Dental Service Coding
HCPCS—Other Health Related Coding
DRG—Diagnosis Related Groups
NDC—National Drug Coding

HIPAA also establishes unique identifiers for healthcare providers, health plans and payers, employers, and eventually patients/individuals. Currently the patient/individual identifier is pending as no consensus could be reached.

Privacy and Security

Administrative simplification also addresses patient privacy and security. Privacy standards exist for disclosure of patient identifiable information (including demographic data), training of healthcare workforce, individual's rights to see records, procedures for amending inaccuracies in medical records, maintenance of privacy when patient information is exchanged between business associates, designation of a privacy officer, procedures for complaints, sanctions for infractions, duty to mitigate, and document compliance. As a component of disclosure, covered entities (providers, payers, and clearinghouses) must make reasonable efforts not to use or disclose more than the minimum amount of protected patient information necessary to accomplish the intended purpose of the use of disclosure. Wide adoption and use of electronic medical record (EMR) systems will present technical challenges to preventing disclosure of this information. In the past, accessibility of paper-based patient records to clinicians was restricted in part by the chart's physical location (i.e., in medical records storage, on the nursing unit in a hospital). However, EMR makes these charts available to a much larger number of clinicians in a wider range of physical locations, making it more difficult to prevent inappropriate uses of this protected information by authorized users of the EMR.

Security standards require establishment of administrative procedures (policies and procedures), physical safeguards (physical access to computers), technical security (individual and network computer access), and electronic signature (optional, but if used, it must be digital). Implementation requirements of the privacy and security standards have received a tremendous amount of debate. While most feel the intent of the standards is good, the cost implications have left many advocating for a longer implementation period. The Department of Health and Human Services estimated the total cost to implement Subtitle II to be approximately $17 billion to $18 billion over the next 10 years. Some industry experts (Kennedy and Blum, 2001) purported implementation and maintenance costs ranging from $50 to $200 million for a large healthcare provider and/or payer. However, these numbers may increase as more healthcare providers implement wireless networks to support the use of EMR and other clinical systems. Use of wireless networks and centralized databases that support EMR and other systems creates a new set of technical security requirements and issues.

HEALTHCARE INFORMATICS

As defined earlier, healthcare informatics comprises the generation, development, application, and testing of information and communication principles, techniques, theories, and technologies to improve the delivery of healthcare with a focus on the patient/consumer, the provider, and, more importantly, the patient–provider interaction. While a number of systems using a number of platforms and technologies have already been developed and are currently being developed, the field itself is still in its infancy. In addition, current systems have been designed to fit within the existing healthcare delivery system and thus are only marginally or superficially impacting the way healthcare is being delivered. The true potential of healthcare informatics is yet to be experienced and will radically transform the way healthcare is delivered and managed in the future. The following sections provide background information on current healthcare applications with an emphasis on two types of users: (a) the consumer/patient and (b) the clinician/provider.

Consumer Health Information

Consumer health information has been defined as "any information that enables individuals to understand their health and make health-related decisions for themselves or their families" (Harris, 1995). Patrick and Koss (1995) listed a variety of organizations and entities that produce and/or disseminate consumer health information including health-related organizations (involved in provision of, or payment for, healthcare services and supporting services), libraries, health voluntary organizations (i.e., American Heart Association, American Cancer, American Lung, etc., and 6000+ other health-interest societies), broadcast and print media, employers, government agencies,

community-based organizations (i.e., churches, YMCA, agencies for the elderly), networked computer health information providers—"virtual" communities. Methods of dissemination are diverse and include informal channels, printed text, broadcast electronic media, dial-up services (telephone), nonnetworked computer-based information (i.e., CD-ROM, Kiosk technology), and networked interactive computer-based information. Harris (1995) discussed a number of problems with health information including how to interpret conflicting or differing information; how to judge reliability; how to choose among many alternatives; and how to deal with the vast quantities of information, much of which is superficial or even inaccurate.

Interactive Health Communication

Print and broadcast dissemination of health information leads to a number of problems such as the timing relative to need, single directionality (difficulty of following up, clarifying, and understanding), timeliness and relevance vis-à-vis updates, and not being unique to individuals (Patrick & Koss, 1995). As Harris (1995) pointed out, electronic sources of health information have the potential to be more timely and complete than other media and can become more accessible to all citizens. Consequently, the use of interactive health communication or consumer health informatics has become increasingly popular. Unfortunately, the advantages have not been fully realized. The science panel on interactive communication and health defined interactive health communication as "the interaction of an individual—consumer, patient, caregiver, or professional—with or through an electronic device or communication technology to access or transmit health information or to receive guidance and support on health-related issues" (Robinson, Patrick, Eng, & Gustafson, 1998, p. 1284). The panel identified six potential functions of interactive health communication applications: (a) relay information, (b) enable informed decision making, (c) promote healthful behaviors, (d) promote peer information exchange and emotional support, (e) promote self-care, and (d) manage demand for health services.

Ferguson (1997) referred to two types of consumer health informatics: community and clinical. Community consumer health informatics includes resources such as online networks, forums, databases, and websites that anyone with a home computer can access. Clinical consumer health informatics includes resources such as programs or systems developed by clinicians, system developers, or HMOs, and is provided only to selected groups of members or patients. With respect to community consumer health informatics resources, there is a growing concern that the barriers to use of online health resources are becoming similar to those for actual health services. These barriers include cost, geography, literacy, culture, disability, and other factors related to the capacity of people to use services appropriately and effectively. Eng et al. (1998) urged public and private sector organizations to collaborate in reducing the gap between the "haves" and "have nots" for health information. They suggested the need for supporting health information technology access in homes and public places, developing applications for the growing diversity of users, funding research on access-related issues, enhancing literacy in health and technology, training health information intermediaries, and ensuring the quality of health information and support.

In a report on consumer health informatics, the U.S. General Accounting Office (1996) described three general categories of consumer health informatics. They include systems that (a) provide health information to the user (one-way communication); (b) tailor specific information to the user's unique situation (customized communication); or (c) allow the user to *communicate and interact* with healthcare providers or other users (two-way communication). The report also cited a number of issues of concern: access, cost, information quality, security and privacy, computer literacy, copyright, systems development (compatibility, infrastructure, and standardization), and the potential for information overload.

Consumer/Patient Web-Based Applications and e-Health

An estimated 95 million Americans seek health information online (Fox 2005). In 1999, the Benton Foundation estimated there were 10,000 or more health-related websites. However, this number has grown tremendously to meet increasing consumer demand for healthcare information. In early 2006, the Yahoo directory of health-related websites (dir.yahoo.com/health) listed over 10,000 websites related to diseases and conditions alone and thousands of other sites addressing topics ranging from pharmacy to alternative medicine to public health. This variety of information available demonstrates that consumers and patients desire a range of information and services including getting disease-treatment information, obtaining report cards on physicians or hospitals, exchanging information with other patients, interacting online with their physicians, and managing their own health benefits. The sheer volume of information on the Internet exceeds most expectations yet raises problems: efficient search for information, information retrieval, information visualization, human-information processing, understanding, and assimilation. Authors of Internet information need to determine how to best structure the information for use by others. e-Health can be described as the transition of healthcare processes and transactions into the Internet-delivered electronic superhighway. Potential problems exist that are specific to patients seeking electronic health-related information (Sonnenberg, 1997), including the lack of editorial control of information, conflict of interest for website sponsors, and unfiltered information presenting an unbalanced view of medical issues. However, referring primarily to self-help groups, Ferguson (1997) believed that the problem of obtaining bad medical information online is not very different from obtaining bad medical information at cocktail parties, in the tabloids, in magazine ads, etc. In addition to the content of information, there are technological problems that may affect both access and use of health information on the Internet: slow modems, poor institutional Internet connections, firewalls that interfere with Internet traffic, malfunctioning message routers, and heavy Internet usage in the immediate geographic area (Lindberg & Humphreys, 1998). Most current concerns about interactive health communication center on the fact that these applications have the potential to both improve health and to cause harm, thus highlighting the need to ensure their accuracy, quality, safety, and effectiveness (Robinson et al., 1998).

Provider Healthcare Informatics

Providers and healthcare provider organizations have long used health-information systems to support both administrative and clinical functions of healthcare delivery and management. However, despite the fact that healthcare is one of the most information-intensive industries, it has very few state-of-the-art information-management systems. Healthcare is fragmented, with hundreds of thousands of payers, hospitals, physicians, laboratories, medical centers, pharmacies, and clinics, each with its own legacy of systems, hardware, software, and platforms. Electronic data interchange and connectivity issues have become critical. Numerous information systems have been developed and implemented, the most noteworthy being integrated EMRs and computerized physician order entry systems. The following is a discussion of some important clinical applications.

Electronic Medical Records (EMRs)

EMRs are increasingly being adopted in both primary-care and inpatient-care environments. EMRs provide functions to document all clinical processes and patient-related information relevant to the delivery of patient care. EMRs are advocated and used to improve the quality, accessibility, and timeliness of patient medical information. However, because EMRs replace traditional patient charts, they affect a wide range of clinical users and clinical processes. As such, they face a number of barriers to adoption, including cost, a lack of tested systems, problems with data entry, inexperienced vendors, confidentiality concerns, and security concerns (Wager, Lee, White, Ward, & Ornstein, 2000). In addition, EMR may have barriers that are specifically related to the practice of medicine. For example, physician use of EMR while with patients could affect patient perceptions of quality of care or quality of physician–patient interactions. While some past studies examining this issue have not shown this to be the case (i.e., Legler & Oates, 1993), a recent survey of pediatric-care primary physicians using EMR indicated that they felt it reduced eye contact with patients and increased the duration of patient visits (Adams, Mann, & Baushner, 2003). Thus, the impact of these technologies on physician–patient interactions is not clear at this point.

A key factor that will contribute to whether EMR systems become more rapidly adopted in primary-care practices is whether or not physicians themselves perceive the systems as improving quality (of the medical records, patient care, and overall performance). A recent study examined this issue using qualitative methods in five community-based practices that had used EMR for at least two years and did not use a duplicate record system (Wager et al., 2000). Results indicated that many physicians and staff members perceived benefits, such as increased access, an ability to search the system, and improved overall quality of medical records. There were, however, several disadvantages mentioned, including the frequency of downtime and the time necessary to develop customized templates.

In a recent study examining the quality of worklife of family physicians, Karsh, Beasley, Hagenauer, and Sainfort (2001) collected quantitative data about EMR in order to compare perceptions of medical records between physicians using EMR and those not using EMR. Specifically, they assessed whether or not the use of EMR was related to perceptions of improved quality of medical records. The results showed that physicians who used EMR perceived their medical records to be more up to date and accessible. Physicians who used EMR were also more satisfied with the overall quality of their medical-records systems. On the other hand, there were no differences in perceptions of whether medical records could be modified to meet individual needs. This suggests that EMR can have positive impacts on medical records (it is possible, though, that the positive responses were caused not by positive traits of the EMR, but rather because of cognitive dissonance).

The results of Karsh et al.'s (2001) study supported those of qualitative studies of physicians who use EMR. Wager et al. (2000) found that physicians in primary practice who had used EMR for at least two years believed EMR to have many benefits over paper-based systems, including increased access and availability of patient information to multiple users, the ability to search the system, improved overall quality of patient records, improved quality of documentation, increased efficiency, facilitated cross-training, and improved communication within the practice. Thus, it is clear that EMR has the potential to improve the quality of patient records and therefore, possibly, the quality of care. However, to fully capitalize on such systems, they need to be designed to maximize usability, connectivity, and portability while guaranteeing privacy and security.

Computerized Physician Order Entry Systems

Computerized physician order entry (CPOE) systems have substantial potential for improving the medication ordering process because they enable physicians to write orders directly online. They ensure complete, unambiguous, and legible orders; they assist physicians at the time of ordering by suggesting appropriate doses and frequencies, by displaying relevant data to assist in prescription decisions, by checking drugs prescribed for allergies and drug–drug interactions. As such, those systems are believed to potentially reduce the incidence of medical errors in general, and medication errors in particular. In a study of the impact of CPOE, Bates, Teich, Lee, Seger, Kuperman, Ma' Luf, Boyle, and Leape (1999) found that such systems substantially decreased the rate of nonmissed-dose medication errors. In addition, other studies have demonstrated that use of CPOE decreases turn-around time for medication orders as well as radiation and laboratory orders (Mekhijan, Kumar, Kuehn, Bentley, Teater, & Thomas, et al., 2002) and improved adherence to clinical guidelines (Overhage et al., 1997).

However, these systems are not a panacea. In some cases, their use has resulted in negative outcomes, including increased physician order entry time (Bates et al., 1994; Shu et al., 2001) and an increased coordination load on clinical-care teams (Cheng et al., 2003).

For computerized physician-order-entry systems to be fully successful, physicians need to use them. However, as the studies previously discussed indicate, the gains achieved come at a cost, frequently imposed on the physicians using the system. This leads to issues of HCI design, usability, and integration within the care delivery processes to ensure that the burden on these users is minimized. The importance of addressing these issues has also been highlighted in several recent studies showing that the potential for new types of prescribing errors can be created by usability problems in the interface and/or lack of integration of the system with clinical work practices (Koppel, Metlay, Cohen, Abaluck, Localio, Kimmel, & Strom, 2005). In certain care situations, such as critical-care environments, this lack of integration between the clinical processes and patient care needs (i.e., urgency of care) can have dire consequences for patients (Han, Carcillo, Venkataraman, Clark, Watson, & Nguyen, et al., 2005). This emphasizes the importance of identifying and resolving potential usability problems prior to implementation, continually monitoring their use, and incrementally improving their usability in the context of the needs of specific clinical-care environments.

Patient Monitoring Systems

Gardner and Shabot (2001) defined patient monitoring as "repeated or continuous observations or measurement of the patient, his or her physiological function, and the function of life support equipment, for the purpose of guiding management decisions" (p. 444). Electronic patient monitors are used to collect, display, store, and interpret physiological data. Increasingly, such data are collected using newly developed sensors from patients in all care settings as well as in patients' own homes. While such data can be extremely useful for diagnosis, monitoring, alerts, as well as treatment suggestions, the amount, diversity, and complexity of data collected present challenges to HCI design.

Imaging Systems

Imaging is a central element of the healthcare process for diagnosis, treatment plan design, image-guided treatment, and treatment-response evaluation. Greenes and Brinkley (2001) noted that the proliferation in number and kind of images generated in healthcare led to the creation of a subdiscipline of medical informatics called "imaging informatics." In their review and summary of the field, Greenes and Brinkley noted that:

as processing power and storage have become less expensive, newer, computationally intensive capabilities have been widely adopted. Widespread access to images and reports will be demanded throughout healthcare delivery networks. . . . We will see significant growth in image-guided surgery and advances in image-guided minimally invasive therapy as imaging is integrated in real time with the treatment process. Telesurgery will be feasible. (pp. 534–536)

They also highlight that such ambitious evolution of imaging systems will be in part dependent on continued advances in user interfaces and software functionality.

Information-Retrieval Systems

Hersh, Detmer, and Frisse (2001) defined information retrieval as the "science and practice of indentification and efficient use of recorded media" (p. 539). In their review of the systems, they addressed four elements of the information-retrieval process: (a) indexing, (b) query formulation, (c) retrieval, and (d) evaluation and refinement. *Indexing* is the process by which content is represented and stored, and *query formulation* is the process by which user information needs are translated in terms of an actionable query. Focusing on clinicians' use, Hersh et al. (2001) pointed out that the evolution of the Internet has posed new and so far unmet challenges regarding the indexing of (essentially loosely and irregularly structured) documents. They call for a unified medical-language system. Thinking of novice users such as consumers and patients, another significant challenge resides in query formulation, since novice users generally have a hard time expressing ill-defined information needs because of their lack of domain knowledge.

Decision-Support Systems

Decision-support systems (DSS) are computer systems used to support complex decision making. The goal of DSS is to improve the efficiency and/or effectiveness of decision-making processes (Shim, Warkentin, Courtney, Power, Sharda, & Carlsson, 2002). As such, many of the complex decision processes in healthcare can benefit from the use of DSS. One class of DSS in healthcare is clinical decision-support systems (CDSS), which provided decision support to clinical users during the process of providing patient care. Another class of DSS, patient decision-support systems (Patient DSS), is designed to educate and support patients as they make decisions that affect their healthcare. The following sections provide a brief overview of these two types of healthcare-related DSS.

Clinical Decision-Support Systems

Musen, Shahar, and Shortliffe (2001) defined a clinical decision support system (DSS) as "any computer program designed to help health professionals make clinical decisions" (p. 575). They characterized clinical decision support systems along five dimensions:

1. system function
2. mode for giving advice
3. style of communication
4. underlying decision-making process
5. HCI

A large number of applications already exists and can be further developed. The opportunities are almost unlimited. Bates et al. (2001) proposed that appropriate increases in the use of information technology in healthcare, especially the introduction of clinical decision-support systems and better linkages in and among systems, could result in substantial reduction in medical errors. For example, studies have demonstrated that implementation of CPOE systems that include CDSS functions (i.e., drug interaction checking, allergy checking, etc.) have reduced medication errors and adverse drug events in both general-care settings (e.g., Bates, Teich, Lee, Seger, Kuperman, Ma'Luf, Boyle, & Leape, 1999) and pediatric-care settings (i.e., Potts, Barr, Gregory, Wright, & Patel, 2004; Cordero, Kuehn, et al., 2004).

However, as noted by Musen et al. (2001), "systems can fail . . . if they require that a practitioner interrupt the normal pattern of patient care" (p. 587). Thus, new technologies (mobile devices, wireless networks, and distance-communication technologies) as well as novel HCIs (based on speech, gestures, and virtual reality) offer huge potential to maximize usability and permit seamless integration of clinical decision-support systems within complex, dynamic work processes.

Patient Decision-Support Systems

The medical industry has increasingly acknowledged the need to enable patients to participate in making health-related decisions. In order to effectively participate in decision making, most patients need to become better informed about the options available and need help assimilating that information to apply it to the decision at hand. These health-related decisions range from choosing a health-insurance plan to working with their doctor to select a treatment for cancer.

Patient DSS are one of several types of decision aids used to help patients participate in health-related decisions. *Decision aids* are interventions provided to assist individuals as they make a deliberative choice between two or more alternatives (Bekker, Hewison, & Thornton, 2003). Patient DSS supports a patient in one or more stages of making a health-related decision. Much of the past and current patient DSS research and development efforts have targeted patients with life-threatening or chronic diseases. Most have focused on supporting decisions regarding treatment options (i.e., medical or surgical therapies), although a few have examined early-detection and other issues (O'Connor, 1999). Patient DSS that support patients faced with treatment decisions provide one or more of the following functions that facilitate patient participation in decision making (O'Connor, 1999; Scott & Lenert, 1998):

Educate the patient. Provide the patient with information about the treatment alternatives and outcomes, especially highlighting risks and benefits associated with each treatment alternative.

Tailor information. Tailor information content and/or presentation based on patient characteristics such as their health and demographics factors.

Assess preferences. Use preference-elicitation methods to assess the patient's values/preferences for the possible intervention outcomes.

Optimize decision. Optimize the decision outcome based on context, heuristics, probabilities of outcomes, and algorithms.

Reviews of patient decision aids including patient DSS (i.e., Molenaar, Sprangers, Postma-Schuit, Rutgers, Noorlander, Hedricks, & De Haes, 2002; O'Connor, 1999; O'Connor, Rostom, Fiset, Tetroe, Entwistle, Llewellyn-Thomas, Homes-Rovner, Barry, & Jones 1999; O'Connor, Stacey, Entwistle, Llewellyn-Thomas, Rovner, Holmes-Rovner, Tait, Tetroe, Fiset, Barry, & Jones 2003b) have revealed that "decision-support strategies have received generally consistent positive ratings by patients in terms of feasibility, acceptability, length, balance, clarity, amount of information, and usefulness in decision making" (O'Connor, 1999, p. 260). Additionally, they demonstrate a number of benefits for patients, including improved decision-relevant knowledge, reduced decisional conflict, improved congruence between values and choice, more realistic expectations, more active patient participation, and fewer patients who are unable to make a decision. Bekker, Hewison, and Thornton (2003) proposed that many observed benefits are a result of decision aids enabling patients to use more effective cognitive and emotional strategies. Note that the reviews do not indicate a consistent, significant effect of decision aids on decision satisfaction, although a generally positive impact on satisfaction with the decision process is indicated. Use of decision aids tends to reduce the number of patients who take a passive role in decision making and the number who are undecided (O'Connor et al., 2003).

In addition to being used to support patients faced with treatment choices, similar support systems can be used to help healthy individuals make health-related choices. For example, DSS can support healthcare consumers as they make the important choice of choosing health insurance. Considerable effort has been made to educate consumers about health insurance. One system available for learning about health insurance plans is the Health Plan Report Card created by the National Committee for Quality Assurance (NCQA) (National Committee for Quality Assurance, 2003). The report card is, in effect, an interactive decision-support tool that helps consumers compare and select a health plan. The Report Card provides information on a variety of attributes for each plan and enables consumers to identify plans that meet their specifications. This system provides access to plan quality information and uses an easily comparable matrix format and a "star" rating system to simplify comparisons of plans. Despite the increasing availability of information on health plans, many consumers still find the information confusing and the decision difficult. By presenting health-plan information in a way that supports and, perhaps, adapts to a consumer's characteristics and their health-plan preferences, we may improve the usability of the information and enable consumers to make more informed decisions about their health-insurance coverage.

Patient Preferences

Patients' preferences and priorities regarding their health and healthcare are applied each time they participate in making health-related choices. However, an individual's stated preferences can be influenced by external factors. For example, researchers have found that the way framing of information about treatments is handled (positive, negative, or neutral) has affected patients' treatment preferences (Llewellyn-Thomas, McGreal, & Thiel, 1995). It is likely that these tendencies for external factors to influence individual preferences will hold for a variety of health-related choices. Therefore, presenting decision makers with information previously not considered in their decision may cause them to change their preference structure. Slovic (1995) viewed "preference construction as an active process" (p. 369). Taking this view, it is important for patient DSS designers to take a patient (user)-centered approach to the design of these systems to ensure that patient decision outcomes are improved and to avoid introducing systematic biases into the decision process. By structuring the education and decision process to help the patient develop a comprehensive picture of relevant facets of the decision, we can help the patient develop a more comprehensive, rational set of preferences to apply to his or her health choices. In this way, DSS can help patients/healthcare consumers make more informed decisions.

HCI Challenges in Patient DSS

The heterogeneity of the patient/healthcare consumer population presents a number of challenges to Patient DSS designers. Patients/consumers vary greatly in age, physical ability, mental ability, computer experience, healthcare/health condition knowledge, and, as discussed previously, decision-related preferences. They also employ different decision strategies when faced with multi-attribute decision tasks such as many of those faced in healthcare. These decision strategies influence what information the patient (decision maker) wants to see and how he or she views that information. Decision makers frequently use either compensatory or noncompensatory strategies to evaluate alternatives and make a decision. Compensatory strategies are those in which the information about every alternative is weighed and compared (Johnson, 1990). Comparatively, in noncompensatory strategies, alternatives may be eliminated after an incomplete search, thereby reducing the cognitive load of the decision task (Johnson, 1990). Selection of a decision strategy is related to a number of factors, including age (Johnson, 1990) and personality type. In addition to the characteristics of the individual, selection of a decision strategy is also influenced by decision-task characteristics such as the complexity of the decision (Payne, 1976) and the type of decision aids available (Todd & Benbasat, 2001).

Because of the differing information and decision-support needs of various user populations, designing effective decision support for a broad range of patients/healthcare consumers is challenging. Therefore, it is imperative that Patient DSS designers model the needs of their target user population and design systems to meet those needs. For example, certain patient populations may suffer from limited physical or cognitive abilities, making it vital to create accessible designs that meet their special needs.

EVALUATING COMPUTERIZED HEALTHCARE APPLICATIONS

While there has been a proliferation of web-based healthcare applications available for consumers and patients, the issue of evaluating applications and guiding users in choosing the best applications has become extremely important. While evaluation is obviously also important for applications targeted at healthcare professionals, the issue is not as critical because healthcare professionals are experts and can exercise their own judgment in the suitability and quality of applications designed to assist them in their work. Consumers and patients, on the other hand, have no or limited basis for exercising such judgment.

Concerned that the growth of the Internet was leading to too much health information with vast chunks of it incomplete, misleading, or inaccurate, Silberg, Lundberg, and Musacchio (1997) proposed four standards for websites:

1. Authorship: authors and contributors, their affiliations, and relevant credentials should be provided.
2. Attribution: references and sources for all content should be listed clearly, and all relevant copyright information noted.
3. Disclosure: website "ownership" should be prominently and fully disclosed, as should any sponsorship, advertising, underwriting, commercial funding arrangements or support, or potential conflicts of interest. This includes arrangements in which links to other sites are posted because of financial considerations. Similar standards should hold in discussion forums.
4. Currency: dates that content was posted and updated should be indicated.

Murray and Rizzolo (1997) pointed out that:

somehow, just the fact that information is traveling quickly through space and being presented on the computer screen lends it an air of authority which may be beyond its due. Sites with official-sounding names can dupe the inexperienced or uncritical into unquestioned acceptance of the content.

In addition to the standards identified by Silberg et al. (1997), Murray and Rizzolo (1997), and others have cited other criteria for evaluating websites. These include the authority of the author/creator, the accuracy of information and comparability with related sources, the workability (user friendliness, connectivity, search access), the purpose of the resource and the nature of the intended users, whether criteria for information inclusion are stated, the scope and comprehensiveness of the materials, and the uniqueness of the resource.

Jadad and Gagliardi (1998) identified a number of instruments used to provide *external* ratings of websites. These ratings are used to produce awards or quality ratings, provide seals of approval, identify sites that are featured as the "best of the web" or "best" in a given category, and/or to declare sites as meeting quality standards. They attempted to determine what criteria were used to establish these ratings and to establish the degree of validation of these rating instruments. However, few organizations listed the criteria behind their ratings and none provided information on interobserver validity or construct validity. Jadad and Gagliardi (1998) also discussed:

• whether it is desirable to evaluate information on the Internet due to concerns over freedom of expression, excessive regulatory control, etc.

• whether it is possible to evaluate information on the Internet due to the lack of a gold standard for quality information and the controversy surrounding its definition.

They also pointed out that information on the Internet is "different from that found in journals—information is produced and exchanged by groups of people (i.e., health professionals, consumers, vendors, etc.) using multiple formats (i.e., text, video, sound) modified at fast and unpredictable rates, and linked within a highly elaborate and complex network of Internet sites." They concluded that with respect to *external independent evaluation* of sites it is not clear:

• whether evaluation instruments should exist in the first place,
• whether they measure what they claim to measure,
• whether they lead to more good than harm,
• whether users may ever notice, or if they notice, whether they will ignore evidence in support or against desirability, feasibility, or benefits of formal evaluations of health information on the Internet.

Robinson et al. (1998) focused more on *internal evaluation* of applications such as self-evaluation by the sponsors or developers of interactive health-communication applications. They pointed out a number of barriers to evaluating these applications. These include the fact that the media and infrastructure underlying applications is in a dynamic state; applications themselves change frequently; many applications are used in situations where a variety of influences on health outcomes exist, few of which are subject to easy assessment or experimental controls; developers lack familiarity with evaluation methods and tools; and developers often believe that evaluation will delay development, increase front-end costs, and have limited impact on sales.

Patrick and Koss (1995) suggested that the effectiveness of consumer health information should be measured by how rapidly and completely desired messages are communicated and how completely intended changes in behavior occur. Saying that Silberg et al.'s criteria (explicit authorship and sponsorship, attribution of sources, and dating of materials) are not enough, Wyatt (1997) provided far more specific direction for evaluating websites. He believes that evaluation of websites should go beyond mere accountability to assess the quality of their content, functions, and likely impact. Evaluating the content should include:

• determining the accuracy of web material by comparing it to the best evidence, i.e., for effectiveness of treatment—randomized trials, for risk factors, cohort studies, or for diagnostic accuracy—blinded comparisons of test with a standard.

• determining timeliness by checking the date on web pages but recognizing that material may not have been current at that time, so need independent comparison with most up to date facts is preferable.

• determining if people can read and understand web material, asking visitors to record satisfaction is unlikely to reveal problems with comprehension since visitors may not realize they misunderstood something or may blame themselves. So, need a minimum reading age for material for the public. However, more accurate to ask users questions based on the web content.

Evaluating the functions of websites should include determining how easy it is to locate a site, how easy it is to locate material within the site, and whether the site is actually used and by whom. For those investing resources in a website, with respect to evaluating the impact, he suggests looking at the impact on clinical processes, patient outcomes, and its cost effectiveness compared to other methods of delivering the same information. He recommended using randomized control trials as the most appropriate method for determining impact. Finally, with respect to evaluation methodology, he pointed out the importance of choosing appropriate subjects (not technology enthusiasts) and the need to make reliable and valid measurements. Wyatt (1997) believed that:

Ideally, investigators would have access to a library of previously validated measurement methods, such as those used for quality of life. However, few methods are available for testing the effect of information resources on doctors and patients, so investigators must usually develop their own and conduct studies to explore their validity and reliability. (p. 1880)

The Science Panel on Interactive Communication and Health (SciPICH) was convened by the Office of Disease Prevention and Health Promotion of the U.S. Department of Health and Human Services to examine interactive health communication technology and its potential impact on the health of the public. The panel was comprised of 14 experts from a variety of disciplines related to interactive technologies and health, including medicine, HCI, public health, communication sciences, educational technology, and health promotion. One of the products of the SciPICH is an evaluation reporting template (Robinson et al., 1998) for developers and evaluators of interactive health communication applications to help them report evaluation results to those who are considering purchasing or using their applications. The template has four main sections: (a) description of the application, (b) formative and process evaluation, (c) outcome evaluation, and (d) background of evaluators. The panel defined the three different types of evaluation as follows:

Formative evaluation. Used to assess the nature of the problem and the needs of the target audience with a focus on informing and improving program design before implementation. This is conducted prior to or during early application development, and commonly consists of literature reviews, reviews of existing applications, and interviews or focus groups of "experts" or members of the target audience.

Process evaluation. Used to monitor the administrative, organizational, or other operational characteristics of an intervention. This helps developers successfully translate the design into a functional application and is performed during application development. This commonly includes testing the application for functionality and may be known as alpha and beta testing.

Outcome evaluation. Used to examine an intervention's ability to achieve its intended results under ideal conditions (i.e., efficacy) or under real world circumstances (i.e., effec-

tiveness), and its ability to produce benefits in relation to its costs (i.e., efficiency or cost effectiveness). This helps developers learn whether the application is successful at achieving its goals and objectives and is performed after the implementation of the application.

Evaluating the effectiveness of web-based applications designed to relay information and/or enable informed decision making is complicated because the "success" of these particular types of applications is (a) not necessarily always related to observable behaviors; (b) based on the quality *and* usability of the information within the application; and (c) a function of the application itself as well as the users. In terms of outcome-evaluation applications, Robinson et al. (1998) gave examples of the types of questions that such evaluation should address:

1. *How much do users like the application?*
2. *How helpful/useful do users find the application?*
3. *Do users increase their knowledge?*
4. *Do users change their beliefs or attitudes (i.e., self-efficacy, perceived importance, intentions to change behavior, and satisfaction)?*
5. *Do users change their behaviors (i.e., risk factor behaviors, interpersonal interactions, compliance, and use of resources)?*
6. *Are there changes in morbidity or mortality (i.e., symptoms, missed days of school/work, physiologic indicators)?*
7. *Are there effects on cost/resource utilization (i.e., cost-effectiveness analysis)?*
8. *Do organizations or systems change (i.e., resource utilization, and effects on "culture")?*

However, for websites designed primarily to provide information or enable informed decision making, only the first three questions apply. Other potential outcomes related to change in behavior *might* be applicable depending on the nature of decisions made. Consequently, the evaluation of these types of web guides needs to focus on the use of the guides (assuming that users who "like" an application will use it more than those who do not), the usefulness of the guides, the usability of the guides, the ability of the guide to increase knowledge, and the contribution of the guide to decision making. These elements can be integrated into a conceptual framework as shown in Fig. 9.1. Part of this framework draws from the Agency for Health Care Policy and Research report on consumer health informatics and patient decision making (1997) as well as Sainfort and Booske (1996).

The framework suggests that to evaluate the effectiveness of web-based health applications, at least three perspectives (at the bottom of the figure) can be taken individually or in combination: (a) the consumer, (b) the site sponsor, and (c) outside experts. The framework posits that the characteristics of the system under evaluation primarily influence accessibility and usability of information. Then, accessibility, in conjunction with consumer/patient characteristics, influence actual *access* to information. In turn, the usability of this information, again in conjunction with consumer characteristics, will influence actual *use* of information by the consumer. Use of information is a complex construct. The framework emphasizes three main uses of information: (a) knowledge (inquiry, verify, learn, augment,

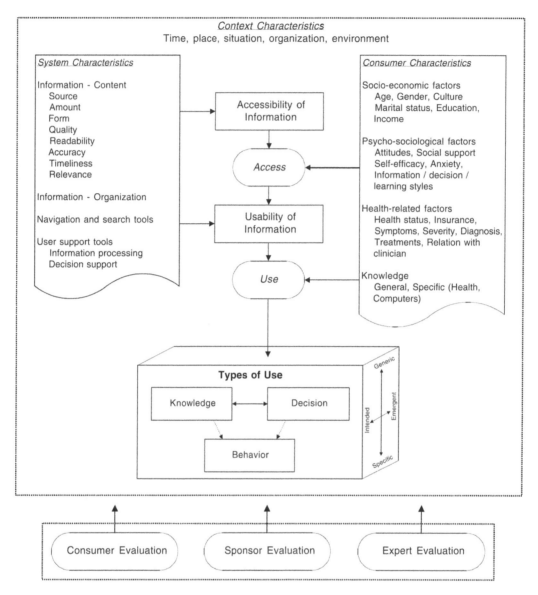

FIGURE 9.1.   Conceptual framework.

etc.), (b) decision making, and (c) behavior (intentions and actual behavior change). All three uses are generally interconnected, with behavior usually following knowledge and/or decision making (whether explicitly or implicitly). Also emphasized in the framework is the fact that "use" can be *generic* (common to most web-based health applications) or *specific* (to the application) and also that use of information accessed can be *intended* (i.e., the information was sought to accomplish a specific purpose or use) or *emergent* (i.e., a piece of information accessed triggers a new use). Finally, the framework shows, surrounding this entire process, that the context (the situation, time, place, organization, etc.) can influence all key elements: accessibility, access, usability, and use. Consumer evaluations are formed because of the experiences they have using the system over time. Expert evaluation usually involves assessment of the system itself and its content, as well as an assessment of its (an-

ticipated) impact on users. Sponsors, evaluation involves both consumer and expert evaluations as well as consideration of organizational resources expended in the design, development, implementation, and operation of the system.

It is important to differentiate information access and quantity from use and usefulness (Booske & Sainfort, 1998). Indeed, while many studies primarily address attempts to measure the "quantity" of information used, others have acknowledged the need to consider the "quality" of information. The introduction of the web as a source of information has introduced a number of additional considerations in evaluation of the use and usefulness of information. The most common measure of information use on the web is the number of "hits" or "visits" to a particular page. Most of these counts do not differentiate between multiple visits by a single user versus multiple users nor do they consider the amount of time spent on a particular page, i.e., whether the page is merely used as a link to elsewhere or as a source of information itself.

In describing three general approaches to information systems evaluation (i.e., the system's output, behavior, and architecture), Orman (1983) defined the quality and quantity of information as the relevant variables in defining an information system's output. However, he went on to say that both quantity and quality of information were "of considerable theoretical interest but of little practical value since neither can be defined or measured with acceptable precision." Orman defined the quality of information produced by a system in terms of "its contributions to the quality of the decisions it aids" and points out that this is "highly influenced by the style and the behavior of the information user and the state of the environment."

This ties in with the concept of the "value" of information, an important part of traditional decision and economic analysis. Within the field of decision science, the primary use of information is to help *make* decisions. However, an examination of literature from other disciplines has shown that there are other ways to conceptualize the value, and thus the effectiveness of information and its usefulness (Orman, 1983). Although it can be thought that the term "information usefulness" could be an analogue for value of information, within the field of HCI, the concept of usefulness includes additional dimensions. The concepts of "perceived usefulness" and "usability" of information *systems* are both receiving more attention as more systems are developed for novice users. For example, Davis, Bagozzi, and Warshaw (1999) developed scales to measure perceived usefulness and ease of use to evaluate specific software in the work setting. They defined perceived usefulness as "a prospective user's subjective probability that using a specific application system will increase his or her job performance within an organizational context." Ease of use was defined as "the degree to which the prospective user expects the target system to be free of effort." Using a seven-point Likert scale, potential users of software responded to six items addressing perceived usefulness and six to assess perceived ease of use.

Of particular relevance in considering the usability, and potentially the usefulness, of information is the manner of display. Although technological advances continue to increase the number of possible methods for disseminating information, the primary *output* media for healthcare information are still print, video, and sound. Print information is often categorized as text, tables, graphs, and figures. There is a quite extensive body of literature comparing the effectiveness of displaying information in tables versus graphs (Jarvenpaa & Dickson, 1988, and Dahlberg, 1991, provided comprehensive summaries of this empirical work). Many of these studies rely on elicited "directed interpretations" where subjects are given specific questions about the information in tables and graphs. In contrast, a more recent study by Carswell and Ramzy (1997) elicited spontaneous interpretation of a series of tables, bar graphs, and line graphs to find out "what information subjects choose to take away from a display rather than their ability to extract information when promoted."

Within the context of decision making, Schkade and Kleinmuntz (1994) found that different characteristics of information display affected aspects of choice processes. They found that organization of information (such as a matrix versus a list) influenced

organization acquisition; form (numeric versus linguistic) influenced information combination and evaluation; while sequence had only a limited effect on acquisition. Johnson, Payne, and Bettman (1988) and others have provided evidence of preference reversals because of different information displays.

Many of the issues surrounding the use of the Internet are issues of usability. Usability is a central notion of the HCI field. Simply put, Shackel (1991) defined usability as the "the capability to be used by humans easily and effectively" while his more formal operational and goal-oriented definition said, "for a system to be usable, the following must be achieved":

Effectiveness. The required range of tasks must be accomplished better than some required level of performance (i.e., in terms of speed and errors); by some required percentage of the specified target range of users; within some required proportion of the range of usage environments.

Learnability. Within some specified time from commissioning and start of user training; based on some specified amount of user training and support; and within some specified relearning time each time for intermittent users.

Flexibility. With flexibility allowing adaptation to some specified percentage variation in tasks and/or environments beyond those first specified.

Attitude. And within acceptable levels of human cost in terms of tiredness, discomfort, frustration, and personal effort; so that satisfaction causes continued and enhanced usage of the system. (p. 48)

In an attempt to place system usability in relation to other system concepts, Shackel (1991) suggested that:

utility (will it do what is needed functionally?), *usability* (will the users actually work it successfully?) and *likeability* (will the users feel it is suitable?) must be balanced in a trade-off against *cost* (what are the capital and running costs and what are the social and organizational consequences?) to arrive at a decision about *acceptability* (on balance the best possible alternative for purchase). (p. 50)

These concepts are all relevant in a decision to purchase or accept an information system, but may not be all relevant when the resource or product under consideration is information. As Orman (1983) pointed out, the value of an information system is different from the value of its information content, "just as the value of a candy machine is different from the value of the candy it dispenses." Furthermore, usability of information is a necessary but not sufficient criterion for information to be useful. (p. 312)

In the context of informing healthcare consumer decisions, Hibbard and Slovic-Jewett (1997) asked how much information is too much. In response to this question, they determined that the critical element is the ability to interpret and integrate information items: integration is a very difficult cognitive process. They suggest that for a consumer, more information is not necessarily better and that the simple provision is not sufficient when the information is complex. In complex-decision situations, it is important to pay particular attention to human-information processing capabilities and differences across individuals. For example, specifically addressing information presentation format, Togo and Hood (1992) found a significant interaction effect between gender and format. Personality differences have also been found to contribute to variation in information processing. The way people gather and evaluate information is at least in part based on their psychological type (Slocum & Hellriegel, 1983). Some people are driven to know details before decisions are made while others feel more comfortable assuming what is not known. Individuals vary on how they receive messages, seek information, organize information, and process information. One approach to categorizing cognitive style, the Myers-Briggs Type Indicator (MBTI), is based on Jung's typology (Myers, 1987).

Other models of learning styles can be used to differentiate among individuals and their preferred methods or strategies for taking in and processing information. Felder (1996) provided a useful summary of these models. For example, Kolb's learning-style model classified people as having a preference for (a) *concrete experience* or *abstract conceptualization* (how they take information in) and (b) *active experimentation* or *reflective observation* (how they internalize information). The Hermann Brain Dominance Instrument (HBDI) classified people in terms of their relative preferences for thinking that are based on the task-specialized functioning of the physical brain, i.e., left brain versus right brain, cerebral versus limbic. The Felder-Silverman Learning Style model classified learners along five dimensions: (a) sensing or intuitive, (b) visual or verbal, (c) inductive or deductive, (d) active or reflective, and (e) sequential or global.

In addition to individual differences, other factors that can affect information processing reflect characteristics of the information itself (John & Cole, 1986), i.e., information quantity, information source, information format (mode of presentation, organization, order), information complexity, nature of access (i.e., voluntary vs. mandatory), instructions in the use of the information, response formats (i.e., recognition, recall, judgment, choice), as well as the interface itself.

Understanding cognitive processes involved in accessing, processing, interpreting, and using healthcare information is critical to the successful design and implementation of web-based healthcare applications. This particular point extends to applications targeted at providers and other healthcare professionals. In looking at the impact of computer-based patient record systems on data collection, knowledge organization, and reasoning, Patel, Kushniruk, Yang, and Yale (2000) indicated that exposure to computer-based patient records was associated with changes in physicians' information gathering and reasoning strategies. They concluded that such technology could have profound influence in shaping cognitive behavior. In such systems, the HCI itself can have a strong influence on information gathering, processing, and reasoning strategies. Recently, Patel, Arocha, and Kaufman (2001) surveyed literature on aspects of medical cognition and suggested that:

cognitive sciences can provide important insights into the nature of the processes involved in human–computer interaction and help improve the design of medical information systems by providing insight into the roles that knowledge, memory, and strategies play in a variety of cognitive activities. (p. 324)

FUTURE OPPORTUNITIES AND CHALLENGES

Information and communication technologies have only begun to impact the healthcare industry. Regarding the use of the Internet, a number of applications will be at the leading edge. Mittman and Cain's 1999 prediction that these applications would include consumer health-information services, online support groups for patients and caregivers, healthcare-provider-information services, provider-patient e-mail, communications infrastructure and transaction services, and EMRs appears to true. Similarly, the barriers they predicted would impede or slow down the development of the Internet in healthcare have largely proven true as well: security concerns, weaknesses inherent to web interfaces (especially browsers, search-engine technology, and inability to interact with legacy systems), mixed or lack of quality of information, physician ambivalence, disarray of current healthcare-information systems, lack of resources for web development, and lack of unified standards for electronic communications and transactions.

For information technology to fully impact healthcare delivery with a focus on both patients and providers, these barriers need to be overcome. We believe that the greatest benefits will be reached in the short term by focusing on the following five areas.

Supporting/Enhancing the Patient–Provider Interaction

The Internet will become a critical element of the physician-patient encounter. Organizational websites will evolve from publishing generic consumer content to providing personalized, online services to all consumers (individuals, patients, providers, etc.). The Internet needs to provide for technology that truly supports two-way interaction between patients and providers and directly impacts care delivery.

Effective technologies need to be developed to fully support, enhance, and extend the patient–physician interaction so as to increase the efficiency of care delivery, increase the quality of care received by patients, as well as increase the effectiveness of the work performed by the physician. This latter point is important since a recent study by Linzer, Konrad, Douglas, McMurray, Pathman, Williams, Schwartz, Gerrity, Scheckler, Bigby, and Rhodes (2000) showed that time stress, defined as reports by the physician that they needed more time for patients than they were allotted, was significantly related to burnout, job dissatisfaction, and patient-care issues. This study was performed with a national sample ($n = 5704$) of physicians in primary care specialties and medical and pediatric subspecialties. It demonstrates that the ability to spend sufficient and quality time with each patient is critical to ensure long-term job satisfaction, avoid physician burnout, and increase quality of patient care.

In developing such technologies, as we mentioned earlier, one would need a full understanding of the various cognitive processes involved in information gathering, knowledge acquisition and organization, reasoning strategies, and decision making. Of particular importance is the recognition of various users with various characteristics potentially using technologies in a variety of situations, contexts, organizations, and environments.

Supporting/Enhancing Collaborative Work Among Providers

The nature of the physician/associate provider (nurse, physician assistant) relationship is evolving from mere interaction to true collaboration with technologies allowing associate providers to increase involvement in clinical decision making and implementation. Technologies to support collaborative work environments in fast-paced, mobile environments are needed.

As mentioned earlier, the healthcare work environment is very complex and involves a variety of healthcare professionals, all with varying needs for information, knowledge, and support. Designing technologies supporting both individual needs as well as collaboration among individuals presents a number of challenges. In a recent study, Jacko, Sears, and Sorensen (2001) showed that different healthcare professionals (physicians, pharmacists, nurses) exhibit very different patterns of use of the Internet for clinical purposes and have very different perceptions of needed enhancements to support their respective needs. The patterns differed in terms of the range and type of information as well as the depth and specificity of information.

Healthcare organizations will be faced with technology integration challenges and will make use of the Internet as well as intranet technologies to manage and organize the variety of applications in use by providers and their patients. Of critical importance for healthcare organizations will be the design of systems to support clinical decisions, knowledge management, organizational learning, administrative transactions, and supply chain.

Developing and Utilizing New Information and Communication Technologies

Wireless, handheld, and nanotechnologies will drive higher adoption rates of the Internet in clinical settings, patient monitoring, and disease management. Both the number of interactive wireless device users and the number of devices will increase significantly in the next few years. New applications for such technologies need to be developed, based on fundamental research conducted to investigate usage of such devices.

Mobile computing and wireless technologies will increasingly become an important part of healthcare's information technology. Turisco and Case (2001) reported the following predictions:

- The number of wireless Internet users will reach 83 million by the end of 2005.
- By the end of 2004, there will be 95 million browser-enabled cellular phones and more than 13 million web-enabled personal digital assistants.
- The wireless LAN market is expected to reach $1 billion in 2001 and this figure will double by 2004.

Turisco and Case (2001) reported that mobile computing applications for healthcare began with reference tools and then moved to transaction-based systems to automate simple clinical and business tasks. They predict that the next step will be to provide multiple integrated applications on a single device. In particular, they mention the following applications: prescription writing, charge capture and coding, lab order entry and results reporting, clinical documentation, alert messaging, and communication, clinical-decision support, medication administration, and inpatient-care solutions.

Designing and Utilizing Adaptive HCIs

The development and integration of information and communication technologies designed to support, facilitate, and enhance patient/provider interactions are critical. In particular, research is needed to design optimal HCIs. A dual focus on (a) increasing job satisfaction and effectiveness for the medical personnel and (b) increasing quality of care as well as safety for patients is needed. To support these objectives, the HCI must possess at least four qualities:

- multimodal, i.e., have the ability to display and accept information in a combination of visual, aural, and haptic modes,
- personalized, i.e., tailored to respond in a manner best suited to the current user and his or her needs,
- multisensor, i.e., have the ability to detect and transmit changes in the user and/or situation,
- adaptive, i.e., have the ability to change its behavior in real time to accommodate user preferences, user disabilities, and changes in the situation/environment.

Optimal interfaces are critical to virtually all applications connecting people to information technologies and people to people via information and communication technologies. Current research is underway to develop intelligent adaptive multimodal interface systems. In the future, interfaces will automatically adapt themselves to the user (capabilities, disabilities, etc.), task, dialogue, environment, and input/output modes in order to maximize the effectiveness of the HCI.

This is especially critical in healthcare for both types of users: consumers/patients and providers. Consumers/patients present a number of challenges regarding the interface. In addition to presenting different personal and sociodemographic characteristics, users in healthcare present varying degrees of health status: healthy consumers, newly diagnosed patients, chronically ill patients, and/or their caregivers will use the same devices in potentially very different ways. Similarly, different providers will have very different characteristics and will perform a multitude of varying and highly dynamic tasks in different contexts and environments.

Moving to e-Health

The opportunities for improving service and decreasing cost via e-commerce technologies and the supply chain are tremendous. In addition, modern information and communication technologies (including sensors, wireless communication, and implant technologies) will enable electronic delivery of health *care*. This is far more comprehensive than the mere electronic delivery of health information to patients and providers and includes new developments such as telemedicine and virtual reality. Krapichler, Haubner, Losch, Schuhmann, Seeman, and

Englmeier (1999) claimed that "virtual environments are likely to be used in the daily clinical routine in [the] medicine of tomorrow" (p. 448). However, a number of barriers will need to be overcome. Organizational barriers to e-Health include infrastructure, organization, culture, and strategy, systems integration, and workflow integration. Technological barriers include integration and security, interface design, connectivity, speed, reliability, and usability issues.

The evolution toward e-Health will involve a number of major changes. For example, organizational websites will evolve from publishing generic consumer content to providing personalized, online interactive services to profitable patient segments. Delivery organizations will focus on direct-to-patient relationship building, migrating to a health system truly centered on patients and communities. Communities of interest will rapidly expand to become a force in healthcare navigation. Wireless and handheld technologies will drive higher adoption rates of the Internet in clinical settings. The Internet will become a critical element of the physician-patient encounter and will support and enhance the patient-provider interaction. These changes will lead to a restructuring of the healthcare industry. Digital health plans will emerge and compete or threaten the traditional healthcare-payer business model. Virtual networks will begin to emerge around specialty services to provide efficient personalized healthcare.

References

Adams, W. G., Mann, A. M., & Bauchner, H. (2003). Use of an electronic medical record improves the quality of urban pediatric primary care. *Pediatrics, 111*(3), 626–632.

Bates, D. W., Boyle, D. L., & Teich, J. M. (1994). *Impact of computerized physician order entry on physician time.* Paper presented at the Annual Symp Comput Appl Med Care.

Bates, D. W., Teich, J. M., Lee, J., Seger, D., Kuperman, G. J., Ma'Luf, N. I., Boyle, D., & Leape, L. (1999). The Impact of Computerized Physician Order Entry on Medication Error Prevention. *J Am Med Inform Assoc,6*(4), 313–21.

Bates, D. W., Cohen, M., Leape, L. L., Overhage, M., Shabot, M. M., & Sheridan, T. (2001). Reducing the frequency of errors in medicine using information technology. *JAMIA, 8*(4), 299–308.

Bekker, H. L., Hewison, J., & Thornton, J. G. (2003). Understanding why decision aids work: Linking process with outcome. *Patient Education and Counseling, 50,* 323–329.

Benton Foundation. (1999). Networking for better care: Health care in the information age. Retrieved December 5, 2001, from http://www.benton.org/Library/health/

BlueCross BlueShield Association. (2006). Medical cost reference guide. Retrieved April 27, 2006, from http://www.bcbs.com/mcrg/chap1/index.html

Booske, B. C., & Sainfort, F. (1998). Relationships between quantitative and qualitative measures of information use. *International Journal of Human–Computer Interaction, 10*(1), 1–21.

Booske, B. C., Sainfort, F., Hundt, A. S. (1999) Eliciting Consumer Preferences for Health Plans. *Health Services Research ;34*(4):839–54.

Carswell, C. M., & Ramzy, C. (1997). Graphing small data sets: Should we bother? *Behaviour and Information Technology, 16*(2), 61–70.

Chassin, M. R., Galvin, R. W., & The National Roundtable on Health Care Quality. (1998). The urgent need to improve health care quality. *JAMIA, 280*(11), 1000–1005.

Cheng, C. H., Goldstein, M. K., Geller, E., & Levitt, R. E. (2003). The effects of CPOE on ICU workflow: an observational study. In: *AMIA Annual Symposium,* pp. 150–44. November 8–12, 2003, Washington, DC.

Cordero, L., Kuehn, L., Kumar, R. R., Mekhjian H. S. Impact of Computerized Physician Order Entry on Clinical Practice in a Newborn Intensive Care Unit. *Journal of Perinatology* 2004;*24*(2):88–93.

Dahlberg, T. (1991). Effectiveness of report format and aggregation: An approach to matching task characteristics and the nature of formats. *Acta Academie Oeconomicae Helsingiensis Series A: 76.* Helsinki, Finland: Helsinki School of Economics and Business Administration.

Davis, F. D., Bagozzi, R. P., & Warshaw, P. R. (1999) User acceptance of computer technology: A comparison of two theoretical models. *Management Science, 35,* 982–1003.

Eng, T. R., Maxfield, A., Patrick, K., Deering, M. J., Ratzan, S. C., & Gustafson, D. H. (1998). Access to health information and support: A public highway or a private road? *Journal of the American Medical Association 280*(15),1371–1375.

Felder, R. M., "Matters of Style." *ASEE Prism, 6*(4), 18–23 (1996 December).

Ferguson, T. (1997). Health online and the empowered medical consumer. *Journal on Quality Improvement, 23*(5), 251–257.

Fox, S. (2005). *Health information online.* Retrieved April 27, 2006, from http://www.pewinternet.org/pdfs/PIP_Healthtopics_May05.pdf

Gardner, R. M., & Shabot, M. (2001). Patient monitoring systems. In E. H. Shortliffe, L. E. Perreault, G. Wiederhold, & L. Fagan (Eds.), *Medical informatics: Computer applications in health care and biomedicine (2nd ed.)* (pp. 443–484). New York: Springer.

Greenes, R. A., & Brinkley, J. F. (2001). Imaging systems. In E. H. Shortliffe, L. E. Perreault, G. Wiederhold, & L. Fagan (Eds.), *Medical informatics: Computer applications in health care and biomedicine (2nd ed.)* (pp. 485–538). New York: Springer.

Han, Y. Y., Carcillo, J. A., Venkataraman, S. T., Clark, R. S. B., Watson, R. S., Nguyen, T. C., Bayir, H., & Orr, R. A. (2005) Unexpected Increased Mortality After Implementation of a Commercially Sold Computerized Physician Order Entry System. *Pediatrics, 116*(6):1506–12.

Harris, J. (1995). Consumer health information demand and delivery: A preliminary assessment. Partnerships for Networked Health Information for the Public. Rango Mirage, California. May 14–16, 1995. Summary Conference Report. Office of Disease Prevention and Health Promotion, U.S. Department of Health and Human Services, Washington, DC.

Hersh, W. R., Detmer, W. M., & Frisse, E. H. (2001). Information-retrieval systems. In E. H. Shortliffe, L. E. Perreault, G. Wiederhold, & L. Fagan (Eds.), *Medical informatics: Computer applications in health care and biomedicine (2nd ed.)* (pp. 539–572). New York: Springer.

Hibbard, J. H., Slovic, P., & Jewett, J. J. (1997). Informing consumer decisions in health care: Implications from decision-making research. *The Milbank Quarterly, 75*(3), 395–414.

Hoffman, C., Rice, D. P., & Sung, H. Y. (1996). Persons with chronic conditions: Their prevalence and costs. *JAMA, 276*(18), 1473–1479.

Institute of Medicine. (2000). *To err is human: Building a safer health system.* Washington, DC: National Academy Press.

Institute of Medicine. (2001). *Crossing the quality chasm: A new health system for the 21st century.* Washington, DC: National Academy Press.

Jacko, J. A., Sears, A., & Sorensen, S. J. (2001). A framework for usability: Healthcare professionals and the Internet. *Ergonomics, 44*(11), 989–1007.

Jadad, A. R., & Gagliardi, A. (1998). Rating health information on the Internet: Navigating to knowledge or to Babel? *JAMA, 279*(8), 611–614.

Jarvenpaa, S. L., & Dickson, G. W. (1988). Graphics and managerial decision making: Research based guidelines. *Communications of the ACM, 31*(6), 764–774.

John, D. R., & Cole, C. A. (1986). Age differences in information processing: Understanding deficits in young and elder consumers. *Journal of Consumer Research, 13*(12), 297–315.

Johnson, M. M. S. (1990). Age differences in decision making: A process methodology for examining strategic information processing. *Journal of Gerontology, 45*(2), 75–78.

Johnson, E. J., Payne, J. W., & Bettman, J. R. (1988). Information displays and preference reversals. *Organizational Behavior and Human Decision Processes, 42*, 1–21.

Karsh, B., Beasley, J. W., Hagenauer, M. E., & Sainfort, F. (2001). Do electronic medical records improve the quality of medical records? In M. J. Smith & G. Salvendy (Eds.), *Systems, Social and Internationalization Design Aspects of Human-Computer-Interaction* (pp. 908–912). Mahwah, New Jersey: Lawrence Erlbaum Associates.

Kaushal, R., Shojania, K. G., & Bates, D. W. (2003). Effects of Computerized Physician Order Entry and Clinical Decision Support Systems on Medication Safety: A Systematic Review. *Archives of Internal Medicine 163*(12), 1409–1416.

Kennedy, J., & Blum, R. (2001, March). HIPAA: The new network security imperative. *Lucent Technologies*, 15.

Koppel, R., Metlay, J. P., Cohen, A., Abaluck, B., Localio, A. R., Kimmel, S. E., & Strom, B. L. (2005). Role of computerized physician order entry systems in facilitating medication errors. *JAMA, 293*(10), 1197–1203.

Krapichler, C., Haubner, M., Losch, A., Schuhmann, D., Seemann, M., & Englmeier, K. H. (1999). Physicians in virtual environments—Multimodal human–computer interaction. *Interacting with Computers, 11*, 427–452.

Kumar, S., & Chandra, C. (2001, March). A healthy change. *IIE Solutions*, 28–33.

Legler, J. D., & Oates, R. (1993). Patients' reactions to physician use of a computerized medical record system during clinical encounters. *The Journal of Family Practice, 37*(3), 241–244.

Levin, I. P., & Jasper, J. D. (1995). Phased narrowing: A new process tracing method for decision making. *Organizational Behavior and Human Decision Processes, 64*(1), 1–8.

The Life and Health Insurance Foundation for Education. (2003). Health insurance. Retrieved November 24, 2003, from http://www.life-line .org/health/index.html

Lindberg, D. A. B., & Humphreys, B. L. (1998). Medicine and health on the Internet: The good, the bad, and the ugly. *JAMA, 280*(15), 1303–1304.

Linzer, M., Konrad, T. R., Douglas, J., McMurray, J. E., Pathman, D. E., Williams, E. S., Schwartz, M. D., Gerrity, M., Scheckler, W., Bigby, J. A., & Rhodes, E. (2000) Managed care, time pressure, and physician job satisfaction: Results from the physician worklife study. *Journal of General Internal Medicine, 15*(7), 441–50.

Llewellyn-Thomas, H. A., McGreal, M. J., & Thiel, E. C. (1995). Cancer patients' decision making and trial-entry preferences: The effects of "framing" information about short-term toxicity and long-term survival. *Medical Decision Making, 15*(1), 4–12.

Mekhjian, H. S., Kumar, R. R., Kuehn, L., Bentley, T. D., Teater, P., Thomas, A., et al. (2002). Immediate benefits realized following implementation of physician order entry at an academic medical center. *JAMA, 9*(5), 529–539.

Mittman, R. & Cain, M. (1999). *The future of the Internet in health care.* Oakland, CA: California HealthCare Foundation.

Molenaar, S., Sprangers M. A. G., Postma-Schuit F. C. E., Rutgers, E. J. T., Noorlander, J., Hedriks, J., & De Haes, H. (2002). Feasability and effects of decision aids. *Med Decis Making, 20*(1), 112–27.

Murray, P. J., & Rizzolo, M. A. (1997). Reviewing and evaluating websites—some suggested guidelines. *Nursing Standard Online, 11*(45). Retrieved July 30, 1997, from http://www.nursing-standard .co.uk/vol11-45/ol-art.htm

Musen, M. A., Shahar, Y., & Shortliffe, E. H. (2001). Clinical decision-support systems. In E. H. Shortliffe, L. E. Perreault, G. Wiederhold, & L. Fagan (Eds.), *Medical Informatics: Computer Applications in Health Care and Biomedicine (2nd ed.)* (pp. 573–609). New York: Springer.

Myers, I. B. (1987). *Introduction to type.* Palo Alto, CA: Consulting Psychologists Press.

The National Committee for Quality Assurance. (2003). *Health plan report card.* Retrieved November 24, 2003, from http://hprc.ncqa .org/index.asp

National Research Council. (2000). *Networking health: prescriptions for the Internet.* Washington DC: National Academy Press.

O'Connor, A. M. (1999). Consumer/patient decision support in the new millennium: Where should our research take us? *Canadian Journal of Nursing Research, 30*(4), 257–261.

O'Connor, A. M., Rostom, A., Fiset, V., Tetroe, J., Entwistle, V., Llewellyn-Thomas, H. A., Homes-Rovner, M., Barry, M., & Jones, J. (1999) Decision aids for patients facing health treatment or screening decisions: Systematic review. *British Medical Journal, 18*, 731–4.

O'Connor, A. M., Stacey, D., Entwistle, V., Llewellyn-Thomas, H. A., Rovner, D., Holmes-Rovner, M., Tait, V., Tetroe, J., Fiset, V., Barry, M., & Jones, J. (2003a) The Cochrane database of systematic reviews: Decision aids for people facing health treatment or screening decisions. The Cochrane Library. Available from: http://gateway1.ovid .com:80/ovidweb.cgi.

O'Connor, A. M., Stacey, D., Entwistle, V., Llewellyn-Thomas, H. A., Rovner, D., Holmes-Rovner, M., Tait, V., Tetroe, J., Fiset, V., Barry, M., & Jones, J. (2003b) Decision aids for people facing health treatment or screening decisions (Cochrane Methodology Review). In: The Cochrane Library, Issue 4, 2003. Chichester, UK: John Wiley & Sons, Ltd.

Orman, L. (1983). Information independent evaluation of information systems. *Information and Management, 6*, 309–316.

Ornstein, S., & Bearden, A. (1994). Patient perspectives on computer-based medical records. *The Journal of Family Practice, 38*(6), 606–610.

Overhage, J. M., Tierney, W. M., Zhou, X-HA, & McDonald, C. J. (1997). A randomized trial of "corollary orders" to prevent errors of omission. *J Am Med Inform Assoc, 4*(5), 364–75.

Patel, V. L., Kushniruk, A. W., Yang, S., & Yale, J. F. (2000). Impact of a computer-based patient record system on data collection, knowledge organization, and reasoning. *JAMIA, 7*(6), 569–585.

Patel, V. L., Arocha, J. F., & Kaufman, D. R. (2001). A primer on aspects of cognition for medical informatics. *JAMIA, 8*(4), 324–343.

Patrick, K., & Koss, S. (1995). Consumer health information "White Paper." Consumer Health Information Subgroup, Health Information and Application Working Group, Committee on Applications and Technology, Information Infrastructure Task Force. Working Draft, May 15, 1995.

Payne, J. W. (1976). Task complexity and contingent processing in decision making: An information search and protocol analysis. *Organizational Behavior and Human Performance, 16*, 366–87.

Potts, A. L., Barr, F. E., Gregory, D. F., Wright, L., & Patel, N. R. (2004). Computerized physician order entry and medication errors in a pediatric critical care unit. *Pediatrics, 113*(1), 59–63.

Robinson, T. N., Patrick, K., Eng, T. R., & Gustafson, D. (1998). An evidence-based approach to interactive health communication. *JAMA, 280*(14), 1264–1269.

Sainfort, F., & Booske, B. C. (1996). Role of information in consumer selection of health plans. *Health Care Financing Review, 18*(1), 31–54.

Schkade, D. A., & Kleinmuntz, D. N. (1994). Information displays and choice processes: differential effects of organization, form, and sequence. *Organizational Behavior and Human Decision Processes, 57,* 319–337.

Schoen, C., Davis, K., Osborn, R., & Blendon, R. (2000, October). Commonwealth Fund 2000 International Health Policy Survey of Physicians' Perspectives on Quality. New York, NY: Commonwealth Fund.

Scott, G. C., & Lenert, L. A. (1998). Extending contemporary decision support system designs to patient-oriented systems. Paper presented at the American Medical Informatics Association, Orlando, FL.

Shackel, B. (1991). Usability—Context, framework, definition, design and evaluation. In B. Shackel & S. Richardson (Eds.), *Human factors for informatics usability* (pp. 376–80). Cambridge, UK: Cambridge University Press.

Shim, J. P., Warkentin, M., Courtney, J. F., Power, D. J., Sharda, R., & Carlsson, C. (2002). Past, present, and future of decision support technology. *Decision Support Systems, 33,* 111–126.

Shu, K., Boyle, D., Spurr, C. D., Horsky, J., Heiman, H., O'Connor, P., et al. (2001). Comparison of time spent writing orders on paper with computerized physician order entry. *Medinfo, 10*(Pt 2), 1207–1211.

Silberg, W. M., Lundberg, G. D., & Musacchio, R. A. (1997). Assessing, controlling, and assuring the quality of medical information on the Internet: Caveant lector et viewor—Let the reader and viewer beware. *JAMA, 277*(15), 1244–1245.

Slocum, J. W., & Hellriegel, D. (1983, July/August). A look at how managers minds work. *Business Horizons,* 58–68.

Slovic, P. (1995). The construction of preference. *American Psychologist 50*(5), 364–371.

Sonnenberg, F. A. (1997, January). Health information on the Internet: Opportunities and pitfalls. *Arch Intern Med, 157,* 151–152.

Todd, P., & Benbasat, I. (1993). An experimental investigation on the relationship between decision makers, decision aids, and decision making effort. *INFOR, 31*(2), 80–100.

Todd, P., & Benbasat, I. (2001). An experimental investigation of the impact of computer based decision aids on decision making strategies. *Information Systems Research, 2*(2), 87–115.

Togo, D. F., & Hood, J. N. (1992). Quantitative information presentation and gender: An interaction effect. *The Journal of General Psychology, 119*(2), 161–167.

Turisco, F., & Case, J. (2001). *Wireless and mobile computing.* Oakland, CA: California Healthcare Foundation.

U.S. Department of Health and Human Services. (1997). *Consumer health informatics and patient decision making* (AHCPR Publication No. 98-N001). Rockville, MD: Author.

U.S. General Accounting Office (1996). *Consumer Health Informatics: Emerging Issues.* Report to the Chairman, Subcommittee on Human Resources and Intergovernmental Relations, House Committee on Government Reform and Oversight. GAO/AIMD-96-86, July 1996. Washington, DC.

Wager, K. A., Lee, F. W., White, A. W., Ward, D. M., & Ornstein, S. M. (2000). Impact of an electronic medical record system on community-based primary care practices. *Journal of the American Board of Family Practice, 13,* 338–348.

Wagner, E. H., Austin, B. T., & Von Korff, M. (1996). Organizing care for patients with chronic illness. *Milbank Quarterly, 74*(4), 511–542.

Workgroup on Electronic Data Interchange. (2000, July/August). HIPAA: Changing the health care landscape. *Oncology Issues,* 21–23.

World Health Organization. (2000). The World Health Report 2000: Health systems: improving performance. Retrieved April 11, 2002, from http://www.who.int/whr/2000/index.htm

Wyatt, J. C. (1997). Commentary: Measuring quality and impact of the World Wide Web. *British Medical Journal, 314*(7098), 1879–1881.

·10·

WHY WE PLAY: AFFECT AND THE FUN OF GAMES

Designing Emotions for Games, Entertainment Interfaces, and Interactive Products

Nicole Lazzaro
XEODesign®, Inc.

WITHOUT EMOTION THERE IS NO GAME

Shakespeare designed the emotional space between characters; game developers design the emotional space between players and game.

Researchers have only just begun to explore the role emotion plays in human activities. In designing emotional responses, most aesthetic disciplines treat audiences as consumers of content and pay little attention to designing emotions from interaction or contribution. Whether it is a movie, a bottle of perfume, or a website advertisement, a broadcast model is used to elicit emotions from observers rather than participants. Scientific research is now beginning to show how emotion influences cognition and behavior offering new opportunities to solicit emotion through action. Emotion emerges from and plays a part in most activities, from following a goal to just goofing around (Damasio, 1994; Ekman, 2003; Norman, 2004; Lazzaro, 2004b). Video games lead the way as interactive products that create emotion. More emotional than software and more interactive than films, games manipulate player affect to create poignant experiences. How they do this provides lessons for the design of games, entertainment interfaces, and other interactive products.

Emotion is essential to maintain player focus, make decisions, perform, learn, and enjoy the process of play. Emotion-rich stimuli grab players' attention, such as a swashbuckling adventure in *Sid Meier's Pirates!* The emotions surrounding swordplay increase players' immersion and negative affect or mindset focuses players on applying effort to overcome obstacles. Meanwhile, positive affect from finding pirate treasure improves exploration of alternatives. Exploring options to get the player's sticky ball up on top of a table in *Katamari Damacy* is made easier by the positive affect created as it squashes and picks up cutely rendered bon-bons, toys, and other items. Strong emotional states also allow easier formation and recollection of memories, especially if the user's emotion matches the emotion of the item to be remembered (Ekman, 2003). Special moves in *Top Spin Tennis* that offer an emotional response in an opponent are easier to remember and motivate the search for more. Games are innovators in the design of emotional responses integrated into the activity to accelerate it or provide friction for game goals. Each affective state, each emotion, carves a unique signature into a player's psychology, physiology, and behavioral state to create a player experience. Hoping that graphic realism alone will create emotions is nearly the same as adding more background color to fix a usability issue.

Lessons about game experiences inform entertainment interfaces and other product experiences. Games are not just entertaining; they are self-motivating activities. User-experience designers for all types of products can take advantage of how games create these emotions from participation. To transition from interface design to user experience design, interfaces need to be more than "transparent." In games, the interface that makes everything easy robs the player of the fun. Pushing a button for a car to drive itself is less thrilling than winning the Grand Prix. In software, the interface cannot do it all because it does not have the user's knowledge. Therefore, not only must interfaces get out of a user's way, they should allow the user to express him or herself by motivating and supporting the cognitive and behavioral functions required for use. Designers can take inspiration from games to fashion emotional responses during interaction; for instance, computers can detect and respond to user emotion, such as "being more helpful" when software detects frustration in a user's face or fingertips. Like interchangeable colored lenses, games employ emotion-producing attributes to support human performance: rose-colored here for a mood boost, yellow for sharper vision later on, green night vision with a distance indicator for the dark corners, and black to win style points from others. Current usability methods (increasing efficiency, effectiveness, and satisfaction) mostly remove frustration points; they do not yet include techniques to measure and craft other emotions. To exaggerate, a 100% usable product would be boring once it eliminates all the challenge. Customers strap software onto their boots like a crampon, but it should not do the job so well that it climbs the mountain for them (Lazzaro, 2004a). It is not a productivity tool if one button-click creates the whole spreadsheet. User experiences should focus on making the process of the task not only easier, but also more enjoyable.

Forget Usability: Make It Fun

"I don't want to feel like the game just wasted two hours of my life."
—A *Might and Magic* player

Wasting time has a whole different meaning to a gamer. Unlike a spreadsheet, the outcome of a game is in the experience of play rather than in the quality of the end result and so is harder to quantify. Traditionally, productivity focuses more on designing a process that creates a better end-product or result. Game design, on the other hand, focuses mostly on designing interactive play that is enjoyable in its own right. Rather than efficiency, game enjoyment requires the pure pleasure of the experience and a fair degree of frustration (Lazzaro, 2004a). Similar to user experiences, player experiences are created when what happens in the game affects the player internally as well as externally. Player experiences are the combination of emotion, thoughts, and other sensations that occur inside and in between players during play. Beyond usability, player experiences design focuses on affect as well as ease of use (Fig. 10.1).

Productivity Software Goals	Game Goal
Task completion	Entertainment
Eliminate errors	Fun to beat obstacles
External reward	Intrinsic reward
Outcome-based rewards	Process is its own reward
Intuitive	New things to learn
Reduce workload	Increase workload
Assumes technology needs to be humanized	Assumes humans need to be challenged

FIGURE 10.1. The goals of productivity and game experiences have several important differences. (Source: Lazzaro & Keeker, 2004)

The interactive entertainment offered by games provides unique opportunities to create emotions in the player and unique challenges for the design professional. A game must be usable enough to play but not so usable as to allow someone to push a button and win. Players crave the illusion of superior control that allows them to accomplish more than others including out-thinking the game designer, but not so much control that they lose their way and do not have a good time.

Emotions Are for More than Entertainment

"Experience is the feeling of what happens."
—Damasio, 1994

Playing games in their discretionary time, gamers mainly play for the emotions the games create. Recent neuropsychology research suggests that two interconnected information-processing systems continually scan the environment to create a person's experience of the world. A person's *cognitive system* interprets and represents the world internally in order to reason, understand, and interact with it. A person's *affective system* interprets external and internal stimuli relative to goals and needs. This affective system kicks in with an emotional and physiological reaction before a cognitive response is ready. Ideas, thoughts, memories, and knowledge are components of cognition; emotions, moods, sentiments, and other internal sensations comprise a person's affective response (Damasio, 1994; Norman, 2004). From a cognitive-psychology perspective, cognition understands the world and affect evaluates it (Norman). On a basic level, we bring items with positive affect (sweet-tasting, soothing voices, warm to the touch) closer to us. We push (generally speaking) objects with negative affect (bitter, bloody, sharp, diseased) away from us. In the context of games, the study of affect must also include the discussion of enjoyment.

Player experiences emerge inside the player from the process of interacting with the game. Player-experience design crafts these cognitive and affective responses in conjunction with user behavior. Therefore, the design of player experiences must refine not only the gamer's cognitive response to a system, for example, by reducing complexity; it must also design the gamer's affective response, for example, by inspiring interest or rewarding success in order to increase engagement and support cognitive tasks. For the purposes of game design, affect supports cognitive as well as behavioral tasks, because emotion has a significant effect on enjoyment, attention, memory, learning, and performance.

Emotion and cognition walk hand in hand. According to Ekman, beyond entertainment, emotions are about goals and the things we care about (2003). This makes the creation of emotions ideally suited for game design because most gameplay offers goals with levels and scores to indicate progress. In films, we feel emotions only if we somehow identify with the characters on the screen and vicariously feel their emotions (Boorstin, 1990). This also happens in games, but more central to interactive entertainment is when players feel emotions from what they accomplish and fail at during the game. In productivity, feedback on effectiveness, such as creating a sales presentation, happens after the task of creating it is complete.

In games, the success-feedback loop is immediate and built into the process of play.

REQUIREMENTS FOR A PLAYER EXPERIENCE FRAMEWORK

Emotion's Five Impacts on Play

"Emotions prepare us to deal with important events without us having to think about what to do."
—Ekman, 2003, p. 20

Games entertain with emotions so players enjoy the ride.

Emotions impact player experiences in five ways. What players like most about games is not the packaging, graphics, or the AI, but the total experience that the game creates for the player. This experience lies in the cognitive, affective, and behavioral changes gamers create for themselves as they play. Emotion generates a big part of the entertainment value, the opportunity for challenge and mastery, the thrill from novelty, the ticket to relaxation, and the opportunity to hang out with friends. Customers buy this ride designed by game designers. These emotions create moving gameplay and make a victory taste sweet (Figure 10.2).

Not only does the act of play produce emotions, but also emotions and affect substantially influence the player and how he or she plays. Unlike user experiences, the primary aim of player experiences is to move the player emotionally along with or counter to the game goal. Such techniques heighten the emotional response in the player. Emotions entertain players, focus their attention, help them decide, aid their performance, and assist and motivate learning. Enjoyable emotions from gameplay increase motivation to play further in order to experience more emotions, which makes games self-motivational. Emotions are there from the first click to final volley.

FIGURE 10.2. Game interaction involves and creates many emotions that are clearly visible on the face during play, as seen here during the tutorial for a popular action game. Copyright© 2004 XEODesign, Inc. All rights reserved.

Emotion During Play Helps Gamers

1. *Enjoy:* Creates entertainment from strong shifts in internal sensations
2. *Focus:* Directs effort and attention
3. *Decide:* Aids decision-making
4. *Perform:* Supports different approaches to action and execution
5. *Learn:* Provides motivation for learning, aids in memory, and rewards progress

Games heighten emotional responses to increase *enjoyment*. *Mario Kart*, for example, adjusts speed of players to keep them together for "close" races; likewise, *Jak and Daxter* keeps players just ahead of rolling boulders. A choice between certain death and escape via a narrow window ledge in *Tom Clancy's Splinter Cell* provides *focus* to fill attention in a way that scrolling through options in a word processor's drop-down menu or walking down an office corridor of options does not. Strong exaggerated affect aids player *decisions* whether it is to attack a goblin in *Worlds of Warcraft* (*WOW*) or grant a Sim bathroom privileges in *The Sims*. Gameplay creates moods that aid players to perform; negative affect during a fire fight in *Battlefield* improves pursuing a narrow course of alternatives such as stopping a sniper, while positive affect from munching letter sounds in *Bookworm* improves identification of new word patterns in this *Scrabble*-like tile game. A player *learns* to keep diners happy in *Diner Dash* when an angry customer empties the tip jar.

1. Enjoy: Emotions create strong shifts in internal sensations to heighten and refresh player experiences

Suspend players over boiling lava or confront them with hideous boss monsters and the game heightens their emotions. Players look for emotional rewards as well as a high score. Much of the enjoyment comes from the player's affective response and ultimately separates player experiences from user experiences. In games, player participation is essential. The emotions come from the players' efforts in accomplishing a task and game enjoyment centers on the experience of strong emotional shifts from their actions. Games provide the environment where the player becomes the central hero to accomplish the extraordinary. The buddy rescued or the enemy vanquished in *Battlefield* is no more real than in film, but the player's role in the achievement is. An interactive medium, the choices offered by the game (including the graphical user interface) must sharply enhance the experience. A poor interface reduces interactivity and harms the game experience. Therefore, a large part of good game design lies in creating effective interfaces that also create emotional responses. Movies, by comparison, invite the audience to share in the joys and sorrows of characters on screen. Where games move beyond film to claim their true power is by rewarding player action with emotions. Movies can never hand the audience a jet ski for the thrill of stopping the impending global thermonuclear war. As part of their unique value proposition, games have to.

2. Focus: Emotion directs effort and attention aids or influences many aspects of cognition by focusing attention, providing immediate feedback and rewards

Emotion supports cognitive tasks by directing, focusing, and holding attention, creating absorbing engagement, and at the same time allowing emotion-laden ideas to dominate thought (Clore & Gasper, 2000 as cited in Brave & Nass, 2003). Events are filtered through mood, we attend more to thoughts that match the current mood, and ideas similar to a user's current mood are remembered better (Ekman, 2003; Thorson & Friestaad, 1985 as cited in Brave & Nass, 2003). Emotional stimuli rewards more detailed analysis.

In games, tight feedback loops reward player actions with immediate visual and audio feedback that motivate the player to want to take another action. This rewarding-stimulus-response loop is a powerful motivating force that reaches a wide massmarket audience. Several game-design techniques magnify the effect emotions already have to focus players on a task. For example, game obstacles sometimes use negative affect to increase player focus and, at other times, they use positive affect to increase creativity and problem solving to provide interesting behaviors or situations to encourage players to explore (using positive affect to increase creativity and problem solving). Emotions reward attention. Game feedback provides new stimuli to interpret and creates new experience. Rewards create positive and negative affect. Failures create negative affect with the hopes that players will double their efforts and try again.

3. Decide: Emotion is vital to decision-making in games

Humans use both thought and emotion to make decisions. Experimental evidence suggests that people with damage to brain structures involved in emotions can generate appropriate logical options and discuss each decision's impacts and tradeoffs with great clarity, but are unable to make the actual choice itself (Damasio, 1994; Norman, 2004). When we select an entree from a restaurant menu, it is what we "feel" like having. There are also logical components to that choice (calories, fiber, who's paying, etc.), but there is something other than logic that happens inside that helps us choose and "feel" good about the selection. Some people even feel that their car drives better after it has been washed or had the oil changed (positive affect). Emotions also let people make snap decisions (Ekman, 2003; Norman, 2004). Emotions help players decide and combine with logic to make these decisions interesting. Because positive and negative affect guides players, it is easier to choose between options with strong emotional stimuli. Game interfaces that supply strong emotional responses also have this effect. Games also use affect to add conflict. Either the emotions support the objective or make achieving it more difficult as players resist the urge to run away. Recognizing this initial role emotion plays in decision-making offers a crucial way to improve decision-support products.

4. Perform: Emotion supports different approaches to action and execution

Emotions are a key component in most tasks. This is easy to see in games, which are often designed to create a particular affective state, which sometimes requires following strict detailed procedures with zero tolerance for error or creative exploration of alternative options. The lesson for human–computer interaction (HCI) is that properly designed emotions support the right affective state to get the job done as well as increase appeal. In productivity software, this is important for both the task (such as struggling to find the right word while writing) and the software (such as struggling to find the right feature), and both create frustration. Like a car's seatbelt alarm, certain levels of frustration create mild negative affect in the user and direct the user toward a certain action. Too much negative affect (such as when players cannot keep cars on the racetrack) makes players feel like they will fail and quit. Software needs aspects that create emotions to support tasks, and allow options that let users balance their own levels of frustration. Like cycles of hard and easy things to do in a game, users need to experience new emotions to refresh their experiences. Emotion creates moods that persist and help the player perform. The emotion the designer chooses should help the actions required. Negative affect narrows attention on aspects relevant to the problem, while positive affect opens it to explore new alternatives. Positive moods influence creativity and flexible thinking for problem solving. Relaxed and happy thought processes expand and become more creative and imaginative, and make players more tolerant of minor difficulties (Norman, 2004). Moods with a negative affect focus attention, while positive moods help players generate new options. Cycling between moods creates variety to refresh the experience, and offers the option of approaching a problem with a different problem-solving strategy related to the new mood.

5. *Learn:* emotion provides motivation and rewards progress

Played for pure enjoyment, game emotion motivates players to pay closer attention and repeat an action, enabling them to master highly complex interfaces and interactions, learn countless features and strategies, and spend hours doing this even though they may fail repeatedly. To increase product mastery, many game methods apply nongame interfaces. While productivity software users prefer to learn the bare minimum number of features to accomplish their work rather than achieving level 42 in spreadsheet wizardryness, game-like motivation may increase exploration of additional product features. All games involve learning. Emotions reward changes and growth inside players as they master what they could not do before. Learning and remembering are easier for emotional stimuli, and when experiencing an emotion or a mood, it is easier to remember thoughts that have a matching emotional context. Improved performance such as learning a new skill or high score is a big part of the enjoyment of play.

Requirements for an Emotion Framework for Games

"The problem with the words 'enjoyment' and 'happiness' is that they're not specific enough; they imply a single state of mind and feeling, in the same way that the terms 'upset' and 'negative' don't reveal whether someone is sad, angry, or disgusted."
—Ekman, 2003, p. 190

A practical methodology for designing emotions should support designers as well as researchers to build better player experiences. More than a model, a practical method for designing and examining player experiences should take into account four perspectives: (a) what players like most, (b) what creates emotions,

(c) what game designers can control, and (d) what researchers can measure. The method's components must be observable, salient to the player, relevant to the player's experience of fun, and apply to a wide variety of game genres and hardware platforms. It must account for how activities offered by best-selling games make them more popular than others with similar features. It should track internal and external aspects of play including complex emotions and their impact on players' lives. A framework must capture emotions' role in enjoyment, maintaining player interest, decision making, performance, and learning. Designing for emotion in games should build on the research and theories of several disciplines. It must cover differences in expectations and play patterns such as Bartle who first classified players as (a) Achievers, (b) Explorers, (c) Socializers or (d) Player Killers (Bartle, 1996). It must measure emotion from play from different game components such as challenge and fantasy (Malone, 1981) and biometrics such as heart rate, control pressure, and facial expression (Mandryk, 2004; Sykes & Brown, 2003; Hazlett, 2003) and specific emotional states such as pleasure and aggression using FMRI (Weber, Ritterfeld, & Mathiak 2006). It should cover emotions from game goals, open-ended play, as well as emotions from game events and those that happen in real life. Most importantly, the method must describe which gameplay mechanisms produce specific emotions that are central components of the player experience (Fig. 10.3).

Evaluating Relevant Frameworks for Emotion, Products, and Entertainment

A limitation shared by current models of emotions and products is that they focus more on positive and negative affect than on how to create specific emotions from interaction; and even less attention is given to how the actions of a game player create specific emotions. To be useful to game and product designers, player-experience methodologies must identify more emotions than simply positive and negative affect. Affective states grow and change during gameplay.

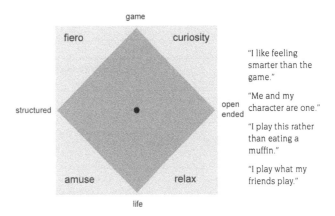

FIGURE 10.3. Players play for "the ride" that games takes them on. Player experiences (PX) generate emotions from reactions along two axes: goal-directed versus open-ended play, what happens inside the game versus how gameplay relates to other aspects of the player's life. Copyright© 2004–2007 XEODesign, Inc. All rights reserved.

Many emotions are enjoyable, such as curiosity, which can also lead to other positive emotions such as wonder and even love. However, players also enjoy play experiences containing negative emotions. The pleasure that comes from many game designs requires mastering difficult situations or experiencing unpleasant emotions. How emotions grow and change during play is a critical part of player experiences. During game testing, designers need more information than whether an action was or not fun. Player observation leads to the design of better games, but this requires a deeper understanding of the player experience and better techniques on how to measure it (Fig. 10.4).

Flow the Optimal Experience: Csikszentmihalyi

The most relevant and influential psychological research outside of games is Csikszentmihalyi's model of on optimal experi-ences or "flow" (Csikszentmihalyi, 1990). He found that people are happiest when engaged in intrinsically motivating activities such as rock climbing, dancing, and gardening. These activities offer long-term, deep, memorable experiences that require concentration and growth. He noted that optimal experiences carefully balance skill with difficulty to create a deeply enjoyable mental and physical state. Without challenge, an activity becomes routine and boredom sets in. With too much challenge, the person becomes too anxious and leaves because he or she feels frustrated (Csikszentmihalyi, 1990). The activity should aim to achieve flow through experience design and create forward movement and personal development through pleasurable engaging experiences. With its balance of difficulty and user skill, flow clearly describes a critical component of play experiences. However, Csikszentmihalyi's model only took into account two emotions: (a) anxiety (frustration) and (b) boredom, and ignored several other emotions essential for gameplay. While these first two emotions

Comparison of Models and Methods to Create Affect from Products and Entertainment Experiences				
XEODesign Four Fun Keys Model	**Hard Fun** Fiero Challenge Game, Goal	**Easy Fun** Curiosity Novelty, Fantasy Game, Open Ended	**Serious Fun** Relaxation Real World Purpose Life, Open Ended	**People Fun** Amusement Social Life, Goal
Bartle's Original 4 Player Types (1996, 2003a, 2003b)	Achiever Player Killer	Explorer		Socializer Player Killer
Boorstin (1990)		Voyeuristic Eye Visceral Eye		Vicarious Eye
Csikszentmihalyi (1990) Ekman (2003)	Enjoyment, flow	Pleasure, microflow Auto appraisal, memory of emotion, imagination, reflective appraisal		Empathy w/ another, Violation of social norm, talking about emotion, making facial expression of emotion
Hassenzahl et al. (2000) Kim (2000)	Ergonomic quality	Hedonic quality		Community
LeBlanc et al. (2004)	Mechanics, dynamics Aesthetics	Aesthetics		
Malone (1981)	Challenge	Curiosity Fantasy		
Norman (2004)	Behavioral		Reflective visceral	Reflective
Piaget (1962)	Formal games with rules	Sensory-motor play Pretend play		
Tiger (1992) Jordan (2000)		Physio-pleasure Psycho-pleasure	Ideo-pleasure	Socio-pleasure
Wright et al. (2003)	Spatial-temporal thread	Compositional thread Sensual thread Emotional thread		
Common Drama and Theater Constructs	Character "motivation", Plot points, objectives 3-act structure	Setting, plot, story, Character, suspension of disbelief	Catharsis, music, Set and costume design	Character Dialog Acting

FIGURE 10.4. Several frameworks describe the emotion resulting from entertainment experience or the use of a product. Comparing their similarities and differences provides interesting insight into the basic requirements for entertainment and product emotions. Copyright© 2004–2007 XEODesign, Inc. All rights reserved.

are important in games, many other emotions play an important role in player engagement. Csikszentmihalyi grouped together physical, mental, and aesthetic challenges, which prevented discussing emotional effects of art and audio separately from the challenge of gameplay. A model for emotions in games should connect the emotions most important to games to how games create them.

Pleasure from Products: Jordan, Norman, and Boorstin

Pleasure comes from different aspects of experiencing a product. To define what he calls the new human factors, Jordan expanded on anthropologist Tiger's Four Pleasures model to create a framework for thinking about pleasure from products. Jordan discussed (a) the ideo-pleasure, (b) physio-pleasure, (c) psycho-pleasure, and (d) socio-pleasures of a product (or emotions from a product's idea, physical, psychological, and social attributes). Each of these Four Pleasures describes enjoyment from a different perspective including how the use, ownership, or identification with a product produces emotion (Tiger, 1992, as cited in Jordan, 2000; Jordan, 2000). Changing an aspect can strengthen a desired emotional response. Norman built on and simplified Jordan's model to three layers of mental processing: (a) the Visceral (an automatic biological response), (b) the Behavioral (learned actions), and (c) the Reflective (involving thought, self-image and relationship to others) (Norman, 2004). Interestingly, filmmaker Jon Boorstin also proposed three ways that films delight audiences: (a) the Visceral Eye (enjoyment on the biological level such as automatic sensory reactions to explosions or speed), (b) Voyeuristic Eye (enjoyment through seeing events unfold), and (c) the Vicarious Eye (enjoyment through identification with people on screen) (Boorstin, 1990). Unfortunately, none of these frameworks go deeply into what actions create specific emotions.

Enter Ekman and Facial Expressions

Through his research of universal facial gestures, Paul Ekman's Facial Action Coding System (or FACS) and a compilation of others' research, Ekman's work has increased understanding of specific human emotions' cross-cultural boundaries. His fascinating and highly accessible book *Emotions Revealed* described how to identify emotions through facial gestures and the important roles emotions play in our lives (Ekman, 2003) (Fig. 10.5).

Because of its specificity, the FACS coding system for facial gestures offers promise as an emotional measure for games and software, and is the basis for many of XEODesign's observational studies. However, Ekman's work focused on the identification of individuals' experience of specific emotions and stops short of discussing how products or their use can create these emotions. What game designers need is a method that connects specific emotions to player actions in the game.

All of these models lay the groundwork for what is ultimately needed by game designers—a method for producing specific emotions from the interactions players most appreciate during goal oriented and open ended play. To do this the game industry needs an expanded method to create specific captivating

Six Plus One Universal Emotions with Universal Facial Gestures	
Emotion	**Example**
Frustration:	Figuring out how to get character off a roof in *Tom Clancy's Splinter Cell* (and all-too-often created by usability issues that detract from the player experience)
Fear:	Falling into boiling lava, fast-moving projectiles aimed at the player in *Doom*
Surprise:	Using *Myst's* linking books for the first time to transport to a new world.
Sadness:	When the young magician *Aerith*, in *Final Fantasy VII*, is murdered
Amusement:	When two Sims get married in *The Sims*, or rolling over and picking up sumo wrestlers in *Katamari Damacy*
Disgust:	Becoming a social outcast (social disgust) after losing the dancing challenge in *Sid Meier's Pirates!*
Curiosity:*	Wanting to know what happens by driving the race track the wrong way in *Project Gotham Racing 3*

*Not all researchers (including Ekman) considered curiosity a universal emotion with a unique facial gesture. I include it here as a seventh emotion because of its importance in games and ease of observation.

FIGURE 10.5. Researchers generally agree that there are at least six emotions with universal facial gestures.

emotions from the best-loved types of gameplay. This is a core focus of XEODesign's independent research.

THE PLAYER EXPERIENCE FRAMEWORK

Why We Play Games: Four Fun Keys Model

XEODesign conducted independent research to identify four key processes that create emotion in best-selling games. (Figure 10.6) Each experience involves a different emotion to create a different Player Experience Profile. By presenting a goal and breaking it into small achievable steps, games create emotions from *Hard Fun*, where the frustration of the attempt is compensated by the feelings of accomplishment and mastery from overcoming obstacles. Outside of goals, games provide novel opportunities for interaction, exploration, and imagination, which create *Easy Fun*. Games that use emotions in play to motivate real-world benefits to help players change how they think, feel, and behave or to accomplish real work create *Serious Fun*. Finally, games that invite friends along get an interpersonal emotional boost from *People Fun* (Lazzaro, 2004b). The Four Fun Keys are a collection of related game interactions (game mechanics) that deliver what players like most about games. Each offers a key to "unlock" unique emotions such as frustration, curiosity, relaxation, excitement, and amusement. Best-selling games provide features that support at least three of these Four Fun Keys to create a wider emotional response in the player. To

The 4 Fun Keys Model for Emotion through Gameplay

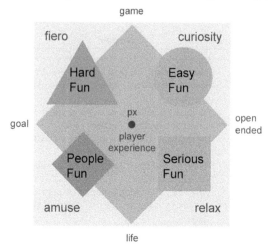

FIGURE 10.6. Each Fun Key is a collection of game mechanics that create a favorite aspect of gameplay. Emotions prepare players for action and reward players. These Four Fun Keys each contribute a unique set of emotions to the game. Changing one of the four key mechanisms will change the Player Experience profile of the game. Copyright© 2004–2007 XEODesign, Inc. All rights reserved.

keep things fresh during a single-play session, gamers move between the four different play styles (Lazzaro, 2004b). Developing each key focuses and rewards the player with emotion from a self-motivating experience that deepens the game's player-experience profile. Designers of products and productivity software can also use these Four Fun Keys to increase emotional engagement for applications outside of games.

Only some of the emotions from playing basketball in the real world come from the Hard Fun of making baskets. Close examination reveals that all four Fun Keys are part of this popular sport. Dribbling the ball or doing tricks like a Harlem Globetrotter offers Easy Fun from novelty and role play. Intentionally blowing away frustration and getting a workout creates Serious Fun. Competition and teamwork make the game even more emotional from People Fun. All four types of fun make basketball's player experience more enjoyable. None of these require story or character. Through examination of how each type of fun creates emotions, designers and researchers can create better and more emotional player experiences (Fig. 10.7).

FIERO FROM HARD FUN GAMEPLAY

Gameplay that rewards effort through challenges to create fiero.

Desire for Fiero Helps Players Master Challenge

"It's easy to tell what games my husband enjoys the most. If he screams 'I hate it. I hate it. I hate it,' then I know two things: a) he will finish it and b) buy version two. If he doesn't say this, he'll put it down after a couple of hours."

—Wife of a hardcore PC gamer

Four Fun Keys to More Emotion through Gameplay

Lead Emotion	Fun Key	Key Game Mechanic
Fiero*	Hard Fun	Affect related to challenge, strategy, and mastery
Curiosity	Easy Fun	Affect related to novelty, ambiguity, detail, fantasy, role-play, and absorbing attention
Relaxation	Serious Fun	Affect related to purposefully changing oneself, learning, or doing real work
Amusement	People Fun	Affect related to competition, cooperation, and socializing with others

*Personal triumph over adversity.

FIGURE 10.7. Game mechanics create emotion from what players like most about play. To this group of easily observable emotions, we add frequently reported emotions collected through verbal descriptions of internal sensations as well as player's body language during play. Copyright© 2004–2007 XEODesign, Inc. All rights reserved.

The most obvious enjoyment in games comes from mastering a challenge and reaching a goal. Hard Fun is a self-motivating activity that keeps the user focused and enthusiastic by providing an obstacle, an objective, and a score. Hard Fun game mechanics challenge a player to overcome an obstacle to achieve a goal. Hard Fun experiences reward mastery, either explicitly with points and bonuses or implicitly through new levels or abilities. Because this type of play requires application of effort, we call this Hard Fun. In our research, we expand on this phrase (Csikszentmihalyi, 1990; Papert) to define Hard Fun as the rewarding process of mastering a challenge that involves the creation and testing of strategies and the application of effort. Hard Fun rewards effort and discourages failed approaches. Hard Fun creates the emotions of frustration and boredom. More importantly, it produces *fiero*, the Italian word Isabella Poggi and Ekman used to describe the personal feeling of triumph over adversity (Poggi, as cited in Ekman, 2003). One of the most important game emotions, fiero is a strong feeling of personally accomplishing something difficult such as defeating the boss monster (Figure 10.8).

Hard Fun requires a high investment of energy from the player. By perfectly balancing player skill with game difficulty, Hard Fun meets many of the characteristics and requirements for flow (Csikszentmihalyi, 1990; Lazzaro, 2004b). For example, *Pac Man's* simple game mechanic (eat dots and avoid ghosts) offers clear long-term and short-term goals, the opportunity to concentrate, achievable tasks, an uncertain outcome, and immediate feedback to player decisions. It creates a deep sense of control through tight feedback loops between player input and action in the game. All of these enhance players' absorption into a challenging activity and improve their ability to perform. Beyond flow's balance of difficulty and skill, game designers do other things to change how players feel about their progress in the game. According to game designer Steve Meretzky, rewards along the path to the goal enhance enjoyment such as power ups, big jumps in score, animations and sounds (Lazzaro, 2005a). For example, the power ups in *Pac Man* create super-charged feelings as players turn the tables to chase ghosts.

Hard Fun

FIGURE 10.8. Similar to Csikszentmihalyi's concept of flow, games must balance difficulty with player skill and provide enough variety of challenge, strategies, and puzzles. If the player gets too frustrated or bored, he or she leaves the game. Charting a player's progress through a game onto Csikszentmihalyi's flow model illustrates that in addition to the requirements for flow, games can be made more emotional from sharp changes in the level of difficulty for game challenges. Game difficulty increases to the end of a level and then resets at the start of the next. When players succeed at the point where they are about to quit, they are more likely to experience fiero. In this figure fiero occurs at the end of level two before continuing to level three of this game (modified from Csikszentmihalyi, 1990). Copyright© 2004–2007 XEODesign, Inc. All rights reserved.

Hard Fun focuses player attention on achieving results by providing opposition and constraints, such as removing an alien threat in *Halo* or aligning puzzle pieces in *Tetris*. Often games provide a choice of strategy, and games with high replay ability offer a choice of goals. For example, *The Sims* offers several winning conditions such as the best-looking house, the most friends, or the most money. The new obstacles, constraints, and tradeoffs from different goals suggest multiple strategies and enhance the challenge. These emotions facilitate achievement of a goal, focus play effort, and reward accomplishment. Negative affect increases focus and concentration (Norman, 2004), such as when the player encounters resistance to the goal through failed attempts. Sometimes Hard Fun emotions run counter to what needs to be done. For example, performing well or moving toward something rather than running away is more challenging when surrounded by negative affect. Other Hard Fun emotions, such as fiero from achievement, put the player into a positive affective state by rewarding the player's efforts. Unlike many emotions, fiero does not require an audience (Poggi as cited in Ekman, 2003) and is a powerful emotion unavailable in film or novels. Players enjoy these emotions from Hard Fun, including the mental focus provided by frustration that helps players concentrate (Fig. 10.9).

While general enjoyment results from emotions that provide motivation, other emotions impact the player's response to the game's stimulus. Decision making is aided or made more challenging by affect, whether it is the positive affect of collecting *BeJeweled*'s colored gemstones or the negative affect from *Bookworm*'s burning letter tiles (which makes their use a priority

and picks up the perceived pace of the game). Emotional states also focus effort and attention on available choices in play. Positive affect, for example, helps to generate new ideas such as what gemstones to match in *BeJeweled*, or negative affect, surrounding an impending invasion in *Civilization,* provides additional focus for decision making on how to respond.

Fiero Enhances a Player's Sense of Progress

"I have to concentrate!"
—Traveler in St. Louis airport on what he likes most about *JamDat Bowling* on his mobile phone. Playing removes distractions from his consciousness and engrosses him in a rewarding activity while he waits.

Players enjoy many other aspects of games with emotions outside of flow's balance of difficulty and skill. To enhance a player's sense of progress as play continues, games offer new tools or abilities along with new obstacles, constraints, and tradeoffs to maintain interest. Levels divide challenge into gradually increasing amounts of difficulty. The difficulty also increases inside each level, with most levels ending in a "boss monster" final challenge similar to a major plot point in a story. Defeating boss monsters or boss puzzles produces fiero, and the start of the next level is often dramatically easier to provide emotional relief. The best-selling game *Halo* uses Hard Fun to enhance a player's performance by the way it breaks a variety of challenges into levels, rewards progress, provides power ups, such as new weapons or armor, and adds new vehicles to offer fresh strategy options (Fig. 10.10).

Hard Fun Creates Challenge with Strategies and Puzzles

Central to the enjoyment of Hard Fun is the creation and testing of strategies, applying creativity, and the development of skills. While some players focus on the goal, many enjoy the process of learning how to win or take pride in how much better they play with each repetition of the game. Unlike real-world sports such as baseball, computer games increase the challenge by changing the rules and winning strategies between levels to keep the emotions from gameplay fresh. Players must strategize to find new ways to play. Rather than requiring more points in less time, best-selling games such as *Collapse* keep players engaged by making a winning strategy obsolete a few levels later (Lazzaro, 2004b). The process of devising new strategies or solving puzzles in new ways is something that players like most about games. The success and failure of these mental tasks and the increase in player skill all create Hard Fun emotion and help players stay engaged and playing.

CURIOSITY FROM EASY FUN GAMEPLAY

Curiosity Retains Attention

"Part of the enjoyment comes from the spy technology . . . cool spy tools are part of the Spy experience."

—Xavier playing *Splinter Cell*

Chain of Hard Fun Emotions Increases Enjoyment

Goal-directed gameplay creates Hard Fun emotions that focus and reward players for overcoming obstacles.

	Emotion	Common Themes and Triggers
fiero! frustration relief boredom	Frustration	Opposition to an important goal, sudden reversal, feeling of being thwarted, physical restraint. Anger prepares the body to remove an obstacle by force. (Ekman, 2003)
Hard Fun emotions involve obstacles, strategy and success and help players accomplish a goal by focusing and rewarding effort.	Fiero* (Italian)	Personal triumph over adversity. (Ekman, 2003) Overcoming difficult obstacles, players raise their arms over their heads. They do not need to experience anger prior to success, but it does require effort and some frustration.
	Boredom**	Repetitive, dull, or tedious tasks. (Ekman, 2003) Lack of interest in the outcome or in playing. Dispelling boredom is also a major reason to play games.

*Fiero is a positive emotion that has a body rather than a facial gesture. It is not yet known what fiero gestures look like across cultural boundaries.
**Ekman does not list boredom as having a universal facial gesture; however, it is frequently seen in games that lack sufficient challenge.
(Source: Lazzaro, 2004b)

FIGURE 10.9.   Hard Fun Emotion Cycle. Copyright© 2004–2007 XEODesign, Inc. All rights reserved.

In fiero, the ultimate game emotion, the sensations are powerful. It is how players feel when they beat the boss monster or make level 20 after difficult struggles, or when they win a tennis match at Wimbledon. As fists punch the air a victorious player screams, "Yes! I did it!"

In XEODesign's play lab, fiero appears as a positive upward gesture of the arms or body after succeeding a challenge. Some players jump their characters up and down to express this emotion. (Source: Lazzaro, 2004b)

FIGURE 10.10.   In fiero a player raises an arm to show their excitement. Copyright© 2004 XEODesign, Inc. All rights reserved.

Less apparent but equally important to Hard Fun, top-rated games offer a lot of gameplay outside of or en route to a goal. Easy Fun is a self-motivating activity that maintains player engagement through novelty beyond an obstacle, goal, or score. Easy Fun offers novel interaction to inspire player curiosity to explore, fantasize, and role play. By balancing what is expected and unexpected, careful game design prevents the player from quitting from either disinterest or disbelief. Easy Fun derives from the ability to explore and create exceptional experiences unavailable in the real world. The emotions of curiosity, surprise, wonder, and awe surrounding Easy Fun capture and retain the player's attention, as opposed to Hard Fun where the emotions of frustration in hopes of fiero help players focus on and apply effort toward attaining a goal (Fig. 10.11).

If Hard Fun revolves around goal achievement, Easy Fun focuses attention by offering opportunities to explore, get lost in a fantasy, role play, or simply horse around. With compelling Easy Fun, the player ignores the goal completely or forgets about keeping score. Easy Fun players find rewarding, open-ended activities on top of the game's main goal. The so-called sandbox play patterns capture player attention and pull players into deep states of immersion through curiosity instead of daring them with challenge. Best-selling games offer opportunities for interaction through unusual yet enjoyable behavior of the controls and novel interaction with the world. Players become fascinated just by interacting with the game, such as being able to flip a car off freeway exit ramps in *GTA* (*Grand Theft Auto*.) Like the fiero

Easy Fun

disbelief
too novel

unexpected

disinterest
too predictable

expected

Game Features
explore
fantasy
fool around
role play
ambiguity
detail
uniqueness

Emotions
curiosity
surprise
wonder
awe

Gameplay that fills attention through novelty to inspire curiosity.

FIGURE 10.11. Easy Fun emotions maintain player attention without challenge through novelty and inspiring fantasy. Similar to Csikszentmihalyi's concept of flow, players will leave a game because of disbelief or disinterest. To maintain player interest, the game design must balance the expected with the unexpected. The player experience profile of Easy Fun includes curiosity, surprise, and wonder. Copyright© 2004–2007 XEODesign, Inc. All rights reserved.

from Hard Fun, Easy Fun has positive peak emotions that reward play such as surprise, wonder, and that "ah-ha" feeling from figuring something out. However, unlike Hard Fun none of these Easy Fun emotions require frustration. Story can often generate the emotions of wonder and surprise, but role-playing and a player's own discovery through exploration create these emotions on a much more personal level. Positive affect encourages creativity and exploration of alternatives (Norman, 2004) that opens the player up to free associations and other possibilities. Easy Fun emotions focus on filling player attention.

Close examination of the conditions, causes, and relationships between emotions provides opportunities for game designers to create even bigger emotional responses in players. It is not that a player cannot feel curious during Hard Fun; but with Easy Fun, curiosity and the sheer joy of interaction drive the player rather than only the score, as it is in Hard Fun. Like improv theater, Easy Fun in games such as *Grand Theft Auto* makes offers to players such as a car and a plate-glass store window. It is up to the player to accept this opportunity and discover how the car and window interact.

Chain of Easy Fun Emotions Increases Enjoyment

Open-ended gameplay creates Easy Fun emotions and increases immersion in an activity (Fig. 10.12).

Easy Fun: the Bubble Wrap of Game Design

"In real life, if a cop pulled me over I'd stop and hand over my driver's license. Here I can run away and see what happens."
—Xavier playing *GTA Vice City*

Easy Fun is the bubble wrap of game design. It is fun without a purpose. Easy Fun provides novel opportunities for interaction that players "discover" alongside of or outside of the main play. The enjoyment and label for Easy Fun comes from the way players goof around, frolic, explore, and play with an ease they do not have when pursuing a specific winning condition as with Hard Fun. Best-selling games such as *GTA, Halo,* and *Myst* offer numerous opportunities for Easy Fun so that when the challenge gets too tough or loses its appeal, the players have many other things to do that create emotional responses. A big role of Easy Fun in best-selling games is to refresh the player between or in the middle of challenges. In *Halo,* for example, players can cycle between the Hard Fun of combat and the Easy Fun of exploring a ring world. The game's battlefield is on the inside of a ring-shaped planet, which inspires curiosity as players investigate as they approach a horizon that instead of dipping down from view, dips up overhead. Easy Fun also provides interest when players pursue the opposite direction of a game goal such as putting the Sims in the pool and removing the ladders in *The Sims* or placing predator and prey animals in the same pen just to see what happens in *Zoo Tycoon* (a flurry of dust and the prey disappears). Through exploring both what's right and what's wrong, games offer more opportunities for emotion, especially from violating norms. In addition to relief from challenge, Easy Fun prevents progress in gameplay from feeling like a skeleton of correct decisions. Easy Fun reinvigorates emotionally and often reinterests the player in the goal. By offering both kinds of fun, the game extends the average play session and lets the player self-regulate the challenge if the Hard Fun becomes too hard.

"The journey is the reward."—Design philosophy at Cyan, creators of *Myst*

The emotions from Easy Fun both inspire and satisfy a player's curiosity. To create the emotions of curiosity, surprise, wonder, and awe in addition to novelty, Easy Fun gameplay uses juxtaposition, where contrast between items or events requires the player to investigate. Like Magritte's surrealist painting, "This is not a pipe," a player of *The Sims* must interpret the Siamese pictograph language spoken by the characters. The player projects in and provides an explanation for any conversation between Sims. The Easy Fun of games provides opportunities for fantasy and role play. Players can take elements of the game and do their own thing with them whether it is wielding an orc's mace in *World of Warcraft* (*WOW*) or a donning a superhero's cape in *City of Heros*. Games also reward player curiosity with details such as in Cyan's *Myst*. In addition to the Hard Fun of puzzles, *Myst* offers Easy Fun gameplay from exploring worlds. To encourage players to slow down and notice, *Myst* provides numerous small details in the environment that reward closer inspection and encourage a slow, careful gameplay style.

Instead of running through at top speed, players spend more time exploring and looking for clues to solve the mystery. This supports the style of interaction needed to play and win the game. In addition to detail, *Myst* captures player attention through ambiguity in the setting, surrealistic juxtapositions such as the boiler room inside a tree, linking books, and faded sketches of other wondrous technology. The conflict that creates emotional tension in Easy Fun is often between what the

---

Chain of Easy Fun Emotions Increases Enjoyment

Unstructured sandbox play creates Easy Fun Emotions that reward players outside of challenge and keeping score

	Emotion	Common Themes and Triggers
wonder / awe / surprise / curiosity / relief	Curiosity*	Unusual, unresolved situation that peaks player's inquisitiveness. (Ekman, 2003) Something that players find strange, odd, or intriguing, such as *Myst's* surreal ship-rock island.
Easy Fun offers emotions surrounding the unique. Games provide a sequence of emotions, often starting with curiosity then creating surprise. Easy Fun can sometimes create wonder or awe if the effect is particularly strong.	Surprise	Sudden change. Briefest of all emotions, does not feel good or bad, after interpreting the event this emotion merges into fear, relief, etc. (Ekman, 2003) such as when matching two blocks clears the whole board.
	Wonder	Overwhelming improbability (Ekman, 2003). Curious items amaze players at their unusualness, unlikelihood and improbability without breaking out of realm of possibilities, such as in *Myst's* linking books.
	Awe	Combination of wonder with a fear and dread (Ekman, 2003), such as a beautiful but impossibly powerful dragon or female warrior in *EverQuest*

*Not all researchers including Ekman recognized curiosity or interest as a universal emotion with a distinct facial gesture. However, those who did recognize it saw the emotion indicated by a lifting and drawing together of the inner eyebrows. In our research, we saw it frequently combined with leaning forward. It was also a feeling reported verbally by players (Lazzaro, 2004b).

FIGURE 10.12.   Easy Fun Emotion Cycle. Copyright© 2005–2007 XEODesign, Inc. All rights reserved.

player knows and does not know. Easy Fun offers detail that rewards player exploration and paying closer attention. Because challenges can feel like a grind, Easy Fun refreshes and provides new and interesting things to do in the game. Easy Fun provides novelty to keep play open ended and interesting not because players wonder about whether they have "the stuff" it takes to reach a goal, but simply to make the player experience interesting, surprising, and far from routine.

RELAXATION FROM SERIOUS FUN GAMEPLAY

Relaxation Changes How Players Think, Feel, and Behave

"Playing helps me blow off frustration at my boss."
—a hard core *Halo* player

In Serious Fun, people play games with a purpose (Fig. 10.13). They play to improve their lives by changing how they think, feel,

Gameplay that changes the self and the real world.

FIGURE 10.13.   Serious Fun creates emotions to engage in activities that players hope will change how they think, feel, and behave or that will accomplish real work. The player experience profile of Serious Fun includes relaxation, excitement, boosts in self-esteem, and the satisfaction from a job well done. Copyright© 2004–2007 XEODesign, Inc. All rights reserved.

or behave. The enjoyment motivates continued engagement with an activity that brings the desired results. The Serious Fun in games reliably relaxes or excites gamers as they play for a purpose to change the player's internal state, develop good habits, improve self-esteem, learn, or do work outside of the game itself. In Serious Fun, the entertainment value captures attention from the emotions surrounding the human values expressed through the act of play. Almost like therapy, players report unparalleled states of concentration and focus during play, making this shift in emotional state one of gaming's biggest takeaways. Whether the activity is fast-action with lots of explosions or slow-paced colored-block matching, players play to experience feelings of excitement, blow away frustration, "get perspective" on their troubles, or create a meditative experience. The stimulation and concentration required drives out bothersome thoughts. Some players simply want to feel more relaxed, excited, or less bored. Players look for a physical and mental workout such as exercise in *Dance Dance Revolution*, heightened reflexes in *Project Gotham Racing*, or an increase in mental acuity from word games such as *Bookworm*. Positive and negative affect from play guides the player toward correct moves, and negative affect increases the perceived pace of the game. Real-world benefits include releasing stress therapeutically, learning new vocabulary or math, improving physical fitness through exercise, and increasing a player's mental agility, which some players believe wards off the effects of aging. Physical movement also creates an emotional release felt after exercise and positive feelings from learning and accomplishing tasks in an engaging way. Games played for any of these reasons offer Serious Fun.

Players play to change themselves. Serious Fun is similar to Easy Fun in that players enjoy immersion rather than challenge.

However, one aspect of enjoyment many players prefer is achieving a purpose outside the game experience itself such as the ability of the game to calm or excite them. As in Easy Fun, players want the game to fill their attention, but Serious Fun focuses on players' affects long after the game is over. The term Serious Fun captures the real effort and intent that players expend to alter their internal states, express their values and beliefs, and improve their real-world skills. Without Serious Fun, a game leaves few long-lasting effects and the play experience feels more like a waste of time. Playing may be hard or easy, but what is important to many players is that the effects of a game last long after the game is over.

Chain of Serious Fun Emotions Increases Emotions

In Serious Fun, many players play to change how they feel, and they take different emotional paths to get there (Fig. 10.14).

Serious Fun: Enjoyment Helps Accomplish Real Work

"I felt better about playing [crosswords] because it's good for me. If someone would tell me Tetris *was good for me I'd feel better about playing that."*
—Ellen on doing crosswords. She believes the memory demands of the game will keep her mentally sharp and delay the onset of Alzheimer's disease.

Serious Fun teaches or accomplishes real work. Learning-type games, from multiplication to database administration, utilize

---

Chain of Serious Fun Emotions Increases Enjoyment

Gameplay that changes how a player feels, thinks, or behaves creates Serious Fun emotions from creating something players value.

	Emotion	Common Themes and Triggers
more frustration → excitement	Excitement	From sudden changes, novelty, and challenge (Ekman, 2003). The unexpected catches attention. Many players enjoy the adrenaline rush. Some want to raise their state of arousal, others like how the emotional intensity makes them more relaxed afterwards.
frustration boredom → zen-like focus → **relaxation**	Relaxation	Relief from negative emotion (Ekman, 2003). Gamers often start a game to attain relief from negative emotions and thoughts prior to playing.

*Self-esteem and knowledge acquisition and the result of exercise are not emotions, but changes in a player's internal states that are reported as desirable results from play (Lazzaro, 2004b).

FIGURE 10.14. Serious Fun Emotion Cycle. A large part of the appeal of games comes from how they change the player inside. This is referred to as Serious Fun. Here a player starts in one of two emotional states and plays a game to end up feeling different. Many gamers appreciate real-world benefits from play. Some play games to become deeply immersed in a process that they hope will change them long after the game is over. Copyright© 2005–2007 XEODesign, Inc. All rights reserved.

emotion to encourage the learning of important content or to provide a mental workout, as in an effort to prevent Alzheimer's. Relaxation and excitement make it easier for the player to focus. Serious Fun provides emotions and opportunities for success unavailable to the player in real life by offering a cheaper, safer, more rewarding experience with the content or activity. Rescue simulations can train firefighters about situations that are too dangerous or expensive to do in real time. In educational games such as *DBA Day* (an Oracle workplace simulation XEODesign worked on) players role play a database administrator, including managing their own time. Gee argued that mastering any simulation requires mastering the content embedded in that simulation. By making decisions for the character in a simulation, he believed that simulation games also teach the morals and values of that character coming from their merged identities. The player makes the decisions and the game character has the special abilities to make changes in the virtual world (Gee, 2003). These provide learning opportunities, whether it is values and actions of a thief in *Sly Cooper and the Thievius Raccoonus* or a restaurant owner's time and people-management skills in *Diner Dash* (Lazzaro, 2005b). Serious Fun provides real benefits from play by using game-like structures to reward concentration and focus attention on an activity that is good for players. Leap Frog's *FLY* pen-top computer helps students with their Spanish vocabulary by offering a verbal translation of words written in English; hearing it spoken in another language motivates practice.

Games offer a new spin on learning by capitalizing on player emotion and interaction to create customized training. *Mavis Beacon Teaches Typing* motivates typists with typing games, such as keeping ants out of a picnic basket or driving a racecar. In addition to explicit learning games (such as Oracle's *DBA Day* or an unrelated project, *Doom DBA*), games also embed techniques to offer players more immersive instruction. The introductory experience in *Halo* not only provides a seamless introduction to the user interface by integrating it into the story, but also adjusts the controls to meet the player's preferences without a dialog box. In *Grand Tourismo*, a detailed car-racing simulation game, the opening time trials break down racing skills, such as cornering, into small, easy-to-master steps that fit together to create a more enjoyable practicing experience. By providing a motivating alternative to accomplishing an otherwise boring or unmotivating task, Serious Fun helps players accomplish real-world objectives such as getting in shape. Someone may play *Dance Dance Revolution* for the excitement of moving to the music, to lose weight by burning calories, or to learn the physical skills and coordination required to dance better. Others play *Karaoke Revolution* to learn how to sing.

Games even make work fun. Players pay for the experience of being a waitress (*Diner Dash*), business owner (*Lemonade Tycoon*), dungeon master (*Dungeon Keeper*), professional sports player (*Madden Football*), or theme-park owner (*Rollercoaster Tycoon*). They even buy the opportunity to sort bugs by color (*Tumblebugs*), pick up their rooms (*Katamari Damacy*), or manage a city (*Sim City*). Games model life problems (*The Sims*) and improve performance during real work. In *The ESP Game*, developed at Carnegie Mellon University, people play a guessing game to enliven the otherwise boring task of providing text labels for images (von Ahn Dabbish, 2004). Idea or prediction

markets use games to beat expert opinion polls. Prediction markets allow participants to express their opinions through buying and selling shares with either virtual or real money. One of the oldest, the University of Iowa's *Iowa Electronic Market*, has allegedly beat expert polls in predicting U.S. presidential elections since 1988. Players of the *Hollywood Stock Exchange* use virtual money to predict what actor, director, or film will receive an Oscar nomination. At *NewsFutures*, players compete for prizes based on their ability to predict news events. Several companies, such as Google, use internal markets to predict launch dates and job openings. Other companies have created internal-prediction markets for sales forecasting (*Newsweek,* 2004), and managing manufacturing capacity (*Harvard Business Review,* 2003). Not without controversy (gambling is illegal in many countries), markets have even been proposed to predict terrorist attacks (*The New York Times,* 2003). Because playing prediction markets accomplishes a real purpose, it changes how participants feel about playing in general. Like offering a prize for a competition to develop a solar car, having real money or reputation on the line increases the excitement. In Serious Fun, players accomplish real work for many reasons, including to meditate, lose weight, label every image on Internet, or beat Wall Street predictions.

AMUSEMENT FROM PEOPLE FUN GAMEPLAY

Amusement Encourages Social Interaction

"It's the people that are addictive, not the game."
—Bob, a hardcore sports game player

One advantage that a computer game has is that it is ready to play when you are. No friend required. Still, for many people it is the experience of playing with friends or family that makes play worthwhile. Group gaming provides a mechanism for social interaction, a quick excuse to invite friends over, and more challenges to gameplay. People Fun creates amusement between gamers as they play to spend time with their friends. People report playing games when their friends do, playing games they do not like, and even playing the types of games that they do not like playing just to spend more time with their friends (Lazzaro, 2004b).

"Since we lost half our guild to Star Wars Galaxies *it's not as fun."*
—A hard core gamer playing *Dark Age of Camelot*

People Fun is a self-motivating activity that maintains player engagement by supporting interaction with other people (Fig. 10.15). People Fun encourages interaction with other players as they establish social hierarchies, joke, and develop social bonds. It creates affect by providing opportunities to cooperate, compete, and watch others play. Some emotions, such as gratitude, generosity, and schadenfreude, the German word for "pleasure at the misfortune of a rival" (Ekman, 2003), cannot occur while playing alone. When people play together they invent new ways to interact, develop house rules, and add their own content to create more pleasure for themselves and their friends.

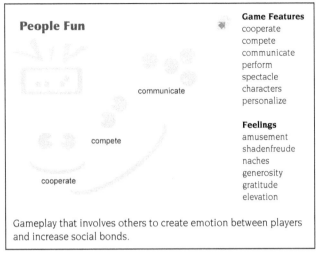

People Fun		Game Features
		cooperate
		compete
		communicate
communicate		perform
		spectacle
		characters
compete		personalize
		Feelings
		amusement
cooperate		shadenfreude
		naches
		generosity
		gratitude
		elevation

Gameplay that involves others to create emotion between players and increase social bonds.

FIGURE 10.15.   People Fun creates emotions that increase a player's enjoyment from social interaction. The player experience profile of People Fun includes many emotions that require two people such as schadenfreude, amusement, naches, and gratitude. Copyright© 2004–2007 XEODesign, Inc. All rights reserved.

Nine Pathways to Emotions (Ekman, 2003)	Example from Games
1. Auto appraisal	Explosions, fire, and boiling lava
2. Reflective appraisal	Thinking about last night's dungeon raid
3. Memory of an emotion	Remembering falling from a cliff
4. Imagination	Thinking about what happens from falling off a cliff
5. Talking about	Discussing game events with other players
6. Empathy or witnessing another's emotion	Another player's facial expression or character emote
7. Instruction by others on how to feel	Another player's assessment of an event
8. Violation of social norm	Driving over other players instead of racing against them
9. Making facial expression of an emotion	Smiling and laughing after defeat in front of friends makes it feel more positive

FIGURE 10.16.   There are nine ways to create emotions in a person.

The most emotion from games arises when people play together in the same room. Whether playing in a living room or cyber café, the frequency, duration, variety, and intensity of emotional displays all increase (Lazzaro, 2004b). The emotions of schadenfreude, amusement, and naches (Yiddish for the pleasure and pride felt when a child or mentee succeeds) occur more frequently when playing with others. Playing over the Internet, even with video or voice chat, elicits fewer emotional displays than when playing in the same room. A major reason behind this increased emotion is that being in the same room allows for additional interaction between players beyond what is available in the game alone. Players exchange insults and outdo each other with witty commentary. They add content and new rules as real time presence allows for more flexibility (Lazzaro, 2004b). Considered in light of Ekman's nine pathways to creating emotions (Fig. 10.16), People Fun uses more of them than Hard Fun, Easy Fun, or Serious Fun. In fact, the last five out of the nine in Fig. 10.16 are enhanced in group face-to-face play.

People Fun: Creates More Emotion When Playing Face to Face

"Enjoyment when your friends blow up."
    —Pat, a PS2 gamer on his favorite game emotion

The involvement of others increases the complexity of the game by creating cooperation and competition, which in turn increases the emotions from play. Different roles for players with shared and opposing objectives increase player interaction. Achieving a goal with a simple rule set becomes exponentially more complex when a player's opponent is another person or group. Not all players like the feelings of rivalry from competition with their friends. Many prefer to team up with their buddies to

compete against computer-controlled characters. To encourage group formation, Massively Multiplayer Online Games (MMOs) often have interdependent classes of players, a design that comes from the *Dungeons and Dragons* paper-based role-playing game. In these games, such as *WOW*, character classes encourage the formation of groups, and treasure quests and dangerous dungeons require players to pool their characters' specializations such as archers, swordsmen, hunters, paladins, wizards, and healers. Improved success rates and score bonuses reward players working together in the game and help form the social bonds that increase a player's emotional attachment to the game (Fig. 10.17).

People Fun keeps players going by providing new ways for players to interact. It provides a variety of player-to-player outcomes and bright spectacles for the audience. It provides the opportunity to perform difficult maneuvers and stunts that reward practice with which to amuse and amaze friends. In *Soul Calibur II*, players interact with each other by fighting rather than racing side by side to beat the clock. The frequency and the variety of player-to-player interaction increase commentary between players and make it more enjoyable for an audience to watch. Spectators can participate throughout with comments and criticisms. People Fun games with hidden aspects, special moves, cheat codes, and Easter Eggs (hidden games, objects, rooms, or animations) offer even more opportunity to impress friends. All of these create admiration and envy between players (Lazzaro, 2004b).

Some Emotions Require Two People

People Fun wins out over other types of gameplay with its exclusive access to emotions that require two people, such as schadenfreude and naches. The most frequent emotion in co-located group gaming is amusement. In a group context, even negative events are sources of laughter. The feeling of naches

Chain of People Fun Emotions Increases Enjoyment between Players

People Fun motivates group interaction, interpersonal relationships, and creates emotions between players.

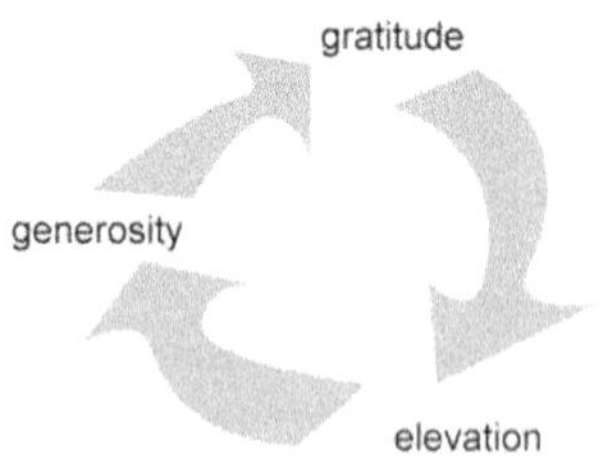

Emotion	Common Themes and Triggers
Amusement	Something that's funny. (Ekman, 2003). Regardless of the emotional tenor of the game, people in groups laugh more than when playing alone.
Schadenfreude*	Pleasure over misfortune of a rival (Ekman, 2003). Competitive players enjoy beating each other, especially a long-term rival. Boasts are made about player prowess and ranking.
Naches*	Pleasure or pride at the accomplishment of a child or mentee (Ekman, 2003). Players enjoy seeing friends succeed as well as the achievements of nonplayer characters they have helped, such as in *Back to Baghdad*, where players can exchange health packs.

*Schadenfreude and naches do not have universal facial gestures; however, they are frequently described by players (Lazzaro, 2004b).

FIGURE 10.17. People Fun Emotion Cycle. In People Fun, some emotions link together like a chain between players, such as offering a healing spell or health pack (generosity) to a fellow player in need (gratitude). A third player feels elevation at witnessing this human kindness and becomes more likely to be generous. Next time a different player may be in need and the links on this chain would rotate. In this way, one game mechanic offers three emotions depending on who starts the chain. Copyright© 2005–2007 XEODesign, Inc. All rights reserved.

and emotional attachment created in *Nintendogs* (a puppy simulation game for the Nintendo DS) are so strong that they not only created a best-selling title, but the title's popularity increased the sales of the hardware platform as well.

People Fun in games is also present in interaction with characters on screen. It enhances emotions with facial expressions for nonplayer characters (NPCs) as well as for player-controlled characters. In *Diner Dash*, players not only have to react to changes in characters' facial expressions but to win the game requires managing customer emotions to prevent them from getting so angry that they leave (Lazzaro, 2005b). Games such as *World of Warcraft* increase emotions between players by offering emotes and chat features. Adding voice and video communication between players increases emotions, and co-located group play takes this one step further by supporting face-to-face collaboration in addition to face-to-screen.

Players Modify Games to Enhance Emotions During Group Play

In People Fun, players like playing together so much that they create house rules and convert single player games to make group play more enjoyable. In order to play a *Buzz Lightyear* game together, one family assigns each person a different key to run, jump, or shoot. To share in a pirate adventure, one set of three college students plays *Sid Myer's Pirates!* game by taking turns. In a bike-racing game where players outnumbered the controllers, the winner is the player who "ran the gauntlet" and beat all challengers in the room, not by winning the race, as that would take too long, but by outdistancing each opponent. This said, some players eschew contact with other people if they mostly care about developing strategy or understanding how to play better. To them, playing a game such as *Hearts* online is better when played alone because they do not want to chat or posture with players that they do not know. Such players prefer the Hard Fun even in multiplayer games. Whether it is *WOW* or *Spades*, and depending on their mood, some end up soloing rather than joining a group.

EMOTION FROM ENTERTAINMENT INTERFACES AND INTERACTIVE PRODUCTS

The Four Fun Keys—Hard Fun, Easy Fun, Serious Fun, and People Fun—play an important role in making games self-motivating. They provide the opportunity for challenge and mastery, prospect for exploration and discovery, give a boost to self-esteem and an excuse to spend time with friends. The gameplay in games capitalizes on the interaction between human emotion, thought, and behavior. From this perspective, productivity applications have the same goal. Productivity applications take for granted that the user provides the motivation for using the tool, such as workplace responsibilities or a boss's deadline. The compelling and enjoyable nature of computer games makes the process itself enjoyable, so game designers spend a great deal of design effort to ensure that their games provide a rich emotional experience during play. Study of these interactions offers clues as to how to make productivity tools more motivating without making it a game. The importance of emotions in design is that emotions can influence when, how, and why to use a product.

Game-Inspired Affect Makes Product Use More Entertaining

Games offer many lessons for product-interaction design outside of entertainment, especially in the relationship between design and emotion. Software should make tasks easy by requiring just a few steps and providing appropriate emotional stimuli to focus and reward the user for completing them. One of the biggest differences between productivity software and software used for entertainment is that the pleasure and motivation come from accomplishing outside tasks, such as completing a spreadsheet or creating a database. Emotions in nongames should create enjoyment, aid decision making, focus effort and attention, and provide motivation. Like games, it is possible to design interactions that are similar to all four types of fun to create a wide spectrum of emotions in nongame software. (Fig. 10.18)

Playful Product Attributes Increase Emotions During Use

Many products employ novel opening techniques to increase customer appeal. Beyond "out-of-the-box" experiences, the experience of the product in use is becoming an important part of a product's competitive advantage. Similar to the way packaging frames consumer expectations and emotional associations with a product, how a product opens or switches on can create emotions and other associations every time the product is used. For example, Danger's *Sidekick* cell-phone lid spins open horizontally with a snap. The *Robo-Book* Light surprises by unfolding like a mechanical arm. The nozzle on *All* liquid laundry soap dispenser features a bright red rubber ball to squeeze for detergent. The designers at IDEO transformed the tiered old carpet sweeper to create the *Swiffer Carpet Flick*. The *Carpet Flick* combines a free-flowing universal joint at the head (Easy Fun), a transparent body so customers miss none of the action (Hard Fun's feedback on goals), and accomplishes real work in an enjoyable way (Serious Fun). The buttons on some productivity software offer Easy Fun if they glow or highlight when moused over. A hard disk defragmenting utility displays progress with an animation that is mesmerizing and game-like to watch, as the colored segments are organized and grouped together. None of these products are games, yet all of them offer Easy Fun with novelty to inspire curiosity, surprise, and a little wonder. Regardless of whether it is a productivity application or a trash can, if it creates enjoyable emotions, it will be used more often and shown off to friends (People Fun).

Breath-relaxation techniques are more effective if practiced daily. To encourage frequent use with Serious Fun the *Stresseraser* (a home-therapy device that provides bio feedback on the user's pulse and breathing exercises) uses both Hard Fun and Serious Fun. The device rewards customers with a point system and game-like chart feedback of their pulse data. The use of a display graph and a gradual increase in number of points earned per session encourages use by offering Hard Fun and Serious Fun at the same time.

"Why not challenge yourself to see how good you can make yourself feel by getting 100 or more points a day for two months?"

"A session of 30 or more points is a great way to start the day off right. Especially on days when you feel like you got up on the wrong side of the bed."

—www.stresseraser.com

The catalog copy on the company's website frames customer expectations in terms of Hard Fun. The game-like intent of the product designers is even more evident in unreleased game modules where customers float a bird around obstacles as a reward for reducing their pulse (Fabricant, 2005). Their approach to relaxation is not about building frustration to eventually succeed through their skill to feel fiero. Instead, with the *Stresseraser*, as in the Interaction Institute's *Brain Ball* game (Hjelm, 2003), people win by relaxing and taking deep breaths. For these products goals and feedback allow players to monitor and change their own emotional state for Hard Fun and Serious Fun.

Creating Affect Through Actions For Nongames

Hard Fun	**Easy Fun**
Affect to increase focus on and enjoyment of a task:	Affect to capture attention and enjoyment:
Spreadsheets	Data mining
Word processing	Searching
Time management	Multivariant problem solving
Decision support	Creativity support

People Fun	**Serious Fun**
Affect to facilitate the interaction and cohesion between individuals:	Affect to help user to accomplish a lot of work consistently over time:
Cell phones	Learning tools
E-mail	Therapies
Human-resource and project-management applications	Medical devices
Groupware	Cleaning products
Presentation software	

FIGURE 10.18. The affect from the Four Fun Keys can increase emotions and enjoyment for several kinds of productivity-software applications and consumer products. Copyright© 2004–2007 XEODesign, Inc. All rights reserved.

Designing Game Affect for Entertainment Access

Playful Interaction and Feedback Facilitates Entertainment Access

Game-inspired affect increases the appeal of entertainment applications such as televisions, game consoles, e-zines, and music players. Because the primary customer goal is to locate content, accessing entertainment options need not be a game in itself. However, entertainment devices can have a game-like feel. They should be even easier to use than productivity applications, because they don't borrow motivation from completing an outside task. They should also create excitement about the entertainment options they provide.

Since the beginning, graphic-user interfaces have employed metaphors such as windows, trashcans, and elevators to explain how features work. Apple's Macintosh OSX operating system established a new trend in modern interface design by going beyond the clear explanation of function to creating an experience. Mac OSX is an interface that is "fun" to use. The dock magnifies icons under the cursor to make them easier to select, at the same time creating pleasurable animation that is pure Easy Fun, even without the goal of opening an application. The genie-out-of-a-bottle animation of document windows as they expand out of a tiny icon on the dock offers more Easy Fun to create curiosity, surprise, and wonder. Operating these features entertains all by itself.

Creating Emotions in Entertainment Access: Apple's iPod Case Study

The most game-inspired device to come out of Apple is the iPod. The iPod's novel combination of hardware and software interfaces creates a new experience full of affect. The novel interaction facilitates entertainment access by offering a customer experience unlike any other. While not technically a game, the iPod uses Hard Fun, Easy Fun, Serious Fun, and People Fun to create emotional experiences for the user.

Central to the user experience of the iPod is the control wheel. It attracts attention and creates emotions of Easy Fun such as curiosity and surprise. Its novel round motion easily scrolls through long play lists. Emotions arising from exploring the free-flowing motion focus attention on finding a song. The quick feedback (including a separate speaker just for the scroll clicks) enhances the sense of mastery (Hard Fun) as a customer gains control. It feels like a game. Even without pressing play, scrolling is enjoyable because using the four directional buttons and wheel mimic a game controller. In addition to the form resembling a stereo speaker, the circular DJ Scratching action takes advantage of an action already associated with making music more exciting and even offers an opportunity to role play for more Easy Fun. The music visualizations on the companion iTunes software offer fascinating animations in time with the music.

A game of *Bricks* (similar to *Breakout*) shipped as an Easter egg (hidden game) in the first edition and later more elaborate games such as Zuma under their own games tab further established the iPod's connection to fun and enjoyment. As easy as the iPod is to use, users must experiment and might experience frustration in learning to understand the menu hierarchy. This offers opportunities for Hard Fun as customers search for their favorite track. Users may experience a feeling of accomplishment rather than fiero when they succeed, but the dancing black silhouette models in the iPod's ad campaigns frolicing with arms overhead clearly evokes fiero and other kinds of joy. That the iPod offers Serious Fun of mood-altering music and videos is a given. For additional Serious Fun, iPods also play business-audio books and offer file storage to do the real work of transporting documents and spreadsheets along with music files. The trademark white headphones offer emotions from People Fun even while separating from others auditorally by sending a clear message about the fun-loving social group to which an iPod customer belongs. These status-symbol icons create People Fun emotions from association with belonging to the "in crowd" as well as the emotions expressed in the ads. Several iPod accessories allow customers to share their music during playback. And iTunes' trading of play lists, creation of custom CDs, and podcasting all provide ways for iPod users to reach out to other people for more People Fun.

Creating Emotions in Entertainment Access: Microsoft's XBOX 360 Case Study

Ready to Inhale. The designers of the XBOX 360 video game platform had a specific customer experience in mind. Whereas the design inspiration for the first XBOX was like the Hulk bulging with muscles to break out of a box, the XBOX 360 was to be more of an inhale, like Bruce Lee drawing breath in preparation for play, poised, and ready to strike. Instead of a file-and-folder metaphor, the main UI for the video-game console is designed to capture a moment of preparation before the challenge. The idea was to avoid fussy animations, yet provide something interesting and intuitive to get the player emotionally ready and mentally focused before play begins. The Easy Fun of navigating the menus increases immersion. The emotions for this interface match the player's goals during this part of the use cycle.

In terms of Easy Fun, the menus for the XBOX 360's main UI curve sideways like rib bones and mimic the gesture of inhalation as they move. The menus slip from side to side in a breathing-like motion. The sounds are organic and whip-like with a satisfying thump at the end as the menu snaps into place. This novel auditory and visual experience makes the menus enjoyable to explore without becoming too complicated or intrusive. Novelty of the sideways menus creates curiosity and encourages exploration. The designers were going for playful (the attitude in games) without being so entertaining as to be a game in itself. It is about providing very simple access and a pleasurable experience that creates excitement but doesn't get in the way.

The most important way the XBOX 360 increases enjoyment of games played on its console is through the use of People Fun. To deliver more People Fun emotions and to foster connections between players, the XBOX 360 offers community features for all its titles. Gamers can create their own persistent player profile (or gamer card) that spans game titles. Personal profiles provide an in-game identity and list in-game accomplishments so friends can compare game-specific achievements, scores, and what game they are currently playing. For the nongamers in the household, the XBOX 360 allows access and display of family photos and video for a different kind of social interaction.

The emotions in these user experiences for iPods and XBOX 360s not only help customers accomplish tasks. The UIs create experiences that make the devices pleasurable to use, put users in the appropriate emotional and cognitive mindset, and capture the next level of the four pleasures that Jordan considers essential in the new human factors: (a) Physio-pleasure, (b) Psycho-pleasure, (c) Ideo-pleasure, and (d) Socio-pleasure (Jordan, 2000). As the focus for HCI shifts from interface design to user experience, more accurate taxonomies for internal experiences and methods to measure these experiences are needed. The development of these tools will facilitate the design of more enjoyable player experiences. While entertainment interfaces such as the iPod and the XBOX 360 have done much to create powerful user experiences on several levels, there is still much more that is possible to do.

NEW DIRECTIONS AND OPEN ISSUES

New Play with Smaller, Smarter, More Flexible Devices

As technology shrinks and integrates itself into more aspects of our lives, people play more games in more places. Devices that are smaller, smarter, mobile, and contextually aware create new opportunities for electronic gaming, which until now has largely been restricted to players at home around a single screen. What happens when the services offered by a laptop become cheap enough to print on a candy wrapper? Mobile and alternate-reality gaming offer the opportunity to make things happen in the real world, which further enhances other experiences such as enjoying the company of friends and dispelling boredom while waiting in line. Games break out into the real world through geo-caching and games of tag through a city. Promotional Alternate Reality Games (ARGs), such as the The Beast for the movie AI and I Love Bees for the game Halo 2, use real-world-web-enhanced gameplay to market products and create communities of millions of players (4ourty2wo, 2005). With ubiquitous computing, everything from mobile phones, to the front door, to a ketchup bottle in a diner could contain enough smarts to offer services. Will they all contain a screen and therefore the potential to host a game? Will we surf the net from our saltshaker or will it provide other opportunities to engage our attention?

Many designers chase faster hardware and better graphics, yet the emotional power of games does not occur on screen; it takes place inside the player's head. The biggest emotions that new technology will create is not through rendering blood, sweat, and tears in molecular detail. The real changes in emotional gaming will happen through supporting a more agile design process and by making gaming devices smaller, sharable, context-aware, and more accessible, such as game controllers with fewer than 12 buttons like Nintendo's Wii. Until we all get personal holodecks and natural language processing, the success of many new kinds of games will come from connecting the cheapest emotion generators around: a player's rivals and friends.

Life is becoming more game-like as games and elements of play inspire new dimensions of product appeal. Like games, products create experiences. Cutting-edge product designers now design customer experiences, not products, as witnessed by the iPod. Designers aim for engagement in addition to making something better/faster/cheaper and easier to use or market. Adding playful elements to goods and services increases the attraction of everything from advertising messages to Southwest Airline's in-flight safety announcements to the design of public spaces. We see the increasing importance of emotions in design of products such as the playful squid shape of the Phillip Stark orange juicer, the surprising opening rotation of Danger's Sidekick mobile phone, and in the pleasing octave chords produced by Segway's acoustically designed motors. The progress being made toward this shift toward more emotional design is even more apparent in games.

Games are already redesigning how we work and shop. Employers use games to screen potential hires for three-dimensional (3D) reasoning skills and train them to solve problems with multiple variables. Games changed consumer processes such as buying and selling on eBay or the dining experience at Dave and Busters (Chuck E. Cheese for adults). Even web software applications like Flikr (www.flikr.com) include more fun to increase appeal by sharing items between friends or by offering features to be "gamed," such as displaying the number of friends a person has in Friendster or LinkedIn (www.friendster .com; www.linkedin.com) social-networking software. Understanding how games create emotions makes products and services more engaging and enjoyable and even improves community and the quality of life.

Open Issues

Better design of the emotions required for and that result from interaction will unlock more human potential from using products and games by priming and rewarding users with appropriate affective states. We know that cognition and emotion walk hand in hand. We are only just beginning to discover how they support each other and how interaction creates emotion. The Four Fun Keys model starts this journey to build and examine player experiences in terms of specific emotions such as fiero from Hard Fun, curiosity from Easy Fun, relaxation from Serious Fun, and amusement from People Fun. Further measurement of specific emotions during different types of cognitive, behavioral, and emotional experiences will help piece together the role individual emotions play in cognition and activities. For example, a surprising event orients an individual's attention to determine whether the source is a benefit or a threat. Surprise does this in addition to creating internal sensations, producing facial expressions, and communicating information about the source to others. Curiosity, the lead emotion from Easy Fun, has a strong cognitive component, which also focuses attention in a pleasurable way. Satisfying curiosity especially when it results in feelings of surprise and wonder produces strong pleasurable sensations that motivate an individual to repeat the activity. Additional research will tell us how to create designs that inspire, maintain, and intensify feelings of curiosity. This is important for applications such as Google to improve searching the Internet or when browsing an e-commerce catalog (www.google.com). Conversely, intensifying fiero from finding the search item also improves the user experience. Informed with a deep understanding of how curiosity and frustration employ different affective states to focus attention, the design of entertainment and productivity products can achieve the next level of engagement by sculpting emotional responses that complement the tasks.

CONCLUSION

Games have the creative freedom to push technical boundaries and be light years ahead of productivity interfaces. They have been champions and early adopters of new interface techniques from joysticks to voice commands. Game interfaces lasso new hardware and experiment with dialogs and menus to deliver novel experiences with a vigor never seen in productivity applications. Mastering these innovative interaction techniques provides richer experiences that inspire a devotion to learning features. Radial menus in *The Sims*, audio menus, and draw-your-own-game interfaces in LeapFrog's *FLY* pen-top computer, the Line Rider web game, the iPod's scroll wheel, and XBOX 360's organic side-scrolling tabs are all examples of original alternatives to navigating file-and-folder hierarchies and dialog boxes. Games boast heads-up displays and voice commands to allow players to multitask (*Tom Clancy's Rainbow Six* on the XBOX) or sing (*Karaoke Revolution*). Still other games use camera interfaces that track the body's movement (*EyeToy* games, *Virus Attack* for camera phones), physical motion (*Dance Dance Revolution* dance pad, Wii Sports) and touch and gestures (*Black and White, Nintendogs, Yoshi's Touch and Go, Electroplankton, Pac-Pix*). Others use positional sensing/ubiquitous computing and motion sensors (*WarioWare: Twisted*). Games offer rich social systems for interpersonal collaboration and communication (*WOW, EverQuest*). Games even prototype future interface technology such as context-sensitive holograms to encourage cursor exploration (*Star Wars Commando*). The interface for each of these technologies supports a fresh experience.

In addition, games offer an emotional punch that has become a cultural force. The automotive industry already consults racing-game designers on how to make more exciting cars. U.S. politicians such as Howard Dean use web games to teach democracy and increase interest in their campaigns. Games are not only innovating interface design, they are also full of emotional experiments. For example, in *Fable*, the moral values expressed by player decisions change the character into a shining, blonde-haired hero or demonic devil with horns. Although not everyone plays computer games, they command a profound influence on other media and culture. Additionally, the generation raised on games (today's college students all played *Oregon Trail* in grade school) will play more games as adults than their parents did.

Games' distinct emotional processes come from what players like most during play, which informs the functional design of product and software. Unfortunately, the sole emotion goal addressed by usability is to reduce frustration. While important, this falls short of offering strong emotional rewards for accomplishing difficult tasks. Crafting these emotions contributes to user success because, when inspired by rich emotional responses, users will explore and learn more of an application's features, making them more efficient at their jobs. Over-engineering a task by making it too predictable, repetitive, and easy to complete increases the likelihood of boredom and actually reduces satisfaction over the long term. Players and workers experience huge emotional rewards for completing complex challenging tasks. It would be unfortunate if these were eliminated from the work we do as humans.

In HCI, the big mistake used to be fixing clumsy complex features with pretty background graphics. Today, the big mistake lies in believing that user experiences will improve if designers remove frustration points (usability), have the interface do all the work, and reduce the task to a series of trivial steps. Customers already have emotional reactions to their software. Designers must learn to speak this language of emotion from interaction.

The role of a good interface is to prepare users to master something difficult, then allow them to give themselves credit for mastering the difficult skill, while leaving enough ambiguity and challenge to make the task fun. What's next for user-experience design goes beyond refining features to logically support how users perform tasks. Design must also address how features motivate users through affective states to support and refresh the task at hand, not by making the software unnecessarily complicated, but by doing what games do: support sequential skill-building to achieve complex goals and encourage players to move beyond their points of failure to feel empowered masters of their own destiny. Experienced designers are already using Hard Fun, Easy Fun, Serious Fun, and People Fun by embedding emotions from goal-directed, open-ended, purposeful, and social play.

By offering life problems in miniature, games provide important clues to the relationship of emotion in human problem solving, goal achievement, and interaction with other people. Researching games clearly defines the relationships between action and emotion as well as between emotions themselves. In games, many emotions are opposites, happen in sequences, have prerequisites, and share links with others. Some emotions require people, relate to goals, or involve the future or the past. Connecting how gameplay leads to specific emotions establishes a framework for the process of creating more emotional user experiences. The Four Fun Keys Model creates effective analysis and design techniques to identify and create a wider range of emotions. It connects emotions to popular types of gameplay and demonstrates how chains of emotions are embedded or released from different activities. By harnessing play, even productivity software and workflow design can take advantage of the motivating force of game emotions.

Games are the new medium of the 21st century. Unlike any other design discipline, player experiences that unfold over time are more interactive than movies, painting, architecture, industrial design, literature, or fashion. Games are dynamic processes that create dynamic experiences. They are much more than a series of static impressions seen in sequence. Games offer a unique set of emotions from accomplishment and failure. At their core is the ability for players to interact with the content. The experiences that come from this interaction, the ability to create emotions, and the way that this interaction moves will separate games as an art form distinct from movies and other visual arts.

Emotions play an essential role in providing the entertainment value that captivates players. From games, we can learn how to improve the emotions that keep users engaged in activities of different types. Through player interaction and control of events, games promise to become more emotional than movies and other entertainment, but whether these are the same emotions remains to be seen. Games now include more detailed storylines to increase player engagement. However, it is clear from XEODesign's research that games create strong emotions even without stories, and it is clear that games already create more emotions through interaction with other players in the game world. Many aspects of the storytelling language of cinema apply to games and they will still move players emotionally. However, emotions from the player's goals and the things he or she cares about will likely prove to be stronger. In games, the mechanic is the moral of the story.

Releasing the full emotional potential of games will not be easy. It requires deep understanding of how emotions work, because more is known about crafting emotional entertainment experiences (such as a movie) by offering the viewer empathy with a protagonist engaged in a predetermined sequence of events. Less is known about creating emotion through being the protagonist. Viewing a prerendered video between game levels is less compelling than having the player's actions create an emotionally rich experience itself, but at the moment these mini-movies are easier to design. Going forward, these dramatic tools for creating affect should inspire rather than dictate game design. Eventually the emotional props from these cut scenes will fade from games just as title cards disappeared from old silent pictures once sound technology allowed the actors and the action to speak for themselves.

Psychology, sociology, theater, literature, film, and only recently games have all studied how entertainment engages audiences. Games are unique in that they entertain by creating emotion from interaction. Before XEODesign's research on emotion and the fun of games, none had studied how entertainment creates specific emotions and uses them to focus attention and motivate play. The Four Fun Keys Model creates the four most important sets of emotions for games. Each fun key is a collection of game interactions (game mechanics) that captivate player attention with different series of emotion that make games self-motivating. These emotion cycles create the player experience and separate best-selling games from their imitators. Based on contextual research of people playing their favorite games cross genre, platform, and gender, the Four Fun Keys describe how emotion comes from what is the most fun about games. Entertainment interfaces and products such as Apple's iPod and Microsoft's XBOX 360 already use interaction from each of the Four Fun Keys to unlock powerful emotions to increase user enjoyment and build stronger brand impressions. Interaction that generates a lot of emotion is more memorable and feels like play.

ACKNOWLEDGMENTS

Jane Booth, Brian "Psychochild" Green, Ron Meiners, Steve Meretzky, and Jeff Pobst provided detailed feedback on drafts of this chapter. Hal Barwood, Lura Dolas, Noah Falstein, Bill Fulton, Lee Gilmore, Russ Glaser, Kevin Keeker, Paolo Malabuyo, Marcos Nunes-Ueno, Randy Pagulayan, Patricia Pizer, Kent Quirk, Brian Robbins, Ann Smulka, Mark Terrano, Gordon Walton, and Richard A. Watson all provided useful guidance, suggestions, and examples.

References

4ourty2wo. (2005). Retrieved December 29, 2005, from http://www .4orty2wo.com/casestudy_ai.html

Ahn, L. & Dabbish, L. (2004). Labeling images with a computer game. *Proceedings Association for Computing Machinery (ACM) Special Interest Group on Computer-Human Interaction. Conference (CHI)*. 319–326. Vienna, Austria. New York: ACM Press.

Bartle, R. (1996). *Hearts, clubs, diamonds, spades: Players who suit MUDs*. MUSE Ltd, Colchester, Essex, UK. Retrieved December 29, 2005, from http://www.brandeis.edu/pubs/jove/HTML/v1/bartle. html

Bartle, R. (2003a). *A self of sense*. Retrieved December 29, 2005, from http://www.mud.co.uk/richard/selfware.htm

Bartle, R. (2003b). *Designing virtual worlds*. New Riders Games. Berkeley, CA: Peach Pit Press.

Boorstin, J. (1990). *Making movies work*. Beverly Hills, CA: Silman-James Press.

Brave, S., & Nass, C. (2002). A. Emotion in human–computer interaction. In J. Jacto & A. Sears (Eds.), *The human–computer interaction handbook: Fundamentals, evolving technologies and emerging applications*, (pp. 81–96) Mahwah, NJ: Lawrence Erlbaum Associates.

Clore, G. C., & Gasper, K. (2000). Feeling is believing: Some affective influences on belief. In N. H. Frijda, A. S. R. Manstead, & S. Bem (Eds.), *Emotions and beliefs: How feelings influence thoughts* (pp. 10–44). Paris/Cambridge: Editions de la Masion des Sciences de l'Homme and Cambridge University Press (jointly published).

Csikszentmihalyi, M. (1990). *Flow: The psychology of optimal experience*. New York: Harper & Row.

Damasio, A. (1994). *Descartes' error: Emotion, reason, and the human brain*. New York: Quill Penguin Putnam.

Ekman, P. (2003). *Emotions revealed*. New York: Times Books Henry Hold and Company, LLC.

Fabricant, R. (2005). *HRV monitor: Creating a guided user experience on handheld devices*. Proc DUX 2005 San Francisco, USA.

Gee, J. (2003). *What video games have to teach us about learning and literacy*. New York: Palgrave Macmillan.

Hassenzahl, M., Platz, A., Burmester, M., & Lehner K. (2000). Hedonic and ergonomic quality aspects determine a software's appeal. *Proceedings Association for Computing Machinery (ACM) Special Interest Group on Computer-Human Interation Conference (CHI)*, 201–208, The Hague, the Netherlands.

Hazlett, H. (2003). Measurement of user frustration: A biologic approach. *Proceedings Association for Computing Machinery (ACM) Special Interest Group on Computer-Human Interaction Conference (CHI)*, 734–735, April, 2003.

Hjelm, I. (2003). BrainBall research + design: The making of Brainball. *Interactions, 10*(1).

Jordan, P. W. (2000). *Designing pleasurable products: An introduction to the new human factors*. London: Taylor & Francis.

Kim, A. J. (2000). *Community building on the Web*. Berkeley, CA: Peach Pit Press.

Lazzaro, N. (2004a, Winter). Why we play games. *User Experience Magazine, 8*.

Lazzaro, N. (2004b). Why we play games: Four keys to more emotion in player experiences. *Proceedings of the Game Developers Conference*, San Jose, California, USA. Retrieved December 28, 2005, from www.xeodesign.com/whyweplaygames.html

Lazzaro, N. (2005a). *Design survey of game designers*. Unpublished manuscript.

Lazzaro, N. (2005b). *Diner dash and the people factor*. Retrieved March 2, 2005, from www.xeodesign.com/whyweplaygames.html

Lazzaro, N., & Keeker, K. (2004). What's My Method? A game show on games. (pp. 1093–1094). *Proceedings Association for Computing Machinery (ACM) Special Interest Group on Computer-Human Interaction Conference (CHI)*, Vienna, Austria.

LeBlanc, M., Hunicke, R., & Zubek, R. (2004). *MDA: A formal approach to game design and game research*. Retrieved March 2, 2005, from http://www.cs.northwestern.edu/~hunicke/pubs/MDA.pdf

Malone, T. (1981). Heuristics for designing enjoyable user interfaces: Lessons from computer games. *Proceedings Association for Computing Machinery (ACM) Special Interest Group on Computer Human Conference (CHI)*, (pp. 63–68).

Mandryk, R. (2004). Objectively evaluating entertainment technology. *Proceedings Association for Computing Machinery (ACM) Special Interest Group on Computer-Human Interaction Conference (CHI)* (pp. 1057–1058). Doctoral Consortium, Vienna, Austria.

Norman, D. A., (2004) *Emotional design: Why we love (or hate) everyday things*. New York: Basic Books.

Papert, S. *Hard Fun*. Retrieved December 29, 2005, from http://www.papert.org/articles/HardFun.html

Piaget, J. (1962). *Play, dreams, and imitation in childhood*. New York: Norton.

Sykes, J., & Brown, S. (2003). Affective gaming: Measuring emotion through the gamepad. *Proceedings Association for Computing Machinery (ACM) Special Interest Group on Computer-Human Interaction Conference (CHI)*, 732–733.

Tiger, L. (1992). *The pursuit of pleasure* (pp. 52–60). Boston: Little, Brown & Company.

Thorson, E., & Friestad, M. (1985). The effects of emotion on episode memory for television commercials. In P. Cafferata & A. Tybor (Eds.), *Advances in consumer psychology* (pp. 131–136). Lexington, MA: Lexington.

Wright, P., McCarthy, J., & Meekison, L. (2003). Making sense of experience. In M. A. Blythe, K. Overbeeke, A. F. Monk, & P. C. Wright (Eds.), *Funology: From usability to enjoyment* (pp. 43–53). Dordrecht, the Netherlands: Kluwer Academic.

Wright, R., Ritterfeld, L., Mathiak, K. (2006). Does playing violent video games induce aggression? Empirical evidence of a functional magnetic resonance imaging study. In *Media Psychology*, Vol 8, No. 1 (pp. 39–60). Mahwah, NJ: Lawrence Erlbaum Associates.

For more articles on emotion and game research, see http://www.xeodesign.com/whyweplaygames

All trademarks are the property of their respective holders.

·11·

MOTOR VEHICLE DRIVER INTERFACES

Paul Green
University of Michigan Transportation Research Institute

INTRODUCTION

This chapter is written for professionals familiar with human—computer interaction (HCI), but not with the issues and considerations particular to motor vehicles. For automotive industry interface designers, this chapter should pull together information dispersed throughout the literature.

HCIs of interest to motor vehicle designers because of the rapid growth of driver information systems that utilize computers and communications, collectively referred to as "telematics." Included under the umbrella of telematics are navigation systems, cell phones, collision-warning systems of various types, lane departure warning systems, and so forth. These systems can (a) potentially enhance public safety, (b) make transport more efficient (saving time and fuel), (c) make driving more enjoyable, and (d) make drivers more productive.

More specifically, this chapter reviews design documents and evaluation methods for driver interfaces in order to promote safety and ease of use. An underlying theme is that the safety-critical nature of driving leads to significant departures from standard HCI practice, and to some methods and measures that are unique to automotive applications. Since this is a reference handbook, engineering practice receives more attention than scientific theory and, furthermore, given its technology focus, the design of traditional (noncomputer) motor vehicle driver interfaces (such as switches for headlights and windshield wipers) is not covered. For information on those interfaces, see Peacock and Karwowski (1993) or the latest edition of the *Society of Automotive Engineers (SAE) Handbook* (Society of Automotive Engineers, 2005). Readers interested in additional research literature on telematics should see Michon (1993), Parkes and Franzen (1993), Barfield and Dingus (1997), Noy (1997), or the proceedings from the 2005 Driving Assessment conference (http://ppc.uiowa.edu/driving-assessment/2005/final/index.htm).

What Kinds of Telematics Products Are Likely in the Near Term?

Over the years, several studies have used expert opinions to predict the future of automotive electronics, specifically telematics applications (e.g., Ribbens & Cole, 1989; Underwood, Chen, & Ervin, 1991; Underwood, 1989, 1992; Richardson & Green, 2000). The most extensive information, however, is often contained in proprietary market surveys not for public distribution. The most recent publicly available projections appear in Green and colleagues (2001), summarizing input from 83 senior executives in the automotive industry. Interestingly, those projections coincide with the time period when this handbook is to be published. Table 11.1 lists the mean value of the year estimated when each feature would be introduced into 10% of new luxury vehicles in 2004–2008.

Since these data were collected in August, 2000, and the planning horizon for a new vehicle is three years, estimates through 2004 should reflect real product plans, not speculation about the future. For some features (built-in personal digital assistant [PDA], electronic toll tag), there was not complete agreement on when introduction would occur, as a significant fraction of respondents thought some features would never be introduced. Nonetheless, the table makes the point that overall, these predictions tend to be slightly optimistic, but not unreasonable. Future products are most likely related to driving support systems (various collision warning and avoidance systems, night vision), information systems (cell phones, e-mail/Internet, PDAs), and the use of voice technology. In about a decade or less, the author believes that wearable computers will be a topic of significant interest for driving. In aggregate, these systems could reshape the driver's task from one of real-time control of the vehicle (steering the vehicle and using the throttle to adjust speed) to information management.

Chapter Organization

How might one organize information on telematics? In their classic paper on usability, Gould and Lewis (1985), identified three key principles to be followed when designing products for ease of use:

1. Early focus on users and tasks
2. Empirical measurement
3. Iterative design

These principles not only apply to office applications and web development, but automotive applications as well. In the automotive context, designers need to understand (a) who drives vehicles (users), (b) what in-vehicle tasks they perform, and, most importantly, (c) the driving task, (d) task context, and (e) the consequence of task failures. These topics are the focus of the first part of this chapter.

Secondly, it is important to be able to measure driver and system performance (empirical measurement). That topic constitutes the second part of this chapter.

Surprisingly, there has been little reporting of how iterative design is used in developing driver interfaces, though the approach is used. A great deal of automotive design relies upon following design guidelines, which is the final focus of this chapter.

WHAT IS THE DRIVING CONTEXT IN WHICH USERS PERFORM TASKS?

Scott McNealy, CEO of Sun Microsystems, once said, "A car is nothing more than a Java technology-enabled browser with tires" (Kayl, 2000). He is dead wrong. The author has never heard of anyone claiming, "A computer came out of nowhere, hit me, and vanished." Yet police officers and insurance adjusters hear such claims about motor vehicles every day. Likewise, the author knows of no one who was ever been killed as a consequence of operating a computer at their desk, but the loss of life associated with crashes arising from normal motor vehicle operation is huge.

TABLE 11.1   Estimated Year of Feature Introduction into Luxury Vehicles from Green et al. (2001)
(Bold = estimated to be in 10 percent of 2006 model year luxury vehicles)

Mean Year Estimate	"Never" Responses*	Feature	Description
2004.3	1	**Built-in wireless phone interface**	
2004.4	0	**GPS navigation**	Uses Global Positioning Satellites for location, e.g., Hertz NeverLost
2004.6	1	**Automatic collision notification**	Calls when air bag deploys, e.g., OnStar in GM vehicles
2004.7	2	**Satellite radio**	Nationwide satellite broadcast of 100 channels, e.g., XM, Sirius
2004.8	4	Removable media for entertainment and data	(e.g., iPod, but for entertainment only)
2004.8	0	Email/internet access	
2004.8	10	Built in PDA (e.g., Palm) docking station	PDA is a personal digital assistant
2004.8	0	Adaptive cruise control (ACC)	Scans for traffic and adjusts speed to maintain driver set separation
2005.0	1	**Rear parking aid**	Several SUVs (e.g., Murano)
2005.2	1	MP3 support	MP3 is an audio file format
2005.2	0	Bluetooth support	Short-range wireless communication allowing direct link of cellular phones
2005.3	0	**Automatic download of traffic/ congestion information**	Navigation systems in several Honda products
2005.3	2	Blind spot detection and warning	Radar and other systems to detect vehicles to the side and rear
2005.5	1	**Voice operation of some controls**	Jaguar, several Nissan vehicles
2005.6	2	Downloadable software features	
2005.6	2	Downloadable software fixes	
2005.6	3	Forward collision warning	Laser radar systems
2005.7	4	Forward parking aid	Sonar-based short range systems to assess distance ahead
2005.7	6	Lane departure warning	Video systems that scan for lane markings, Infiniti SUVs
2005.7	4	Dual voltage (42/12 volt)	Current power is 12-volt
2005.8	12	Built-in electronic toll and payment tag	Allows travel through toll booth without stopping
2005.8	0	General purpose text/data speech capability	
2005.9	7	Large general-purpose display	
2005.9	6	Offboard applications via data link	
2005.9	9	Night vision systems	Infrared-based systems to show objects ahead on a head-up display (HUD), Cadillac
2006.0	16	Black box crash recorder	
2006.0	7	Active suspension	
2006.1	8	Forward collision braking only	Active, not just a warning
2006.2	9	General purpose computer (e.g., AutoPC)	
2006.3	62	Karaoke	Most respondents said never
2006.3	9	Drowsy driver detection	
2006.7	1	42 v electrical system	
2006.8	10	Open electronics bay with utilities	
2006.8	7	Fingerprint or voice-controlled entry	
2006.9	11	Interface to wearable computer	
2007.1	14	Large area HUD	For speedometer, navigation, back up camera, etc.
2007.1	25	Alcohol-impaired driver detection	Most respondents said never
2007.1	1	Brake by wire	
2007.1	4	Drive by wire	
2007.2	6	Hybrid drive-train (electric/combustion)	
2007.6	15	Automatic lane control	Takes over steering
2008.0	20	Forward collision braking and steering	
2008.2	26	All-electric drive-train	

*Respondents had the option of identifying a particular year or selecting never.

According to the World Health Organization (Peden et al., 2004), 0.75 to 1.8 million people die in road traffic crashes each year, or about 3,000 per day. (See also http://www.unece.org/trans/main/welcwp1.html.) Traffic crashes have been ranked ninth in terms of the percentage of disability-adjusted life years lost (Murray & Lopez, 1996), well ahead of war (16th), violence (19th), and alcohol use (20th). If the current trends continue, by the year 2020, traffic crashes will become the third largest cause of death and disability after clinical depression and heart disease. In 1990, traffic crashes ranked as the ninth biggest killer in the world.

Additional insights come from crash data for the United States, for which reliable, detailed crash statistics are available. According to the U.S. Department of Transportation (2005a), 42,636 deaths occurred in traffic crashes in the United States in 2004. Of them, 19,091 were passenger car occupants; 5,801 were pickup truck occupants; 4,641 were pedestrians; 4,735 were in SUVs; 4,008 were motorcyclists; 2,036 were in vans; 761 were in larger trucks; 725 were bicyclists; and 808 were in other categories. (For additional information, see the latest edition of *Injury Facts* [National Safety Council, 2004] and the U.S. Department of Transportation Fatality Analysis Reporting System [FARS] website [http://www-fars.nhtsa.dot.gov/].)

What Kinds of Crashes Are Associated With Telematics?

It is widely accepted that some telematics tasks are distracting and that distraction can lead to crashes. The most commonly cited statistics concerning distraction-related crashes (Table 11.2) were based on data from the Crashworthiness Data System (CDS). CDS is an annual probability sample of approximately 5,000 police-reported crashes involving at least one passenger vehicle that was towed from the scene (out of a population of almost 3.4 million tow-away crashes). Minor crashes (involving property damage only) are not in CDS. CDS crashes are investigated in detail by specially trained teams of professionals who provide much more information than is given in police reports.

Note that distraction-related crashes primarily involved a single vehicle (41%), though rear-end crashes (moving, 10%; stopped, 22%) were also common. Intersection crashes represented another 18% of the total. Crashes tended to peak in the morning rush hour and, to a much lesser extent, in the evening rush. The overwhelming majority of the crashes occurred in good weather (86% clear, 10% rain, 4% snow/hail/sleet), and many occurred at lower speeds (40% at 0–35 mph, 40% at 40–50 mph, 17% at 55–60 mph, and 4% at over 65 mph). Thus, these data suggest that device test scenarios should emphasize situations in which single-vehicle crashes (often run-off-the-road), as well as those involving rear-end collisions into stopped vehicles, are likely.

Other recent evidence appeared in Stutts, Reinfurt, Staplin, and Rodgman (2001) and Eby and Kostyniuk (2004). Table 11.3 shows summary statistics from Stutts and colleagues (2001; CDS

files from 1995, the first year for which driver attention was examined, until 1999) concerning the types of distraction involved.

These data, however, should be used with some caution. First, there is a high percentage of unknown and missing data in the set, an outcome typical when distraction is coded. For example, the driver attention status (1 of 360 variables included) was "unknown" for 36% of the cases. Several years have elapsed since the midpoint of this study, and many driver interfaces have changed considerably since the data were collected. For example, according to the Cellular Telecommunications and Internet Association (CTIA; http://files.ctia.org/pdf/CTIA_Survey_Year_End_2006_Graphics.pdf), the number of cell phones in use grew by more than a factor of four between 1997 and 2006. Given that increase in exposure, the percentage of such crashes is likely to be underestimated by that factor using the data from Stutts and colleagues (2001). Unlike external distractions, distractions due to vehicle hardware can be reduced or eliminated by appropriate engineering on the part of manufacturers and suppliers.

The Scope of the Cell Phone Problem

Because telematics devices are new, the data on device-related crashes is limited. This situation has hampered progress in understanding the risks of such devices, especially cell phones, which receive the most attention. At the time this chapter was written, 22 states record whether cell phones are causal factors in crashes, with most just starting to collect this information in the last few years (Sundeen, 2005). The National Conference of State Legislatures data suggested that cell phones were involved in approximately 0.18% of all crashes. Currently, 6% of all drivers are using cell phones at any moment (U.S. Department of Transportation, 2005b); however, given increased feature content (MP3 support, text messaging, broadcast TV), cell-phone use is likely to increase even more than over the last few years.

In terms of fatalities, the data from the National Police Agency (NPA) in Japan is often cited. Admittedly, there is less evidence to assess the quality of that data than there is for U.S.

TABLE 11.2   Distraction/Inattention Crashes by Crash Type from Knipling and Goodman (1996)

Crash Type	Sleepy	Distracted	Looked But Did Not See	Unknown	Attentive	Total
Single Vehicle	5.8	18.1	0.2	31.8	44.1	100.0
	66.2	41.2	0.7	20.6	45.9	30.0
Rear-End/Lead Vehicle Moving	12.7	21.3	3.4	48.3	14.3	100.0
	27.9	9.6	2.0	6.4	2.9	5.9
Rear-End/Lead Vehicle Stopped	—	23.9	11.4	52.6	11.8	100.0
	—	21.9	13.8	14.1	4.9	12.1
Intersection/Crossing Path	—	7	17.9	52.8	22.3	100.0
	—	18.1	63.6	39.8	26.6	34.3
Lane Change/Merge	—	5.6	17.2	41.8	35.3	100.0
	—	1.6	6.7	3.4	4.6	3.8
Head-On	1.0	7.0	8.1	46.4	37.5	100.0
	1.7	2.2	3.5	4.3	5.4	4.2
Other	—	7.3	9.7	53.5	28.9	100.0
	—	5.4	9.7	11.4	9.7	9.7
Total Crashes	2.6	13.2	9.7	45.7	28.8	100.0
	100.0	100.0	100.0	100.0	100.0	100.0

TABLE 11.3 Crashes for Which Distraction Was a Known Factor
(8.3 percent of the crashes)
(Bold = equipment-related distraction)

Distraction	Estimate (%)	Error (%)
Outside person, object, or event	29.4	±4.7
Adjusting radio/cassette/CD	**11.4**	**±7.2**
Other occupant	10.9	±3.3
Moving object in vehicle	4.3	±3.2
Other device/object	2.9	±1.6
Adjusting vehicle/climate controls	**2.8**	**±1.1**
Eating and/or drinking	1.7	±0.6
Using/dialing cell phone	**1.5**	**±0.9**
Smoking-related	0.9	±0.4
Other distractions	25.6	±6.0
Unknown distraction	8.6	±5.3

TABLE 11.4 Mobile Phone and Navigation Systems-Related
Crashes in Japan

System	Result	1997	1998	1999
Cell Phones	Injuries	2,095	2,397	2,418
	Deaths	20	28	24
Navigation Systems	Injuries	117	131	205
	Deaths	1	2	2

TABLE 11.5 Cell Phone and Navigation Crashes
per Year in Japan

Crashes	1997	1998	1999	2000	2001	2002	2003	2004
Cell phones	2297	2648	2583	1453	3040	2847	2597	1868
Navigation systems	127	146	219	190	1415	1307	1098	1253

federal or state data. Further, it is suspected that even when investigated, mobile phone use is underreported, because reporting involves drivers admitting to something inappropriate. Tables 11.4 and 11.5 show the annual crash statistics for phones and, for perspective, navigation systems. The decreases in the number of cell-phone crashes after 1999 and 2003 reflect new legal restrictions on cell phone use.

Using these data, cell-phone-related deaths in the United States for the year in which this chapter was written (2005) can be estimated. (See Wierwille, 1995.) The average number of cell-phone-related deaths in Japan was about 25 per year for 1997–1999. In determining this estimate, one must realize that a random variation exists in deaths per year, and the number increased over that period. On average, the number of traffic deaths in the United States is about four and a half times that of Japan. Assuming everything else is equal, this would suggest at least 112 (=25 × 4.5) cell-phone-related deaths per year in the United States for 1997–1999. If the number of calls per driver and the mean call duration are assumed unchanged (they have actually increased by 30%), then the change in exposure of the population between 1998 and 2005 is simply due to the increase in the number of phones in existence, roughly 3.19 times. Using the CTIA data to determine the number of users and 112 deaths in 1998 as the base estimate, the total for 2005

cell-phone-related crashes in the United States is 357 deaths (=112 × 3.19); this number should continue to increase at 15% or so per year, the current growth rate of cell phone subscribers. Other sources suggest the number of cell-phone-related deaths is much higher (Pena, 2007).

By way of comparison, about 120 children and small-stature women were killed in front-passenger air-bag crashes, and about 200 people in Firestone/Explorer rollover crashes over several years. Both of these situations have led to major changes in federal requirements and product engineering changes, as well as significant financial losses to several manufacturers (Green, 2001c). In response to the Firestone situation, the U.S. Senate passed the TREAD Act (http://www.driveusa.net/treadact.htm) three weeks after it was introduced. Hence, designing telematics interfaces to assure safety and usability is a topic worthy of significant effort.

Supporting the crash statistics, the most-often-cited study of cellular-phone use is Redelmeier and Tibshirani (1997; see also Redelmeier & Tibshirani, 2001). They examined data for almost 700 drivers who were mobile-phone users and were involved in motor-vehicle crashes that resulted in substantial property damage. Each driver's mobile-phone records for the day of the crash and the previous week were examined. Redelmeier and Tibshirani reported the risk of a crash was 4.3 times greater when a mobile phone was used than when it was not. Interestingly, hands-free units had a greater risk ratio (though not significant) than hand-held units (5.9:1 versus 3.9:1). Other data (Koushki, Ali, & Al-Saleh, 1999; McEvoy et al., 2005; Violanti & Marshall, 1996) suggested similar risk ratios. Other human factors data show drivers taking longer to respond to brake lights of lead vehicles, departing from the lane more often, and exhibiting other undesired characteristics (Horrey & Wickens, 2004) while using a cell phone.

If phone use is a concern, which tasks lead to crashes? Table 11.6 shows task-specific crash statistics for cell phones and, for comparison, navigation systems. These NPA data identify four primary tasks: (a) receiving a call, (b) dialing, (c) talking, and (d) other. What other represents is unknown. It is also unknown if talking includes voice services (voice mail, bank by phone, etc.). Notice that NPA reported the number one cause of crashes was receiving a call. Most people, either at work or at home (and while driving, even in hazardous traffic situations), immediately (within 1–4 seconds) answer the phone when it rings, almost regardless of the situation (Nowakowski, Friedman, & Green, 2001). (An exception is at dinnertime in the United States, when telemarketing calls are common.) For example, one could be in his or her office talking to a very important person such as the

TABLE 11.6 Driver Tasks and Crashes
(January–November, 1999) from NPA Data

Cell Phone		Navigation System	
Task	Crashes	Task	Crashes
Receiving Call	1077		
Dialing	504	Looking	151
Talking	350	Operating	46
Other	487	Other	8
Total	2418		205

President of the United States. If the phone rang, many people might ask the President to wait a moment while they answered the incoming call, though it is highly unlikely someone more important might be calling. In fact, many people now have call waiting so they can accept additional incoming calls while on the phone. Logically, answering the phone should not usually be more important than driving, but people behave otherwise out of habit.

Not only is the answering task a problem because of its immediacy, but also sometimes because of its visual demands. Many drivers use portable phones that might be in a jacket pocket on the seat or in a purse. Thus, in answering the phone, the driver's first task might be to search for it. Although hands-free systems can reduce the duration of the disruption, they do not eliminate the disruption itself or its incorrect prioritization.

In the Japanese data, the second most common cause of crashes was dialing. Clearly, going to a hands-free interface can reduce the visual and manual demands, though not the cognitive demands of dialing; however, the current evidence is that there is no difference in crash risk when hands-free phones are provided. It could be that providing a hands-free phone makes it easier to dial, so hands-free phone users engage in more conversations while driving. Another possibility is that seeing a phone being held is a warning to other drivers of a potentially distracted driver.

Finally, a substantial number of calls were associated with just talking on the phone. While some distractions associated with talking on the phone are manual (such as holding the phone), most distractions are cognitive, with drivers focusing on the conversation, rather than on driving. It is this aspect of cell phone use that is commonly observed by drivers. Current wisdom in the U.S. is that if a person is weaving between lanes and it is night, they are drunk. If they are weaving and it is during the day, they are on the phone.

Talking on the phone is different from talking to a passenger, especially an adult in the front seat. Often, that adult behaves as a co-driver, looking both ways at intersections and checking the mirrors during lane changes, and speaking less in higher risk situations (Drews, Pasupathi, & Strayer, 2004; Crundall, Bains, Chapman, & Underwood, 2005). When drivers move their heads to scan an intersection or expressway entrance or exit, passengers often stop talking. People on the phone have no knowledge of the driving situation, and just keep talking. Technology could assist in reducing the scope of this problem, for example by providing cues to inform callers of the driving situation and drivers of the call duration or their driving performance.

Recognizing the risks of cell phone use, many countries (Australia, England, Israel, Italy, Switzerland, and Spain) permit only hands-free operation. As was noted earlier, however, hands-free use is essentially no different in risk from handheld operation. Many states are also considering restrictions on cell phone use while driving, with the exception of making 911 and other emergency calls. Obtaining passage of such laws has proven difficult.

The author believes that a complementary approach is to implement an in-vehicle workload manager (Michon, 1993; Green, 2000b, 2004; Hoedemaeker, de Ridder, & Janssen, 2002; Piechulla, Mayser, Gehrke, & König, 2003). As initially conceived, such systems would use data from four sources: (a) the navigation system (such as lane width and radius of curvature), to assess the demands due to road geometry, (b) the adaptive cruise control system (headway and range rate to vehicles ahead), to assess traffic demands, (c) the traction control system, to assess road surface friction, and (d) the wipers, lights, and clock, to assess visibility. This information—along with information on the driver (e.g., age) and the specific visual, auditory, cognitive, and motor demands of each in-vehicle task—could be used to schedule the occurrence of in-vehicle tasks. Thus, when driving on a curving road in heavy traffic in a downpour, incoming mobile phone calls could be directed to an answering machine and 30,000-mile maintenance reminders could be postponed. When the driving task demand is low, drivers could have access to a wide range of functions. Being able to reliably predict the momentary workload of driving, however, has proven to be very difficult.

As an alternative, it is useful to think more broadly about when the driver is likely to be overloaded and, as some vehicle engineers have realized, that often occurs when the driver is maneuvering or about to maneuver. Maneuvering includes (a) changing lanes, (b) merging onto an expressway, (c) turning at an intersection, (d) parking, (e) braking in response to a lead vehicle, (f) accelerating from a traffic light, and so forth. These situations are much easier to identify than the overload situations described earlier and, furthermore, are situations when drivers are less likely to desire additional tasks (such as responding to an incoming call) and would likely lead to much greater market acceptance of a workload manager (Eoh, Green, Schweitzer, & Hegedus 2006).

What Kinds of Trips Do People Make and Why?

Every five years, the U.S. Department of Transportation conducts the National Household Travel Survey to obtain travel data for the United States (Hu & Reuscher, 2004). In the most recent data from 2001, an average person in the United States traveled about 14,500 miles per year, making four trips per day. On average, people drove about 40 miles per day, with most of the miles (about 35) covered in a personal vehicle. Keep in mind that these are averages, and that public transit (including school buses) prevalent in urban areas accounts for only 2% of all trips. The travel situation is likely to be different for more urbanized countries (Japan, most of Europe) as well as where public transit is more prevalent.

According to the 2001 data (Table 11.7), the most common reason for travel is family and personal business, which includes shopping, running errands, and dropping off and picking up others, accounting for almost half of the trips.

These and other data (on trip distances, travel speeds, time of day, etc.) in the National Household Travel Survey provide

TABLE 11.7 Summary of Trip Purposes

Purpose	% Person Trips
Family & personal business	44.6
Work	14.8
Social & recreational	27.1
School & church	9.8
Work-related	2.9
Other	0.8

background information on the context for which telematics devices should be designed and may suggest scenarios that should be used in evaluating safety and usability.

In contrast to the emerging understanding of the driving task, less is known about the real use of in-vehicle devices while driving, in particular the frequency and duration of use.

Who Are the Users?

Almost any adult has the potential of driving, with the determination being made by completing simple driver licensing requirements. Thus, in some ways, the driving population represents the population of candidate users for office computer systems. For most states, the percentage of drivers of either gender is within 1% of being equal (Highway Statistics 2003, 2004a). Data with regard to age are shown in Fig. 11.1. Notice that the percentage of the population that is licensed hits 80% at age 21 and remains at that level until the mid-70s. Thus, in designing in-vehicle systems for motor vehicles, few adults can be excluded, which differs from the design of office computer systems, where the emphasis is on the working population (generally less than 65 years old). Further, because of a wide age range and other differences, significant differences in individual performance can be expected. For example, in UMTRI telematics studies, older drivers typically required one and a half to two times longer to complete tasks than younger drivers (Green, 2001d). This fact, along with the requirement to design and test for the reasonable worst-case drivers, makes testing drivers over age 65 imperative.

In contrast to users of computers, operators of motor vehicles must be licensed. In the United States, the process of becoming a licensed driver begins with obtaining a copy of the state driving manual and learning the state's traffic laws. Candidates must also pass vision tests (see http://www.lowvision care.com/visionlaws.htm) and take a test of rules of the road to obtain a learner's permit, often on their 16th birthdays. Consistent with the increasingly common practice of graduated driver licensing, the learner can drive at restricted times with adult supervision, must generally complete a driver's education class (which often includes gory crash movies to convince teenagers not to drink and drive and to wear seat belts), and after a few years and an on-the-road test, is licensed to drive. (For details, see http://www.highwaysafety.org/laws/state_laws/grad_license.html,

http://www.insure.com/auto/teenstates.html, and http://www .hsrc.unc.edu/pubinfo/grad_overview.htm.) The rationale for graduated licensing is to provide new drivers with more experience under less risky conditions. To put this in perspective, Mayhew, Simpson, and Pak (2003) show that crash rates per 10,000 novice drivers drop dramatically with time, being about 120, 100, and 70 after one, three, and six months of being licensed. Similarly, recognizing the increased risk of elderly drivers, some states have special renewal procedures for them (IIHS, 2006).

In the United States, obtaining a commercial driver's license, which is needed to drive buses, large trucks, and other vehicles, is a more complex process. Most candidates either obtain (a) on-the-job training, (b) training integrated into their lifestyle (using machinery on a farm), or (c) training at truck driving schools (Sloss & Green, 2000). That population tends to be older than the working population as a whole, and is predominantly male.

Driver licensing practice varies from country to country. In Japan, for example, the failure rate for the basic licensing exam is much higher than in the United States, and there is much greater use of special schools to train drivers.

What Kinds of Vehicles Do People Drive (The Platform Question)?

For computers, people are concerned about the (a) brand, (b) amount of memory, (c) processor speed, (d) capacity of the hard drive, (e) type and version of operating system (Windows, Mac, or Linux), (f) type and version of browser, and so forth. The hardware rarely moves very far from its initial location; however, the content of individual computers is in a state of flux, being constantly updated over a life span of just a few years. Fortunately, the physical interface is fairly consistent—a QWERTY keyboard, mouse, and often a 17-inch or 19-inch monitor. The on–screen "desktop" is a more flexible space than the motor vehicle instrument panel.

Unlike personal computers, a motor vehicle is almost completely identified by its make, model, and year. Updates over an average 11-year life span are rare. In most countries (at least where there is left-hand drive), many aspects of driving are fairly consistent: the (a) input devices (steering wheel, brake, and throttle), (b) method of operation, (c) location, and (d) primary displays (windshield and mirrors). In contrast, there is no consistency in the controls or displays for telematics interfaces. Furthermore, although new motor vehicle models are offered once per year, major changes typically occur every five years. Computer software and hardware model upgrades occur almost continually, with a three-year-old computer generally being considered obsolete. Thus, the hardware life cycles of the two contexts are quite different.

According to the U.S. Department of Transportation, 237 million motor vehicles were registered in the United States in 2003 (Highway Statistics 2003, 2004b). Of them, 57.3% were automobiles, 40.1% were trucks, 2.3% were motorcycles, and 0.3% were buses. In less-developed countries, motorcycles make up a much larger fraction of the vehicle fleet.

For 2004, North American car and truck production was 16.9 million vehicles (Tierney, 2005). Many people do not realize that the United States now produces more trucks than cars. The

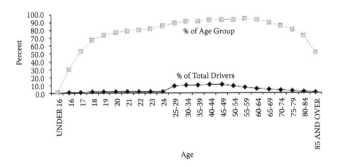

FIGURE 11.1.   Distribution of driver age groups developed from U.S. Department of Transportation data.

TABLE 11.8 Top 20 Vehicles in U.S. (Sales in 2004) from
Wards Auto World
(Bold = trucks)

Rank	Make	Model	Sales
1	**Ford**	**F-Series**	**901,931**
2	**Chevrolet**	**Silverado**	**680,768**
3	Toyota	Camry/Camry Solara	426,990
4	**Dodge**	**Ram Pickup**	**426,289**
5	Honda	Accord	386,770
6	Honda	Civic	309,196
7	Chevrolet	Impala	290,259
8	**Ford**	**Explorer**	**276,303**
9	Chevrolet	Malibu	268,017
10	Ford	Taurus	248,148
11	Toyota	Corolla	243,208
12	**Dodge**	**Caravan**	**242,307**
13	Nissan	Altima	235,889
14	**GMC**	**Sierra**	**213,756**
15	Ford	Focus	208,339
16	Chevrolet	Cavalier	195,275
17	**Chevrolet**	**Trail Blazer**	**193,862**
18	**Chevrolet**	**Tahoe**	**186,161**
19	**Ford**	**Escape**	**183,430**
20	**Jeep**	**Grand Cherokee**	**182,313**

TABLE 11.9. Some Driving-Specific Usability Measures

Category	Measure
Lateral	Number of lane departures
	Mean and standard deviation of lane position
	Standard deviation of steering wheel angle
	Number of larger steering wheel reversals
	Time to line crossing
	Steering entropy
Longitudinal	Number of collisions
	Time to collision
	Headway (time or distance to lead vehicle)
	Mean and standard deviation of speed
	Speed drop during a task
	Heading entropy
	Number of braking events over some g threshold
Visual	Number of glances
	Mean glance duration
	Maximum glance duration
	Total eyes-off-the-road time

predominance of trucks is due the popularity of minivans, sport utility vehicles (SUVs), and light trucks for personal use, not the production of heavy trucks for commercial use. Table 11.8 summarizes U.S. sales, including vehicles made in Japan, which comprise a significant fraction of U.S. sales. Trucks predominate sales. Thus, as designers think about packaging telematics applications, trucks are a likely platform.

WHAT MEASURES OF SAFETY AND USABILITY ARE OF INTEREST? (THE EMPIRICAL MEASUREMENT ISSUE)

As was noted earlier, usability is difficult to achieve without empirical measurement. First impressions suggest that the measurement of usability of office computer and web applications and the measurement of the usability of telematics applications are quite similar. In an office, one measures task completion time, errors, and ratings of ease of use. A typical usability lab has (a) a one-way mirror, (b) cameras, (c) video editing equipment, (d) audio mixers, and (e) at least two rooms, one for the subject and one for an experimenter.

In a typical laboratory for examining telematics, the same measures may be obtained; however, other driving-specific measures, as listed in Table 11.9, may also be obtained, especially in driving simulators and on-the-road tests. (See Green, 1995a, b, c.) In addition, a host of other performance measures often used in non-HCI contexts are also often collected: (a) ratings of workload (Cooper-Harper ratings, NASA TLX), (b) measures of situation awareness, and (c) measures of object and event detection (pertaining to traffic). (See Roskam et al., 2002; Tijerina, Angell, Austria, Tan, & Kochhar, 2003; Johansson et al., 2004.) Because the user is engaged in concurrent performance

of a manual control and information management tasks, the situation is in some ways similar to that described by Landry in the aerospace chapter. (See chapter 12.)

A major challenge in assessing the safety and usability of telematics is dealing with the trade-offs that drivers naturally make. The impression is that when preoccupied with an in-vehicle task, drivers lose awareness of the driving context—that is, situation awareness. Drivers attempt to compensate by slowing down (to make driving easier), allowing for larger headways and, if very preoccupied, paying less attention to steering (so lane variance and the number of lane departures increase); however, drivers can respond in strange ways. For example, if asked to use two different in-vehicle systems, one of which is not well designed, they might attempt to maintain equal performance on both: slow down more for the more difficult interface, but compensate by having better steering performance for the poorer interface. Assessment is difficult because the trade-off functions for all of these measures are unknown. One strategy used to overcome the trade-off problem is to minimize the opportunity for trade-offs. For example, this might include using cruise control to fix the speed (and in some cases, headway) and provide incentives and feedback to maintain driving performance, so only task time and errors might trade off.

HOW ARE TELEMATICS EVALUATED?

The assessment of mobile applications often occurs in contexts other than simulators and real roads as suggested above. Table 11.10 provides a summary of the contexts and their strengths and weaknesses. (See Green, 1995a, for additional details.)

Over the last decade, enormous improvements have been made in the quality of the tools available for human factors evaluations of in-vehicle evaluations. These include:

1. Reductions in the size of video cameras (to that of a postage stamp) and their cost.

TABLE 11.10. Evaluation Contexts

	Method and Equipment	Advice and Comments
Focus groups	• Groups of eight to 12 people demographically similar to customers sit around a table and discuss a product or service guided by a facilitator • Camera is often behind one-way mirror • Generally done in multiple cities (one or two groups/city) • Often conducted by a marketing firm	• Useful in getting ideas for product concepts, but not predictions of the safety or usability of new products because the products are never used • Approach might be used by manufacturers when a usability test might be more appropriate • Generally no quantitative data • Essential to report actual quotes from participants, not what the facilitator recalls
Clinics	• Customers in various cities are given the opportunity to experience a new product and its competition, often two or three vehicles, side by side • Customers say which product or feature they prefer	• Only exposes users to a limited number of options • Performance data often not collected • Approach is commonly used by industry • Because the results are highly proprietary, published studies are rare
Part task simulation	• A sample of users operate the device (e.g., computer-simulation of a new radio) and user task times, errors, and comments are recorded • Test facility is not sophisticated	• Not done that often • Relatively less common now than in the past as simulators have improved and instrumented vehicles have become more common
Driving simulator	• Typically driving simulators are fixed base (no motion) and cost $25,000 to $250,000 each • Simulators at manufacturers tend to be in 1–3 million dollar range, though some are much more (e.g., Ford is about $10,000,000) • 1–5 projectors with total 40–210 degree forward field of view, real vehicle cab, steering system with torque feedback, and realistic sound • Rear image may be projected or mirrors may be replaced with small LCDs • For advice, see Green, Nowakowski, Mayer, and Tsimhoni, 2003; Green, 2005	• Operation requires considerable experience • Simulator sickness is a major problem • Each experiment requires construction of a test road/world and scripting the behavior of vehicles and pedestrians • Facility can require considerable space (e.g., 1000 square feet) • Generally requires large number of fixed small (lipstick or smaller) cameras • Best-known vendors in U.S. are Systems Technology Inc. ($5,000–$25,000, http://www.systemstech.com/content/view/23/39/), GlobalSim Corporation ($75,000–$150,000, http://www.drivesafety.com/), and OKTAL in Europe (http://www.scaner2.com/php/)
Instrumented vehicle on test track or public roads	• Production vehicle (usually a car and often in the past, a station wagon) is fitted with cameras aimed at driver, forward scene, instrument panel, and lane markings, and with sensors for steering wheel angle, brake pressure, speed, and headway • Eye fixation system may also be provided • System of interest is also installed • In a typical experiment, the driver is asked to follow a test route while a back seat experimenter operates the test equipment	• Typical cost is at least $100,000–$250,000 though some low-cost systems may utilize a single camcorder on a vertically-mounted curtain rod in the back seat aimed at the instrument panel • Problem in the past has been finding enough space and power for the equipment, which has been solved by laptops and lipstick cameras • Commonly, instrumented research test vehicles are used, though there is increasing use of systems that can be installed in the subject's own vehicle
Operational field test	• Compact instrumentation is installed in a fleet of vehicles (10–50) • Each vehicle is borrowed by a potential user for a week, a month, or even a year • Driving performance is surreptitiously recorded by the vehicle • Unlike an instrumented car, continuous video is not recorded • In addition to data recorded by the instrumented vehicle, GPS-determined location is also recorded • Vehicles are periodically polled for data (and data is automatically dumped) by an independent digital cellular phone • Test is confined to a single metropolitan area	• Each test requires unique instrumentation • Tests are very expensive ($10,000,000 to $40,000,000) and can only be conducted with significant government support • Experiment generally lasts several years • Planning stage for experiments takes several years • At any given time, there may only one operational field test in progress in the U.S.

2. Significant improvements in the fidelity, along with significant reductions in the cost, of wide field-of-view driving simulators.
3. Electronic innovations that allow packaging compact instrumentation systems in vehicles and that reduce power consumption (and heat generation).
4. GPS systems for precise tracking of vehicle location.
5. Digital cellular phones for remote downloading of vehicle data.

Development of better, lower-cost eye-fixation systems is still needed, though the leading vendors of hardware for driving research, Seeing Machines, the manufacturers of FaceLab (http://seeingmachines.com/), and Smart Eye (http://www.smarteye.se) are making progress. Challenges include their accuracy and ability to hold calibration, dealing with glasses (common for older drivers), and operation in bright sunlight.

WHAT DESIGN DOCUMENTS EXIST FOR TELEMATICS?

What Types of Documents Are There?

To establish that a product is safe and easy to use, empirical measurement is important. Beyond verification, findings from such measurements can be fed back to the design team to improve the design.

Design is also guided by product and service specifications. In the case of office applications, style guidelines exist for various platforms and applications. For telematics applications, publicly available, product-specific interface guidelines do not exist but there are other types of important written materials (Green, 2001a, 2001b; Schindhelm et al., 2004). Automotive design documents fall into five general classes: (a) principles, (b) information reports, (c) guidelines, (d) recommended practices, and (e) standards. Principles give high-level recommendations for design and are similar to those found in office HCI applications, such as "design interfaces to minimize learning."

Information report is a term used by SAE to refer to a compilation of engineering reference data or educational material useful to the technical community. Information reports do not specify how something should be designed, but provide useful background information.

Guidelines give much more specific advice about how to design an interface element. For example, guideline 9 in chapter 7 of Green (1995) stated, "Turn displays should show two turns in a row when the turns are in close proximity" (p. 41), where close proximity means 0.1 miles apart or less. The impact of guidelines can depend on the issuing organization. For example, automotive design guidelines written by research organizations have no real authority. Guidelines written by the International Organization of Standardization (ISO), while technically voluntary, can become requirements, because in some countries, type approval (approval for sale) requires compliance with ISO guidelines. For vehicle models sold worldwide, global manufacturers find building common vehicle systems that comply with ISO standards to be less costly than building noncompliant, country-specific systems. In Japan, the Japan Automobile Manufacturers Association (JAMA, 2004) has a set of guidelines for navigation systems. Although theoretically voluntary, "requests" from the National Police Agency make the JAMA guidelines a requirement for all original equipment manufacturers (OEMs).

Recommended practice, a term used by SAE, refers to methods, procedures, and technology that are intended as guides to standard engineering practice. The content may be of a general nature, or may present data that have not yet gained broad engineering acceptance. Common practice is to follow a recommended practice, if possible. It is recognized that a product liability action against a product (especially in the United States) is extraordinarily difficult to defend if the product design deviates from recommended practice.

A standard specifies how something must be done and includes broadly accepted engineering practices, or specifications for a material, product, process, procedure, or test method. In the case of SAE, a standard is technically voluntary because SAE has no enforcement powers. Even more so than with a recommended practice, a product not complying with an SAE standard is almost not defendable in a product liability action, and is unlikely to be purchased from a supplier by an OEM.

Non-ISO Documents

Table 11.11 provides a summary of the design document activities to date excluding those of the ISO (described later). As indicated in the table, the EU guidelines are quite brief and are merely statements of very general principles (for example that interfaces should be simple to operate). Others contain a limited set of specifics (Alliance, JAMA, SAE J 2364, and J2365, TRL), and others contain significant design details (Battelle, HARDIE, UMTRI). Readers should not take the view that the newer or longer sets of guidelines are necessarily better. Representatives from industry (with academic involvement for SAE efforts) developed the Alliance, EU, JAMA, and SAE guidelines. Contractors (Battelle and UMTRI) working for the U.S. Department of Transportation developed the guidelines bearing their names. The TRL and HARDIE guidelines were contracted efforts in Europe.

See http://www.umich.edu/~driving/guidelines/guidelines.html. for unofficial electronic copies. (To avoid copyright problems, only draft and not final versions have been posted for SAE and ISO documents.)

All of these guidelines are voluntary except for the JAMA guidelines, as noted earlier. In the United States, the most important guidelines from the perspective of regulation are SAE Recommended Practices J2364 and J2365, especially J2364, and the Alliance guidelines. They are all discussed later in this chapter. Many manufacturers and suppliers are complying with them.

The guidelines developed by the Alliance, EU, and JAMA are being revised and that process is expected to continue for several years. For updates, readers are therefore advised to see either the authoring organization or the previously noted UMTRI website.

TABLE 11.11. Major Non-ISO Telematics Guidelines and Recommended Practices

Common Document Name	Reference	Size (pages)	Comments
Alliance guidelines	Alliance of Automobile Manufacturers (AAM) (2003 June 17); version 3	67	Restatement of EU principles, plus considerable details and rationale, to be used by almost all manufacturers in U.S.; key sections are principles 2.1 and 2.2, which still need development; still being updated by AAM.
Battelle guidelines	Campbell, Carney, and Kantowitz, (1997)	261	Voluminous document with references to interface design, heavy on trucks. User interface has been said to have a Windows OS flavor, includes physical ergonomics information (e.g., legibility, control sizes) which are not included in the UMTRI guidelines.
EU guidelines	Commission of the European Communities (1999)	2	Mostly "motherhood" statements. Some revisions are expected.
HARDIE guidelines	Ross, Midtland, Fuchs, Pauzie, Engert, Duncan, et al., (1996)	480	Early set of European guidelines, less data than UMTRI or Battelle.
JAMA guidelines	Japan Automobile Manufacturers Association (2004)	15	First set of detailed design guidelines for driver interfaces. These guidelines are voluntary in Japan but followed by all OEMs there and sometimes by aftermarket suppliers. Some aspects are particular to Japan. Device location restrictions are important.
SAE J2364 ("15-second rule")	Society of Automotive Engineers (2004, August)	13	Specifies the maximum allowable task time and test procedures for navigation system tasks performed while driving for systems with visual displays and manual controls; also describes an interrupted vision (visual occlusion) method as well; See also SAE J2678.
SAE J2365 (SAE calculations)	Society of Automotive Engineers (2002, May)	23	Method to compute total task time for tasks not involving voice used early in design to estimate compliance.
TRL checklist	Quimby (1999)	18	Simple check list.
UMTRI guidelines	Green, Levison, Paelke, and Serafin (1995)	111	First set of comprehensive design guidelines for the U.S. Includes principles, general guidelines, and specific design criteria with an emphasis on navigation interfaces.

ISO Documents

Much of the ISO activity over the last decade has occurred under the auspices of the International Organization of Standardization Technical Committee 22, Subcommittee 13 (ISO TC 22/SC 13-Ergonomics Applicable to Road Vehicles, in particular, Working Group 8 [WG8—Transport, Information, and Control Systems or TICS]; Green, 2000a). WG8 has about 50 delegates from the major vehicle producing nations, with the most 15 active members appearing at meetings held 2 to 3 times per year, usually in Europe.

Table 11.12 shows the standard and technical reports developed (or in progress) by Working Group 8 that pertain to telematics. (For the complete list, see ISO, 2007.) Most of the standards can be quite general, sometimes not containing the detail found in the Battelle, HARDIE, or UMTRI guidelines. For a variety of reasons, they tend to emphasize measurement methods and organization over specifications and safety limits. To pro-

mote international harmonization, national standards organizations, technical societies (e.g., SAE), and government organizations (e.g., U.S. Department of Transportation) often permit ISO standards to supersede their own standards, so ISO standards are very important.

Note: ISO documents follow a very-well-defined, three-year process through several stages (Preliminary Work Item [PWI], Committee Draft [CD], Draft International Standard [DIS], Final Draft International Standard FDIS], and International Standard [IS]) as they are passed from the working group to the subcommittee to the technical committee, and finally, to the secretariat for review and approval. The major hurdles are the working group and subcommittee, where passage requires two-thirds of the nations participating. The emphasis of this process is on building a voluntary consensus (ISO, 2003). Some items that are more informational in nature become technical reports instead of standards. Because of the limited number of experts available, WG8 is very selective in adding items to its work program.

TABLE 11.12.   ISO TC 22/SC 13/WG 8 Work Program

Effort	Summary	Status
Dialogue management principles and compliance procedures	Provides high-level ergonomic principles (compatibility with driving, consistency, simplicity, error tolerance, etc.) to be applied in the design of dialogues that take place between the driver of a road vehicle and an in-vehicle information system while the vehicle is in motion. Provides general directions on how to test for compliance.	Std 15005
Specifications and compliance procedures for in-vehicle auditory presentation	Provides requirements for auditory messages including signal levels, appropriateness, coding, etc., along with compliance test procedures.	Std 15006
Measurement of driver visual behavior	Generally describes video-based equipment (cameras, recording procedures, etc.) and procedures (subject descriptions, experiment design parameters, tasks, performance measures, etc.) used to measure driver visual behavior.	Std 15007, part 1 and TS part 2
Legibility (visual presentation of information)	Provides requirements for character size, contrast, luminance, etc. and specifies how they are to be measured.	Std 15008
Warning system messages	This state-of-the-art literature review (circa 2002) of warning systems covers topics such as alarm theories; the design of visual, auditory, and tactile warnings; redundancy; etc. The report contains summaries of a significant number of studies.	TR 16352
Message priority	Provides two methods for determining a priority index for in-vehicle messages (e.g., navigation turn instruction, collision warning, low oil) presented to drivers while driving. For one method, priority is based on criticality (likelihood of injury if the event occurs and urgency (required response time)), determined on four-point scales by experts.	TS 16951
Suitability of TICS while driving	Generally describes a process for assessing whether a specific TICS, or a combination of TICS with other in-vehicle systems, is suitable for use while driving. It addresses: a user-oriented TICS description and context of use, TICS task description and analysis, assessment, and documentation.	Std 17287
Occlusion method to assess distraction	Describes how the visual demand of a display can be assessed by periodically blocking (occluding) the driver's view of the display. Includes requirements for number and training of subjects, test hardware, viewing and occluded periods, and two metrics for data analysis.	Std 16673 (in press)
Lane change test	Proposes a procedure for testing the demand of telematics devices using a PC-based driving simulator. Subjects perform a number of lane changes, some of which occur while using an in-vehicle device. The test is based on results of the Advanced Driver Attention Metrics (ADAM) project. Complements SAE and AAM procedures.	NWI 26022

SAE J2364 (The 15-Second Rule)

Of the design documents in the literature, few have generated as much discussion as SAE Recommended Practice J2364, also known as "The 15-Second Total Task Time Rule" or "15-Second Rule." SAE J2364 establishes two procedures for determining if a navigation-system-related data entry task involving visual displays and manual controls is excessive while driving (Society of Automotive Engineers, 2004a, b; Green, 1999c). The practice applies to OEM and aftermarket products. The practice does not and should not apply to voice interfaces or passenger operation because the task demands are fundamentally different. There is no reason why the requirements do not make sense for similar tasks for other systems given its performance basis.

In developing this practice, criteria were selected to be related to crash risk, likely to lead to design improvements, and easy to measure. Some suggested this document should contain design criteria, such as a specification for a maximum number of items on a menu. It became clear, however, that such a design practice needed a performance basis, and specifying a design standard

would constrain innovation, so a performance-based practice was developed instead. Many measures were then proposed, with eyes-off-the-road time being most popular. The logic behind this measure is simple. If drivers are not looking at the road while driving, they are more likely to crash, with the likelihood of a crash increasing with increased eyes-off-the-road time. As shown in Green (1999d), this is reflected in the following equation:

$$\# \text{ U.S. deaths in 1989} = *\text{market.penetration.fraction}*$$
$$[-.133 + (.0447 * (\text{mean glance time})^{1.5} (\# \text{ of glances}) (\text{frequency}))]$$

where

market.penetration.fraction is the fraction of vehicles with a system (10% ≥ .1), *mean glance time* is the number of seconds drivers looked away from the road, *# of glances* is the number of times the device is looked at for each use sequence, *frequency* is the number of use sequences per week.

As an example, a task—say, entering an address—could have a mean glance duration of 2.7 seconds, require an average of 27.5 glances per entry, and be performed twice per week.

Unfortunately, eyes-off-the-road time is time-consuming, expensive to measure, and requires specialized equipment and skilled personnel. It also requires a fully functional system installed either in a simulator or test vehicle, something that is only available late in design. Invariably, problems are identified so close to production that few, if any, changes can be made. One of the key lessons from the Gould and Lewis design principles from HCI is that feedback from early on in the design process is critical.

A review of the literature (Green, 1999d) found that task time while driving was highly correlated with eyes-off-the-road time. Looking at in-vehicle systems is time shared with driving, so the more the driver needs to look, the longer the task will take. Further, dynamic (on-the-road) task time and eyes-off-the-road time are correlated with static task time, the time to complete the task when the vehicle is parked (e.g., Green, 1999d; Farber et al., 2000; Young et al., 2005). Static task time is easy to measure and can be done using prototypes available early in design.

The current SAE J2364 test procedure (Society of Automotive Engineers, 2004a) requires that 10 subjects between the ages of 45 and 65 be tested. Each subject completes five practice trials and three test trials in a *parked* vehicle, simulator, or laboratory mockup. The test cases to be examined (addresses for destination entry) are to be representative of what is planned for production.

In the static method, the subject performs the task, with the duration being from when the subject is told to start until the goal is achieved. Timing is continuous except for computational interruptions equal to or greater than 1.5 seconds, a time period during which the device is computing (for example, a route). If feedback is provided to the driver, that period is excluded from the 15-second task time limit. The interface complies with J2364 if the mean of the log of the task times is less than the log of 15 seconds. (Logs were used to reduce the influence of long outliers.)

It must be emphasized that the 15-second limit is for a static test. On-the-road drivers will take 30% to 50% longer overall, and furthermore alternate between looking inside the vehicle and looking at the road. The test procedure does not suggest drivers can safely look away from the road continuously for 15 seconds.

Some have argued that use of static task time fails to identify interfaces requiring long glance durations. However, analysis of real products shows the primary risk is from tasks that take too long to complete. In fact, it is very difficult to think of driver tasks for navigation systems that have short total task times but very long glance durations. In real driving, people truncate glances to the interior when the glances become too long but tend to complete tasks, even if they are unacceptably long. In practice, eliminating tasks with long completion times (the worst tasks) also eliminates many of the tasks with long glance durations.

Nonetheless, J2364 provides an alternative method involving visual occlusion. In that method, either the subjects wear special LCD glasses, or vision to the device is otherwise periodically interrupted, simulating looking back and forth between the road and the device. (Unlike driving, though, subjects do nothing in the occlusion interval.) The device is visible for 1.5 seconds and occluded for 1 to 2 seconds, with 1.5 seconds being recommended. Compliance is achieved if the sum of the log of the viewing times is less than the log of 20 seconds.

Alliance Principles

The Alliance of Automobile Manufacturers, the trade association of most major manufactures of automobiles in the United States (except for Nissan, Honda, Subaru, and Isuzu), has devoted considerable effort in developing design guidelines. Although the document scope states that it applies only to advanced information and communication systems, it should apply to traditional information or communications systems (e.g., entertainment systems), since the low-level tasks (reading displays, pressing buttons, etc.) and the manner in which those tasks interfere with driving are the same for all systems.

Their design principles document is a detailed elaboration of the 24 principles in the EU guidelines (e.g., "The system should be located . . . in accordance with relevant . . . standards. . . ." "No part . . . should obstruct any vehicle controls or displays. . ."), which seem obvious at a high level; however, the important content of the principles is in how they are achieved. Each principle has four parts: (a) rationale, (b) criterion/criteria, (c) verification procedures, and (d) examples.

The most important principle is 2.1: "Systems with visual displays should be designed such that the driver can complete the desired task with sequential glances that are brief enough not to adversely affect driving." Two alternative sets of criteria are offered. Alternative A says that the 85th percentile of single-glance durations should generally not exceed 2 seconds, and task completion should not require more than 20 seconds of total glance time. Verification can be achieved by a visual occlusion procedure (1.5 second viewing time, 1.0 second occlusion time), or by monitoring eye fixations directly using either a camera aimed at the face or an eye fixation monitoring system in either a divided-attention or on-road test.

Alternative B requires that the number of lane exceedances do not exceed the number associated with a reference task such as manual radio tuning, and that cars following headway should not degrade under those conditions, either. The verification procedure is stipulated to be driving on a divided road (either real or simulated) at 45 mph or less in daylight, on dry pavement, with low to moderate traffic. Additional details are provided describing the location of the radio, the stations to choose among, what constitutes a trial, subject selection (equal numbers of men and women between the ages of 45 and 65), and so forth.

Although both procedures seem well described on the surface, additional details and constraints are needed to make those procedures repeatable. For example, the differences in performance between driving in "low" and "medium" traffic could be quite considerable. One source of ideas is Lai's (2005) dissertation, which specifically proposes using variations in lateral position and variations in speed maintenance as assessment measures. Nonetheless, the principles represent a reasonable first step in developing a test protocol. On the other hand, on-road and simulator tests are extremely expensive and time-consuming, often impractical, and occur too late to have a useful impact. It is therefore the author's preference to rely on simpler evaluation procedures such as J2364 and the J2365 calculation procedure described in the next section.

WHAT TOOLS AND ESTIMATION PROCEDURES CAN AID TELEMATICS DESIGN?

SAE J2365 Calculations

SAE J2365 (Green, 1999a) was developed to allow designers and engineers to calculate completion times early in design, when it is still a concept that can easily be modified. As with J2364, J2365 is for in-vehicle tasks involving visual displays and manual controls evaluated statically—that is, while parked (or in a bench-top simulation). SAE J2365 applies to both OEM and aftermarket equipment. Though intended for navigation systems, J2365 should provide reasonable estimates for most in-vehicle tasks involving manual controls and visual displays.

The calculation method is based on the Goals, Operators, Methods, and Selection rules (GOMS) model described by Card, Moran, and Newell (1980) with task time data from several sources. The keystroke data was drawn from UMTRI studies of the Siemens Ali-Scout navigation system (Steinfeld, Manes, Green, & Hunter, 1996; Manes, Green, & Hunter, 1998). Search times were based on Olson and Nilsen (1987–1988) and the mental time estimates were drawn from the Keystroke-Level Model (Card, Moran, & Newell, 1983) and UMTRI Ali-Scout studies. Thus, the times shown in Table 11.13 have been tailored for the automotive context. (See also Nowakowski, Utsui, & Green, 2000.)

The basic approach involves top-down, successive decomposition of a task. The analyst divides the task into logical steps. For each step, the analyst identifies the human and device task operators. Sometimes analysts get stuck using this approach because they are not sure how to divide a task into steps. In those cases, utilizing a bottom-up approach may overcome such roadblocks. For each goal, the analyst identifies the method used.

The analyst is advised to use paragraph descriptions of each method to document them and then convert those descriptions into pseudo code. All steps are assumed to occur in series; multiple tasks cannot be completed at the same time. Furthermore, most drivers are assumed to use only visible, noncognitively loading shortcuts. Invisible shortcuts are likely to be used only by experts.

Next, the pseudo code task description is entered into an Excel spreadsheet. The analyst looks up the associated time for each operator listed in Table 11.13 and sums them to determine total task time. To assist in understanding the process, the practice provides a step-by-step example of entering a street address into a PathMaster/NeverLost navigation system, a popular U.S. product. For background on the calculation method, see Green (1999b).

The J2365 approach shares a number of assumptions, many of which are also shared with the basic GOMS model. For example, the model assumes error-free performance, which while not likely, can be adjusted for (say by increasing the computed value by 25%). Further, activities are assumed to be routine cognitive tasks, with users knowing each step and executing them in a serial manner. Again, adjustments in computed time can account for users sometimes forgetting what is next.

Though many of these assumptions are not true, adjustments can be made for them and many times the adjustments are small. Furthermore, violations of assumptions tend to affect all interfaces equally, so decisions about which of several interfaces is best still hold. As a practical matter, the estimates are good enough for most engineering decisions. Readers should keep in mind that J2364 only requires the use of 10 subjects at most, so there is some error in those estimates. Those errors are likely to be as large as variability among analysts and among J2365 estimates.

IVIS Estimates

A more complex estimation procedure, the IVIS DEMAnD model (In-Vehicle Information System Design Evaluation and Model

TABLE 11.13   Operator Times (seconds)

Code	Name	Operator Description	Young Drivers (18–30)	Older Drivers (55–60)
			Time (s)	
Rn	Reach near	From steering wheel to other parts of the wheel, stalks, or pods	0.31	0.53
Rf	Reach far	From steering wheel to center console	0.45	0.77
C1	Cursor once	Press a cursor key once	0.80	1.36
C2	Cursor 2 times or more	Time/keystroke for the second and each successive cursor keystroke	0.40	0.68
L1	Letter or space 1	Press a letter or space key once	1.00	1.70
L2	Letter or space 2 times or more	Time/keystroke for the second and each successive cursor keystroke	0.50	0.85
N1	Number once	Press the letter or space key once	0.90	1.53
N2	Number 2 times or more	Time/keystroke for the second and each successive number key	0.45	0.77
E	Enter	Press the Enter key	1.20	2.04
F	Function keys or shift	Press the function keys or Shift	1.20	2.04
M	Mental	Time/mental operation	1.50	2.55
S	Search	Search for something on the display	2.30	3.91
Rs	Response time of system-scroll	Time to scroll one line	0.00	0.00
Rm	Response time of system-new menu	Time for new menu to appear	0.50	0.50

Note 1: The keystroke times shown in Table 11.13 include the time to move between keys.
Note 2: System response times to display new menus may be empirically determined.

of Attention Demand, described by Hankey, Dingus, Hanowski, Wierwille, & Andrews, 2000a, 2000b). The model, which runs under Windows 98 or Windows NT, allows analysts to calculate a wide range of performance characteristics for proposed user interfaces. (The CD can be obtained from the U.S. Department of Transportation, Federal Highway Administration, Turner-Fairbank Highway Research Center in McLean, Virginia.)

Consistent with common understanding, the model assumes there are five basic human resources: (a) visual input, (b) auditory input, (c) central processing, (d) manual output, and (e) speech output. Overload of any one of these resources will affect task performance. The more demanding an in-vehicle task, the greater the likelihood of a crash. The model does not include a haptic component because haptic displays are rare in contemporary vehicles. Although many have developed models of human performance that partition human resources more finely, a five-component model is sufficient for most in-vehicle analyses. The data in the model were based on four experiments: (a) Gallagher (2001), (b) Blanco (1999), (c) Biever (1999), and (d) research conducted by Westat, a consulting company. These experiments concentrated on reading visual displays while driving, though there was work on auditory information as well.

To use the model, analysts need to create a description of each task drivers perform. Generally, that involves selecting the task in question from a large library of tasks in the database, modifying an existing description, or creating one from scratch. Tasks are grouped into seven categories: (a) conventional, (b) search, (c) search-plan, (d) search-plan-interpret, (e) search-plan-compute, (f) search-compute, and (g) search-plan-interpret-compute. Analysts need to select the driver age category (or specify all ages), the traffic density, the road complexity, the reliance on symbols/labels, the location of the display, and other characteristics.

The output of the model includes about 20 parameters, such as (a) the expected number of glances, (b) total task time, (c) ratings of mental demand and frustration, (d) total task demand, and so forth. In addition, the model output specifies if design thresholds are exceeded. The model proposes two sets of thresholds: (a) yellow-line and (b) red-line. Yellow-line thresholds were sets of points at which there was a measurable degradation in driving performance ($p < .05$) from baseline driving in the research conducted to support model development. Red-line thresholds indicated that a composite group of surrogate safety measures of driving performance was substantially affected. The red-line values were determined primarily from the literature and expert opinion. Table 11.14 shows those thresholds.

Independent validation data of the model has yet to appear in the literature.

TABLE 11.14. IVIS Yellow and Red Line Thresholds from Hankey, Dingus, Hanowski, Wierwille, and Andrews (2000a)

Measure	Affected (yellow)	Substantially Affected (red)
Single glance time	1.6 s	2.0 s
Number of glances	6 glances	10 glances
Total visual task time	7 s	15 s
Total task time	12 s	25 s

CLOSING THOUGHTS

This chapter makes the following key points:

1. Driving is quite different from sitting at a desk in an office because of the concern for crash risk.
2. A large number of new telematics applications will appear in vehicles over the next few years. They should radically re-shape the driver's task, providing the driver with a flood of information.
3. This flood of information has the potential to distract drivers from driving. Distraction is associated with specific types of crashes (single vehicle run off road, rear end) that are most common under generally good driving conditions.
4. Relative to other distractions, the number of telematics-related crashes has been low in the recent past; however, current projections are that in excess of 200 people are killed per year in the U.S. in cell-phone-related crashes, and the number is growing. This number of fatalities exceeds that from front air bags and Firestone/Explorer tire rollovers, both of which had a major financial impact on industry and led to federal regulatory changes.
5. Workload managers may be a long-term alternative to legislation to reduce distraction that leads to crashes.
6. Trips are made for a wide variety of purposes that need to be considered in assessing telematics applications.
7. Telematics need to be developed for a wide range of driver ages. Even well into their retirement years, most people still drive and will need to be able to use telematics.
8. Though people commonly think of passenger cars as the primary form of personal transportation in the United States, more trucks (especially pickups, SUVs, and vans) are produced in the United States than passenger cars, and they will be the most common platform in the future.
9. In assessing safety and usability, a wide variety of measures of longitudinal and lateral control are used in addition to task completion time, errors, and subjective ratings of ease of use. A major challenge in interface evaluation is dealing with performance trade-offs between measures.
10. Over the last few years, there have been significant advances in driving simulators and instrumented vehicles that have improved their quality and reduced their cost for safety and usability evaluations; however, the cost of these systems may be out of the range of most laboratories, especially those in academic settings.
11. The Alliance, Battelle, EU, HARDIE, JAMA, SAE, TRL, and UMTRI, along with ISO, have developed key design documents. The SAE recommended practices (especially SAE J2364, the 15-second rule) and the Alliance guidelines have the most impact in the United States, as does JAMA in Japan. ISO is in the process of developing a large set of standards and other organizations are updating theirs.
12. SAE J2365 and IVIS can assist in predicting task performance time, a simple measure for evaluating telematics safety and usability.

Thus, while the HCI literature provides a framework for test methods and evaluation, a great deal is specific to the motor vehicle context because of the safety-critical nature of the context and the timesharing not found in office activities. To meet the needs of the future, the cost of the methods needs to be reduced, and reliable tools, especially for recording eye fixa-

tions, are needed. Significant research is needed to support the development driver performance models (and workload managers) and understand how drivers use real telematics applications.

References

Alliance of Automobile Manufacturers. (2003, June 17). Statement of principles on human-machine interfaces (HMI) for in-vehicle information and communication systems (draft). Washington, DC: Author. http://www.umich.edu/~driving/guidelines/guidelines.html).

Barfield, W., & Dingus, T. A. (1997). *Human factors in intelligent transportation systems.* Mahwah, N. J.: Lawrence Erlbaum Associates.

Biever, W. J. (1999). *Auditory based supplemental information processing demand effects on driving performance.* Unpublished master's thesis, Virginia Polytechnic Institute and State University, Blacksburg.

Blanco, M. (1999). *Effects of in-vehicle information systems (IVIS) tasks on the information processing demands of a commercial vehicle operations (CVO) driver.* Unpublished master's thesis, Virginia Polytechnic Institute and State University, Blacksburg.

Campbell, J. L., Carney, C., & Kantowitz, B. H. (1997). *Human factors design guidelines for advanced traveler information systems (ATIS) and commercial vehicle operations* (CVO; Technical Report FHWA-RD-98-057). Washington, DC: U.S. Department of Transportation, Federal Highway Administration.

Card, S. K., Moran, T. P., & Newell, A. (1980, July 23). The keystroke-level model for user performance time with interactive systems. *Communications of the ACM, (7),* 396–410.

Card, S. K., Moran, T. P., & Newell, A. (1983). *The psychology of human-computer interaction.* Hillsdale, NJ: Lawrence Erlbaum Associates.

Commission of the European Communities. (1999). Statement of principles on human machine interface (HMI) for in-vehicle information and communication systems. (Annex 1 to Commission Recommendation of 21 December 1999 on safe and efficient in-vehicle information and communication systems: A European statement of principles on human machine interface). Brussels, Belgium: European Union.

Crundall, D., Bains, M., Chapman, P., & Underwood, G. (2005). Regulating conversation during driving: a problem for mobile phones? *Transportation Research Part F: Traffic Psychology and Behaviour, 8,* 197–211. Elsevier: Amsterdam.

Drews, F. A., Pasupathi, M., & Strayer, D. L. (2004). Passenger and cellphone conversations in simulated driving. In *Proceedings of the Human Factors and Ergonomics Society 48th Annual Meeting,* 2210–2212.

Driving Assessment 2005: Proceedings of the Third International Symposium on Human Factors in Driver Assessment, Training and Vehicle Design. (2005). Iowa City, Iowa: University of Iowa. Retrieved April 3, 2007, from http://ppc.uiowa.edu/driving-assessment/2005/final/index.htm

Eby, D. W., & Kostyniuk, L. P. (2004). *Distracted driving scenarios: A synthesis of literature, 2001 Crashworthiness Data System (CDS) data, and expert feedback* (SAVE-IT project technical report, Task 1). Washington, DC: U.S. Department of Transportation. Retrieved April 3, 2007, from (http://www.volpe.dot.gov/hf/roadway/saveit/docs/dec04/finalrep_1.pdf).

End-of-Year 2006 Top-Line Survey Results. (2006). Cellular Telecommunications and Inernet Association, Washington, D.C. Retrieved April 5, 2007, from http://files. ctia.org/pdf/CTIA_Survey_Year_End_2006_Graphics.pdf

Eoh, H., Green, P. A., Schweitzer, J., & Hegedus, E. (2006). *Driving Performance Analysis of the ACAS FOT Data and Recommendations for a Driving Workload Manager.* (Technical Report UMTRI-2006-18). Ann Arbor, Michigan: University of Michigan Transportation Research Institute.

Farber, E., Blanco, M., Foley, J. P., Curry, R., Greenberg, J., & Serafin, C. (2000), Surrogate measures of visual demand while driving. In *Proceedings of the IEA/HFES 2000 Congress* [CD-ROM]. Santa Monica, CA: Human Factors and Ergonomics Society.

Fatality Analysis Reporting System (FARS) Web-Based Encyclopedia. (2007). National Center for Statistics & Analysis, U.S. Department of Transportation. Retrieved April 3, 2007, from http://www-fars.nhtsa .dot.gov/

Gallagher, J. P. (2001). *An assessment of the attention demand associated with the processing of information for in-vehicle information systems (IVIS),* Unpublished master's thesis, Virginia Polytechnic Institute and State University, Blacksburg.

Gould, J. D., & Lewis, C. (1985, March). Designing for usability: Key principles and what designers think. *Communications of the ACM, 28*(3), 300–311.

Green, P. (1995a). Automotive techniques. In Weimer, J., (Ed.), *Research techniques in human engineering* (2nd ed., 165–208). New York: Prentice-Hall.

Green, P. (1995b). *Measures and methods used to assess the safety and usability of driver information systems* (Technical report FHWA-RD-94-088). McLean, VA: U.S. Department of Transportation, Federal Highway Administration.

Green, P. (1995c). *Suggested procedures and acceptance limits for assessing the safety and ease of use of driver information systems* (Technical report FHWA-RD-94-089). McLean, VA: U.S. Department of Transportation, Federal Highway Administration.

Green, P. (1999a). Estimating compliance with the 15-second rule for driver-interface usability and safety. In *Proceedings of the Human Factors and Ergonomics Society 43rd Annual Meeting* [CD-ROM]. Santa Monica, CA: Human Factors and Ergonomics Society.

Green, P. (1999b). *Navigation system data entry: Estimation of task times* (Technical report UMTRI-99-17). Ann Arbor, MI: University of Michigan, Transportation Research Institute.

Green, P. (1999c). The 15-second rule for driver information systems. In *ITS America Ninth Annual Meeting Conference Proceedings* [CD-ROM]. Washington, DC: Intelligent Transportation Society of America.

Green, P. (1999d). *Visual and task demands of driver information systems* (Technical report UMTRI-98-16). Ann Arbor, MI: University of Michigan Transportation Research Institute.

Green, P. (2000a). *The human interface for ITS display and control systems: developing international standards to promote safety and usability.* Paper presented at the International Workshop on ITS Human Interface, Utsu, Japan. June 8, 2000.

Green, P. (2000b). Crashes induced by driver information systems and what can be done to reduce them (SAE paper 2000-01-C008). In *Convergence 2000 Conference Proceedings* (27–36), Warrendale, PA: Society of Automotive Engineers.

Green, P. (2001a). *Safeguards for on-board wireless communications.* Paper presented at the Second Annual Plastics in Automotive Safety Conference, Troy, Michigan. February 5, 2001.

Green, P. (2001b). *Synopsis of driver interface standards and guidelines for telematics as of mid-2001* (Technical report UMTRI-2001-23). Ann Arbor, MI: University of Michigan, Transportation Research Institute.

Green, P. (2001c). *Telematics: Promise, potential, and risks.* Presented as the Management Briefing Seminar panel session—Traverse City Conference. Ann Arbor, MI: University of Michigan, Transportation Research Institute, Office for the Study of Automotive Transportation. August 6, 2001.

Green, P. (2001d, February 19–20). *Variations in task performance between younger and older drivers: UMTRI research on telematics.* Paper presented at the Association for the Advancement of Automotive Medicine Conference on Aging and Driving, Southfield, MI.

Green, P. (2004). Driver distraction, telematics design, and workload managers: safety issues and solutions (SAE paper 2004-21-0022). In *Proceedings of the 2004 International Congress on Transportation Electronics* (pp. 165–180). Warrendale, PA: Society of Automotive Engineers.

Green, P. (2005). How driving simulator data quality can be improved. *Proceedings of the Driving Simulation Conference North America 2005 [CD-ROM].* November, 2005, Orlando, Florida. Retrieved April 3, 2007, from http://www.umich.edu/~driving/publications.html

Green, P., Flynn, M., Vanderhagen, G., Ziomek, J., Ullman, E., & Mayer, K. (2001). *Automotive industry of trends in electronics: year 2000 survey of senior executives* (Technical Report UMTRI-2001-15). Ann Arbor, MI, University of Michigan, Transportation Research Institute.

Green, P., Levison, W., Paelke, G., & Serafin, C. (1995). *Preliminary human factors guidelines for driver information systems* (Technical report FHWA-RD-94-087). McLean, VA: U.S. Department of Transportation, Federal Highway Administration.

Green, P., Nowakowski, C., Mayer, K., & Tsimhoni, O. (2003). Audiovisual system design recommendations from experience with the UMTRI driving simulator. *Proceedings of the Driving Simulator Conference North America 2003 [CD-ROM].* October 2003, Dearborn, Michigan. Retrieved April 3, 2007, from http://www.umich.edu/~driving/publications.html

Hankey, J. M., Dingus, T. A., Hanowski, R. J., Wierwille, W. W., & Andrews, C. (2000a). *In-vehicle information systems behavioral model and design support final report* (Technical report FHWA-RD-00-135). McClean, VA: U.S. Department of Transportation, Federal Highway Administration. Retrieved April 3, 2007, from (http://www.tfhrc.gov/humanfac/00-135.pdf).

Hankey, J. M., Dingus, T. A., Hanowski, R. J., Wierwille, W. W., & Andrews, C. (2000b). *In-vehicle information systems behavioral model and design support: IVIS DEMAnD prototype software user's manual* (Technical report FHWA-RD-00-136). McClean, VA: U.S. Department of Transportation, Federal Highway Administration. Retrieved April 3, 2007, from http://www.tfhrc.gov/humanfac/00-136.pdf

Highway Statistics 2003. (2004a). Section III: Driver Licensing. Office of Highway Policy Information, Federal Highway Administration: Washington, D.C. Retrieved April 3, 2007, from http://www.fhwa.dot.gov/policy/ohim/hs03/dl.htm.

Highway Statistics 2003. (2004b). Section II: Motor Vehicles. Office of Highway Policy Information, Federal Highway Administration: Washington, D. C. Retrieved April 3, 2007, from http://www.fhwa.dot.gov/policy/ohim/hs03/htm/mv1.htm.

Hoedemaeker, M., de Ridder, S. N., & Janssen, W. H. (2002). *Review of European human factors research on adaptive interface technologies for automobiles* (TNO report TM-02-C031). Soesterberg, The Netherlands: TNO Institute for Perception.

Horrey, W. J., & Wickens, C. D. (2004). Cell phones and driving performance: A meta-analysis. In *Proceedings of the Human Factors and Ergonomics Society 48th Annual Meeting* (2304–2308). Santa Monica, CA: Human Factors and Ergonomics Society.

Hu, P. S., & Reuscher, T. R. (2004). Summary of Travel Trends: 2001 National Household Travel Survey. U. S. Department of Transportation: Washington, D. C. Retrieved April 3, 2007, from http://nhts.ornl.gov/2001/pub/STT.pdf

Insurance Institute for Highway Safety (IIHS). (2006). *US driver licensing procedures for older drivers.* Arlington, Virginia. Retrieved April 3, 2007, from http://www.iihs.org/laws/state_laws/older_drivers.html

International Organization of Standardization. (2003). *How are ISO standards developed?* Geneva, Switzerland. Retrieved April 3, 2007, from http://www.iso.ch/iso/en/stdsdevelopment/whowhenhow/how.html

International Organization of Standardization. (2007). *List of technical committees: TC 22/SC 13 Ergonomics applicable to road vehicles.* Geneva, Switzerland. Retrieved April 4, 2007, from http://www.iso.org/iso/en/CatalogueListPage.CatalogueList?COMMID=869&scopelist=PROGRAMME.

International Organization of Standardization. (2001). *Road vehicles—Measurement of driver visual behaviour with respect to transport information and control systems—Part 2: Equipment and procedures* (ISO Technical Specification 15007-2: 2001). Geneva, Switzerland: Author.

International Organization of Standardization. (2002a). *Road vehicles—Ergonomic aspects of transport information and control systems—Dialogue management principles and compliance procedures* (ISO Standard ISO 15005:2002). Geneva, Switzerland: Author.

International Organization of Standardization. (2002b). *Road vehicles—Measurement of driver visual behaviour with respect to transport information and control systems—Part 1: Definitions and parameters* (ISO Standard 15007-1: 2002). Geneva, Switzerland: Author.

International Organization of Standardization. (2003a). *Road vehicles—Ergonomic aspects of transport information and control systems—Procedure for assessing suitability for use while driving* (ISO Standard 17287:2003). Geneva, Switzerland: Author.

International Organization of Standardization. (2003b). *Road vehicles—Ergonomic aspects of transport information and control systems—Specifications and compliance procedures for in-vehicle visual presentation* (ISO Standard 15008: 2003). Geneva, Switzerland: Author.

International Organization of Standardization. (2004a). *Road vehicles—Ergonomic aspects of transport information and control systems (TICS)—Procedures for determining priority of on-board messages presented to drivers* (ISO Trial Standard 16951: 2004). Geneva, Switzerland: Author.

IInternational Organization of Standardization. (2004b). *Road vehicles—Ergonomic aspects of transport information and control systems—Specifications and compliance procedures for in-vehicle auditory presentation* (ISO Standard 15006: 2004). Geneva, Switzerland: Author.

International Organization of Standardization. (2005). *Road vehicles—Ergonomic aspects of in-vehicle presentation for transport information and controls systems—Warning systems* (ISO Technical Report 16352: 2005). Geneva, Switzerland: Author.

International Organization of Standardization. (2006). *Road vehicles—Ergonomic aspects of transport information and control systems—Occlusion method to assess visual distraction due to the use of in-vehicle information and communication systems* (ISO Draft Standard 16673). Geneva, Switzerland: Author.

Japan Automobile Manufacturers Association. (2004). *JAMA guideline for in-vehicle display systems, version 3.0.* Tokyo, Japan: Author. Retrieved April 3, 2007, from http://www.jama.or.jp/safe/guideline/pdf/jama_guideline_v30_en.pdf.

Johansson, E., Engstrom, J., Cherri, C., Nodari, E., Toffetti, A., Schindhelm, R., et al. (2004). *Review of existing techniques and metrics for IVIS and ADAS assessment* (AIDE deliverable 2.2.1). Brussels, Belgium: European Union.

Kayl, K. (2000). The networked car: Where the rubber meets the road. Retrieved April 5, 2007, from http://sun.systemnews.com/articles/32/1/ja/2746.

Koushki, P. A., Ali, S. Y., & Al-Saleh, O. I. (1999). Driving and using mobile phones: Impacts on road accidents. *Transportation Research Record, 1694*, 27–33. National Research Council, Washington, D.C.

Lai, F. C. H. (2005). *Driver attentional demand to dual task performance.* Unpublished doctoral dissertation, University of Leeds, Institute for Transport Studies, United Kingdom.

Manes, D., Green, P., & Hunter, D. (1998). *Prediction of destination entry and retrieval times using keystroke-level models* (Technical report UMTRI–96–37). Ann Arbor, MI: University of Michigan, Transportation Research Institute.

Mayhew, D. R., Simpson, H. M., & Pak, A. (2003). Changes in collision rates among novice drivers during the first months of driving. *Accident Analysis and Prevention, 35*(5), 683–691. Elsevier: Amsterdam.

McEvoy, S. P, Stevenson, M. R., McCartt, A. T., Woodward, M., Haworth, C., Palamara, P., et al. (2005, July 12). Role of mobile phones in motor vehicle crashes resulting in hospital attendance: A case-crossover study. *British Medical Journal, 1–5.* Retrieved April 3, 2007, from http://bmj.bmjjournals.com/cgi/content/abstract/bmj.38537.397512.55v1

Michon, J. A. (Ed.) (1993). *Generic intelligent driver support.* London: Taylor & Francis.

Murray, C. J. L., & Lopez, A. D. (1996b). *The global burden of disease.* Boston: Harvard School of Public Health.

National Safety Council. (2004). *Injury facts.* Itasca, IL: Author.

Nowakowski, C. Friedman, D., & Green, P. (2001). *Cell phone ring suppression and HUD caller ID: Effectiveness in reducing momentary driver distraction under varying workload levels* (Technical report 2001-29). Ann Arbor, MI: University of Michigan, Transportation Research Institute. Retrieved April 3, 2007, from (http://www.umich.edu/~driving/publications.html).

Nowakowski, C., & Green, P. (2000). *Prediction of menu selection times parked and while driving using the SAE J2365 method* (Technical report 2000-49). Ann Arbor, MI, University of Michigan, Transportation Research Institute.

Nowakowski, C., Utsui, Y., & Green, P. (2000). *Navigation system evaluation: the effects of driver workload and input devices on destination entry time and driving performance and their implications to the SAE recommended practice* (Technical report UMTRI-2000-20). Ann Arbor, MI, University of Michigan, Transportation Research Institute.

Noy, Y. I. (Ed.). (1997). *Ergonomics and safety of intelligent driver interfaces.* Mahwah, NJ: Lawrence Erlbaum Associates.

Olson, J. R., & Nilsen, E. (1988). Analysis of the cognition involved in spreadsheet software interaction. *Human-Computer Interaction, 3,* 309–349.

Parkes, A. M., & Franzen, S. (1993). *Driving future vehicles.* London: Taylor and Francis.

Peacock, B., & Karwowski, W. (1993). *Automotive ergonomics.* London: Taylor & Francis.

Peden, M., Scurfield, R., Sleet, D., Mohan, D., Hyder, A. A., Jarawan, E., et al. (Eds.). (2004). *World report on road traffic injury prevention.* Geneva, Switzerland: World Health Organization. Retrieved April 3, 2007, from http://www.who.int/world-health-day/2004/infomaterials/world_report/en/

Pena, P. (2007). *An Education on Common Objections to Cell Phone Legislation.* Retrieved April 3, 2007,. from http://www.geocities.com/morganleepena/rebuttal.htm

Piechulla, W., Mayser, C., Gehrke, H., & König, W. (2003). Reducing Drivers' mental workload by means of an adaptive man-machine interface. *Transportation Research Part F: Traffic Psychology and Behaviour, 6,* 233–248. Elsevier: Amsterdam.

Quimby, A. (1999). *A safety checklist for the assessment of in-vehicle information systems: scoring performance* (Project Report PA3536-A/99). Crowthorne, UK: Transport Research Laboratory.

Redelmeier, D. A., & Tibshirani, R. J. (1997). Association between cellular-telephone calls and motor vehicle collisions. *New England Journal of Medicine, 336,* 453–458.

Redelmeier, D. A., & Tibshirani, R. J. (2001). Car phones and car crashes: some popular misconceptions. *Canadian Medical Association Journal, 164*(11), 1581–1582.

Ribbens, W. B., & Cole, D. E. (1989). *Automotive electronics Delphi.* Ann Arbor, MI: Office for the Study of Automotive Transportation, University of Michigan, Transportation Research Institute.

Richardson, B., & Green, P. (2000). *Trends in North American intelligent transportation systems: A year 2000 appraisal* (Technical Report UMTRI–2000-9). Ann Arbor, MI, University of Michigan, Transportation Research Institute.

Roskam, A. J., Brookhuis, K. A., deWaard, D., Carsten, O. M. J., Read, L., Jamson, S., et al. (2002). *Development of experimental protocol* (HASTE Deliverable 1). Brussels, Belgium: European Commission.

Ross, T., Midtland, K., Fuchs, M., Pauzie, A., Engert, A., Duncan, B., et al. (1996). *HARDIE design guidelines handbook: human factors guidelines for information presentation by ATT systems.* Luxembourg: Commission of the European Communities.

Schindhelm, R., Gelau, C., Keinath, A., Bengler, K., Kussmann, H., Kompfner, P., et al. (2004). Report on the review of available guidelines and standards (AIDE deliverable 4.3.1). Brussels, Belgium: European Commission. Retrieved April 4, 2007, from http://www.aide-eu.org/pdf/sp4_deliv/aide_d4-3-1.pdf

Sloss, D., & Green, P. (2000). National automotive center 21st century truck (21T) dual use safety focus (SAE Paper 2000-01-3426). Warrendale, PA: Society of Automotive Engineers (published in National Automotive Center Technical Review, Warren, Michigan, U.S. Army Tank-Automotive and Armaments Command, National Automotive Center, 63–70).

Society of Automotive Engineers. (2002, May), *Calculation of the time to complete in-vehicle navigation and route guidance tasks* (SAE recommended practice J2365). Warrendale, PA: Author.

Society of Automotive Engineers. (2004a, May). *Navigation and route guidance function accessibility while driving* (SAE recommended practice 2364). Warrendale, PA: Author.

Society of Automotive Engineers (2004b, May). *Rationale document for SAE J2364* (SAE information report J2678). Warrendale, PA: Author.

Society of Automotive Engineers. (2005). *SAE handbook 2005.* Warrendale, PA: Author.

Steinfeld, A., Manes, D., Green, P., & Hunter, D. (1996). *Destination entry and retrieval with the Ali-Scout navigation system* (Technical report UMTRI–96-30, also released as EECS-ITS LAB FT97-077). Ann Arbor, MI: University of Michigan, Transportation Research Institute.

Stutts, J. C., Reinfurt, D. W., Staplin, L., & Rodgman, E. A. (2001). *The role of driver distraction in traffic crashes.* Washington, DC: AAA Foundation for Traffic Safety. Retrieved April 3, 2007, from (http://www.aaafts.org/pdf/distraction.pdf).

Sundeen, M. (2005). *Cell Phones and Highway Safety: 2005 State Legislative Update.* National Council of State Legislatures. Denver, Colorado. Retrieved April 3, 2007, from http://www.ncsl.org/programs/transportation/cellphoneupdate05.htm

Tierney, C. (2005. Big 3 market share dips to all-time low. *The Detroit News,* January 5, 2005. MediaNews Group: Detroit, Michigan. Retrieved April 4, 2007, from http://www.detnews.com/2005/autosinsider/0501/06/A01-50668.htm

Tijerina, L., Angell, L., Austria, A., Tan, A., & Kochhar, D. (2003). *Driver workload metrics literature review.* Washington, DC, U.S. Department of Transportation, National Highway Traffic Safety Administration.

U.S. Department of Transportation. (2005a). *2004 Traffic Safety Annual Assessment—Early Results* (Technical Report DOT HS 809 897). Washington, DC: Author. Retrieved April 3, 2007, from http://www-nrd.nhtsa.dot.gov/pdf/nrd-30/NCSA/RNotes/2005/809_897/

U.S. Department of Transportation. (2005b). Driver Cell Phone Use in 2005—Overall Results (Technical Report DOT HS 809 967). Washington, D.C.: Author. Retrieved April 3, 2007, from http://www-nrd.nhtsa.dot.gov/pdf/nrd-30/NCSA/RNotes/2005/809967.pdf

Underwood, S. E. (1989). Summary of preliminary results from a Delphi survey on intelligent vehicle-highway systems (technical report). Ann Arbor, MI: University of Michigan.

Underwood, S. E. (1992). *Delphi forecast and analysis of intelligent vehicle-highway systems through 1991: Delphi II. Ann Arbor, Program in Intelligent Vehicle-Highway Systems* (IVHS Technical Report-92-17), Ann Arbor, MI: University of Michigan.

Underwood, S. E., Chen, D., & Ervin, R. D. (1991). Future of intelligent vehicle-highway systems: A Delphi forecast of markets and socio-technological determinants. Transportation Research Record No. 1305, 291–304.

Violanti, J. M., & Marshall, J. R. (1996). Cellular phones and traffic accidents: An epidemiological approach. *Accident Analysis & Prevention*, 28, 265–270. Elsevier: Amsterdam.

Wang, J.-S., Knipling, R. R., & Goodman, M. J. (1996). The role of driver inattention in crashes: New statistics from the 1995 crashworthiness data system. *Association for the Advancement of Automotive Medicine 40th Annual Conference Proceedings* (377–392). Des Plaines, IL: Association for the Advancement of Automotive Medicine.

Wierwille, W. (1995). Development of an initial model relating driver in-vehicle visual demands to accident rate. *Proceedings of the Third Annual Mid-Atlantic Human Factors Conference* (1–7). Blacksburg, VA: Virginia Polytechnic Institute and State University.

Wooldridge, M., Bauer, K., Green, P., & Fitzpatrick, K. (2000). *Comparison of workload values obtained from test track, simulator, and on-road experiments.* Paper presented at the Transportation Research Board Annual Meeting, Washington, DC.

Young, R., Aryal, B., Muresan, M., Ding, Z., Oja, S., & Simpson, N. (2005). Road-to-Lab: Validation of the static load test for predicting on-road driving performance while using advanced in-vehicle information and communication devices. *Driving Assessment 2005: Proceedings of the Third International Symposium on Human Factors in Driver Assessment, Training and Vehicle Design,* 240–254. Iowa City, Iowa: University of Iowa. Retrieved April 3, 2007, from http://ppc.uiowa.edu/driving-assessment/2005/final/index.htm

·12·

HUMAN–COMPUTER INTERACTION IN AEROSPACE

Steven J. Landry
Purdue University

HUMAN–COMPUTER INTERACTION IN AEROSPACE

Aviation's interest with human–computer interaction (HCI) concepts began with the need to understand how pilots could interact with mechanical displays that indicated such things as airspeed, altitude, aircraft orientation, heading, and bearing from a radio navigation aid. As aircraft technology progressed and displays became more numerous and more complex, the interface issues became more pronounced. As travel by aircraft was commercialized, the consequences of these interface issues became more severe. As modern air-traffic control was introduced, a new arena for human interface issues opened. By the 1970s, computers were in commercial aircrafts, and air-traffic control was utilizing radar displays and automated flight data processing. This progression has not stopped, as higher levels of automation and new types of interfaces have been steadily added to flight decks and air-traffic control rooms over the succeeding decades.

This continuous progression of displays and automation has taxed the capabilities of researchers to fully investigate the vast and varied challenges associated with aviation HCI. This chapter discusses the efforts that have been made, are being made, and those issues which are just now on the horizon.

PAST CHALLENGES, SOLUTIONS, AND PRINCIPLES

Human factors issues related to aviation displays were recognized as early as 1923, as indicated by the following quotation from the National Advisory Committee for Aeronautics (the predecessor to NASA):

The reaction of the aviator to his instruments has to be considered, as well as the operation of the instruments themselves. This is evident enough in the case of appliances such as oxygen apparatus, intended solely for the comfort and efficiency of the aviator, or in the case of complicated instruments such as bombsights. It is equally true, however, with the more simple, direct-reading instruments. It is not enough for such instruments to be mechanically correct; they must be, in the case of service instruments, readily intelligible to the pilot. The manipulation of the instrument must not make an appreciable demand on his time or attention. The visibility must be satisfactory both day and night. Finally, service instruments must be "foolproof." While much can be accomplished by technical instruction courses for aviation personnel, still the personal prejudice of the average pilot has to be reckoned with. If the instrument for any reason fails to appeal to the individual pilot, he will take great chances rather than trouble to look at it. On the other hand, if the instrument pleases his fancy, he may grow so attached to it that he will claim he could not fly safely without it, even though the instrument is scientifically known to be incorrect. Curious examples of this circumstance were found in the popularity of the earlier liquid type Pitot tube among the British pilots and the spinning-top inclinometer among the French. (Hersey, 1923, pp. 8–9)

Another early example of aviation cognitive engineering research was in 1939, when an Applied Psychology Unit at Cambridge University in England (headed by Frederic Bartlett) began studying the design of aviation equipment (Bartlett, 1943). Bartlett's group studied manual control performance, the interrelationship between controls and displays, and human vigilance. Also in 1939, the National Research Council in the United States created a Council on Aviation Psychology. The U.S. Army and Navy created programs on aviation psychology in 1940 and 1945, respectively.

World War II provided a wealth of aviation experience, and after the war was over, aviation-savvy graduate students entered universities on the GI Bill. At the same time, advances were being made in navigation systems technology, cathode ray tube (CRT) displays, aircraft instrumentation, and control technology. The rapid increase in civil aviation made the development of air-traffic control a priority as well, and the advent of radar displays peaked interest in cognitive engineering issues for air-traffic controllers.

These developments led to an explosion of research in the 1950s, both in the United States and in Europe. One of the more influential reports was a 1951 blue ribbon committee report titled "Human Engineering for an Effective Air-Navigation and Traffic-Control System" (Fitts, 1951b). This paper summarized the state of research as it concerned air transportation, and identified key areas of future research. As such, it provides a convenient milepost for a discussion of the history of aviation cognitive engineering research.

This chapter will begin from a historical perspective, covering the major issues conceived of in the 1950s and how they were addressed in the succeeding decades, if at all. Following that will be a discussion of modern concerns, which grew out of the computerization of flight decks and innovative display concepts such as heads-up displays. Lastly will be a discussion of the future challenges for aviation and air-traffic control displays.

One concern for researchers in the 1950s regarded the display of information on "transitory displays," mainly dials and gauges. As such, the main focus was on (a) whether (and how) to integrate information on a single display, (b) how to arrange displays, (c) how displays, controls, and the comprehension of spatial information were related, and even (d) what types of displays were best for certain functions.

Many of these questions have been successfully addressed, providing today's engineers with some of the most concrete guidance that cognitive engineering has to offer. Some of them continue to be fascinating topics of research to this day, and of course, many new topics related to information display and assessment have arisen over the years.

Single-Sensor Single-Instrument (SSSI) Displays versus Integrated Displays

The problems inherent in SSSI displays have become increasingly obvious as the number and complexity of sensors (and other information sources) have increased. This has been particularly noticeable in aviation, where flight decks have undergone significant changes since 1951, but has also been true for nuclear plants, which were just beginning to be tested in 1951.

Interestingly, air-traffic controllers have only recently begun to encounter this problem as the number of displays (and elements on those displays) provided to a controller (and in the Traffic Management Unit [TMU]) has substantially increased.

SSSI displays have been gradually replaced (or incorporated into multifunction displays) as display technology has improved, although there are still examples of traditional SSSI displays on older aircraft and plants (it is an expensive venture to replace them with graphical displays). It seems clear from the research, however, that fewer displays are better than many displays (although the former is not without its own problems).

The reason that SSSI displays are generally considered to be worse than an integrated display seems to be related to the fact that a person's ability to monitor dynamic information is very low, even when only keeping track of a few items (Monty, 1973). Also, aircraft operators have demonstrated that they have difficulty integrating a large amount of displayed information (Murphy, McGee, Paler, Paulk, & Wempe, 1978). These findings mean that it is difficult for an operator to maintain comprehension of a dynamic situation when forced to scan from display to display. A study of the operation of manipulators showed subjects performed 30% to 40% better using two displays as opposed to four displays, despite a greater amount of information in the four-display case (Bejczy & Paine, 1977). This result extended a much earlier finding showing that fewer detections would be missed detections if there were, for example, five times the error rate on one display as compared to five displays each with one fifth the error rate (Conrad, 1951).

The presence of large numbers of displays also increases the amount of irrelevant data through which the operator must sort, leading to the problem of clutter. If the SSSI displays must, at every moment, display all the information that may be needed in any state of the system, then there must be irrelevant information present at any given time, and the operator must filter out this irrelevant information. As the quantity of irrelevant information increases, operator performance decreases, particularly if the information is similar (Hodge & Reid, 1971; Well, 1971; Dorris, Connolly, Sadosky, & Burroughs, 1977).

Irrelevant data is a problem, even assuming the operator knows what data is relevant (which is, and continues to be, its own problem). The relevant data must then be located from the different (and often, physically separated) instruments, and then mentally integrated to obtain the higher order representation. This operation can be cognitively intensive, and during situations of stress or short time constraints, it may not be possible (Vicente & Rasmussen, 1992).

In the next section, these issues—general design principles, arrangement of displays, and clutter—will be discussed in more detail.

Basic Display Principles

In examining the question of whether fewer displays containing integrated information were better, and how information may be integrated, researchers have developed a number of principles for the design of displays. In particular, researchers

early on studied problems in discerning the direction of motion of a pointer, and in the use of multipointer displays (which were being used in altimeter designs; Fitts, 1951a). Since then, a number of design principles with regard to such quantitative displays have been put forth (Sanders & McCormick, 1993).

One study compared digital speedometers to dials and curvilinear displays (all electronically generated). For accuracy and speed of reading, the digital display performed consistently better (Simmonds, Galer, & Baines, 1981). In general, digital displays have been found superior when a precise numeric value is required, and when the values presented remain visible long enough to be read.

Fixed-scale displays with moving pointers are better in the opposite cases—where precision is not required or the values change quickly. They are also better in cases where direction and rate of change of the values are important. For the design of analog displays, the following guidelines (Heglin, 1973) have been generally accepted:

- A moving pointer against a fixed scale is preferred.
- Small changes in quantity are more visible with a moving pointer display.
- If the quantity to be read is analogous to some physical interpretation, horizontal or vertical fixed scales with moving pointers should be used in order to provide a zero reference.
- Do not mix multiple moving-element indicators when they are related to the same function, as this can cause reversal errors.
- Moving scale displays can be used where the scale is too great to be displayed on the face of an instrument.
- If a control changes the quantity on the display, it is less ambiguous for the pointer to move than the scale.

In addition, for quantitative displays the numeric progression and length of scale units affect the speed and accuracy of quantitative reading (Whitehurst, 1982), and scale progressions should be in increments of 1s or 5s (to ease determination of readings in multiples of 1 and 5). Unusual progressions and decimals should be avoided. It was also found that the distance between graduation markers should be no less than 0.05 to 0.07 inches apart (greater under low illumination conditions), that scale markers should be used, and they should have graduation marks for the lowest unit to be read, but in no case lower than one fifth or one tenth (Cohen & Follert, 1970). Bar-type displays should only show one segment of the bar, should not have the bar extended to zero, and should not be used where an indication at zero is needed (Green, 1984).

Altimeter designs were of particular concern in the late 1940s, due to a spate of accidents related to misreading them. One of the early altimeters responsible for some of the problems violated one of the principles above: it had three pointers on the same scale. The operator would have to read three pieces of information and integrate them to get altitude. In 1968, a study determined that an integrated vertical display of altitude was best in terms of time and accuracy of reading (Roscoe, 1968). In explaining the results, the researcher indicated that key factors were the analogy of the vertical instrument to altitude and the integration of all elements of altitude into one display.

The distinction between quantitative and check-reading instruments has survived to this day, with the added distinction of qualitative and situational awareness instruments (Sanders & McCormick, 1993). In fact, despite the prevalence of digital computers driving them, graphical displays often show representations of analog-like displays for instruments designed for qualitative or check-reading. This is because analog displays were found to be more quickly read for this purpose, and contributed less to workload, when compared to digital displays (Hanson, Payne, Shively, & Kantowitz, 1981).

Check-reading display research. Researchers demonstrated that for a number of instruments configured together for check reading, the pointers on round instruments were best positioned at 9 or 12 o'clock (Sanders & McCormick, 1993). When all pointers show nearly identical readings, the grouping of the instruments and positioning of the pointers appeal to the similarity gestalt of human cognition. Violation of the gestalt is immediately and reliably recognized. Further research indicated that the addition of a line joining the instruments at their null position improved or added to the gestalt and improved performance (Dashevsky, 1964).

Qualitative instrument research. The distinction does not appear to have been made between quantitative and qualitative (perhaps a subset of which is check reading) instruments before 1960. Qualitative instruments are distinct from quantitative in that the specific numerical value is irrelevant; instead the approximate value, its value in relation to some standard, its rate, or its trend are of interest. The difference may have first been noticed in a study that examined two types of readings of an instrument using three types of scales. In that study, subjects had to either read the quantitative value, or read high, low, or OK. Open-window displays (see Fig. 12.1) were read the quickest for quantitative values, but vertical scales were read quickest for the qualitative assessment (Elkin, 1959).

Qualitative scales also typically have some identification of ranges. This can be done through the use of color, or by the addition of coded markings that have some association to the meaning of the range (Sabeh, Jorve, & Vanderplas, 1958).

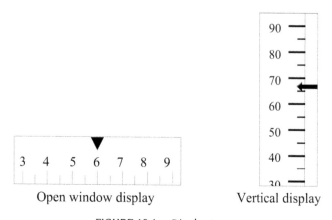

Open window display   Vertical display

FIGURE 12.1.   Display types.

Arrangement of Displays

A second question brought up in the Fitts (1951b) report regarded the organization of the display elements. A number of advances in relation to this arrangement problem were made shortly after its publication. Extensive studies (most notably one in which Fitts himself participated) were underway examining visual scanning patterns to determine the optimal layout of instruments (Fitts, Jones, & Milton, 1950). These studies led the Civil Aeronautics Board (later to become the FAA) to establish a standard arrangement for flight instruments, commonly referred to as the "basic T," in 1953 (Civil Aeronautics Board, 1953). It included six basic instruments, namely, (a) airspeed, (b) attitude, (c) altitude, (d) direction, (e) "climb" (which is taken here to mean vertical speed), (f) flight path deviation, and (g) their required arrangement. This was later reduced to just the first four instruments (Civil Aeronautics Board, 1957).

The analysis used in relating visual scanning patterns to display positioning was referred to as "link analysis." In this analysis, a "link" was sequential fixations on two items. The more occurrences of sequential fixations between two items, the stronger the link. Items strongly linked should then be located next to one another, a technique and principle used in other fields (such as in operation centers aboard naval vessels) (Chapanis, 1959). Other studies of eye movements determined that the best location for important information is just above or below the glareshield of a vehicle (Cole, Milton, & McIntosh, 1954). Since then, advances have been made in automating information organization (Mendel & Sheridan, 1986) and instrument-panel layout (Pulat & Ayoub, 1979), although methods of this sort are not widely used.

The layout of the instruments not only serves to reduce the time to scan the instruments, but can also affect workload. A principle of display design called the "proximity compatibility principle" states that, when attempting to integrate two sources of information from two locations, greater workload is induced by greater separation, as the physical proximity of the sources should mimic the cognitive proximity of the information (Wickens & Carswell, 1995).

This principle relates to the requirement of the operator to integrate different sources of information. There are two basic types of integration covered by this principle. The first involves locating a source of information (such as the heading indicator) next to related sources of information (such as bank indication and turn coordinator). The second involves the display itself integrating more than one piece of information, such as the attitude indicator showing both pitch and bank at the same time.

Multifunction graphical displays (MFDs) overcome these two limitations. These displays provide, in one viewport, a number of elements that may need to be integrated frequently (although not always), satisfying the proximity compatibility principle (Seidler & Wickens, 1992). The content of the MFD can be manually controlled, providing the opportunity to present a wealth of information for which there previously was no physical space.

Two problems arise, however. First, information that is not currently displayed is unavailable to the pilot. If that information changes, the pilot will be unaware of the change. Second, the "depth" added to the display presents the problem of navigation, although this can be mitigated through good organization (and minimal depth) of the pages (Allen, 1983; Schneiderman, 1998, Francis, 2000). If these principles are not met, however, there is the possibility that the operator will become lost in the database. In addition, there is workload (and heads-down time) associated with the controls for the display.

How Much Information Can Be Displayed Effectively?

The problem of how much information can be fit on a display, has significantly evolved since 1951, but is still a topic of debate and research. A large body of work on visual processing has been amassed over the years, and many of the theories are still hotly debated. What is certain is that distinguishability of items on a display is overwhelmed by the somewhat limited perceptual processing capacity of the human. Without going into a great amount of detail, some of the findings with respect to this issue will be summarized.

There are two fundamental limits on extracting information from a visual scene: (a) processing and (b) attention. At any time, only a small amount of the information available to the retina can be processed (although all of it is perceived). The mechanics of processing are still being determined, but it appears likely that some high-level representation is quickly obtained, followed by a "weeding out" of some portions of the image (VanRullen & Thorpe, 2001). Scene processing allows a variety of information to be extracted, but specific objects of interest may or may not be extracted in sufficient detail. Focused attention captures this detail, but objects can only be processed serially, and subsequent processing of these objects is inhibited (referred to as "negative priming"; Tipper & Driver, 1988). The combination of these effects yields an effective capacity of 4 to 6 items (VanRullen & Koch, 2003). This number represents how many items can be represented in visual awareness at any one time. Objects/information may then be transferred to long-term memory, a process that has also been researched extensively.

How much information can be fit on one display, then, is dependent on the task and environmental factors. If one is reading a depiction of an instrument approach with the intention of memorizing details, and there is no time pressure, a great deal of information can be displayed and recalled effectively. If an air-traffic controller has a fraction of a second to glean track information from a radar display, he or she is unlikely to extract more than 4 to 6 pieces of information from that display. Moreover, what information would be extracted from that display is unclear, although factors of the displayed information (salience, size, color, etc.) can influence selective attention.

One further question remains in regards to information display that we will address here—the rate at which information can be extracted. This is also complex, dependent upon physiological factors, features of the display, complexity of the information to be extracted, and other factors. Simple responses to visual stimuli can occur within 30 to 60 msec (Iwasaki, 1996), but more complex objects (such as text) can require a great deal more time.

In summary, the answer to the question of how much information can be fit on a display is, "It depends." The question has

become even more relevant as electromechanical displays have been almost universally replaced by computer-driven electronic displays, which can display a great deal of information in one place. These displays are also three-dimensional (3D) in the sense that they can contain pages of information (adding a new task—navigation of a display), and their contents can be modified by the user (or automation).

Clutter

Clutter is a significant issue in two main classes of displays: (a) multifunction displays (such as air-traffic control displays and flight deck displays that show route, weather, and traffic), and (b) heads-up displays (HUDs). The source of the clutter is different in the two cases. In the former case, a large quantity of data is being overlaid on the display, whereas in the latter, the display is overlaid on a source of great clutter (the outside world). In both cases, the problem is the same: response time for a stimulus is increased with the proximity of nonstimulus items (Eriksen & Eriksen, 1974). Additionally, clutter can increase the time it takes for the pilot to locate an item, or even obscure items of interest (Wickens, 2003). Clutter has also been mentioned by pilots in subjective ratings for some time (Abbott et al., 1980).

The solution to this problem is to provide the operator with decluttering capability—that is, the ability to either manually or automatically remove information from the display; however, if the information is no longer present, it cannot be integrated with other information, and changes to the information, which could be important, will of course not be noticed. This is in addition to the workload required to remove or add information to the display, which could be considerable, particularly in the case of navigating a multifunction display.

Displaying Spatial Relationships

The Fitts (1951b) report suggested that "studies should be made to determine principles governing the effective display of information about the relative position of aircraft and ground objects in tri-dimensional space" (p. xvii). Specifically, the committee felt that research should be directed at determining the best types of projections for displaying spatial relations, and how symbolic and pictorial displays are best used for this purpose. In addition, more studies were called for to determine what display characteristics were required when control manipulation resulted in changes in the display (e.g., turning a control yoke results in changes in bank and heading indications).

Aviation display is a natural place for interest in the display of spatial relations to develop. Air-traffic controllers must gauge both absolute and relative position and motion of aircraft in at least two dimensions (often three), and provide spatial control guidance to pilots. A pilot must, through use of instrumentation, determine relative (and sometimes absolute) position in 3D coordinates, and keep track of 3D orientation. This is accomplished through the use of (a) an attitude indicator (ADI), which shows pitch and bank, (b) a heading indicator, which shows absolute orientation, (c) a turn indicator, which shows yaw rate, and (d) one or more instruments showing relative position and bearing (e.g., altimeters, radio magnetic indicator, course deviation indicator, vertical deviation indicator, distance measuring equipment).

One problem that almost certainly drove the recommendation in the Fitts (1951b) report regarding the display of orientation was the problem of the ADI. In particular, what should move, the aircraft or the earth? Several principles related to this question (Roscoe, 1981) have been put forth in consideration of the body of research into interfaces:

- Principle of pictorial realism: The display should be pictorially analogous to the real world.
- Principle of compatible motion: The part that moves in the real world should be the part that moves on the display.
- Principle of integration: Related information should be integrated on the same display.
- Principle of pursuit presentation: A display should allow pursuit tracking versus just compensatory tracking—that is, if a target moves both absolutely and relatively, both sources of movement should be displayed.

So a moving aircraft on an ADI or an air-traffic controller's radarscope would be compatible with the principle of compatible motion; however, while the radarscope also complies with the principle of pictorial realism, from the pilot's perspective, the moving aircraft would (loosely) violate it; from the pilot's perspective, it appears that the horizon is banked as opposed to the aircraft. Although designers have settled on the moving earth model for ADI design, the debate was never satisfactorily resolved. Experiments based on (actually, an extension of) a proposal to integrate the two models (Fogel, 1959) showed improved performance (Beringer, Willeges, & Roscoe, 1975). This topic continues to be discussed to this day, with researchers arguing that the moving earth ADI has continued to be used only because of the lack of interest on the part of pilots to change it, despite evidence that it is has been causing accidents for decades (Bryan, Stonecipher, & Aron, 1954).

In addition to the orientation question addressed mainly by the ADI, there are questions about how best to portray relative position. This is a common problem for both air-traffic controllers, who must assess relative bearing and range between aircraft, and pilots, who must frequently assess their relative bearing and range from some navigational fix. More recently, the question of vertical navigation has received attention.

A typical horizontal-navigation problem for a pilot is common to anyone familiar with basic instrument flying. From any given position, a pilot could be asked to fly to a "fix," whose position is defined by bearing and range relative to a radio navigation aid (e.g., a VOR). Although this can be calculated from current heading and the ownship's bearing and range to the VOR, it is not a trivial task. A simpler task is to fly inbound on a course to a VOR or outbound on a radial from a VOR. Another example is the task of determining the position of the ownship given access to several radio navigational aids.

Recent research has qualified the difficulties inherent in these operations as "representational" problems (Zhang, & Nor-man, 1994). The theory follows from the argument that the information to accomplish many tasks is distributed between the external stimuli (the interface) and internal cognition (in this case, of the pilot; Norman, 1993). Furthermore, the information representations may have different scales: (a) ratio, (b) interval, (c) ordinal, or (d) nominal. Efficient external representations reflect all the categories of information present in the "real-world" object. Deficiencies in information would require that information be distributed to the internal cognition of the operator, while surplus information (information present in the display that is not in the object) could be misleading.

Application of this theory to the problem of horizontal navigation shows that different equipment requires different levels of internal cognition. A simple ADF or VOR system requires internal computation of angular differences and range. A map type display, however, explicitly represents these, simplifying the task.

An extreme example of this representational problem is the difficulty in determining relative position from terrain (or weather) on a paper map (or even presented aurally!) from their knowledge of absolute position as given by the flight instruments. It is difficult and time-consuming to accurately comprehend this situation. This was apparently recognized, resulting in the display of "minimum safe" altitudes, which changes the nature of what needs to be comprehended (the pilot only must know his or her position with respect to that altitude, a one-dimensional, relative task).

INTRODUCING AUTOMATION TO THE FLIGHT DECK

In addition to research concerning information display and assessment, much was being learned about automation in general. The Fitts report (1951b) mentioned the need to understand how tasks should be allocated between human and machine, and this was the subject of some effort on the part of Fitts himself. Computer technology was in its infancy, and little was known in the 1950s about the possible limits on the ability of machines to assist humans. Many thought that these machines would gradually replace all human functions. Today, researchers better understand the different ways in which we can describe the relationship between human and operator, although this relationship is changing as new types of automated assistance are being added to aircraft and air-traffic systems.

This section will discuss the concept of allocating roles to automation and humans, beginning with some of the early concepts, and then progressing to more recent characterizations of the roles of humans and automation. Following this will be a discussion of how cognitive engineering has helped shape early systems, consisting mainly of control automation, and how it is significantly involved in newer types of systems, consisting mainly of information automation.

Autopilots

In a sense, automatic control capability, such as an autopilot, neatly satisfied the prominent conception of the proper roles of automation and human operators. The automation took over the routine work of simultaneously maintaining course, altitude, and speed. Humans had a new system to deal with, but could leverage their long-term memory capability, reasoning, and judgment. However, Fitts (1951b) also noted that humans are poor monitors, and even suggested a specific proper role of automation may be to monitor humans. The addition of a highly reliable system to monitor for long periods of time violates that principle.

The autopilots, in fact, were a source of a number of accidents. A much-discussed accident in which an L-1011 crashed into the Everglades because the autopilot was inadvertently disconnected is an example of the "graceful" degradation of the autopilot system. The automation fails with little or no notice to the flight crew, who are not good at monitoring it to start with. Several other incidents demonstrated that pilots can overrely on automation to perform its tasks (Parasuraman & Riley, 1997).

The autopilot's development began in 1914, when Lawrence Sperry created a gyroscopic device capable of stabilizing an aircraft's flight without input from the pilot. This led to the design of a fully automatic aircraft by 1915, and a fully automatic pilot by 1916, although it would not be until 1931 that an autopilot was licensed to operate with passengers on board (Davenport, 1978). An automatic landing system was invented in 1937 by Capt. Carl Crane, and since, numerous refinements in control theory have been applied to autopilots. One particularly interesting addition to control theory was the finding that the behavior of human–machine systems with simple feedback control systems (such as an aircraft under nominal flight conditions) can be modeled using a simple control law (McRuer & Graham, 1965).

While changes to the underlying control theory remain transparent to pilots, changes in the capabilities of the autopilots have not. Pilots interfacing with the autopilot often face "mode confusion," a term that indicates when the pilot's understanding of the mode in which the autopilot is operating differs from its actual mode. When this occurs, incidents such as those described above are more prevalent.

In general, however, autopilots have been amazingly successful. They relieve pilots of a great deal of manual workload, particularly during routine phases of flight such as cruise. Autopilots are also capable of increased accuracy at navigation, which is reflected in their ability to land aircraft in poor weather, including when poor visibility completely obstructs the pilot from seeing the runway.

Role Allocation

The "Fitts list" shown in Table 12.1 served as the launching point for studies on the allocation of tasks to humans and automation.

TABLE 12.1. Fitts List

Humans Are Good At	Machines Are Good At
Sensory functions, detection	Speed and power
Perceptual ability	Routine, repetitive work
Flexibility, improvisation	Computation
Judgment and selective recall	Short-term storage
Reasoning	Simultaneous operations
Long-term memory	Short-term memory

Automation for the control of aircraft had been around for decades by 1951. Not much more was automated, however, until the late 1960s and early 1970s, when some systems, navigation, and automated alerting were added. The early 1980s saw the advent of graphical displays, flight management systems, and the removal of flight engineers. The late 1980s introduced the concept of glass cockpits, replacing nearly all electromechanical "steam gauges" with graphical displays (Billings, 1997).

Over the decades, several different attempts at methodologies for allocating function have been made. This section will be organized around those efforts: (a) the Fitts list, (b) the "automate everything" philosophy, (c) supervisory control, (d) guidelines for automation, and (e) levels of automation.

Fitts list. In order to discuss the advantages and disadvantages of this approach, but without rehashing Fitts' discussion of them, we will first discuss the contents of the list, and findings surrounding the truth or fallacy of the claims. Then we will give examples of applying the Fitts list to automation, examining the strengths and weaknesses of the approach. Finally, I will give some examples of criticism levied at this method of function allocation.

Many of the items in the list still appear to be true. Very little argument can be made with the assertion that machines can generate more power than humans can, and they can perform many tasks faster than a human. On the other side of the list, humans are still much more capable of improvisation and making judgments, and computers are not able to reason in any meaningful way, despite significant advances in computer technology.

Yet some comparisons appear to be more complicated than suggested by the Fitts list. While a human's perceptual ability is quite impressive, some machines are capable of remarkable feats of detection: thermal imaging systems that can "see" through walls, satellites that can read individual license plates, microscopes that can image individual atoms, explosive detection equipment that can "smell" tiny amounts of specific chemicals, and so on. Despite this, it still requires a human to interpret the results. Machines still have only primitive (and brittle) abilities to deal with recognition (particularly of symbol systems, faces, and expressions, etc.).

In general, it appears humans are "wired" for rapid operation in a highly uncertain and diverse environment, whereas machines are still generally relegated to rather specific operations in a well-regulated and defined environment. Therefore, while machines may be better at routine and repetitive work, they are generally unable to deal with events or circumstances outside of the expected operating regime. Humans, however, are able to operate in the face of such events, leading to the notion that humans are suited for "supervising" automation (which will be discussed in a later section).

In practice, Fitts lists have been difficult to apply in allocating function (Sheridan, 1998). One likely reason is that a Fitts list approach requires automation to be assigned to a particular function. In reality, most functions require that some aspects are done by human operators and some by automation. Since the functions are shared between humans and automation, any distinction of human or machine for a function is likely to be suboptimal.

The "automate everything" philosophy. Although this method of allocation seems unjustifiable on its face, it has had its proponents, and its reasons. Humans have typically been seen as the source of a majority of error in the system. Replacing them with "trustworthy" automation, which also came with the benefits of lower cost and more efficient operation, did not seem quite so ridiculous several decades ago. Officials of two different aircraft manufacturers have been quoted as adhering to some form of this approach even within this last decade (Billings, 1997). In addition, these thoughts came at a time when the growth of technology appeared to be facilitating the design of automation.

By the late 1970s, it was becoming apparent that humans and automation had trouble getting along. Among the problems frequently cited (Wiener & Curry, 1980) were:

- Crew errors exacerbating automation errors
- Improper setup of automated systems
- Improper response to alerts
- Failure to monitor
- Loss of proficiency

In truth, however, the hurdles for an entirely automated aircraft have been, so far, too great to be surmounted. For one, the current social and political climate would have to undergo significant changes. Furthermore, since that time it has become apparent that automation is not as reliable as had been believed at the time. Examples of failures in automatic equipment have been common and persistent (Wiener & Curry, 1980).

In addition, under this strategy the operator would be left with any functions that could not be automated (Parasuraman, Sheridan, & Wickens, 2000). The result is that ill-defined, difficult tasks may be assigned to an operator that is significantly detached from the operation of the system. It is unlikely that the operator could perform well under these circumstances.

In such a system, the operator may be left with the task of monitoring the system with the resumption of control should the automation fail. Unfortunately, the skills necessary to do so would likely be degraded from nonuse. In addition, retrieval from long-term memory is more difficult if information goes unused for long periods of time, and the model of the system required for diagnosis and action is unlikely to be present in an operator who rarely actively controls it. These last few points have been referred to as "ironies of automation," since the more advanced the automation, the more valuable may be the operator's function, but the less able the operator may be to fulfill that function (Bainbridge, 1983).

Human supervisory control. About the same time as the automate-everything approach, a concept of interaction between humans and automation was being discussed, called human supervisory control (Ferrell & Sheridan, 1967). This was an offshoot of research into how humans could control a remote vehicle (specifically, vehicles on the moon). It became apparent that under the three-second delay for information to be sent to and return from the moon, some control loops would have to be closed by the automation, with the operator acting as a supervisor of the automation (Sheridan, 1992).

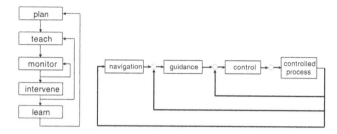

FIGURE 12.2. Sheridan's supervisory and control loops. (Adapted from Thomas B. Sheridan. *Telerobotics Automation, and Human Supervisory Control,* Figures 1.2 and 1.3, Massachusetts Institute of Technology, published by The MIT Press).

The concept reflected the growing role of automation, and the changing of the human's role from operator to manager of automation. As shown in Fig. 12.2, several levels of supervisory and control loops exist, with the human closing the supervisory loops, and each control loop being able to be closed by either the automation or the human.

Designers could now employ control theory concepts to the human-automation problem. The role of humans was essentially defined by this paradigm, and the designer could then look at lower level tasks and turn to questions such as (a) what human resources are required for each task, (b) how much time and effort is required, (c) how much time and energy is available, and (d) what the consequences are of accomplishing or not accomplishing an action (Sheridan, 1992).

Sheridan goes on to describe the limitations of humans and automation, and the implications of this for applying the concept. These admonishments went unheeded, however, and automation was steadily added to the flight deck and control rooms, with the expectation that the operators would handle the additional managerial workload.

Levels of automation. Parasuraman, Sheridan, and Wickens (2000) recently proposed a new model for allocation of automation functions across four classes and along a continuum of low automation to high automation (Parasuraman, Sheridan, & Wickens, 2000). The four classes are (a) information acquisition, (b) information analysis, (c) decision and action selection, and (d) action implementation. The level of automation can be allocated for each of these four levels, and can be adapted "on the fly" during operational use. The overriding consideration in determining level of automation is the consequences for human performance in the wake of that automation decision.

Information acquisition automation relates to the "sensing and registration of input data." This includes such things as moving sensors (low level of automation), organizing information (moderate), highlighting information (moderate), and filtering information (could be a high level of automation).

Information analysis automation relates to the requirement to integrate and/or transform information. Low-level examples include showing track predictions or trends. At higher levels, data may be integrated, displaying only high-level information to the operator.

Decision and action selection automation assists in determining courses of action and in making decisions. Low-level automation in this class may simply recommend courses of action. At higher levels the automation may perform the action that it feels is appropriate. The difference is demonstrated by the ground-proximity warning system (GPWS), which recommends an action and is at a fairly low level, and the ground collision avoidance system, which is at a significantly higher level as it automatically takes control to avoid a ground impact if the pilot does not take action.

Action implementation automation performs actions for the operator. The level is determined by the amount of manual control left to the human operator. For example, an autopilot on heading hold may still require heading changes using a turn controller and speed control using throttles. This would be a lower level of action implementation than a fully coupled autopilot.

The appropriate level of automation for different classes should then be decided upon by examining possible solutions at different levels of automation and applying primary and secondary criteria in a possibly iterative fashion. The primary criteria are based upon aspects of human performance such as mental workload, situation awareness, complacency, and skill degradation. The secondary criteria are based upon more practical concerns, such as reliability of the automation and the costs versus benefits of automation decisions.

Ecological allocation. Advocates of this approach (although they do not use this term to describe it) regard function-based representations of work as inadequate representations of the actual workplace. The workplace is complex, and there is interplay between technology and work—that is, just as the workplace is changed by automation, the use of the automation may be changed by the workplace. The claim is then that one cannot adequately allocate function without fully comprehending the work environment and the situations that arise therein. Suchman (1987) stressed that work is "situated," and as a result the system must "incorporate both a sensitivity to local circumstances and resources for the remedy of troubles in understanding that inevitably arise" (p. 28).

Moreover, the situations that arise can fundamentally change the work environment. In a discussion of previous work, Wright, Dearden, and Fields (1998) related three themes that emerged from their (and other related) studies:

The first is that divisions of labor are often dynamically re-allocated on the basis of local contingencies on an occasion of practice. The second is that the processes by which such working divisions of labor are achieved are a significant component of that work. The third theme is that even where divisions of labor are pre-computed and reified in such mechanisms as standard operating procedures, there is still work to be done in making those procedures work. (Wright et al., 1998, p. 339)

Often, due to the changing nature of work, organization, and even situation, the underlying functions of automation can change. In a study of two incidents in the London Underground (subway system) in which trains left their drivers behind, it was discovered that the operators had altered one of the two controls required to operate the train (Wright et al., 1998). One automated control requires all doors of the train to be shut before the train will move. A second control is simply

a button for the operator to push to start the train moving. The operators, apparently believing that the second control was redundant, taped the button in the on position. The two incidents happened when operators left their station to fix stuck doors. When the doors were closed, the "circuit" was closed, and the trains began moving.

Wright et. al. (1998) argued that the operators had unknowingly changed the allocation of functions. Their function had been removed—the automation now had sole control to operate the vehicle. Yet these types of changes to function are not only the result of errors. Pilots make use of automation in ways that change their function as well (e.g., stall warning systems as angle of attack indicators, situational awareness instruments as navigation sensors).

For this reason, it may be desirable to allocate functions dynamically, as it is done in normal social work contexts. In this concept, the allocation of function would depend on the context of each occasion of use (Scallen, Hancock, & Dudley, 1996). In this case, designers would have to examine the consequences of improper configuration of the automation and how to mitigate the likelihood of that occurring (Dearden, Harrison, & Wright, 1998).

One method of obtaining the richer representations of the work context called for in this concept would be the use of scenarios (Dearden et al., 1998). Other methods include ethnographic approaches (Hughes, Randall, & Shapiro, 1992) and contextual design methods (Beyer & Holtzblatt, 1998). The former finds researchers immersing themselves in the work environment under study to attempt to understand the social organization of the work. In the latter, the work environment is viewed from different perspectives (e.g., user, cultural relationships, sequences of actions, information artifacts, physical layout) to define roles and tasks.

Guidelines for automation. Over the years, a number of efforts at establishing guidelines for automation have been described. Several of these, by notable authors, are discussed here.

Wiener and Curry first presented guidelines for automation in 1980. They advised designers to create automation that was comprehensible to the operator, so that the operator could detect and properly react to failures in the system. They also advised that the system should appear to perform the task in much the same way as the operator, allowing for individual styles of operation while not affecting overall system performance. They suggested that operators should be well trained on the setup of the system, as well as nominal and off-nominal operation, emphasizing its use as an aid rather than as a replacement. Operators should also be trained on the alerts of the systems, which should be able to be validated by the operator, and should clearly indicate both the source of the alarm and the severity of the situation. Wiener and Curry also suggested that monitoring may be an important task for the operator, who should be trained and motivated to monitor, and whose workload should be kept at a moderate level to avoid both high- and low-workload situations.

Billings, in 1997, agreed with many of these findings, but stressed that the automation should never have control over the system. The operator should be in control, and even should not be automatically prevented from exceeding normal operating limitations. Billings also warned against overautomation, the tendency of complex automation to become incomprehensible, and the resulting system being too complex for the operator to control in the case of automation failure.

Parasuraman and Riley (1997) added to these guidelines by pointing out the possibility that in certain situations operators can "over-rely" on automation, while in other situations operators can "under-rely" on the automation (Parasuraman et al., 2000). These situations generally occur when the operator does not have sufficient knowledge about the automation or the situation, or when confidence in the automation is misplaced (either high or low). They found that sudden alerts do not adequately allow the operators to prepare to respond, and suggested the use of preparatory, or "likelihood," alerts. They warned that in many cases, automation replaced operator error with designer error, which can occur if designers do not incorporate knowledge of operational practice into their design, and do not allow for some nonintended uses of the automation.

Early Air-Traffic Control

Efforts at controlling air traffic began quickly after the advent of aircraft. These early efforts consisted of drafting procedures for safe operation of aircraft, and were undertaken by private companies until the 1930s (Gilbert, 1973). As aviation grew, dedicated professionals increased their efforts to track aircraft and ensure their separation from one another.

This process was, and to a surprising degree still is, a highly manual process. Initially, aircraft had contact via radio with their companies, who relayed position and timing information to a central control facility. The controllers in that facility would record this information on a large blackboard and manipulate aircraft markers (also called "shrimp boats") on a map table. They would analyze this information in an attempt to predict potential conflicts between aircraft. This ability to mentally project aircraft states is still relied on today when radar displays fail.

Progress from these modest beginnings was slow compared to the progress of aircraft technology. The two most significant advances in air-traffic technology occurred after World War II. At least partially due to their experiences in the Berlin Airlift, air-traffic personnel adopted the practice of speaking directly to aircraft. In addition, the development of radar allowed controllers to track aircraft in real time.

Early radar displays. Although the application to air-traffic control is obvious, early efforts at controlling aircraft with radar were troubled. Early radar displays showed the position of each aircraft relative to the radar station, but did not display other critical information, such as aircraft identification, speed, or altitude. Because of this, associating targets with radar returns was difficult, and projecting their status into the future was still a largely cognitive process. The recognition of this helped bring about the introduction of secondary surveillance radar, in which a transponder on board each aircraft transmits an identification code and altitude to a ground receiving station (Gilbert, 1973). This information can then be correlated with the radar returns, eliminating the need for map tables.

Flight strips and strip bays. Controllers gradually replaced the blackboard with small strips of paper on which information about each flight was recorded. These would be placed in a rack (or bay) in temporal order, and even passed from controller to controller as the flight progressed. As computers were added in the 1960s and 1970s, one of their first tasks was to automate this flight data processing, including printing flight strips.

MODERN CHALLENGES, SOLUTIONS, AND PRINCIPLES

The first 50 years of aviation saw significant changes and innovations in both aircraft and air-traffic control. Recent developments such as (a) complex flight management systems, (b) multifunction displays, (c) highly integrated information systems, (d) alerting and warning systems, and (e) sophisticated automatic control capability have presented even more complex challenges for aviation professionals and researchers.

Glass Cockpits, Flight Management Systems, Mental Models, and Mode Confusion

Beginning in the 1970s, airliners were equipped with flight management computers, which would assist pilots in numerous tasks, including flight path management, navigation, and fuel burn management. The interface to the flight management computer is through a control display unit, which consists of (a) a small, alphanumeric keyboard, (b) several function keys, and (c) a small display screen, all of which fits into about a 6-inch by 10-inch space. On this unit, pilots must enter navigation information, select information to be displayed, and navigate through menus. By the 1980s, flight-management computers were integrated with cathode ray tube (CRT) displays, creating flight decks mostly devoid of gauges, also called "glass cockpits." An example of a glass cockpit (from an MD-11 aircraft) is shown in Fig. 12.3.

Such systems integrated the operation of the aircraft with navigation, resulting in a "fundamental shift in aircraft automation" (Billings, 1997). Although such displays provided the benefits of a highly integrated display, there was a cost in terms of workload, crew coordination, error management, and vigilance (Wickens, Fadden, Merwin, & Verves, 1998). In addition, a number of accidents have been associated with pilots misunderstanding the state of the automation. These accidents have been attributed to the pilots lacking an accurate "mental model" of the automation.

Mental models are a conceptual representation of a system posited cognitively. The precursors to mental model research date back nearly a century. These early inquiries (Wittgenstein, 1922; Craik, 1943), and a significant body of recent work, argue that cognition is inherently imagistic, so that mental models are a theory of how people think. This theory is a topic of contentious debate in the psychological community, with many researchers vehemently denying that people think in these "pictures in the head," but rather that thought consists of formal rules of inference (Pylyshyn, 2003).

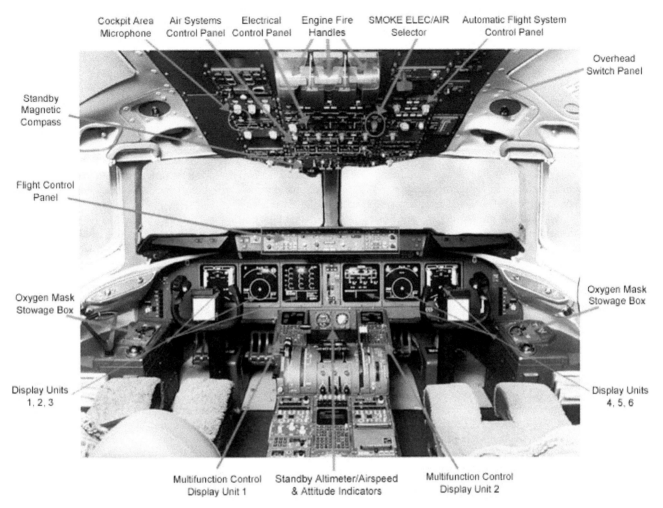

Cockpit Area Microphone
Air Systems Control Panel
Electrical Control Panel
Engine Fire Handles
SMOKE ELEC/AIR Selector
Automatic Flight System Control Panel
Overhead Switch Panel
Standby Magnetic Compass
Flight Control Panel
Oxygen Mask Stowage Box
Display Units 1, 2, 3
Oxygen Mask Stowage Box
Display Units 4, 5, 6
Multifunction Control Display Unit 1
Standby Altimeter/Airspeed & Attitude Indicators
Multifunction Control Display Unit 2

FIGURE 12.3.   MD-11 flight deck instruments. (Photo from Transportation Safety Board of Canada Report Number A98H0003, page 32, Transportation Safety Board of Canada, 1998. Reproduced with the permission of the Minister of Public Works and Government Services, 2007).

Undeniably, however, people at least think about such imagistic models. In doing so, operators reason about their system based on these models. There is convincing evidence that such models correlate well with certain operator errors, and point to ways in which good design can mitigate these errors. As such, we will try to skirt the contentious issue of whether mental models are theories or models, and stick to how the concept has been deployed in support of cognitive engineering.

Mental model research was born out of the observation that many people use analogy to understand complex phenomena. This idea became more formalized in the cognitive science literature, highlighted by a flood of papers and books on mental models in the early 1980s (Gentner & Stevens, 1983; Johnson-Laird, 1983). If one believed that a person's knowledge of a complex system was a product of that person's model of the system, then an understanding of the limitations of these models was crucial to explaining behavior and errors.

Nowhere did mental models seem to fit in better than in explaining a number of human–automation interaction failures. Flight deck automation was becoming increasingly complex,

taxing the abilities of pilots to accurately model it. Error is likely where such inaccuracies exist. Several particularly prominent examples of this have occurred in the last 15 years. In 1988, a Boeing 767 was inadvertently put into a vertical speed mode, nearly resulting in a crash. A similar situation occurred in an Airbus 320, when a crew member mistakenly selected a descent rate of 3,300 feet per minute instead of on a 3.3° downward slope and flew into terrain. In 1994, a China Airlines Airbus 300 aircraft crashed after pilots tried to continue a landing while the autopilot had been inadvertently put into a go-around mode. Also in 1988, an Airbus 320 was flown into terrain by an aircrew during an air show; the crew believed that the envelope protection of the aircraft would prevent a stall, but the aircraft was in a mode where that protection was not provided.

These incidents indicated that one piece of flight deck automation particularly prone to such mental model inaccuracies is the flight management system, which interacts closely with the autopilot. Researchers had been aware of mode problems in HCI for some time, and in due time these issues arose in the interaction of the pilot with the computer that operates the autopilot. This aspect of the problem—as a computer interface issue—has significantly affected the research on mental models in aviation.

In addition to calling for training for pilots to better understand their systems, much effort has been put into creating models of the pilot's interaction with the flight management system in an effort to understand how systems might be designed to mitigate these model inaccuracies.

Heads-Up Displays

In order to minimize the need to view instruments inside the cockpit, "heads-up" displays (HUDs) were superimposed on the windscreen, providing at least the basic T information. The same method can be used to superimpose images on the inside of the visor of a helmet (known as "helmet mounted displays" [HMDs]). HUDs have moved from the military, where they found their first application, to commercial aircraft and even automobiles.

One major difference between the instrumentation on a HUD and the same instrumentation on a traditional instrument panel is that the world can be viewed through the HUD. This allows for the possibility of integrating real external images with the instrumentation without violating the proximity principle. The flip side of this, however, is that the world peeking through the HUD provides a great deal of clutter on the display, although some research indicates that this is not a significant increase in difficulty (Ververs & Wickens, 1998). Overall, the HUD has a number of advantages that make it superior to conventional positioning of displays. The HUD enhances the operator's ability to switch between near and far domains (Levy, Foyle, & McCann, 1998), although strangely this reverses to a detriment for the detection of unexpected events (Wickens & Prevett, 1995).

Information Automation

The rapid rise in computer and display technology has allowed engineers to develop systems that alleviate some of the information burden on the operator. These systems perform automated monitoring (such as conflict alerting and prevention), provide information to assist the pilot's decision making, and help provide coordination between different operators.

Alerting systems. In the late 1970s, aircraft were undergoing a rapid expansion in terms of warnings and alerting systems. One survey found general agreement that there were too many warnings and that standards and guidelines were needed for warning system design (Cooper, 1977). Another researcher determined that the number of visual alerts and flags had doubled, while the number of aural alerts had increased by 50% since the beginning of the commercial jet age (Veitengruber, 1977). Based on the data at the time, researchers were recommending that alerts be reduced, prioritized, and inhibited during critical phases of flight.

Since then, alerting systems have grown significantly in capability and complexity. Current alerting systems may utilize multiple sensors, ground communications, large knowledge databases, and may have sophisticated algorithms for alert threshold and guidance recommendations (Pritchett, 2001).

One of the most widely analyzed problems in alerting system technology relates to problems with conformance to the ground-proximity warning system. The ground-proximity warning system (GPWS) was a complex alert designed to warn of something extremely hazardous: proximity to or closure to terrain. It was complex in that closure to terrain can be caused by a number of things, including landing, and the system had to discriminate these events. In part due to this complexity, and in part due to poor design, the system frequently produced false alarms. This had the effect of reducing operator trust in the automation and causing underreliance (the pilots would not believe the alert).

High false alarm rates can also be caused by improper setting of thresholds for alerts, a process that is analogous to signal detection. The threshold for an alert can be based upon the system operating characteristics curve, which is in turn a function of the sensors (Kuchar, 1996). Setting the threshold is then a tradeoff between missed detections and false alarms. Typically, this means that the fewer the false alarms, the more the missed detections. Interestingly, some researchers have warned against having too low a false alarm rate (Farber & Paley, 1993). If the false alarm rate is too low, not only will the occurrence be infrequent and unexpected, but also it will likely require a later alert, compounding the difficulty of reacting to the alert.

The problem of false alarms is further exacerbated by the low base rate of incidents requiring an alert (Parasuraman & Riley, 1997). In these cases, the relative rate of false alarms is increased. Yet in many cases, the cost of a false alarm is considered much lower than a missed detection, leading designers to err on the side of economy and/or safety.

These (generally) are cases where the alerting system is functioning as a signal detector or a hazard detector, the latter being a more complex version of the former. Alerting systems have also added hazard-resolving logic, such as in TCAS II (Pritchett, 2001) and NASA's Advanced Airspace Concept (Erzberger, 2004). The benefit of such an approach is that the resolving maneuver can be incorporated into the alerting logic, resulting in a reduction of false alarms; however, such a system relies heavily on the pilot appropriately performing the maneuver. It is also possible that the maneuver may actually induce a collision by creating a problem where none actually existed.

Decision support systems. In addition to alerting systems, there are also systems on board aircraft and, to an even greater extent, in air-traffic control which are considered decision support tools. These tools do not warn or alert, but provide information to support decision making. This information is often integrated from several sources, or enhanced in some way.

One simple example of a decision support system on a flight deck is the weather radar system. This system provides no discrete alert, but rather shows the pilot where hazardous weather exists to help pilots determine how to deviate to avoid it. With no weather radar, pilots had to rely on visual means, which can only detect gross features such as height, darkness, and the presence of lightning. Weather radar, however, not only shows rates of precipitation (an indication of severity), but can also show the lateral and vertical extent of the weather.

Another decision support system is the Engine Indication and Crew Alerting System. While some aspects of this system are more correctly considered an alerting system, much of the system is simply decision support. This display shows pilots the status of most of the systems on board. It can be configured to depict engine instruments or other graphical depictions of the operation of the systems on board the aircraft.

As mentioned, there are a number of decision-support tools being used in air-traffic control. In fact, the number of such displays is increasing at such a rate that the FAA is attempting to integrate these systems to reduce the sheer number of separate systems operating in air-traffic facilities.

Modern Air-Traffic Control

The current air-traffic control system, while reliant on computer resources, has been designed to be resilient to the loss of such resources. Computers are still used for display of primary and secondary radar information, as well as flight data processing. The upgrading of these capabilities from the 1970s has been slow, impacted by the widespread replacement of experienced controllers following the air-traffic controller strike of 1981 and the failure of the Advanced Automation System (AAS) program of the late 1980s (Lee & Davis, 1996). On the heels of the AAS system, however, have come a number of advances, including attempts to automate flight strips and the introduction of automation aids.

Automated flight strips and situation awareness.
Flight strips have proven remarkably resilient to change. Many controllers still use flight strips, and flight strip bays are still used in air-traffic control facilities. Recently, however, manual flight strips are being replaced with electronic flight strips, most notably in the User Request and Evaluation Tool (URET), which took 25 years to implement (Arthur & McLaughlin, 1998). Similar efforts to replace flight strips have been undertaken in Europe (Berndtsson & Normark, 1999). The reasons for their slow demise are highly instructive of some of the problems with automating some forms of human work.

Despite the significant workload of the manual process of dealing with flight strips, controllers have been reluctant to part with them, or to use suggested replacements. This is, in large part, due to the flight strips contribution to the controller's situation awareness. For a controller, whose job is still to a large degree cognitive, their situation awareness is crucial (Whitfield & Jackson, 1983). Researchers have found that controllers indicate that flight strips are a primary means of establishing and maintaining their situation awareness (Harper & Hughes, 1991).

It is a notable theme that modern research attention has turned from information display, as described above, to information assessment. As more complex automation is added to flight decks and air-traffic facilities, and as more information is provided to operators, accessing and comprehending that automation and information became problematic. Researchers began to understand that not only did they need to know how to display information, but also how operators represented and utilized that information, in order to provide a better match between technology and the human.

Interest in information assessment has lead to a significant body of research into situation awareness (SA). This has been a particularly important concept in aviation, since safe and successful operation in aviation requires a great deal of knowledge about the environment outside of an individual aircraft. Many aviation accidents can be attributed to a lack of knowledge of the environment external to the aircraft (e.g., collisions with aircraft and the ground). Air-traffic controllers' knowledge of the "big picture" includes not only the states of the aircraft in their own sector, but those of neighboring sectors, airports, weather, and much more. Attempts to understand how pilots and controllers obtain and maintain this "big picture" have been the focus of SA research.

Since as far back as the first World War, military aviators have understood the importance of SA and have instructed their pilots in developing and maintaining good SA. One of Germany's top World War I fighter aces, Oswald Boelcke, listed among his "dicta" that "the pilot must acquire the habit of "taking in" unconsciously the general progress of the whole multi-aircraft dogfight going on around the individual combat in which the pilot will become involved . . . (so that) no time (is) wasted in assessment of the general situation after the end of an individual combat" (Hacker, 1984, p. 1). Boelcke also prescribed knowledge of one's own machine, the enemy's machine, and navigational fixes. SA has been a significant part of aeronautical training since that time.

As a research topic, however, the concept was mostly ignored by researchers until the 1980s. Due to increasing flight deck and air-traffic automation, pilots' and air-traffic controllers' role as supervisor of these systems was increasing, reducing the time they could spend in developing SA. At the same time, a great deal of new sensor information was becoming available to designers of aviation automation, information that could be used to reinforce the controllers' and pilots' SA. Researchers began looking into what SA is, what affects an operator's ability to construct SA and to keep it, and how it might be measured.

Many definitions of SA exist, with most agreeing that SA is, at least in part, the comprehension of elements of the environment that have (or may have) some bearing on the task being accomplished. Some efforts have viewed SA as a static, information-driven product, some have viewed it as a dynamic process, while others have viewed SA as a high-level description of certain aspects of task behavior. A significant body of work has also gone toward determining how to measure SA.

One of the most widely quoted definitions of SA is "the perception of elements in the environment within a volume of time and space, the comprehension of their meaning, and the projection of their status in the near future" (Endsley, 1988, p. 97). Another researcher has called SA "an integrated understanding of factors that will contribute to the safe flying of the aircraft under normal or nonnormal conditions" (Regal et al., 1988, p. 65). These definitions describe SA as a product, something that an operator either has or does not have. As such, errors resulting from a lack of SA can be studied. Researchers have found that controllers, viewing recorded air-traffic files, were unable to recall many details about traffic when asked about them during the scenario, including call signs, control level (who has control of the aircraft), altitude, speed, and heading; controllers often even failed to report that some aircraft were present at all (Endsley & Rodgers, 1996). Other researchers studying error report databases have found that in only about 20% of serious operational air-traffic errors were controllers aware of the problem developing (Gosling, 2002).

Researchers who view SA as a product often defer the process of obtaining and maintaining SA to a separate process, often referring to it as "situation assessment". This type of SA has been described as "(an) adaptive, externally-directed consciousness" (Smith & Hancock, 1995, p. 137), and "the integration of knowledge resulting from recurrent situation assessments" (Sarter & Woods, 1991, p. 45). Here, the sources of information important to SA (visual, auditory, tactile, other sensory input, knowledge of procedures and regulations, etc.) are diverse and often clearly identifiable. The operator's ability to obtain SA begins with the most fundamental cognitive processes (detection), and progresses to very sophisticated concepts (comprehension and projection).

Some researchers have called into question the utility of the SA concept. Many feel it is a high-level concept that does not have sufficient granularity to really explain anything. SA has been referred to as a description of observations of humans operating complex systems in a dynamic environment (Billings, 1995) and as simply equivalent to expertise (Crane, 1992). As such, SA is a description of a set of cognitive processes that are used together, and is really only useful in categorizing or grouping behaviors and errors. Some researchers feel that considering SA as a causal agent can be counterproductive to understanding operator behavior (Flach, 1995).

No one set of measures has been clearly identified for SA. Typical measures can be broken into three categories: (a) explicit (or knowledge-based), (b) implicit (or performance-based), and (c) subjective (Vidulich, 1992). Examples of explicit methods include participants' recall of a situation (post hoc), an ongoing narrative provided concurrently with the task, or freezing a scenario and questioning the participant about the decisions, events, or the task environment. These measures can be compared to actual state of the system to provide a better measure, but have been criticized as too subjective (in the case of recall; Fracker, 1991), or too intrusive (in the case of freezing a scenario; Sarter & Woods, 1991).

Implicit measures examine task performance, and correlate that with SA. These measures are generally unobtrusive and objective, and (as mentioned earlier) can be used in conjunction with explicit measures. It has been suggested that these measures can succeed where explicit methods cannot, particularly in situations where a determination of the timing of events is important and where the subject may be unaware of his or her deficiency (Pritchett, Hansman, & Johnson, 1995). These measures may examine performance at the task overall, or alter the task to determine whether the subject notices the change.

Subjective ratings are assessments of SA made either by an observer or by the subject. These ratings can use a number of different scales, and can be either direct or relative (such as by comparing SA in one situation to SA in another situation). Although these ratings can be affected by a number of factors, including task performance, one technique that has been studied extensively is the Situation Awareness Rating Technique (SART; Taylor, 1989. This technique has the operator rate 10 constructs on a seven-point scale. The constructs are grouped into three categories: (a) attentional demand (which includes the instability, variability, and complexity of the situation), (b) attentional supply (including arousal, spare mental capacity, concentration, and division of attention), and (c) understanding (which includes information quality and quantity, and familiarity). SART has been found to be more sensitive than overall subjective measures of SA (Selcon & Taylor, 1989), although some researchers feel that SART confounds SA with workload (Jones & Endsley, 2000).

Decision support tools in air-traffic control. One of the most significant areas of improvement within air traffic over the last two decades has been in decision support tools. Over the years, a number of automation tools have been slowly added to the air-traffic control system, and others have been considered and failed. These advancements have included the introduction of two conflict prediction systems, the failed attempt to provide sequencing and runway selection aids, and the introduction of centrally scheduled time-based metering tools.

The first conflict prediction system was created in the early 1980s as a quality assurance program. The "operational error detection program" (OEDP) was designed to identify when two aircraft had come closer to one another than allowed under FAA regulations. If the OEDP was set off, a record was made of the transgression, and investigation took place. This is obviously highly undesirable for controllers, who have nicknamed the system "the snitch patch." As a result of the OEDP, controllers routinely pad separation between aircraft to ensure that the snitch patch does not activate (Cotton, 2003).

The OEDP also gives a three-minute warning that a violation is going to occur, although the system is not considered highly reliable by controllers. Part of the reason for the system's unreliability is that its predictions are based on simple extrapolations of the current states of the aircraft involved, without consideration of intentions (such as its intention to stop climbing or to change course). The system's unreliability has led to several efforts to replace it. The URET system mentioned previously includes a "conflict probe," which examines the predicted flight path of aircraft and gives an indication of the potential for collision (Brudnicki & McFarland, 1997).

In the 1990s, NASA undertook an effort to provide automation tools for controllers and traffic managers. The suite of tools, called the Center-TRACON Automation System (CTAS), was based upon combining radar position and speed data, aircraft flight plans, weather data, and models of aircraft flight characteristics. The models were based upon those used in the flight management systems in aircraft, which are known to be highly accurate. One of the first tools proposed was the Final Approach Spacing Tool (FAST). This system was to be used by controllers handling aircraft close to busy airports. These controllers transition aircraft from intermediate altitudes about fifty miles out from an airport and coordinate their arrival to the runway. This is a very complex task, requiring controllers to predict the proper sequencing and merging of aircraft from several directions. FAST generated advisories for sequencing, runway assignment, headings, and speeds to ease the merging and spacing of aircraft (Lee

& Davis, 1996). Although deemed acceptable by controllers (Lee & Sanford, 1998), the system was ultimately not implemented. One of the likely factors was that although the system was extremely accurate in its predictions, it was not perfectly so, and controllers may have felt that the workload and distraction of the few inaccurate predictions outweighed its benefits.

Another part of the CTAS suite of tools is the Traffic Management Advisor (TMA), which provides a similar service as FAST for controllers handling aircraft further out from the airport (out to about 300 miles). This system provides delay advisories for each aircraft approaching a busy airport. Controllers are then responsible for slowing aircraft to meet the assigned delays, utilizing the same techniques they use to space aircraft normally. The main shift between the current method and TMA is that the spacing TMA provides is time-based, versus the distance-based method currently used. Time-based metering in this fashion has been shown to be more efficient than distance-based (Sokkappa, 1989; Vandevenne & Andrews, 1993), and TMA has shown significant benefits where it has been adopted (Knorr, 2003). Similar efforts focused on runway sequencing and scheduling have been developed in Europe, Canada, and Australia (Ljungberg & Lucas, 1992; Robinson, Davis, & Isaacson, 1997; Barco Othogon, 2003; NavCanada, 2003). More advanced systems are also being developed which move some separation authority to the flight deck (Prevot et al., 2003).

FUTURE CHALLENGES

Aviation has undergone a great deal of change since its inception, although predictions of its development path have not always been accurate. Forty or 50 years ago, predictions of where the air transport system would be would probably have included much more automation of air-traffic control and high reliance on satellite systems, with larger and faster aircraft than were then utilized. These aspects have largely not been realized to the extent expected. The aviation system of the future will undoubtedly hold other unexpected developments, but a few challenges for researchers, engineers, and practitioners are clear.

Three-Dimensional Displays

To overcome the limitations of two-dimensional (2D) displays, several alternatives to providing three dimensions in a display have been investigated. Air-traffic and navigation displays are both trying to represent three dimensions in a 2D display. Inevitably, some information is lost in the display in this case. This can be accomplished through binocular, stereoscopic, or holographic devices (Mountford & Somberg, 1981), but practical problems make this unlikely. Also unlikely due to the cognitive demands of integration is using time frame compressions (a display of sequential images; Roscoe, 1981); however, significant research has been accomplished in using perspective illusion displays, where the illusion of three dimensions is given through the use of perspective.

Three-dimensional displays can portray a variety of viewpoints, to different effect (Wickens & Prevett, 1995). In testing such displays, 2D displays, with either some analog display of altitude or an additional profile view, are used for comparison. This adds complexity to testing these displays; often, additional information has to be added to current displays in order to isolate the 3D presentation from the additional information provided by the display.

In addition, dual panel 2D displays, although they may contain the same information as 3D displays, have an additional cost—that of scanning and integrating. Therefore, it would seem that 3D displays adhere to several tenets of good display design: proximity, compatibility, integration, and pictorial realism.

If the airplane is not navigating vertically, however, the 3D display may impose a clutter cost, as the operator must sort through the irrelevant altitude information. This is in addition to the problem of ambiguity of position of objects along the line of sight of the display (McGreevy & Ellis, 1986). Lastly, if the 3D perspective is immersed (the perspective is as the world looks from the cockpit), no information concerning the world to the side or behind the aircraft is presented, resulting in a "keyhole" view of the world (Woods, 1984).

For judgments requiring mapping between the world and the display, a 3D display with a viewpoint outside the cockpit seems to provide the best performance (Olmos, Wickens, & Chudy, 1997). This benefit is again tied to the principles of realism and integration, and is generally viewed as providing better overall situation awareness than immersed 3D views.

Tunnels in the sky. Enhancements to the immersed viewpoint of a 3D display can present the operator with a highly integrated pathway, or "tunnel in the sky." Superior performance in navigation and tracking has been demonstrated using this display (Wickens & Prevett, 1995), although they are currently still in development. The biggest obstacle to the deployment of such automation seems to be the cost of certification, rather than any technical or cognitive issues. It appears that these displays only suffer from effects similar to those found on other perspective displays (confounding of distance along the line of sight) and on HUDs (difficulty in detecting unusual events due to the compelling nature of the display; Wickens, Mavor, Parasuraman, & McGee, 1998).

The scale problem in air-traffic control. Increases in computational ability have made 3D displays even more practical, but many issues surrounding them have not been resolved. In some ways, aviation is a 3D task, with vertical and horizontal navigation; however, the scales of these navigation tasks are different, one expressed in feet, the other in nautical miles. It is extremely difficult to portray these two axes on a single display without distorting one or the other, with significant repercussions for judgments and decisions of the operator using the display. In addition, air-traffic control is only marginally a 3D task, with aircraft separated for much of their flight through the use of altitude stratification. This separation method places aircraft at discrete altitudes, with climbing and descending transitions between these strata. As such, separation in the vertical axis can be assured for much of the aircraft simply by comparing two numbers. Additional information about vertical position is superfluous (except during climbs and descents).

Coordination Support Systems

One type of information system which is likely to become prevalent in the future is what is referred to here as a coordination support system. It is becoming more common that work, including in the air-traffic system, is distributed across organizations. For example, alerts generated by an aircraft's TCAS system have to be coordinated between the two aircraft in the conflict pair to avoid having the aircraft's resolutions conflict. In addition, coordinating arrivals into airports involves affecting aircraft many hundreds of miles away from the airport, across air-traffic control organizational and even facility boundaries.

One approach to enabling such cooperative work is to actively share information between organizations through procedural means. Such an approach means that one organization, with localized information of interest, distributes that information to other parties. In the air-traffic system, however, the great majority of the information needed for collaborative work is present in the task environment, or relates to the impact of other organizations' decisions on the traffic situation. If such ecological information could be provided, then collaborative work would happen naturally.

Such an approach is supported by the concept of a common information space (CIS) (Schmidt & Bannon, 1992). A CIS is a source of information shared within and across organization boundaries in order to conduct coordinated work for common purposes. Instead of passing information from one organization to another, information is made available through the CIS, which can then be accessed and manipulated by any interested party. Moreover, objects can be created within the CIS which serve as translators across different communities. These objects, referred to as "boundary objects," are entities in which decontextualized information resides in a particular form (Starr & Griesemer, 1989). The distinction which defines boundary objects is their purpose in acting as translators, or, as defined by Bannon (2000), they are a "means for sharing items in a common information space" (p. 5). Information, devoid of context, is presented in a form that allows users from different communities of interest to comprehend it within their own contexts. This object is accessed by users in different communities of interest in order to conduct their particular portion of the work. The information in boundary objects is contextualized only within each community of interest, and this context differs in each separate community.

Technologies Needed for 3× Air Traffic

The rate of growth of air traffic has been consistent for a number of decades, and is not expected to change significantly in the next few decades. Such a rate of growth would double or triple air traffic, a level which some researchers feel is unsupportable by the current method of air-traffic control.

Such a conclusion is consistent with the historical development of the air-traffic system. In order to grow beyond its capabilities in the 1940s, radar was added, reducing the workload of the controllers, which enabled them to deal with more traffic and

higher aircraft speeds. To continue to grow, computerized flight data processing was added, further reducing workload and allowing controllers to handle more traffic. Increases in staffing, reduction in sector sizes, strategic control, and other innovations have allowed further increases in air-traffic levels. To further continue these increases, researchers argue that a more fundamental shift is needed, by sharing some of the air-traffic responsibilities between the current authority (air-traffic controllers) and the flight deck, by allowing automation to accomplish some current tasks of the controllers, or some combination of these.

Datalink and the loss of the party line.   One high-workload task for controllers is communication. For some air-traffic control tasks, the capacity of the radio frequency is a limiting factor for how many aircraft can be handled within a sector. This is also a limiting factor for how small sectors can be, since controllers must still talk to each aircraft at least twice (once upon entering, once upon leaving the sector), and small sectors reduce the time between these communications.

One solution to this communication workload is to allow communication directly between the data processing system and the flight, through a system known as "datalink." This technology has been suggested for decades (Schmidt & Saint, 1969), but its use has been stifled by several issues.

One of these issues is that pilots' situation awareness is affected by the availability of information on the communication channel. Pilots hear the communication of other aircraft with the controller, and can glean information that is relevant to them, such as expectations of future clearances, the impact of weather on aircraft ahead of them, and positions and intentions of other aircraft. If this "party line" information is passed over a data channel, pilots would no longer have access to it, reducing their situation awareness (Midkiff & Hansman, 1993).

4D contracts: Future changes in the roles of pilots and controllers?   The flight management system used by the current generation of commercial aircraft is capable of meeting arrival times over points in its flight path, subject to the constraints of the flight envelope. The logic behind these flight management systems is also available to ground-based systems, and is in fact the heart of the Center-Tracon Automation System, a suite of air-traffic management tools developed by NASA (Erzberger & Nedell, 1989).

These ground-based systems can use the flight management algorithms to make predictions of arrival times for all aircraft in the National Airspace System (NAS). This has been accomplished for flights arriving at a number of the busiest airports in the United States in the Traffic Management Advisor system (Swenson et al., 1997). This capability is being expanded to regional or national airspace, and for scheduling aircraft over any congested point in the NAS (Landry, Farley, & Hoang, 2005).

One proposal for increasing the capacity of the air-traffic system is to allow such a system to negotiate "4D contracts," indicating navigational waypoints, altitudes, and times of arrival at those points for all aircraft. The flight management systems would then be responsible for adhering to the contract. These contracts would ensure conflict-free routes, and coordinate arrivals into congested resources, including airports.

Such a system would radically alter the roles of pilots and controllers in relation to their automation. Pilots and controllers would be monitors of a highly automated system, in comparison to their current roles of actively controlling the flight paths of aircraft. As such, this presents enormous challenges to the aviation HCI community.

Automating separation assurance. The primary responsibility of an air-traffic controller is to provide separation assurance between aircraft. Even if datalink and 4D contracts were established, the controller is still limited by the cognitively intensive process of monitoring for conflicts. This task is also under consideration for automated assistance, either on the flight deck or through a centralized system.

One proposal is to equip flight decks with a Cockpit Display of Traffic Information (CDTI) system, and utilize an advanced form of the current Traffic Collision and Avoidance System (TCAS). A second proposal is to utilize a centralized system for providing alerts (Erzberger, 2004). Such a system would automatically coordinate resolutions, and provide for an independent backup in case of failures to resolve. In both cases, automation would detect conflicts, identify resolutions, and (perhaps) automatically perform the resolutions. Such systems overcome the latencies in the human detection, communication, and action cycles; however, as the human operators are pushed back further from the actual operation of the systems, their ability to intervene in the case of automation failures wanes. Because of this, such automation must be highly reliable for the system to succeed.

UAVs in the Airspace System

The rapid development of uninhabited aerial vehicles (UAVs) will also greatly impact the air-traffic system of the future. To date such vehicles have been mostly constrained to special uses, such as the military. It seems likely, however, that these vehicles will need to be integrated into the air-traffic control system as their uses expand.

Recently, efforts have begun to identify the issues related to integrating UAVs into the airspace system (Wegerbauer, 2005).

Such vehicles, while not piloted in the traditional sense, are controlled by a remote pilot. The issues related to their use of the airspace system concern collision danger and safe extraction of the vehicle in the case of a communication (or other system) failure. These are but the first issues to be dealt with, particularly as one considers more advanced UAVs with increasing amounts of autonomy.

CONCLUSIONS

Fitts (1951b) and his colleagues described the research challenges in visual displays as of 1951. The document is remarkable in that it spurred (or foresaw) a great deal of research that continues to this day. Some of that can be attributed to the general nature of the questions, but many of the specific research questions and methods have been examined and used over the last forty years. What was noticeably (and understandably) absent from the discussion of transitory displays was any reference to higher cognition. In some ways, this is indicative of the progress made in aerospace human–machine system research. Many of the difficulties associated with these concepts originated as higher levels (and greater amounts) of automation were added to the flight deck.

The progression of research has also followed this model. Early research resulted in a number of principles for integrating displays, for determining the form of displays for different purposes, and for the positioning of displays. Later work has concentrated on higher level cognitive issues such as mental models and situation awareness. In addition, work has been done on understanding how to apply automation, resulting in a number of principles regarding automation use and allocation of function. Most recently, new types of displays and automation such as warning systems and decision support systems have been introduced, yielding a new set of problems for researchers to tackle. As the air transportation system matures and the demand for ever-greater levels of capacity increases, it seems that greater reliance on automated systems will be required. For such a system to be successful, such automation must be based on sound principles.

References

Abbott, T. S., Mowen, G. C., Person, L. H., Jr., Keyser, G. L., Jr., Yenni, K. R., & Garren, J. F., Jr. (1980). *Flight investigation of cockpit-displayed traffic information utilizing coded symbology in an advanced operational environment.* Hampton, VA: NASA Langley Research Center.

Allen, R. B. (1983). Cognitive factors in the use of menus and trees: An experiment. *IEEE Journal on Selected Areas in Communication, 1*(2), 333–336.

Arthur, W. C., & McLaughlin, M. P. (1998). *User request evaluation tool (URET): Interfacility conflict probe performance assessment.* McLean, VA: MITRE.

Bainbridge, L. (1983). Increasing levels of automation can increase, rather than decrease, the problems of supporting the human operator. *Automatica, 19*, 775–779.

Bannon, L. J. (2000, August). *Understanding common information spaces in CSCW.* Workshop on Cooperative Organisation of Common Information Spaces, Delft, Technical University of Denmark.

Barco Orthogon. (2003). *Osyris.* Retrieved January 27, 2005, from http://www.barco.com/barcoview/downloads/BVW_Osyris_new_6p.pdf

Bartlett, F. C. (1943). *Instrument controls and displays—Efficient human manipulation.* London, UK: Medical Research Council.

Bejczy, A. K., & Paine, G. (1977). Displays for supervisory control of manipulators. In *MIT Proceedings of the 13th Annual Conference on Manual Control* (pp. 275–284), Cambridge, MA: Massachusetts Institute of Technology.

Beringer, D. B., Willeges, R. C., & Roscoe, S. N. (1975). The transition of experienced pilots to a frequency-separated aircraft altitude display. *Human Factors, 17*, 401–414.

Berndtsson, J., & Normark, M. (1999, November). The coordinative functions of flight strips: Air traffic control work revisited. *ACM Group99: International Conference on Supporting Group Work*, Phoenix, AZ.

Beyer, H., & Holtzblatt, K. (1998). *Contextual design: Defining customer centred system*. San Francisco: Morgan Kaufman.

Billings, C. E. (1995). Situation awareness measurement and analysis: A commentary. D. Garland & M. Endsley (Eds.), *Proceedings of the International Conference on Experimental Analysis and Measurement of Situation Awareness*, Daytona Beach, FL, Embry Riddle University Press.

Billings, C. E. (1997). *Aviation automation: The search for a human-centered approach*. Mahwah, NJ: Lawrence Erlbaum.

Brudnicki, D. J., & McFarland, A. L. (1997). *User request evaluation tool (URET) conflict probe performance and benefits assessment*. McLean, VA: MITRE.

Bryan, L. A., Stonecipher, J. W., & Aron, K. (1954). 180-degree turn experimen. *University of Illinois Bulletin, 54*(11), 1–52.

Chapanis, A. (1959). *Research techniques in human engineering*. Baltimore: The Johns Hopkins Press.

Civil Aeronautics Board (1953). *Civil Air Regulations* Part 4b. 4b.611 (b).

Civil Aeronautics Board (1957). *Civil Air Regulations* Part 4b. Amendment 4b-7.

Cohen, E., & Follert, R. L. (1970). Accuracy of interpolation between scale graduations. *Human Factors, 31*(5), 481–483.

Cole, E. L., Milton, J. L., & McIntosh, B. B. (1954). Eye fixations of aircraft pilots. IX. Routine maneuvers under day and night conditions, using an experimental panel (Tech. Rep. No. 53-220). Wright Patterson AFB, OH, USAF WADC.

Conrad, R. (1951). Speed and load stress in sensory-motor skill. *British Journal of Industrial Medicine, 8*, 1–7.

Cooper, G. E. (1977). *A survey of the status of and philosophies relating to cockpit warning systems*. Moffett Field, CA: NASA Ames Research Center.

Cotton, B. (2003, July). *For spacious skies—the twenty-first century promise of free flight*. Paper presented at AIAA International Air and Space Symposium and Exhibition: The next 100 years. Dayton, OH.

Craik, K. (1943). *The nature of explanation*. Cambridge: Cambridge University Press.

Crane, P. M. (1992). Theories of expertise as models for understanding situation awareness (Tech. Rep. No. ADP006943). Armstrong Lab Williams, AZ.

Dashevsky, S. G. (1964). Check-reading accuracy as a function of pointer alignment, patterning, and viewing angle. *Journal of Applied Psychology, 48*, 344–347.

Davenport, W. W. (1978). *Gyro! The life and times of Lawrence Sperry*. New York: Charles Scribner.

Dearden, A., Harrison, M., & Wright, P. (1998). Allocation of function: scenarios, context and the economics of effort. *International Journal of Human Computer Studies, 52*(2), 289–318.

Dorris, A. L., Connolly, T., Sadosky, T. L., & Burroughs, M. (1977). More information or more data? Some experimental findings. *Proceedings of the Human Factors Society 21st Annual Meeting*. Santa Monica, CA: Human Factors Society.

Elkin, E. H. (1959). *Effect of scale shape, exposure time and display complexity on scale reading efficiency* (Tech. Rep. No. 58-472). Wright Patterson AFB, OH, USAF WADC.

Endsley, M. R. (1988, October). *Design and evaluation for situation awareness enhancement*. Paper presented at the Human Factors Society 32nd Annual Meeting, Santa Monica, CA.

Endsley, M. R., & Rodgers M. D. (1996, September). *Attention distribution and situation awareness in air traffic control*. Paper presented at the 40th Annual Meeting of the Human Factors and Ergonomics Society, Santa Monica, CA.

Eriksen, C. W., & Eriksen, B. A. (1974). Effects of noise letter upon the identification of a target letter in a nonsearch task. *Perception and Psychophysics, 16*, 143–149.

Erzberger, H. (2004, August). *Transforming the NAS: The next generation air traffic control system*. Paper presented at the 24th International Congress of the Aeronautical Sciences, Yokohama, Japan.

Erzberger, H., & Nedell, W. (1989). *Design of automated system for management of arrival traffic*. Moffett Field, CA: National Aeronautics and Space Administration.

Farber, E., & Paley, M. (1993, April). *Using freeway traffic data to estimate the effectiveness of rear-end collision countermeasures*. Paper presented at the Third Annual IVHS America Meeting, Washington, DC.

Ferrell, W. R., & Sheridan, T. B. (1967). Supervisory control of remote manipulation. *IEEE Spectrum, 4*(10), 81–88.

Fitts, P. M. (1951a). Engineering psychology in equipment design. In S. Sevens (Ed.), *Handbook of experimental psychology*. New York: Wiley.

Fitts, P. M. (1951b). *Human engineering for an effective air-navigation and traffic-control system*. Washington, DC: National Research Council Committee on Aviation Psychology.

Fitts, P. M., Jones, R. E., & Milton, J. L. (1950). Eye movements of aircraft pilots during instrument-landing approaches. *Aeronautical Engineering Review, 9*(2), 1–6.

Flach, J. M. (1995). Situation awareness: Proceed with caution. *Human Factors, 37*, 149–157.

Fogel, L. (1959). A new concept: The kinalog display system. *Human Factors, 1*(1), 30–37.

Fracker, M. L. (1991). Measures of situation awareness: Review and future directions (Tech. Rep. No. 1991-0127). Wright-Patterson AFB, OH, USAF Armstrong Laboratory.

Francis, G. (2000). Designing multifunction displays: An optimization approach. *International Journal of Cognitive Ergonomics, 4*(2), 107–124.

Gentner, D., & Stevens, A. L. (1983). *Mental models*. Hillsdale, NJ: Lawrence Erlbaum.

Gilbert, G. A. (1973). Historical development of the air traffic control system. *IEEE Transactions on Communications, 21*(5), 364–375.

Gosling, G. D. (2002, January). *Analysis of factors affecting the occurrence and severity of air traffic control operational errors*. Paper presented at the 2002 TCB Annual meeting, Washington, DC.

Green, P. (1984). *Driver understanding of fuel and engine gauges*. Warrendale, PA: Society of Automotive Engineers.

Hacker, E. W. (1984). Learning from the past: A fighter pilot's obligation. Retrieved July 29, 2005, from http://www.globalsecurity.org/military/library/report/1984/HEW.htm

Hanson, R. H., Payne, D. G., Shively, R. J., & Kantowitz, B. H. (1981). *Process control simulation research in monitoring analog and digital displays*. Paper presented at the Human Factors Society 25th Annual Meeting, Rochester, NY.

Harper, R., & Hughes, J. (1991). What a F-ing system! Send 'em all to the same place and then expect us to stop 'em hitting: Making technology work in air traffic control. In G. Burton (Ed.), *Technology in working order: Studies of work, interaction and technology*. Cambridge, England, Rank: Xerox Cambridge EuroPARC.

Heglin, H. J. (1973). *NAVSHIPS display illumination design guide: Human factors*. San Diego: Naval Electronics Laboratory Center.

Hersey, M. D. (1923). Aeronautic instruments, Section I: General classification of instruments and problems including bibliography (Tech. Rep. No. 105). Langley, VA: NASA.

Hodge, M. H., & Reid, S. R. (1971). The influence of similarity between relevant and irrelevant information upon a complex identification task. *Perception and Psychophysics, 13*, 193–196.

Hughes, J. A., Randall, D., & Shapiro, D. (1992). Faltering from ethnography to design. In *Proceedings of The ACM Conference on Com-*

puter Supported Cooperative Work, Toronto, Ontario, ACM Press, pp. 115–122.

Iwasaki, S. (1996, April). Speeded digit identification under impaired perceptual awareness. Presented at Toward a Science of Consciousness 1996. Tucson, AZ, University of Arizona.

Johnson-Laird, P. N. (1983). *Mental models: Towards a cognitive science of language, inference, and consciousness.* Cambridge, MA: Harvard University Press.

Jones, D. G., & Endsley, M. R. (2000, October). *Can real-time probes provide a valid measure of situation awareness?* Paper presented at Human Performance, Situation Awareness and Automation: User Centered Design for the New Millennium Conference, Savannah, GA.

Knorr, D. (2003, December). *Free flight program performance metrics to date.* Washington, DC: Federal Aviation Administration.

Kuchar, J. K. (1996). Methodology for alerting-system performance evaluation. *Journal of Guidance, Control, and Dynamics, 19*, 438–444.

Landry, S. J., Farley, T., & Hoang, T. (2005, June). *Expanding the use of time-based metering: Multi-center traffic management advisor.* Paper presented at the 6th USA/Europe ATM 2005 R&D Seminar, Baltimore.

Lee, K. K., & Davis, T. J. (1996). The development of the final approach spacing tool (FAST): A cooperative controller-engineer design approach. *Journal of Control Engineering Practice, 4*(8), 1161–1168.

Lee, K. K., & Sanford, B. D. (1998). *Human factors assessment: The passive final approach spacing tool (pFAST) operational evaluation.* Moffett Field, CA: National Aeronautics and Space Administration.

Levy, J. L., Foyle, D. C., & McCann, R. S. (1998, October). *Performance benefits with scene-linked HUD symbology: An attentional phenomenon?* Paper presented at the 42nd Annual Meeting of the Human Factors and Ergonomics Society, Santa Monica, CA.

Ljungberg, M., & Lucas, A. (1992, September). *The OASIS air traffic management system.* Paper presented at the 2nd Pacific Rim International Conference on Artificial Intelligence, Seoul, Korea.

McGreevy, M. W., & Ellis, S. R. (1986). The effect of perspective geometry on judged direction in spatial information instruments. *Human Factors, 28*, 439–456.

McRuer, D., & Graham, D. (1965). Human pilot dynamics in compensatory systems (Tech. Rep. No. 65-5). Air Force Flight. Dynamics Lab, Wright-Patterson AFB, OH.

Mendel, M. B., & Sheridan, T. B. (1986). *Optimal combination of information from multiple sources.* Arlington, VA: Office of Naval Research.

Midkiff, A. H., & Hansman, R. J. (1993). Identification of important "party line" information elements and implications for situational awareness in the datalink environment. *Air Traffic Control Quarterly, 1*(1), 5–30.

Monty, R. A. (1973). Keeping track of sequential events: Implications for the design of displays. *Ergonomics, 16*, 443–454.

Mountford, S. J., & Somberg, B. (1981). Potential uses of two types of stereographic display systems in the airborne fire control environment. *Proceedings of the Human Factors Society, 25*, 235–239.

Murphy, M. R., McGee, L. A., Paler, E. A., Paulk, C. H., & Wempe, T. E. (1978). Simulator evaluation of three situation and guidance displays for V/STOL aircraft zero-zero landing approaches. *IEEE Transactions on Systems, Man, and Cybernetics, 8*, 19–29.

NavCanada. (2003). *SASS: Sequencing and scheduling system.* Retrieved January 27, 2005, from http://www.navcanada.ca/contentdefinition files/TechnologySolutions/products/StandAlone/sass/SASSen.pdf

Norman, D. A. (1993). *Things that make us smart.* Reading, MA: Addison-Wesley.

Olmos, O., Wickens, C. D., & Chudy, A. (1997, April). *Tactical displays for combat awareness: An examination of dimensionality and*

frame of reference concepts, and the application of cognitive engineering. Paper presented at the 9th International Syposium on Aviation Psychology, Columbus, OH, Ohio State University.

Parasuraman, R., & Riley, V. (1997). Humans and automation: Use, misuse, disuse, abuse. *Human Factors, 39*(2), 230–253.

Parasuraman, R., Sheridan, T. B., & Wickens, C. D. (2000). A model for types and levels of human interaction with automation. *IEEE Transactions on Systems, Man, and Cybernetics, 30*(3), 286–297.

Prevot, T., Shelden, S., Mercer, J., Kopardekar, P., Palmer, E., & Battiste, V. (2003, October). *ATM concept integrating trajectory-orientation and airborne separation assistance in the presence of time-based traffic flow management.* Paper presented at the 22nd Digital Avionics Systems Conference, Indianapolis, IN.

Pritchett, A. R. (2001). Reviewing the role of cockpit alerting systems. *Human Factors in Aerospace Safety, 1*(1), 5–38.

Pritchett, A. R., Hansman, R. J., & Johnson, E. N. (1995). Use of testable responses for performance-based measurement of situation awareness. In D. J. Garland & M. Endsley (Eds.), *Experimental analysis and measurement of situation awareness.* Daytona Beach, FL: Embry-Riddle Press.

Pulat, B. M., & Ayoub, M. A. (1979). A computer-aided instrument panel design procedure. *Proceedings of the Human Factors Society, 23*, 191–192.

Pylyshyn, Z. (2003). Mental imagery: In search of a theory. *Trends in Cognitive Science, 7*(3), 113–118.

Regal, D. M., Rogers, W. H., & Boucek, G. P. (1988). *Situational awareness in the commercial flight deck—definition, measurement, and enhancement.* Warrendale, PA: Society of Automotive Engineers.

Robinson, J. E. III, Davis, T. J., & Isaacson, D. R. (1997, August). *Fuzzy reasoning-based sequencing of arrival aircraft in the terminal area.* Paper presented at the AIAA Guidance, Navigation and Control Conference, New Orleans, LA.

Roscoe, S. N. (1968). Airborne displays for flight and navigation. *Human Factors, 10*(4), 321–332.

Roscoe, S. N. (1981). *Aviation psychology.* Ames: The Iowa State University Press.

Roscoe, S. N. (1997). Horizon control reversals and the graveyard spiral. *CSERIAC Gateway 7*(3), 1–4.

Sabeh, R., Jorve, W. R., & Vanderplas, J. M. (1958). Shape coding of aircraft instrument zone markings (Tech. Rep. no 57-260). Wright-Patterson AFB, USAF WADC.

Sanders, M. S., & McCormick, E. J. (1993). *Human factors in engineering and design.* New York: McGraw-Hill.

Sarter, N. B., & Woods, D. D. (1991). Situation awareness: A critical but ill-defined phenomenon. *The International Journal of Aviation Psychology, 1*(1), 45–57.

Scallen, S. F., Hancock, P. A., & Dudley, J. A. (1996). Pilot performance and preference for short cycles of automation in adaptive function allocation. *Applied Ergonomics, 26*(6), 397–403.

Schmidt, H. P., & Saint, S. P. (1969, December). *Human bottleneck in air traffic control.* Paper presented at the ASME Winter Annual Meeting, Los Angeles.

Schmidt, K., & Bannon, L. (1992). Taking CSCW seriously: Supporting articulation work. *Computer Supported Cooperative Work, 1*(1), 7–40.

Schneiderman, B. (1998). *Designing the user interface.* Cambridge, MA: Addison-Wesley.

Seidler, K., & Wickens, C. D. (1992). Distance and organization in multifunction displays. *Human Factors, 34*, 555–569.

Selcon, S. J., & Taylor, R. M. (1989). Evaluation of the situational awareness rating technique (SART) as a tool for aircrew systems design. In *Proceedings of the NATO AGARD Conference on Situational Awareness in Aerospace Operations* (AGARD-CP-478). Springfield, VA: National Technical Information Service.

Sheridan, T. B. (1992). *Telerobotics, automation, and human supervisory control.* Cambridge, MA: The MIT Press.

Sheridan, T. B. (1998). Allocating functions rationally between humans and machines. *Ergonomics in Design, 6*(3), 20–25.

Simmonds, G. R., Galer, M., & Baines, A. (1981). *Ergonomics of electronic displays.* Warrendale, PA: Society of Automotive Engineers.

Smith, K., & Hancock, P. A. (1995). Situation awareness is adaptive, externally directed consciousness. *Human Factors, 37*(1), 137–148.

Sokkappa, B. G. (1989). *The impact of metering methods on airport throughput.* McLean, VA: MITRE.

Starr, S. L., & Griesemer, J. R. (1989). Institutional ecology, "translations," and boundary objects: Amateurs and professionals in Berkeley's Museum of Vertebrate Zoology. *Social Studies of Science, 19,* 387–420.

Suchman, L. (1987). *Plans and situated actions: The problem of human-machine communication.* Cambridge, UK: Cambridge University Press.

Swenson, H. N., Hoang, T., Engelland, S., Vincent, D., Sanders, T., Sanford, B., et al. (1997 June). *Design and operational evaluation of the traffic management advisor at the Fort Worth air route traffic control center.* Paper presented at the 1st USA/Europe Air Traffic Management R&D Seminar, Saclay, France.

Taylor, R. M. (1989). Situational awareness rating technique (SART): The development of a tool for aircrew systems design. In *Proceedings of the NATO AGARD Conference on Situational Awareness in Aerospace Operations* (AGARD-CP-478). Springfield, VA: National Technical Information Service.

Tipper, S. P., & Driver, J. (1988). Negative priming between pictures and words in a selective attention task: Evidence for semantic processing of ignored stimuli. *Memory and Cognition, 16*(1), 64–70.

Vandevenne, H. F., & Andrews, J. W. (1993). Effects of metering precision and terminal controllability on runway throughput. *Air Traffic Control Quarterly, 1*(3), 277–297.

VanRullen, R., & Koch, C. (2003). Competition and selection during visual processing of natural scenes and objects. *Journal of Vision, 3*(1), 75–85.

VanRullen, R., & Thorpe, S. J. (2001). The time course of visual processing: from early perception to decision-making. *Journal of Cognitive Neuroscience, 13*(4), 454–461.

Veitengruber, J. E. (1977). Design criteria for aircraft warning, caution, and advisory alerting systems. *Journal of Aircraft, 15*(9), 574–581.

Ververs, M. P., & Wickens, C. D. (1998). Head-up displays: Effects of clutter, symbology, intensity, and display location on pilot performance. *The International Journal of Aviation Psychology, 8*(4), 377–403.

Vicente, K. J., & Rasmussen, J. (1992). Ecological interface design: Theoretical foundations. *IEEE Transactions on Systems, Man and Cybernetics, 22,* 589–606.

Vidulich, M. A. (1992, October). *Measuring situation awareness.* Paper presented at the Human Factors Society 36th Annual Meeting, Atlanta, GA.

Wegerbauer, C. (2005 May/June). The Access 5 project—Enabling direct routine UAS operations in the U.S. National Airspace System. *Unmanned Vehicles Magazine, 10*(3).

Well, A. (1971). The influence of irrelevant information on speeded classification tasks. *Perception and Psychophysics, 13,* 79–84.

Whitehurst, H. O. (1982). Screening designs used to estimate the relative effects of display factors on dial reading. *Human Factors, 24*(3), 301–310.

Whitfield, D., & Jackson, A. (1983). The air traffic controller's picture as an example of a mental model. In *Proceedings of the IFAC Conference on Analysis, Design, and Evaluation of Man-Machine System.* G. Johannsen & J. E. Rijnskorp (Eds.). London: Pergamon.

Wickens, C. (2003). Aviation displays. In P. Tsang & M. Vidulich (Eds.), *Principles and practice of aviation psychology* (pp. 147–199). Mahwah, NJ: Lawrence Erlbaum.

Wickens, C. D., & Carswell, C. M. (1995). The proximity compatibility principle: Its psychological foundation and its relevance to display design. *Human Factors, 37*(3), 473–494.

Wickens, C. D., & Prevett, T. (1995). Exploring the dimensions of egocentricity in aircraft navigation displays. *Journal of Experimental Psychology: Applied, 1*(2), 110–135.

Wickens, C. D., Fadden, S., Merwin, D., & Ververs, P. M. (1998, November). *Cognitive factors in aviation display design.* Paper presented at the 17th Digital Avionics Systems Conference, Belleview, WA.

Wickens, C. D., Mavor, A. S., Parasuraman, R., & McGee, J. P. (1998). *The future of air traffic control: Human operators and automation.* Washington, DC: National Academy Press.

Wiener, E. L., & Curry, R. E. (1980). Flight-deck automation: Promises and problems. *Ergonomics, 23*(10), 995–1011.

Wittgenstein, L. (1922). *Tractatus logico-philosophicus.* London: Routledge & Kegan Paul.

Woods, D. D. (1984). Visual momentum: A concept to improve the cognitive coupling of person and computer. *International Journal of Man-Machine Studies, 21,* 229–244.

Wright, P., Dearden, A., & Fields, B. (1998). Function allocation: A perspective from studies of work practice. *International Journal of Human Computer Studies, 52*(2), 335–355.

Zhang, J., & Norman, D. A. (1994). Representations in distributed cognitive tasks. *Cognitive Science, 18,* 87–122.

·13·

USER-CENTERED DESIGN IN GAMES

Randy J. Pagulayan, Kevin Keeker, Thomas Fuller,
Dennis Wixon, and Ramon L. Romero
Microsoft Game Studios

INTRODUCTION

The intent of this chapter is to review principles and challenges in game design improvement and evaluation and to discuss user-centered techniques that address those challenges. First, we will present why games are important, followed by the definitions and differences between games and productivity software. In the next section, we discuss the principles and challenges that are unique to the design of games. That discussion provides a framework for what we believe includes the core variables that should be measured to aid in game design testing and evaluation. The chapter will conclude with some examples of how those variables can be operationalized through methods used by the Games User Research group at Microsoft Game Studios.

WHY GAMES ARE IMPORTANT

Computers that appeared commercially in the 1950s created a technological barrier that was not easy to overcome for the greater population. Only scientists, engineers, and highly technical persons were able to use these machines (Preece, Rogers, Sharp, Benyon, Holland, & Carey, 1994). As computers became less expensive, more advanced, and more reliable, the technology that was once available only to a small group of people was now permeating itself throughout the entire population and into everyday life. In order to ease this transition, the need for a well-designed interface between the user and the technology became a necessity. Computer games come from similar origins, and potentially may head down a similar path. Some early attempts at making commercial video games failed due to unnecessarily high levels of complexity. Nolan Bushnell, cofounder of Atari, stated the issue quite succinctly: "No one wants to read an encyclopedia to play a game" (Kent, 2000, p. 28). In retrospect, some of the most successful video games were the ones that were indeed very simple.

Entertainment in general is a field of great importance to everyone. Today, the video games industry is one of the fastest growing forms of entertainment to date. According to the Entertainment Software Association (ESA), revenue from computer and console games has grown from 3.2 billion dollars in 1994 (Interactive Digital Software Association, 2000) to 7.3 billion dollars in 2004 (Entertainment Software Association, 2005). Approximately 248 million games were sold in 2004 (Entertainment Software Association, 2005). To put this into perspective, according to the ESA, 47% of Americans either purchased or planned to purchase one or more games in 2005 (Entertainment Software Association, 2005). These statistics do not even consider the international importance of video games.

Video games are quite important to the software development industry due to their roles in pushing research and technology forward. Games have acted as catalysts for hardware innovations (e.g., graphics cards, processors, and sound cards; e.g., Kroll, 2000), research on the relationships between artificial intelligence and synthetic characters or computer graphics (e.g., Funge, 2000; Laird, 2001), and physics (e.g., Hecker, 2000). These are all areas that previously were confined to engineering and computer science communities.

Players Are as Diverse as the Games That They Play

With annual sales reaching 8 billion dollars across a broad range of game genres and platforms, it is no surprise that the players are also quite diverse. Video games are no longer associated with male adolescents and young adults. According to the ESA (2005), 75% of American heads of household, over half of all Americans over the age of 9 have played, and 19% of Americans over the age of 50 years old have played an electronic game. The average age of those who play video games is 30, and 43% of Americans who play games are female (ESA, 2005). In addition, the extremely popular casual online games market is dominated by older females.

Assumptions that games are limited to a narrow market may stem from claims that the electronics industry has traditionally been marketed to boys (e.g., Chaika, 1996). Controversial beliefs about males possessing an innate proclivity towards video gaming or an innate skill advantage over females are commonly sported in casual conversation and sometimes offered by scholarly publications (e.g., Williams, 1987). While Nintendo's positioning has been demonstrated by previous marketing campaigns and promotions on such things as cereal boxes (Herz, 1997; Provenzo, 1991), genetic claims tend to be driven by speculation and stereotype.

DEFINING GAMES

As stated, the gaming industry is becoming as widespread and popular as computers and televisions. As the industry matures, games become more complex, making games difficult to define narrowly. Articulating a clear and succinct set of principles that capture the essence of games is not straightforward. However, shedding some light on the difference between games and productivity applications may help. When comparing games to productivity software, principles and methods can be successfully applied to both games and productivity applications. Regardless of the domain, the same techniques would be applied to understand a misleading button label or a confusing process model.

At the same time, there are fundamental differences between productivity applications and gaming applications. Some of the differences can be clear. Games are usually more colorful, wacky, and escapist than productivity applications, with the inclusion of interesting story lines, or animated characters.

User-centered design principles have not reached game makers to the degree that they have influenced other electronic applications. The following sections will describe some differences that have important implications for user research on games.

The Goals Are Different

At their roots, productivity applications are tools. Similarly, the design intentions behind productivity applications are to make task performance easier, quicker, less error prone, more quality driven, and accessible to populations of interest. In this sense, a word processor or spreadsheet differs only from a powered woodworking tool in terms of its complexity and domain of application. One can argue that better word processors allow a wider set of people to make documents better, faster, and easier and with fewer errors. Similarly, a power mitre box cuts wood with angles that are more precise, faster, and more reliable than a hand mitre box or a hand saw. The focus of design and usability is to produce an improved product or result.

Games at their roots are different. Games are intended to be a pleasure to play. Ultimately, games are like movies, literature, and other forms of entertainment. They exist in order to stimulate thinking and feeling. Their outcomes are more experiential than tangible. This is not to say that word processors or other tools cannot be a pleasure to use or that people do not think or feel when using them, but that is rarely their primary design intention. In a good game, both the outcomes and the journeys are rewarding. This fundamental difference leads us to devote more of our effort to measuring perceptions, while productivity applications focus more on task completion.

The Rules Are in the Game World

Games create an artificial world with artificial objectives. Capturing the king has a chess-specific meaning that is different from the precise meaning outside of the chess world. In some sense, this simplifies the problem of the game designer and of the usability practitioner. User goals are almost completely defined by the game designer. The resulting concern is to ensure that user interpretations match the game design's expectations (or differ in a creative and entertaining manner).

Industry Competition Is Intense

Competition within the games industry is more intense than other software domains. Games compete with each other and many other forms of entertainment for your attention. Pretend that you have to write a book chapter. For that, there are a limited number of viable tools: handwriting, dictation, a typewriter, or a limited set of software applications. However, imagine that you have a few hours to kill and that you want to avoid productivity at all costs. You could consult the *New York Times* crossword puzzle, argue politics with your best friend, watch a mystery film, or stock up with health potions and attack demons in *Diablo* (2000).

Games Must Challenge

Perhaps the most important distinction between games and productivity applications is the complex relationship between challenge, consistency, and frustration. To illustrate this difference, consider the following example. It has been said that the easiest game to use would consist of one button labeled "Push." When you push it, the display would say, "YOU WIN." This game would have few, if any usability issues. However, it would not be fun either. Games deliberately impose constraints, whereas productivity applications attempt to remove them. Games deliberately evolve, whereas productivity applications strive for consistency. At their hearts, the fun in most games comes from

learning how to overcome ever-increasing challenges. Ironically, some educators, practitioners, and researchers think it comes from fantasy and bright colors.

Many of the characteristics mentioned above are not unique to games. All of these characteristics are relevant to the discussion of the ways in which user research must adapt to be useful for game development. Later we will discuss how these differences create particular challenges for both the design and evaluation of games.

TYPES OF GAMES

In the following sections, we review some common gaming platforms and game types.

Platforms

One of the simplest classifications of games is by the platform or hardware on which they are played. Cassell and Jenkins (2000) differentiated between games played on a PC and those played on a console. Crawford (1982) divided games into even finer categories; arcade/coin operated, console, PC, mainframe, and handheld. Different gaming platforms can be differentiated by their technical capabilities, physical control interface, visual interface, and the context in which they are played.

PC. A useful distinction can be made between PC games that are normally acquired through retail outlets on CDs, persistent world games in which much of the content lives on the Internet, and casual, web-based games. While there are many technical differences between these kinds of games, the important distinction is the business model and level of investment required by the players. Most retail games need to rely heavily on flash, reputation, and recognizable novelty to attract users to a relatively large investment. Continued investment in the game is useful only to build a reputation or to convince customers with higher thresholds that they will get their moneys' worth from the game. Massively multiplayer online persistent worlds (MMOs) require continued user investment to obtain monthly subscription fees. The long learning curves and reinforcement schedules in MMOs are tailored to make players invest more time, money, and effort for long-term rewards. Many casual gamers are attracted to free, familiar games. For these casual users, gameplay is squeezed between activities that are more important to them. As a result, games with very little learning investment are very appealing to casual gamers. Removing penalties for setting aside the game (for minutes, hours, days, or forever) can mean the difference between successful and unsuccessful casual games (see the casual games section for more details).

Console. For user research purposes, the most important unique characteristic of console video game systems is the advantage of a fixed set of hardware. By contrast with PC games, there is very little game setup, minimal maintenance efforts (e.g., few software patches, video card conflicts, etc.), and a consistent set of input device capabilities. Some game genres are more popular on console than PC. This difference in popularity can usually be attributed to differences in the input devices that are typically associated with each platform. It is hard to enter text using a gamepad. Most popular MMOs are still published for PCs in spite of the fact that fast Internet connections and hard drives are becoming more popular additions to console systems. This may change as it becomes easier to attach other devices such as keyboards or voice-command hardware that would make communication more robust.

More recently, casual games are also available for all console systems, including Xbox 360, where consumers can download, try, and purchase casual games directly from the console via a broadband connection.

Portable devices. Portable devices include portable gaming systems (handhelds), mobile phones, and personal digital assistants (PDAs). Handheld devices are a standardized platform, with screens and controls designed primarily for gaming. Mobile phones on the other hand provide greater challenges to creating meaningful gameplay experience, including limited screen resolution and small controls (Hyman, 2005). Mobile games generally must support games that can be played in shorter blocks of time and that can work well with the limitations of the handsets. The unlimited combination of handset controls, processor and network capabilities, and runtime environments create even more challenges to developing quality games (IDGA, 2005; Hyman 2005).

Consumers primarily view mobile phones as communication devices, which are not meant to replace the gaming experience available on handhelds such as the Nintendo DS and Sony PSP (Moore & Rutter, 2004). Unlike handheld gaming devices, mobile phones are generally carried at all times. This creates an opportunity to provide gaming at any time, anywhere (Hyman, 2005; IDGA, 2005). However, with mobility comes the greater need to design for time, location, social, emotional, and motivational context (Sidel & Mayhew, 2003).

Cross platform. Cross-platform convergence is an emerging trend in the games industry that complicates the classification of games by platform and that creates new challenges in user-centered game design. Developers are now creating multiple versions of a game that interact with one another across platforms. For example, Nintendo games such as *Animal Crossing* (2002), *The Legend of Zelda: The Wind Waker* (2003), and *Harvest Moon* (2003) allow users to connect a Game Boy portable device to the GameCube console to access additional features and content or to use the Nintendo DS as a touch screen controller. Another example of cross-platform games would be a title where the player may execute specific planning or strategic tasks on a mobile device at any time, then have those actions affect the full game experience the next time it is played on a console or PC. This cross-platform design effect has been referred to as a reciprocation effect (Yuen, 2005).

The next generation of cross-platform experiences may involve matching players for real time or asynchronous competitive and co-operate multiplayer experiences regardless of what platform they are on (portable, PC, console, etc.). As these trends become realized, the promise of interesting challenges and opportunities for game developers continues to increase.

Game Types

Retail games (NPD classifications). The NPD Group (a marketing company) uses a fine-grained classification scheme for game type that is referred to quite often in the games industry. They offer the following classes and subclasses as seen in Table 13.1. Games can also be categorized in a more granular fashion by splitting each genre up into a subgenre. For instance, action games can be divided into Action Driving Hybrid (*True Crime, GTA3,* etc.), Action Combat (*Max Payne, Tomb Raider,* etc.), and Platformer (*Jak and Daxter, Super Mario Sunshine,* etc). In general, the genres represent types of gameplay mechanics, themes, or content.

Casual games. Casual games are a rapidly growing segment of the video games market, producing an estimated $500 million in online sales in 2005 (IDG Entertainment, 2005). Casual games as a whole are appealing to a broad range of users, can be found on all platforms, and represent several genres.

TABLE 13.1 The NPD Group Super Genre Classification Scheme for Games

Category	Description
Action	Control a character and achieve one or more objectives with that character.
Fighting	Defeat an opponent in virtual physical hand-to-hand or short-range combat.
Racing	Complete a course before others do and/or accumulate more points than others while completing a course.
Shooter	Goal is to defeat enemies in combat with ranged weapons (first-person shooters, third-person shooters).
Strategy	Strategically manage resources to develop and control a complex system in order to defeat an opponent or achieve a goal. The goal is to defeat opponent(s) using a large and sophisticated array of elements.
Role-Playing	Control a character that is assuming a particular role in order to achieve a goal or mission. Rich story and character development are common.
Family entertainment	The primary objective is to interact with others and/or to solve problems. This genre includes puzzles and parlor games.
Children's entertainment	Same as family entertainment, but geared to a younger audience.
Sports	Manage a team or control players, to either win a game or develop a team to championship status. Involves individual, team, and extreme sports.
Adventure	Control a character to complete a series of puzzles that are required to achieve a final goal.
Arcade	Games on coin-operated arcade machines, or games that have similar qualities to classic arcade games. Generally, they are fast paced, action games.
Flight	Plan flights and pilot an aircraft in a realistic, simulated environment.
All other games	Educational, compilations, non flight simulators, rhythm games, etc.

For these reasons, they do not fit into either of the previous categorization schema. Casual games generally meet the following requirements: (a) are easy to start and control; (b) can be enjoyed in small time intervals; (c) do not require an extensive investment in time to enjoy or progress, and therefore usually do not have deep linear story lines; and (d) have discreet goals and rules that are intuitive or easy to learn. Casual games often have significantly lower development costs, can be distributed digitally (or via physical media), offer trial versions and lower retail prices (or are free), and utilize lower system requirements and storage space. These requirements and attributes make these games well suited for platforms such as PC distributed via the web and mobile devices, where they are most common.

Websites such as Real Arcade, Pogo.com, Yahoo! Games, and MSN Games provide free online casual games, pay per play casual games, subscription services, downloadable games that include free trials, and games that can be purchased for $5 to $20. Mobile phone carriers and a variety of websites provide casual games for mobile devices. Although there are feature rich, 3-D noncasual games available for mobile devices, the top selling titles are casual games such as *Tetris, Bejeweled, Bowling*, and *Pacman* (Hyman, 2005).

PRINCIPLES AND CHALLENGES OF GAME DESIGN

Having differentiated games from other applications, we can look at some of the unique issues in game design and evaluation.

Identifying the Right Kind of Challenges

Games are supposed to be challenging. This requires a clear understanding of the difference between good challenges and frustrating usability problems. Most productivity tools struggle to find a balance between providing a powerful enough tool set for the expert and a gradual enough learning curve for the novice. However, no designer consciously chooses to make a productivity tool more difficult. The ideal tool enables users to experience challenge only in terms of expressing their own creativity. For games, learning the goals, strategies, and tactics to succeed is part of the fun.

Unfortunately, it is not always clear which tasks should be intuitive (e.g., easy to use) and which ones should be challenging. Input from users becomes necessary to distinguish good challenges from incomprehensible design. Take a driving game for example. It is not fun having difficulty making your car move forward or turn. However, learning to drive is still a fundamental part of the challenge in the game. While all cars should use the same basic mechanisms, varying the ways that certain cars respond under certain circumstances may be fun. It should be challenging to identify the best car to use on an icy, oval track as opposed to a rally-racing track in the middle of the desert. The challenge level in a game must gradually increase in order to maintain the interest of the player.

Addressing Different Skill Levels

Unfortunately, not all players start from the same place in terms of gaming experience or talent. Obviously, frequent failure can be a turn off. Success that comes too easily can also become repetitive. Games must address the problem of meeting all players with the correct level of challenge. Tuning a game to the right challenge level is called "game balancing."

There are many ways to balance the difficulty of the game. The most obvious way is to let players choose the difficulty themselves. Many games offer the choice of an easy, medium, or hard difficulty level. While this seems like a simple solution, it is not simple to identify exactly how easy the easiest level should be. Players want to win, but they do not want to be patronized. Too easy is boring and too hard is unfair. Either experience can make a person cease playing.

Another approach to varying skill levels is to require explicit instruction that helps all users become skilled in the game. You might imagine a tutorial in which a professional golfer starts by explaining how to hit the ball and ends by giving instruction on how to shoot out of a sand trap onto a difficult putting green. Instruction, however, need not be presented in a tutorial. It could be as direct as automatically selecting the appropriate golf club to use in a particular situation with no input from the user, similar to the notion of an adaptive interface, where the interface provides the right information at the right time.

The environments, characters, and objects in a game provide another possibility for self-regulation. Most games will offer players some choices regarding their identities, their opponents, and their environments. The better games will provide a variety of choices that allow users to regulate the difficulty of their first experiences. With learning in mind, it is not uncommon for novice players to choose champion football teams to play against weak opponents. As long as players can distinguish the good teams from the bad ones, and the teams are balanced appropriately, users will be able to manage their own challenge levels.

Some games take it even further by identifying the skill level of the player and regulating the instruction-level appropriately. In this situation, instruction can be tuned to the skill level of the player by associating it with key behavioral indicators that signify that the player is having difficulty. If the game does not detect a problem, it does not have to waste the player's time with those instructions. In *Halo 2* (2004), the game detects difficulties that a player may have with certain tasks. For example, to get into a vehicle, the player must press and hold the X button on his or her controller when standing next to the driver's seat of that vehicle. If the player is standing in the right position, but taps the X button repeatedly (instead of holding the button down), the game will present a more explicit instruction to press and hold the button. This is just one of many dynamic instructions that appear throughout *Halo 2* based on behavioral indicators of player difficulty.

Productivity tools have implemented similar problem-identification features, but often with mixed success due to the open nature of tasks in most productivity applications. Good game tutorials have succeeded by setting clear goals and completely constraining the environment. Doing so focuses the user on the specific skill and simplifies the detection of problem behavior. Other lessons from game tutorial design will be described in later sections of this chapter.

Another in-game approach to auto-regulating the difficulty level requires adjusting the actual challenge level of the opponents during the game. Evaluating the success of the player and adjusting the opponent difficulty during the game is often called "dynamic difficulty adjustment," or "rubber banding." When players perform very skillfully, their performances are moderated by computer-generated bad luck and enhanced opponent attributes. In a football game, the likelihood of fumbling, throwing an interception, or being sacked may increase as the player increases his or her lead over their opponent. Even though this may seem like a good solution, there can be a downside. Most people would prefer to play a competitive game (and win) than to constantly trounce a less-skilled opponent. However, overdeveloped rubber banding can cheat a skilled player out of the crucial feeling of mastery over the game.

A final approach focuses on providing tools that maximize the ability of the trailing player to catch up with the leading player. The key for the game designer is to think of ways to maintain challenge, reward, and progress for the unskilled player without severely hampering the skilled player. One interesting and explicit example of this is found in *Diddy Kong Racing* (1997). In this game, the racer can collect bananas along the roadway. Each banana increases the top speed of your car. The player can also collect missiles to fire forward at the leading cars. Each time you hit a car with a missile, it not only slows the car's progress, but it jars loose several bananas that the trailing player can pick up. Thus, trailing players have tools that they can use to catch the leaders even if the leaders are not making any driving mistakes. The chief distinction between this and dynamic difficulty adjustment is that the game is not modifying skills based on success. Instead, the rules of the game provide the trailing player with known advantages over the leader.

Players Must Be Rewarded Appropriately

Explicit or slow reinforcement schedules may cause users to lose motivation and quit playing a game. Because playing a game is voluntary, games need to quickly grab the user's attention and keep him or her motivated to come back repeatedly. One way to accomplish this is to reward players for continued play. Theories of positive reinforcement suggest behaviors that lead to positive consequences tend to be repeated. Thus, it makes sense that positive reinforcement can be closely tied to one's motivation to continue playing a game. However, it is less clear which types of reinforcement schedules are most effective.

Although the model should not necessarily be used for all games, research suggests that continuous reinforcement schedules can establish desired behaviors in the quickest amount of time (Steers & Porter, 1991). Unfortunately, once continuous reinforcement is removed, desired behaviors extinguish very quickly. Use of partial reinforcement schedules takes longer to extinguish desired behaviors but may take too long to capture the interest of gamers. Research suggests that variable ratio schedules are the most effective in sustaining desired behaviors (Jablonsky & DeVries, 1972). This kind of schedule is a staple of casino gambling games, in which a reward is presented after a variable number of desired responses. Overall, there is no clear answer. Creating a game that establishes immediate and

continued motivation to continue playing over long periods is a very complex issue.

Another facet of reinforcement systems that may impact enjoyment of a game is whether the player attributes the fact that they have been playing a game to extrinsic or intrinsic motivations. Intrinsic explanations for behavior postulate that the motivators to perform the behavior come from the personal needs and desires of the person performing the behavior, whereas extrinsically motivated behaviors are those that people perform in order to gain a reward from or please other people. In research on children's self-perceptions and motivations, Lepper, Greene, and Nisbett (1973) discovered that children who were given extrinsic rewards for drawing were less likely to continue drawing than those who had only an intrinsic desire to draw. The conclusion that they drew is that children perceived their motivations to draw as coming from extrinsic sources and thus discounted their self-perceptions that they liked to draw.

The same may be true of reward systems in games (Lepper & Malone, 1987; Malone, 1981). To a certain degree, all reinforcement systems in games are extrinsic because they are created or enabled by game developers. Some reward systems are more obviously extrinsic than others are. For instance, imagine the following rewards that could be associated with combat in a fantasy role-playing game (RPG). The player who slays a dragon with the perfect combination of spell casting and swordplay may acquire the golden treasure that the dragon was hoarding. In this situation, the personal satisfaction comes from being powerful enough to win and smart enough to choose the correct tactics. The gold is an extrinsic motivator. The satisfaction is intrinsic. By analogy from Lepper, Greene, and Nisbett's (1973) research, feelings of being powerful and smart (intrinsic motivators) are more likely to keep people playing than extrinsic rewards.

Collecting and Completing

The chief goal of many games is to acquire all of the available items, rewards, and knowledge contained in the game. In games such as *Pokemon Crystal* (2000), the central challenge is to acquire as many of the *Pokemon* characters as you can and learn all of their skills well enough to outsmart your opponent at selecting the right characters for a head-to-head competition. Not coincidentally, the catch phrase for the *Pokemon Crystal* game is "Gotta catch 'em all!"

This game mechanic is also used by numerous games to add depth and repeat play. Though this is not the primary mechanic in *Madden NFL 06* (2005), the ability to collect electronic player cards and use them strategically in games provides incentive for gamers to experiment with much of the content that they may not experience if playing through a standard season.

Story

Characters and narrative help gamers attach meaning and significance to action sequences. Those in the games industry propose that many games neither have nor need a story. It is our contention that the key to understanding narrative in games is to realize that story lines may be both embedded in the game, or they may emerge in the course of playing a game.

When most consumers think about story, they think about embedded story lines (Levine, 2001). *Final Fantasy X* (2001) tells its story by cutting action sequences with a series of full-motion cut scenes, real-time cut scenes, and player-driven character interactions. The embedded story forms a central part of the appeal of the game. Levine (2001) pointed out that much of the narrative in a game emerges in the successes and failures experienced by players throughout the course of the game. This is especially true of multiplayer games, in which story is often generated exclusively by the interactions of the participants. As Levine described it, the story is generated by replaying an abstract narrative structure with a strict set of rules and possibilities within a novel set of circumstances. Sporting events both within and outside of the video game world provide an excellent example of this type of narrative. No author scripted the result of the last World Cup tournament, but each such event has the potential to be an epic narrative for both participants and viewers.

Technological Innovation Drives Design

There is a great deal of pressure on designers to utilize new technologies that may break old interaction models. The desire to experience ever more realistic and imaginative games has pushed game developers into engineering and computer-science research domains. Likewise, technology often drives game design in order to showcase new capabilities. The constant demand for novelty can be strong enough incentive for game makers to try untried designs, "spruce up" familiar interfaces, and break rules of consistency. For example, the original *NFL2K* series (1999) sported a new interface model in which users selected interface areas by holding the thumbstick in a direction while pressing a button. It is possible that Sega chose the new design primarily because it was new and different from existing interfaces. It required the somewhat new (at that time) mechanics of the thumbstick on the controller, it minimized movement because one could point directly at any given item in the menu, and it was cleverly shaped like a football (which made more sense for *NFL2K* than for another sport). However, errors due to more error-prone targeting and the hold and click interaction metaphor made the system harder for first-time players to use.

Perceptual-Motor Skill Requirements

The way that functions are mapped onto available input devices can determine the success or failure of a game. A crucial part of the fun in many games comes from performing complex perceptual-motor tasks. While today's arcade-style games include increasingly more sophisticated strategic elements, a core element of many games is providing the ability to perform extremely dexterous, yet satisfying physical behaviors. These behaviors are usually quick and well-timed responses to changes and threats in the environment. If the controls are simple enough to master and the challenges increase at a reasonable difficulty, these mostly physical responses can be extremely satisfying (Csikszentmihalyi, 1990).

Problems can arise when games require unfamiliar input devices. This is a common complication in console game usability research because new console systems usually introduce new input device designs unique to that system (see Fig. 13.1). Furthermore, game designers often experiment with new methods for mapping the features of their games to the unique technical requirements and opportunities of new input devices. Unfortunately, this does not always result in a better gameplay experience.

Balancing Multiplayer Games

As we have seen, different strategies can be used to support a broad range of player skill levels. While this is complicated in single-player games, it becomes even more daunting when players of different skills make up both the opponents and the allies. By far the most common strategy for regulating the challenge level of online multiplayer games is to provide strong matchmaking tools. *Internet Backgammon* (2001) automatically matches players from around the world based on their self-reported skill level. Other games use algorithms that count player winning percentages and strength of opponent.

Another strategy seeks to solve skill balance problems outside of the game by offering players a wide array of arenas and game types. Rather than actively connecting players of like skill, this approach provides a broad array of places to play and allows players to self-select into game types that suit their style and attract the players with whom they want to associate. A final approach, used frequently by instant messaging clients (e.g., Yahoo IM, MSN Messenger), is to make it easy to start a game with friends. Though skill levels of friends may not always match, you are presumably less likely to perceive the match as unfair against people that you know.

Game designers also employ a variety of in-game strategies to balance out the competition. The most common way to allow less skilled players to compete effectively with players that are more skilled is to play team games. Many games allow players to

FIGURE 13.1. This is a sample of different types of input devices used in games, from the traditional keyboard and mouse, to different console gamepads.

take on a variety of roles. *Capture the Flag* is a common back-yard tag-style game that forms the basis for a game type in many first-person shooters. Some players go on offense to capture the opponent's flag, some stay back to defend their flag. Likewise, some players may take a long-range sniper weapon to frustrate the enemy, while others take weapons that are more effective at close range. The defending and sniping roles can be more comfortable for some novice players because they allow the player to seek protective cover, they require less knowledge of the play field, and one-on-one combat can be avoided.

Some first-person shooters also allow the game host to set handicaps for successful players and bonuses for less successful players. For example, one version of *Unreal Tournament* increased the size of the player's character with every point that they won, and decreased the size of the character every time that the character died. The size differences made successful players easier targets and unsuccessful players more difficult targets.

Enabling Social Networking

Skill level is not the only dimension of importance. Recent systems, such as the Xbox Live service, seek to group players of similar attitudes and behaviors. The goal is to place more antagonistic players with like-minded others who appreciate tough talk, while protecting players who do not want to or should not be exposed to aggressive or offensive language.

This is just one of many tools that games have developed to promote group play and improve communication between online players. Most online games include some form of in-game messaging. Often this messaging is tailored for fast and efficient communication of key functions or timely game events. Most massively multiplayer online RPGs have negotiation systems that help people trade objects safely and efficiently. Most first-person shooters present automatic messages telling you which players have just been killed by whom. Real-time strategy (RTS) games allow gamers to set flares or markers on the map to notify their allies of key positions. Each game genre has a key set of in-game communication tools to support the game play.

More frequently, game designers are starting to embed cooperative tools into the environment itself. Much of the benefit of vehicles in first-person shooting games comes from their use as cooperative tools. Used in concert, these tools can often be extremely effective. From a game design perspective, they provide a great incentive for people to come together, strategize, and work co-operatively. These shared successes can be a huge part of the fun in online games.

Massively multiplayer games employ a wide variety of incentives for players to form groups. Success comes from taking on missions that are far too dangerous for any single player to accomplish on their own. Players choose a role and learn to complement the strengths of the other members of their group in order to overcome fearful enemies and accomplish great quests. Recent massively multiplayer games have invested even more in the creation of large-scale guilds. Most massively multiplayer games intentionally keep large areas of information secret from their players—to provide the need and the opportunity for symbiotic relationships between experts and novices. In addition, modern MMOs often provide pyramid-scheme style bonuses in

wealth and experience to the leaders and captains of guilds. In return, members receive physical, material, educational, and social protection from their leaders and peers.

IMPORTANT FACTORS IN GAME TESTING AND EVALUATION

Most game genres are subtly different in the experiences that they provoke. It may seem obvious that the point of game design is making a fun game. Some games are so fun that people will travel thousands of miles and spend enormous amounts of money to participate in gaming events. However, we would like to propose a potentially controversial assertion. That the fundamental appeal of some games lies in their ability to challenge, to teach, to bring people together, or simply to experience unusual phenomena. Likewise, the definition of fun may be different for every person. When you play simulations games, your ultimate reward may be a combination of learning and mastery. When you play something like the *MTV Music Generator* (Codemasters, 1999), your ultimate reward is the creation of something new. When you go online to play card games with your uncle in Texas, you get to feel connected. *Flight Simulator 2000* (2000) lets people understand and simulate experiences that they always wished they could have. While these may be subcomponents of fun in many cases, there may be times when using "fun" as a synonym for overall quality will lead to underestimations of the quality of a game. While a fun game is often synonymous with a good game, researchers are warned to wisely consider which measures best suit the evaluation of each game that they evaluate.

Game Designer Intent

As stated earlier, the goal of a game is to create a pleasurable experience. Goals in a game are not necessarily derived from external user needs as in productivity applications. In games, goals are defined in accordance with the game designer's vision, which is a novel position because historically, success has been defined by the accomplishment of user tasks and goals (Pagulayan, Gunn, Romero, 2006). When approaching a game for user-centered design or testing, it is best to assume the role of facilitating the designer's vision for the game (Pagulayan & Steury, 2004; Pagulayan, Steury, Fulton, & Romero, 2003) because many times, only the designer can recognize when the player experience is not being experienced as intended. In traditional usability testing, it is often very recognizable when there is user error, but not so in games.

Davis, Steury, and Pagulayan (2005) discussed a case study in the game *Brute Force* which revealed that players were not encountering certain gameplay features early enough in their gameplay experience. The players were not failing, but players were taking much longer to play through the second mission than intended. By understanding the design vision, the authors were able to work with the designers to provide feedback that resulted in shortening the second mission and in reordering other missions to match the design intent of the game.

Ease of Use

The ease of use of a game's controls and interface is closely related to fun ratings for that game. Think of this factor as a gatekeeper on the fun of the game. If the user must struggle or cannot adequately translate their intentions into in-game behaviors, they will become frustrated. This frustration can lead the user to perceive the game as being unfair or simply inaccessible (or simply not fun). Thus, it becomes very clear why usability becomes very important in games. Ease of use should be evaluated with both usability and attitude-measurement methodologies, which are discussed later in the chapter.

Starting a game. Starting the kind of game that the user wants is an easily definable task with visible criteria for success. This is something one can measure in typical usability laboratories. Though designers often take game shell (the interface used to start the game) design for granted, a difficult or confusing game shell can limit users' discoveries of features and impede their progress toward enjoying the game. The immediate concern for users can be starting the kind of game that they want to play. Games often provide several modes of play. When the game shell is difficult to navigate, users may become frustrated before they have even begun the game. For example, we have found that many users are unable to use one of the common methods that sports console games use to assign a game controller to a particular team. This has resulted in many users mistakenly starting a computer versus computer game. Depending on the feedback in the in-game interface, users may think that they are playing when, in fact, the computer is playing against itself. In these cases users may even press buttons, develop incorrect theories about how to play the game, and become increasingly confused and frustrated with the game controls. The most effective way to avoid these problems is to identify key user tasks and test their usability.

Pagulayan, Steury, Fulton, et al. (2003) discussed a case study where they found issues with difficulty settings in the game shell. In early usability testing, participants were having problems with setting the difficulty level of opponents in *Combat Flight Simulator* (2000). This is a case where the users' gameplay experiences would have been quite frustrating because of a usability error in the game shell if it were not addressed.

Tutorials or instructional gameplay. As mentioned earlier, tutorials are sometimes necessary to introduce basic skills in order to play the game. In this situation, instructional goals are easily translated into the usability labs with comprehension tasks and error rates.

One of the risks of not testing tutorials or instructional missions is inappropriate pacing, which can often result from an ill-conceived learning curve at the start of the game. Many games simply start out at too difficult a challenge level. This is an easy and predictable trap for designers and development teams to fall into because when designers spend months (or even years) developing a game, they risk losing track of the skill level of the new player. A level that is challenging to the development team is likely to be daunting to the beginner.

Unfortunately, the reverse can also be troubling to the new user. Faced with the task of addressing new players, designers

may resort to lengthy explanations. Frequently, developers will not budget time to build a ramped learning process into their game. Instead, they may realize late in the development cycle that they need to provide instruction. If this is done too abruptly, the learning process can end up being mostly explanation, and to be frank, explanation is boring. The last thing that you want to do is to bore your user with a longwinded explanation of what they are supposed to do when they get into your game. It is best to learn in context and at a measured pace or users may just quit the game.

A very positive example is the first level of *Banjo Kazooie* (2000). At the start, the player is forced to encounter a helpful tutor and listen to a few basic objectives. Then, they must complete some very basic objectives that teach some of the basic character abilities. Much of the tutorial dialogue may be skipped, but the skills necessary to continue must be demonstrated. In this way, the game teaches new skills but never requires tedious instruction. The player learns primarily by doing. All of this is done in the shadow of a very visible path onto the rest of the game so the user never loses sight of where they need to go.

In-game interfaces. In-game interfaces are used primarily to deliver necessary status feedback and to perform less-frequent functions. We measure effectiveness with more traditional lab usability testing techniques and desirability with attitude measurements such as surveys (see next section for example).

Some PC games make extensive use of in-game interfaces to control the game. For example, simulation and RTS games can be controlled by keyboard and mouse presses on interface elements in the game. Usability improvements in these interfaces can broaden the audience for a game by making controls more intuitive and reducing tedious aspects of managing the game play. In-game tutorial feedback can make the difference between confusion and quick progression in learning the basic mechanisms for playing. In this situation, iterative usability evaluations become a key methodology for identifying problems and testing their effectiveness (see next section for example).

Many complex PC and console video games make frequent use of in-game feedback and heads-up displays (HUD) to display unit capabilities and status. For example, most flight combat games provide vital feedback about weapons systems and navigation via in-game displays. Without this feedback, it can be difficult to determine distance and progress toward objectives, unit health, and attack success. This feedback is crucial for player learning and satisfaction with the game. With increasing game complexity and 3-D movement capabilities, these displays have become a crucial part of many game genres. Usability testing is required to establish whether users can detect and correctly identify these feedback systems (see Pagulayan, Steury, Fulton, et al., 2003, for detailed example of usability testing the HUD in *MechWarrior 4: Vengeance,* 2000).

Mapping input devices to functions. A learnable mapping of buttons, keys, or other input mechanisms to functions is crucial for enjoying games. We measure effectiveness with usability techniques and desirability with attitude measurements. Without learnable and intuitive controls, the user will make frequent mistakes translating their desires into onscreen actions. We have seen consistently that these kinds of mistakes are enormously frustrating to users, because learning to communicate one's desires through an eight-button input device is not very fun. The selection of keys, buttons, and other input mechanisms to activate particular features is often called "control mapping." Players tend to feel that learning the control mapping is the most basic part of learning the game. It is a stepping-stone to getting to the fun tasks of avoiding obstacles, developing strategies, and blowing things up.

By contrast to other ease of use issues, evaluating the control mapping may involve as much subjective measurement as behavioral observation. Button presses are fast, frequent, and hard to collect automatically in many circumstances. Furthermore, problems with control mappings may not manifest themselves as visible impediments to progress, performance, or task time. Instead, they may directly influence perceptions of enjoyment, control, confidence, or comfort. Due to differences in experience levels and preferences between participants, there may also be significant variation in attitudes about how to map the controls.

Dissatisfaction with the controller design can also be a central factor that limits enjoyment of all games on a system. For example, the results of one set of studies on the games for a particular console game system were heavily influenced by complaints about the system's controller. Grasping the controller firmly was difficult because users' fingers were bunched up and wrists were angled uncomfortably during game play. Ratings of the overall quality of the games were heavily influenced by the controller rather than the quality of the game itself.

Because of these concerns and the importance of optimizing control mappings, we recommend testing control mappings with both usability and attitude-assessment methodologies.

Challenge

Challenge is distinct from ease of use and is measured almost exclusively with attitude-assessment methodologies. This can be a critical factor to the enjoyment of a game, and obviously can be highly individualized and is rightly considered subjective.

Consumers may have difficulties distinguishing the appropriate kinds of challenge that result from calculated level and obstacle design, from the difficulty that is imposed by inscrutable interface elements or poor communication of objectives. In either case, the result is the same. If not designed properly, the player's experience will be poor. Thus, it is up to the user-testing professional to make measurement instruments that evaluate the appropriateness of the challenge level independent of usability concerns. In one example, Pagulayan, Steury, Fulton, et al. (2003) used attitude-assessment methodologies to determine the final design of the career mode in *RalliSport Challenge* (2002). In this situation, finding a solution to the design problem was not necessarily related to ease of use issues, or other usability-related issues. The final design was based on what was most fun and appropriately challenging for users.

Pace

We define pace as the rate at which players experience new challenges and novel game details. We measure this with attitude-measurement methodologies.

Most designers will recognize that appropriate pacing is required to maintain appropriate levels of challenge and tension throughout the game. You might think of this as the sequence of obstacles and rewards that are presented from the start of the game to the end. However, the way a designer will address pace will depend on a variety of issues, including game type, game genre, and their particular vision for the gameplay experience. In a tennis game, pace can be affected by a number of things, including the number of cut-scenes in between each point, to the actual player and ball movement speed (Pagulayan & Steury, 2004).

One group at Microsoft uses a critical juncture analogy to describe pacing. As a metaphor, they suggest that the designer must attend to keeping the user's attention at 10 seconds, 10 minutes, 10 hours, and 100 hours. The player can always put down the game and play another one, so one must think creatively about giving the user a great experience at these critical junctures. Some games excel at certain points but not others. For example, the massively multiplayer game may have the user's rapt attention at 10 seconds and 10 minutes. The fact that hundreds of thousands pay $10 per month to continue playing indicates that these games are very rewarding at the 100-hour mark. Anyone who has played one of these games can tell you that they are extremely difficult and not too fun to play at the 10-hour mark. At this point, you are still being killed repeatedly. That is no fun at all. Arcade games obviously take this approach very seriously. Though they may not scale to 100 hours, good arcade games attract you to drop a quarter and keep you playing for long enough to make you want to spend another quarter to continue.

Pacing may also be expressed as a set of interwoven objectives much like the subplots of a movie. Again, *Banjo Kazooie* (2000) provides an excellent example of good pacing. Each level in *Banjo Kazooie* contains the tools necessary to complete the major objectives. Finding the tools is an important part of the game. New abilities, objectives, skills, and insights are gradually introduced as the player matures. While progressing toward the ultimate goal (of vanquishing the evil witch and saving the protagonist's sister), the player learns to collect environmental objects that enable them to fly, shoot, become invincible, change shape, gain stamina, add extra lives, and unlock new levels. This interlocking set of objectives keeps the game interesting and rewarding. Even if one is unable to achieve a particular goal, there are always sets of sub goals to work on, some of which may provide cues about how to achieve the major goal.

Summary

Attitude methodologies are better apt to measure factors such as overall fun, graphics, sound, challenge, and pace. The typical iterative usability is an exploratory exercise designed to uncover problem areas where the designers intentions do not match the users expectations; as a result, we typically choose not to use a usability test to assess fun or challenge. When attempting to assess attitudinal issues as overall fun and challenge, we make use of a survey technique that affords testing larger samples. Internally, we have adopted the term *Playtest* or sometimes *Consumer Playtest* for this technique. At the same time, we use typical iterative usability methods to determine design elements that contribute to or detract from the experience of fun.

USER RESEARCH IN GAMES

Introduction to Methods—Principles in Practice

In the following section, we propose various methodologies and techniques that attempt to accurately measure and improve game usability and enjoyment. Many of the examples are taken from techniques used and developed by the Games User Research group at Microsoft Game Studios.

Our testing methods can be organized by the type of data being measured. At the most basic level, we categorize our data into two types: behavioral and attitudinal. Behavioral refers to observable data based on performance, or particular actions performed by a participant that one can measure. This is very similar to typical measures taken in usability tests (e.g., time it takes to complete a task, number of attempts it takes to successfully complete a task, task completion, etc.). Attitudinal refers to data that represent participant opinions or views, such as subjective ratings from questionnaires or surveys. These are often used to quantify user experiences. Selection of a particular method will depend on what variables are being measured and what questions need to be answered.

Another distinction that is typically made is between formative and summative evaluations, which we apply to our testing methods as well. Formative refers to testing done on our own products in development. Summative evaluations are benchmark evaluations, done either on our own products or on competitor products. It can be a useful tool for defining metrics or measurable attributes in planning usability tests (Nielsen, 1993), or to evaluate strengths and weaknesses in competitor products for later comparison (Dumas & Redish, 1999).

While these methods are useful, they do not allow us to address issues with extended gameplay, that is, issues that may arise after playing the game for a couple of days or more. This is problematic, because one of the key challenges in game design is longevity. With the competition, the shelf life of a game becomes very limited.

Usability techniques. Traditional usability techniques can be used to address a portion of the variables identified as important for game design. In addition to measuring performance, we use many standard usability techniques to answer how and why process-oriented questions. For example, how do users perform an attack, or why are controls so difficult to learn? We use (a) structured usability tests, (b) rapid iterative testing and evaluation (RITE; Medlock, Wixon, Terrano, & Romero, 2002), and (c) other variations and techniques, including open-ended usability tasks, paper prototypes, and gameplay heuristics. For clarity of presentation, each technique will be discussed separately, followed by a case study. Each case study will only contain information pertinent to a specific technique; thus examples may be taken from a larger usability test.

Structured usability test. A structured usability test maintains all the characteristics that Dumas & Redish (1999) proposed as common to all usability tests: (a) goal is to improve usability of the product, (b) participants represent real users, (c) participants do real tasks, (d) participant behavior and verbal comments are observed and recorded, and (e) data are analyzed, problems are diagnosed, and changes are recommended. We have found that issues relating to expectancies, efficiency, and performance interaction are well suited for this type of testing. Some common areas of focus for structured usability testing are in game shell screens, or control schemes. The game shell can be defined as the interface where a gamer can determine and or modify particular elements of the game. This may include main menus and options screens (e.g., audio, graphics, controllers, etc.).

An example that uses this method is in the *MechCommander 2* (2001) usability test. Portions of the method and content have been omitted.

Case Study: MechCommander 2 Usability Test. MechCommander 2 (MC2) is a PC RTS game where the gamer takes control of a unit of *mechs* (e.g., large giant mechanical robots). One area of focus for this test was on the 'Mech Lab, a game shell screen where mechs can be customized (see Fig. 13.2). The gamer is able to modify weaponry, armor, and other similar features and is limited by constraints such as heat, money, and available slots.

The first step in approaching this test was to define the higher-order goals. Overall, the game shell screens had to be easy to navigate, understand, and manipulate, not only for those familiar with mechs and the mech universe, but also for RTS gamers who are not familiar with the mech universe. Our goal was for gamers to be able to modify and to customize mechs in the 'Mech Lab.

As mentioned, one of the most important steps in this procedure is defining the participant profiles. Getting the appropriate users for testing is vital to the success and validation of the data since games are subject to much scrutiny and criticism from gamers. To reiterate, playing games is a choice. For

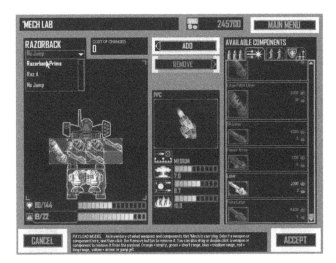

FIGURE 13.2. Screenshot of the 'Mech Lab in MechCommander 2.

MechCommander 2 (2001), we defined two participant profiles which represented all of the variables we wanted to cover. The characteristics of interest included those who were familiar with RTS games (experienced gamers) and those who were not RTS gamers (novice gamers). We also wanted gamers that were familiar with the mech genre, or the mech universe. Overall, we needed a landscape of gamers that had some connection or interest that would make them a potential consumer for this title, whether through RTS experience, or mech knowledge.

Tasks and task scenarios were created to simulate situations that a gamer may encounter when playing the game. Most importantly, tasks were created in order to address the predefined higher order goals. Participants were instructed to talk aloud, and performance metrics were recorded (e.g., task completion, time). The following are examples from the usability task list.

1. Give the **SHOOTIST** jumping capabilities.

This task allowed us to analyze participant expectations. To succeed in this task, one had to select a "CHANGE WEAPONS" button from a different game shell screen, which brought them into the *'Mech Lab*. If the task was to change a weapon on a mech, the terminology would probably have been fine. Thus, this task had uncovered two main issues: (a) could they get to the *'Mech Lab* where you modify the mech, and (b) were they able to discover the process of modifying the mech. It was accurately predicted that gamers would have difficulties with the button terminology. Thus, that button was changed to "MODIFY MECH."

To change the components (e.g., add the jump jets), participants could either select the item and drag it off the mech or select the item and press the "REMOVE" button (see Fig. 13.2). One unexpected issue that arose was that participants unknowingly removed items because the distance required for removing an item was too small. The critical boundary that was implemented was too strict. In addition, participants had difficulties adding items by dragging and dropping because the distance required for adding an item was too large (e.g., the item would not stay on the mech unless it was placed exactly on top of the appropriate location). Appropriate recommendations were made and implemented.

2. Replace the **MG Array** with the **Flamer**.

One of the constraints presented for modifying a mech was heat limit. Each weapon had a particular heat rating. For example, if the heat limit for a mech is 35 and the current heat rating is 32 only weapons with a rating of 3 or fewer could be added. In this task, the "Flamer" had a heat rating much larger than the "MG Array," thus making it impossible to accomplish this task without removing more items. The issues here were the usability of the heat indicator, heat icons, and the discoverability of heat limit concept. None of the participants figured this out. Recommendations included changing the functionality of the Heat Limit Meter and adding better visual cues to weapons that exceed the heat limit. Both of these changes were implemented.

Rapid iterative testing and evaluation method (RITE).
Medlock et al. (2002) documented another common usability method used by the Games User Research Group at Microsoft Game Studios, which they refer to as the RITE method. In this method, fewer participants are used before implementing changes, but more cycles of iteration are performed. With RITE, it is possible to run almost two to three times the total sample size of a standard usability test. However, only one to three participants are used per iteration with changes to the prototype immediately implemented before the next iteration (or group of one to three participants).

The goal of the RITE method is to address as many issues and fixes as possible in a short amount of time in hopes of improving the gamer's experience and satisfaction with the product. However, the utility of this method is entirely dependent on achieving a combination of factors (Medlock et al., 2002). The situation must include (a) a working prototype; (b) the identification of critical success behaviors, important, but not vital behaviors, and less important behaviors; (c) commitment from the development team to attend tests and immediately review results; (d) time and commitment from the development team to implement changes before the next round; and (e) the ability to schedule and run new participants as soon as the product has been iterated. Aside from these unique requirements, planning the usability test is very similar to more traditional structured usability tests.

It is very helpful to categorize potential usability issues into four categories: (a) clear solution, quick implementation, (b) clear solution, slow implementation, (c) no clear solution, and (d) minor issues. Each category has implications for how to address each issue. In the first category, fixes should be implemented immediately and should be ready for the next iteration of testing. In the second category, fixes should be started in hopes that they can be tested by later rounds of testing. For the third and fourth category, more data should be collected.

The advantage of using the RITE method is that it allows for immediate evaluation and feedback of recommended fixes that were implemented. Changes are agreed upon and made directly to the product. If done correctly, the RITE method affords more fixes in a shorter period. In addition, by running multiple iterations over time, we are potentially able to watch the number of usability issues decrease. It provides a nice, easily understandable, and accessible measure. In general, more iterations of testing are the better. However, this method is not without its disadvantages. In this situation, we lose the ability to uncover unmet user needs or work practices, we are unable to develop a deep understanding of gamer behaviors, and we are unable to produce a thorough understanding of user behavior in the context of a given system (Medlock et al., 2002).

The following example demonstrates how the RITE method was used in designing the *Age of Empires II: The Age of Kings* (1999) tutorial. Again, portions of the method and content have been omitted (see Medlock et al., 2002, for more details).

Case Study: Age of Empires II: The Age of Kings Tutorial. Age of Empires II: The Age of Kings (*AoE2*; 1999) is an RTS game for the PC in which the gamer takes control of a civilization spanning over a thousand years, from the Dark Ages through the late medieval period. In this case study, a working prototype of the tutorial was available, and critical concepts and behaviors were defined. Also, the development team was committed to attending each of the sessions. They were committed to quickly implementing agreed upon changes. Finally, the resources for scheduling were available. The key element in this situation for

TABLE 13.2. Age of Empires Example Concepts and Behaviors Categorized into Three Concepts Using the RITE Method (Medlock, Wixon, Terrano, & Romero, 2002)

Essential Concepts/Behaviors	Important Concepts/Behaviors	Concepts/Behaviors of Lesser Interest
• movement • multiselection of units • "fog of war" • scrolling main screen via mouse	• queuing up units • setting gathering points • garrisoning units • upgrading units through technology	• using hotkeys • using mini-map modes • using trading • understanding sound effects

success was the commitment from the development team to work in conjunction with us.

In the *AoE2* tutorial, there were four main sections: (a) marching and fighting (movement, actions, unit selection, the fog of war[1]), (b) feeding the army (resources, how to gather, where to find), (c) training the troops (use of mini-map, advancing through ages, build and repair buildings, relationship between housing and population, unit creation logic), and (d) research and technology (upgrading through technologies, queuing units, advancing through ages). Each of these sections dealt with particular skills necessary for playing the game. In essence, the tutorial had the full task list built in. In the previous case study, this was not the case.

At a more abstract level, the goals of the tutorial had to be collectively defined (with the development team). In more concrete terms, specific behaviors and concepts that a gamer should be able to perform after using the tutorial were identified, then categorized into the three levels of importance: (a) essential behaviors that users must be able to perform without exception, (b) behaviors that are important, but not vital to product success, and (c) behaviors that were of lesser interest. Table 13.2 lists some examples of concepts and behaviors from each of the three categories. This is an important step because it indirectly sets up a structure for decision rules to be used when deciding what issues should be addressed immediately, and what issues can wait.

The general procedure for each participant was similar to other usability tests we often perform. If participants did not go to the tutorial on their own, they were instructed to do so by the specialist.

During the session, errors and failures were recorded. In this situation, an error was defined as anything that caused confusion. A failure was considered an obstacle that prevented participants from being able to continue. After each session, a discussion ensued among the specialist and the development team to determine what issues (if any) warranted an immediate change at that time.

In order to do this successfully, certain things had to be considered. For example, how can one gauge how serious an issue is? In typical usability tests, the proportion of participants experiencing the error is a way to estimate its severity. Since changes are made rapidly here, the criteria must change to the intuitively estimated likelihood that users will continue to experience the error. Another thing to consider is clarity of the issue, which was assessed by determining if there is a clear solution. We have often found that if issues do not have an obvious solution then the problem is not fully understood. Finally, what errors or failures were essential, important, or of lesser interest. Efforts of the development team should be focused on issues related to the essential category when possible.

At this point, we broke down the issues into three groups. The first group included issues with a solution that could be quickly implemented. Every issue in this group was quickly implemented before the next participant was run. The second group consisted of issues with a solution that could not be quickly implemented. The development team began working on these in hopes they could be implemented for later iterations of the test. Finally, there were issues with no clear solutions. These issues were left untouched because more data was needed to assess the problem at a deeper level (e.g., more participants needed to be run). Any fixes implemented in the builds were kept as each participant was brought in. Thus, it was possible that many of the participants experienced a different version of the tutorial over the duration of testing.

Overall, seven different iterations were used across 16 participants. Fig. 13.3 represents the number of errors and failures recorded over time. The number of errors and failures gradually decreased across participants as new iterations of the build were introduced. By the seventh and final iteration of the build, the errors and failures reliably were reduced to zero.

Although we feel that the *AoE2* tutorial was an enormous success, largely due to the utilization of the RITE method, the method does have its disadvantages and should be used with caution. Making changes when issues and solutions are unclear may result in not solving the problem at all, while creating newer usability problems in the interface. We experienced this phenomena a couple of times in the *AoE2* study.

Also, making too many changes at once may introduce too many sources of variability and create new problems for users. Deducing specifically the source of the new problem becomes very difficult. A related issue is not following up changes with enough participants to assess whether or not the solution really addressed the problem. Without this follow up, there is little evidence supporting that the implementations made were appropriate (which is a problem with traditional usability methods as well). The last thing to consider is that other important usability issues that may surface less frequently are likely to be missed. Using such small samples between iterations allows for the possibility that those less occurring issues may not be detected.

Variations on usability methods. Now that we have presented two general types of usability testing, it is worth mentioning some variations on these methods: (a) open-ended tasks, (b) paper prototyping, and (c) gameplay heuristics.

---

[1]The *fog of war* refers to the black covering on a mini-map or radar that has not been explored yet by the gamer. The fog of war lifts once that area has been explored. Use of the fog of war is most commonly seen in RTS games.

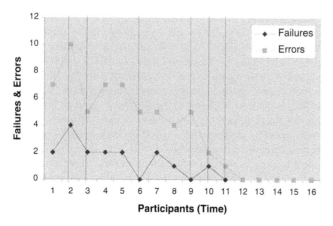

FIGURE 13.3. A record of failures and errors over time for the Age of Empires Tutorial when using the RITE method (Medlock et al., 2002).

In general, it is often recommended that tasks in usability tests be small, with a specified outcome (e.g., Nielsen, 1993). However, we have found situations where the inclusion of an open-ended task yields important data as well. In many usability studies, we often include an open-ended task where participants are not instructed to perform or achieve anything in particular. In other words, there is no specified outcome to the participant. These tasks can be used to analyze how gamers prioritize certain tasks or goals, in a nonlinear environment. These tasks are also useful in situations where structured tasks may confound the participant experience or situations where we are interested in elements of discovery. An example of an open-ended task is as follows.

1. Play the game as if you were at home. The moderator will tell you when to stop.

This example was taken from a usability test on *Halo: Combat Evolved* (2001; Pagulayan, Steury, Fulton, et al., 2003). Participants were presented with a mission with no instruction other than playing as if they were at home. Traditional usability metrics were not tracked or used. Instead, the focus was watching players and the tactics and strategies they employed while playing through the game. Results demonstrated that novice players would start firing at enemies as soon as they were visible, which was not how the designers intended combat to occur. The design intent was for combat to occur at closer ranges.

By allowing participants to play through the mission with no structured task, designers were able to detect the strategies players would employ. As a result, several changes were made to the gameplay to encourage players to engage in much closer combat (see Pagulayan, Steury, Fulton et al., 2003, for more details).

Prototyping, heuristic evaluations, and empirical guideline documents are other techniques we often use when more time-consuming testing cannot be done. In practice, these techniques do not differ when used on games. Nielsen (1993) categorized prototyping and heuristic evaluations as "discount usability engineering," and we would agree. We also tend to view empirical guideline documents in a similar manner. Empirical guideline documents are essentially lists of usability principles for partic-

ular content areas based on our collective experience doing user-testing research. Examples of some of these content areas include console game shell design, PC game shell design, PC tutorial design principles, movement, aiming, and camera issues in first and third person shooter games, and online multiplayer interfaces. Desurvire, Caplan, and Toth (2004) developed a list of heuristics targeted at computer and video games. These have been broken down into four general categories: game play, game story, game mechanics, and game usability. For the full list of gameplay heuristics, see Desurvire et al., (2004).

Survey Techniques

The use of surveys has been explored in great depth (e.g., Bradburn & Sudman, 1988; Couper, 2000; Labaw, 1981; Payne, 1979; Root & Draper, 1983; Sudman, Bradburn, & Schwarz, 1996) and is considered a valid approach for creating an attitudinal data set as long as you ask questions that users are truly capable of answering (see Root & Draper, 1983). We conduct surveys in the lab and over the Internet in order to collect attitudinal data regarding gameplay experiences, self-reported play behaviors, preferences, styles, and motivations.

Formative Playtest surveys. We combine hands-on play with surveys to create a formative lab method called "Playtest." In nearly all instances, questioning follows the pattern of a forced-choice question (sometimes a set of forced-choice questions) followed by a more open-ended question encouraging the participants to state the reasons behind their response. The goal of a Playtest is to obtain specific information about how consumers perceive critical aspects of a game and provide actionable feedback to game designers (Davis et al., 2005). The basic Playtest method is used formatively to compare a product at time one and time two during its development, while a modified version of the basic method is used cumulatively to compare a summative evaluation to other, relevant games.

Playtests have several advantages and characteristics that differentiate them from usability studies and online survey research. For one, they focus mainly on perceptions. A larger sample size is required for statistical power and generalization of the results to the target audience. In most cases, 25 to 35 participants are used as a pragmatic trade-off between confidence and resource constraints. Specialized studies may require more. Statistical representations of the attitudinal measures are used to identify potential strengths and weaknesses that need to further be explored, provide information about the extent or severity of an attitude, or to describe trends of interest that come to light.

Conducting the studies in the lab as opposed to online allows us to test in-progress games during early development. It also allows us to take more control of the testing situations and monitor a limited amount of play behaviors that can be paired with the attitudinal data.

On the other hand, there are a few notable disadvantages. First, structured lab studies limit the amount of gameplay that can be tested in a single study. An artificial lab scenario or time line may prevent the engineer from collecting critical information about the game experience. This is a classic validity trade-off present in most lab studies. Second, the amount of preparation that is required to design a Playtest questionnaire takes some time. The study plan cannot be adjusted during the course of the study. The engineer must be intimately aware of the development team's questions to include all of the relevant questions in the survey. This requires domain knowledge in games and experience in survey design.

Second, the data can sometimes be difficult to interpret if the problem is not major or if the game facet is novel. Sometimes the data does not provide specific causes for the problem, making it difficult to nail down the best solution. Follow-up studies (Playtest or usability) are often recommended (or necessary) to parse out these issues. Finally, the approachable nature of the data is ripe for misinterpretation by team members who may not be familiar with the current state of the game, design details, or the methods used in Playtest. It has been rightfully asserted that usability tests do not necessarily require a formal report upon completion (Nielsen, 1993). We believe that may be less true for using this kind of technique. Careful and consistent presentation formats are required. Instructions and contextual information that will help with interpretation should be included, along with some form of qualitative data (e.g., open-ended comments). Metrics from games in development should focus on problem detection and often cannot be used for predictive purposes.

Development of the Playtest surveys. Several steps have been taken to increase the effectiveness and efficiency of Playtest. First, core questions for each genre, play style, and core game feature have been developed, using a multistep iteration and evaluation process. This process included construct development, survey item development, formative evaluation with subject matter experts, and formative evaluation of question validity using Cognitive Interviewing techniques and statistical validation. While many survey items have been validated, most Playtests require new items to be created in order to meet the specific needs of that particular game and study goal. Therefore, a repository of previously used customized questions has been created, along with guidelines for writing new survey items.

Type of Formative Studies

There are essentially three types of formative studies that are conducted during development: (a) the critical facet test, (b) the initial experience test, and (c) the extended Playtest. We also run tests that do not easily fit into any of these categories, including subtle variations on the above, a few cases of large sample observationally based studies, and studies that focus on games from a more conceptual level.

As in the usability section, for clarity of presentation, each technique will be discussed separately, followed by a case study. Each case study will only contain information pertinent to a specific technique; thus examples may be taken from a larger Playtest.

Critical facet Playtest. Games often take the form of repeating a core experience within an array of different contexts and constraints. A driving game is always about aiming an object that is hurtling through space. The critical facet Playtest focuses

directly on that core experience and making sure that it is fun. While the core experience can often be assessed in usability testing, Playtesting is necessary to assess attitudes and perceptions about the core experience.

The following example demonstrates how a critical facet test was applied to the critical facets of *Oddworld: Munch's Oddysee* (2001), an Xbox game. *Oddworld: Munch's Oddysee* is a platform/ adventure game that allows you to switch back and forth between two main characters as they proceed through the increasingly difficult dangers of Oddworld on a quest to save Munch's species from extinction. The core gameplay is exploring the realm by running, jumping, and swimming through the environment. In this case, there were concerns about the user's visual perspective—which we typically call the camera. Does the camera support exploration of the environment?

Case study: Munch's Oddysee, Camera. Previous usability testing with the *Oddworld: Munch's Oddysee* (2001) determined that while some users indicated dissatisfaction with the behavior of the camera, other participants chose not to mention it at all while engaged in the open-ended usability tasks. The camera's behavior was programmed to create maximal cinematic effects (e.g., sometimes zooming out to show the size of an area) and attempt to enhance gameplay. The camera would often show a specific view with the intent of showing you what was behind the next door or on the other side of the wall while still keeping the main character in view. While many users liked the look, style, and behavior of the camera, users often wanted more control over the behavior of the camera. Indeed some participants would actively say things such as, "That looks really cool right now [after the camera had done something visually interesting] but I want the camera back pointing this way now." Because feedback from the usability lab contained both positive and negative feedback, the development team did not see the usability data as conclusive. Further, changing the camera would be a major cost to the game in terms of redesign and redevelopment time.

After having played the game for an hour, 25 participants were asked for general perceptions of the game. More specific questions followed. Questions related to the camera were asked in the latter portion of the questionnaire because previous experience in the usability lab had shown that merely mentioning the camera as part of a task would often cause participants previously silent on the subject to vociferously criticize aspects of the camera's behavior. With the knowledge that we wanted to factor out any priming-related effects, two analyses were conducted.

The first analysis was based on the response to the questions related to the behavior of the camera itself. Nearly half of the participants (46%) indicated that the camera did not give them enough flexibility of control.

The second analysis went back through individual responses to determine the attitudes of those participants who mentioned the camera before the survey first broached the subject. Forty-three percent of the participants were found to have mentioned the camera in a negative fashion prior to being asked specifically about the camera questions.

Based on this data and other anecdotal evidence, the development team chose to give the players more flexibility of camera control. The result was more frequent use of a camera behavior we termed a *third-person follow camera*. The behavior of this camera had the double advantage of being more easily controlled by users and of conforming to a set of behaviors more often expected by users. It maintained focus on the main character without major adjustments to point of view (e.g., to "look over a wall" or "behind a door"). Other camera behaviors (e.g., still camera behaviors that pan with the character) are still a part of the game but have been localized to areas where these alternative camera behaviors can only create an advantage for the user.

Initial experience Playtest. As with many things, first impressions are a key component of overall satisfaction with a game. Given that many games are a linear experience/narrative, there is a lot of value in obtaining attitudinal data related to the first portions of gameplay. Lessons learned at first are often applied throughout the game. Obviously, the later portions of the game will never be experienced unless the first portions of the game are enjoyed.

The following example explains how a set of formative initial experience tests were run for *MechCommander 2*, the RTS game described earlier in the chapter. The earlier usability test focused on the choices that users could make prior to starting a mission, whereas this test focused on in-game components, such as users' ability to take control of their squad and lead them through battles, while experiencing satisfaction and motivation to continue playing.

Case study: MechCommander 2, Initial Missions. Twenty-five participants, who were representative of the target market, were brought onsite and were asked to begin playing from the first mission of the game. After each mission, participants were asked to stop playing and report their impressions of the fun, the excitement, and the clarity of objectives. Participants also indicated their comfort level with the basics of controlling the game. The participants were able to play through as many as three missions before the session ended. For purposes of brevity, this case study will focus on the results related to the first mission.

Although the first mission had been designed to offer tutorial elements in a brief and fun way, the experience was generally not satisfying to participants. Approximately a third of the participants had a poor initial impression of the game. They felt that the challenge was too easy and that the mission was not exciting. Furthermore, there were some problems related to clarity of goals. By combining the survey results with opportunistic observation, it was noted that some participants moved their units into an area where they were not intended to go. This disrupted their experience and limited their ability to proceed quickly. Responses to more open-ended questions indicated that some of the core gameplay components and unique features did not come across to users. Some users complained about the standard (predictable or commonplace) nature of the game. Finally, several participants complained about the fact that they were being taught everything and wanted to turn the tutorial off.

A number of actions were selected from team insight and user testing recommendations. First, the team decided to separate the tutorial missions from the required course of the game—in order to save experienced players from the require-

ment of completing the tutorial. Second, the scope of the first mission was expanded so that users would have more time in their initial experience with the game. Third, the clarity of objectives was improved via minor interface changes. Fourth, addressing the same issue, the design of the map on which the mission took place was revamped to require a linear approach to the first mission that limited the likelihood of users becoming lost. Fifth, the amount and challenge of combat were increased. Finally, one of the unique components of the game was introduced. The new mission included the ability to call in support from off-map facilities. Despite all these changes, the mission was still targeted to be completed within approximately 10–15 minutes.

A follow-up Playtest was intended to verify the efficacy of the changes. Twenty-five new participants were included in the study. Results from this test indicated that the design changes had created a number of payoffs. Far fewer participants felt the mission was too easy (13% vs. 33%), only 3% indicated that the mission was not exciting, measures of clarity of objectives improved, and surprisingly, there was no drop off on ratings of comfort with basic controls because of the tutorial aspects being improved. Results were not a total success, as some participants were now rating the mission as too hard while others in response to open-ended questions complained about the overly linear nature of the mission (e.g., there was only one way to proceed and few interesting decisions related to how to proceed through the mission). In response to these results, the development team decided to retune the first mission to make it a little easier, but not to address the comments related to the linearity of the mission. The second mission of the game was less linear, and so it was hoped that, with the major constraints to enjoyment removed, most participants would proceed to the second mission and find the mission goals to be more engaging. The data from the second mission was rated far more fun than the first mission, validating their assumption.

Extended Playtests. While the initial experience with a game is very important, the success or failure of a game often depends on the entire experience. In an attempt to provide game developers with user-centered feedback beyond the first hour of gameplay, we conduct extended Playtests that test several hours of consecutive gameplay. This is done by conducting Playtest studies that run for more than two hours, or by having participants participate in more than one study. Attitudinal data is taken at significant time or experience intervals (e.g., after a mission). Participants can also provide self-directed feedback about a specific point in the game at any time during the session. Basic behaviors and progress (e.g., completed level 4 at 12:51 p.m.) can be recorded by Playtest moderators and used in conjunction with the attitudinal data.

Focus groups. In addition to usability and survey techniques, we sometimes employ other user-centered design methods that are qualitative in nature. Focus groups provide an additional source of user-centered feedback for members of the project team. Our Playtest facilities offer us the opportunity to precede a candid user-discussion with hands-on play or game demonstration. In this setting, participants can elaborate on their experience with the product, provide feedback about the user-testing process, speculate about the finished version of the product, generate imaginative feature or content ideas, or find group consensus on issues in a way that models certain forms of real-world shared judgments.

For example, a focus group discussion regarding a sports game revealed that users had varying opinions about what it meant for a game to be a simulation versus an arcade-style game. Additionally, they had many different ideas as to what game features made up each style of game. From this conversation, we learned a few of the Playtest survey questions were not reliable and needed revision.

Besides the general limitations inherent in all focus group studies (Greenbaum, 1988; Krueger, 1994), the focus groups described here typically include practical limitations that deviate from validated focus group methods. Generally, focus group studies comprise a series of 4 to 12 focus groups (Greenbaum, 1988; Krueger, 1994); however, we typically run one to three focus groups per topic. This makes the focus group more of a generative research tool rather than a confident method for identifying real trends and topics in the target population. Fewer group interviews make the study more vulnerable to idiosyncrasies of individual groups.

A Combined Approach

As you can see, we have provided a number of different techniques that we use in game development and evaluation. The presentation has been structured in such a way as to clearly demonstrate the differences between the techniques we use. The important thing to realize is that no one method exists independent of other methods. The following is a brief case study on an Xbox game called *Blood Wake* (2001) that demonstrates how some of these techniques can be used together.

Case Study: Blood Wake. Blood Wake (2001) is a vehicular combat console game using boats and a combination of powerful weapons. Our initial main concern was the aiming mechanism and the controls—users must be able to control their boat and destroy the enemy. A secondary concern was related to the initial missions, goals, gameplay, and general look and feel of the game

The first problem we addressed was the aiming mechanism. Users were presented with two different types of aiming schemes: auto-aim and manual-stick aim. In the auto-aim scheme, the target reticule automatically snapped to an enemy that was within a reasonable distance allowing the player to maneuver their boat while shooting the enemy. In the manual-stick aim scheme, the user had to manually aim the reticule using the right stick on the gamepad with no assistance. In previous usability studies, we have found difficulties with the manual aiming mode. However, this method is used quite often in popular games; thus the general belief was that it was better and preferred. Many felt the auto-aim method would be too easy.

We decided to run a within-subjects design Playtest to assess which scheme users liked better. Our results showed that 73% of participants liked the auto-aim, whereas only 38% liked the manual-stick aim. This was proof enough to convince the developers to focus their efforts on the auto-aim method.

Although we solved the aiming controversy, unoptimized navigation controls made it difficult to assess the fun and chal-lenge of the game. Poor attitudinal ratings could be explained by the gameplay or the poorly tuned boat controls. In order to optimize these controls, we headed to the usability labs. There we presented participants with two different control schemes. In one scheme, throttle and steering were both mapped to the left analog stick (push forward/back for throttle; push left/right for steering). This enabled the user to drive and steer with one finger. In another scheme, the throttle was mapped to the right trigger, and steering was on the left analog stick. Separating the steering from the throttle would map more directly onto familiar mental models for driving and steering cars. Previous data indicated that neither scheme was clearly superior, so the goal of this test was to optimize the controls for both schemes.

Participants were asked to perform common maneuvers in the game. The following are examples taken from the usability task list:

1. Once the game starts, you should see a bunch of boxes that form a path in the distance. As fast as you can, try to follow the trail of items and collect as many boxes as you can without turning around if you get off course.
2. Somewhere on this map, there is a straight line of boxes. Find the straight line of boxes then try to slalom (like a skier) through the line of the boxes. Try to do this as fast as you can without collecting any of the boxes.

As you can see from the two examples, the tasks were very performance based. Participants were instructed to think aloud while we measured task performance and task time.

By the end of the study, we were able to provide detailed information to the team about how to optimize the controls. Some of the issues that were addressed included the sensitivity of the steering, controlling boat speed, and difficulties moving the boat in reverse. Iteration on the controls was followed by multiple Playtests that focused on user perceptions of five of the initial missions. The results of the Playtest validated our earlier recommendations and allowed us to tweak the controls even more.

By running multiple Playtests, we were also able to get data and feedback on gameplay. After each mission, we elicited feedback on the perceived difficulty, clarity of objectives, attitudes about the environment, and general fun ratings. This allowed us to pinpoint where participants were running into problems, such as not knowing where to go next and not understanding how to complete objectives. This feedback led to dramatic changes in the design of the missions.

CONCLUSION

The need for the continued development of user-centered design methods in video games has indeed arrived. Games drive new technologies and affect an enormous amount of people. In addition, games represent a rich space for research areas involving technology, communication, attention, perceptual-motor skills, social behaviors, virtual environments, just to name a few. It is our position that video games will eventually develop an intellectual and critical discipline, like films, which would result in continually evolving theories and methodologies of game design. The result will be an increasing influence on interface design and evaluation. This relationship between theories of game design and traditional HCI evaluation methods has yet to be defined, but definitely yields an exciting future.

As mentioned earlier, user-centered design methods are beginning to find their way into the video games industry. Commercial game companies, such as Ubisoft Entertainment, Electronic Arts, Inc., and Microsoft Corporation, employ some level of user-centered methodologies in their game development processes, and this presumably will grow. However, this is not to say that video games are qualitatively different from other computer applications. There are just as many similarities as differences to other computer fields that have already benefited from current user-centered design methods. Thus, it makes sense to utilize these methods when applicable, but also to adapt methods to the unique requirements that we have identified in video games.

In this chapter, we emphasized the difference between games and productivity applications in order to illustrate the similarities and differences between these two types of software applications. We also chose to reference many different video games in hopes that these examples would resonate with a number of different readers. Case studies were included to try to demonstrate in practice how we tackle some of the issues and challenges mentioned earlier in the chapter. It is our inten-

tion that practitioners in industry, as well as researchers in academia, should be able to take portions of this chapter, and adapt them to their particular needs when appropriate, similar to what we have done in creating the actual methods mentioned in this chapter. That said, we are upfront that most, if not all of our user-testing methods are not completely novel. Our user-testing methods have been structured and refined based on a combination of our applied industry experience, backgrounds in experimental research, and of course, a passion for video games. This allows us to elicit and utilize the types of information needed for one simple goal and to make the best video games that we can for as many people as possible.

ACKNOWLEDGMENTS

We would like to thank the Games User Research Group at Microsoft Game Studios. In addition, we would like to express our gratitude to Paolo Malabuyo, Michael Medlock, and Bill Fulton for their insights and input on the creation of this chapter, and Kathleen Farrell and Rally Pagulayan for their editing assistance and reviewing early drafts. The views and opinions contained in this chapter are those of the authors and do not necessarily represent any official views of Microsoft Corporation.

References

Age of Empires II: Age of Kings [Computer software]. (1999). Redmond, WA: Microsoft Corporation.

Animal Crossing [Computer software]. (2002). Redmond, WA: Nintendo of America.

Banjo Kazooie [Computer software]. (2000). Redmond, WA: Nintendo of America.

Bradburn, N. M., & Sudman, S. (1988). *Polls and surveys: Understanding what they tell us.* San Francisco: Jossey-Bass.

Blood Wake [Computer software]. (2001). Redmond, WA: Microsoft Corporation.

Brute Force [Computer software]. (2003). Redmond, WA: Microsoft Corporation.

Cassell, J., & Jenkins, H. (2000). Chess for girls? Feminism and computer games. In J. Cassell & H. Jenkins (Eds.), *From Barbie to Mortal Kombat: Gender and computer games* (pp. 2–45). Cambridge, MA: The MIT Press.

Chaika, M. (1996). *Computer games marketing bias.* ACM Crossroads. From http://www.acm.org/crossroads/xrds3-2/girlgame.html

Combat Flight Simulator [Computer software]. (1998). Redmond, WA: Microsoft Corporation.

Couper, M. P. (2000). Web surveys: A review of issues and approaches. *Public Opinion Quarterly, 64,* 464–494.

Crawford, C. (1982). *The art of computer game design.* Berkeley, CA: Osborne/McGraw-Hill.

Csikszentmihalyi, M. (1990). *Flow—The psychology of optimal experience.* New York, NY: Harper & Row.

Davis, J. P., Steury, K., & Pagulayan, R. J. (2005). A survey method for assessing perceptions of a game: The consumer playtest in game design. *Game Studies: The International Journal of Computer Game Research, 5*(1). From http://www.gamestudies.org/0501/davis_steury_pagulayan/

Desurvire, H., Caplan, M., & Toth, J. (2004). Using heuristics to improve the playability of games. *CHI 2004: Conference on Human Factors in Computing Systems,* Vienna, Austria. ACM's Special Interest Group on Computer-Human Interaction.

Diablo [Computer software]. (2000). Paris, France: Vivendi Universal.

Diddy Kong Racing [Computer software]. (1997). Redmond, WA: Nintendo of America.

Dumas, J. S., & Redish, J. C. (1999). *A practical guide to usability testing* (Rev. ed.). Portland, OR: Intellect Books.

Entertainment Software Association. (2005). 2005 Sales, demographics and usage data: Essential facts about the computer and video game industry. Washington, DC: Entertainment Software Association.

Final Fantasy X [Computer software]. (2001). Tokyo, Japan: Square Co., Ltd.

Flight Simulator 2000 [Computer software]. (2000). Redmond, WA: Microsoft Corporation.

Funge, J. (2000). Cognitive modeling for games and animation. *Communications of the ACM, 43*(7), 40–48.

Greenbaum, T. L. (1988). *The practical handbook and guide to focus group research.* Lexington, MA: DC Heath and Company.

Halo: Combat Evolved [Computer software]. (2001). Redmond, WA: Microsoft Corporation.

Halo 2 [Computer software]. (2004). Redmond, WA: Microsoft Corporation.

Harvest Moon [Computer software]. (2003). Redmond, WA: Nintendo of America.

Hecker, C. (2000). Physics in computer games. *Communications of the ACM, 43*(7), 35–39.

Herz, J. C. (1997). *Joystick nation: How videogames ate our quarters, won our hearts, and rewired our minds.* New York, NY: Little, Brown and Company.

Hyman, P. (2005). State of the industry: Mobile games. *Game Developer Magazine, 12* (14), 11–16.

IDG Entertainment, (2005, August). *IDG Entertainment casual games market report.* Oakland, CA: IDG Entertainment.

International Game Developers Association. (2005). *IDGA 2005 Mobile games whitepaper.* Paper presented at the Game Developers Conference 2005 by the Mobile Games SIG, San Francisco, CA.

Interactive Digital Software Association (2000). State of the industry report (2000–2001). Washington, DC: Interactive Digital Software Association.

Internet Backgammon, Windows XP [Computer software]. (2001). Redmond, WA: Microsoft Corporation.

Jablonsky, S., & DeVries, D. (1972). Operant conditioning principles extrapolated to the theory of management. *Organizational Behavior and Human Performance, 7,* 340–358.

Kent, S. L. (2000). *The first quarter: A 25-year history of video games.* Bothell, WA: BWD Press.

Kroll, K. (2000). Games we play: The new and the old. *Linux Journal, 73es.* From http://www.acm.org/pubs/articles/journals/linux/2000-2000-73es/a26-kroll/a26-kroll.html

Krueger, R. A. (1994). *Focus groups: A practical guide for applied research.* Thousand Oaks, Ca: Sage.

Labaw, P. (1981). *Advanced questionnaire design.* Cambridge, MA: Abt Books Inc.

Laird, J. E. (2001). It knows what you're going to do: Adding anticipation to a Quakebot. *Proceedings of the Fifth International Conference on Autonomous Agents, Canada,* 385–392.

Lepper, M., Greene, D., & Nisbett, R. (1973). Undermining children's intrinsic interest with extrinsic rewards. *Journal of Personality and Social Psychology, 28,* 129–137.

Lepper, M. R., & Malone, T. W. (1987). Intrinsic motivation and instructional effectiveness in computer-based education. In R. E. Snow & M. J. Farr (Eds.), *Aptitude, Learning and Instruction III: Conative and Affective Process Analyses.* Hillsdale, NJ: Lawrence Erlbaum Associates.

Levine, K. (2001). *New opportunities for Storytelling.* Paper presented at the Electronic Entertainment Exposition, Los Angeles, CA.

Madden NFL '06 [Computer software]. (2005). Redwood City, CA: Electronic Arts Inc.

Malone, T. W. (1981). Towards a Theory of Intrinsic Motivation. *Cognitive Science, 4,* 333–369.

MechCommander 2 [Computer software]. (2001). Redmond, WA: Microsoft Corporation.

MechWarrior 4: Vengeance [Computer software]. (2000). Redmond, WA: Microsoft Corporation.

Medlock, M. C., Wixon, D., Terrano, M., & Romero, R. (2002, July). *Using the RITE method to improve products: a definition and a case study.* Orlando, FL: Usability Professionals Association.

Microsoft Word 2003 [Computer software]. (2005). Redmond, WA: Microsoft Corporation.

Moore, K., & Rutter, J. (2004). Understanding Consumers' Understanding of Mobile Entertainment. In K. Moore & J. Rutter (Eds.), *Proceedings of Mobile Entertainment: User-centred Perspectives* (pp. 49–65). *CRIC,* University of Manchester.

MTV Music Generator [Computer software]. (1999). Warwickshire, United Kingdom: The Codemasters Software Company Limited.

NFL2K [Computer Software]. (1999). San Francisco, CA: Sega of America, Inc.

Nielsen, J. (1993). *Usability engineering.* San Francisco, CA: Morgan Kaufmann.

Oddworld: Munch's Oddysee [Computer software]. (2001). Redmond, WA: Microsoft Corporation.

Pagulayan, R., Gunn, D., & Romero, R. (2006). A Gameplay-Centered Design Framework for Human Factors in Games. In W. Karwowski (Ed.) *2nd Edition of International Encyclopedia of Ergonomics and Human Factors* (pp. 1314–1319). London: Taylor & Francis.

Pagulayan, R. J., & Steury, K. (2004). Beyond usability in games. *Interactions, 11*(5), 70–71.

Pagulayan, R. J., Steury, K., Fulton, B., & Romero, R.L. (2003). Designing for fun: User-testing case studies. In M. Blythe, A. Monk, K. Overbeeke, & P. Wright (Eds.), *Funology: From usability to enjoyment* (pp. 137–150). Netherlands: Kluwer Academic Publishers.

Payne, S. L. (1979). *The art of asking questions.* Princeton, NJ: Princeton University Press.

Pokemon Crystal [Computer software]. (2000). Tokyo, Japan: Nintendo Japan.

Preece, J., Rogers, Y., Sharp, H., Benyon, D., Holland, S., & Carey, T. (1994). *Human-computer interaction.* Reading, MA: Addison-Wesley.

Provenzo, E. F., Jr. (1991). *Video kids: Making sense of Nintendo.* Cambridge, MA: Harvard University Press.

RalliSport Challenge [Computer Software]. (2003). Redmond, WA: Microsoft Corporation.

Root, R. W., & Draper, S. (1983, December 12–15). *Questionnaires as a software evaluation tool.* Paper presented at the ACM CHI, Boston, MA.

Sidel, P. H., & Mayhew, G. E. (2003). The Emergence of Context: A Survey of MobileNet User Behaviour. From http://www.mocobe.com/pdf/EmergenceofContext1.pdf

Steers, R. M., & Porter, L. W. (1991). *Motivation and work behavior* (5th ed.). New York, NY: McGraw-Hill, Inc.

Sudman, S., Bradburn, N. M., & Schwarz, N. (1996). *Thinking about answers: The application of cognitive processes to survey methodology.* San Francisco, CA: Jossey-Bass.

The Legend of Zelda: The Wind Waker [Computer Software]. (2003). Redmond, WA: Nintendo of America.

Williams, J. H. (1987). *Psychology of women: Behavior in a biosocial context (3rd ed.).* New York: W. W. Norton.

Yuen, M. (2005, September). The Tipping Point: The convergence of wireless and console/PC game design. *Game Developers Magazine.*

AUTHOR INDEX

Gordon-Salant, S., 24, 31
Gosling, G. D., 210, 214
Gould, J. D., 178, 189, 192
Gould, S. J., 53, 67
Graber, M. A., 115, 122
Graf, P., 119, 122
Gragoudas, E. S., 96, 108
Graham, D., 203, 215
Gray, D. B., 77, 79, 90, 91
Gray, J., 21, 30
Gray, P., 79, 91, 99, 109
Green, P., 177–196, 178, 179, 181, 182, 183, 184, 185, 186, 187, 188, 189, 190, 192, 193, 194, 195, 199, 214
Greenbaum, T. L., 233, 234
Greenberg, J., 189, 192
Greenberg, R., 103, 107
Greene, D., 222, 235
Greenes, R. A., 143, 152
Greenfield, P., 12, 15
Gregor, P., 51–68, 57, 59, 62, 63, 64, 65, 66, 67, 68, 112, 116, 117, 119, 122
Gregory, D. F., 144, 153
Gribble, S., 35, 37, 47
Gribbons, W., 111, 112, 122
Gribbons, W. M., 111–124
Griesemer, J. R., 212, 216
Griffin, P., 42, 45, 48
Griffith, E. R., 72, 91
Griffiths, V., 116, 121
Gross, C. R., 21, 25, 28, 30
Guerrier, J. H., 20, 25, 27, 30
Guggiana, V., 102, 107
Guha, M. L., 39, 47
Gullo, D. F., 44, 47
Gunn, D., 218, 224, 225, 226, 230, 235
Gustafson, D., 21, 31, 141, 142, 146, 147, 153
Gustafson, D. H., 141, 152, 153
Gutwin, C., 94, 104, 107
Guyer, D. R., 96, 108
Guzdial, M., 37, 42, 44, 46, 47, 48

H
Haag, Z., 5, 14
Hacker, E. W., 209, 214
Hagenauer, M. E., 143, 153
Halgren, S., 35, 36, 47
Hall, J., 8, 9, 13, 14
Hallett, M., 74, 90
Hamalainen, P., 40, 41, 47
Hamilton, H., 130, 133
Hamilton, S., 130, 133
Hammond, K., 25, 26, 27, 30
Hammond, M., 74, 91, 92
Han, Y. Y., 143, 152
Hancock, P. A., 205, 210, 215, 216
Hankey, J. M., 191, 193
Hanna, L., 36, 37, 40, 47
Hanowski, R. J., 191, 193
Hansen, J., 83, 86, 90, 92
Hansen, P. K., 83, 92
Hansman, R. J., 208, 210, 212, 215
Hanson, R. H., 200, 214
Hanson, V. L., 125–134, 126, 127, 128, 132, 133
Harkreader, A., 77, 84, 91
Harper, R., 209, 214

Harper, S., 94, 105, 108
Harris, J., 141, 152
Harris, T. A., 4, 5, 14
Harrison, M., 205, 214, 216
Hassenzahl, M., 160, 175
Haubner, M., 151, 153
Haworth, C., 181, 194
Hay, K. E., 42, 44, 48
Hayes, B. C., 27, 32
Hayes, K. H., 21, 30
Hazen, M., 4, 14
Hazlett, H., 159, 175
Heck, H., 104, 107
Hecker, C., 218, 235
Hedriks, J., 145, 153
Hegedus, E., 182, 192
Heglin, H. J., 199, 214
Heider, F., 10, 14
Heiman, H., 143, 154
Hellriegel, D., 150, 154
Henderson, V., 130, 133
Hernandez, M., 20, 32
Hernandez-Rebollar, J. L., 130, 133
Hersey, M. D., 198, 214
Hersh, W. R., 144, 152
Hertzog, C., 19, 20, 25, 26, 30
Herz, J. C., 218, 235
Hewison, J., 144, 145, 152
Hewitt, J., 86, 90
Hibbard, J., 149, 152
Hickey, D. T., 43, 44, 47
Hicks, J. E., 72, 90
Hinterberger, T., 81, 89, 91
Hjelm, I., 171, 175
Hludik, F. C. J., 81, 91
Hoang, T., 197, 200, 204, 212, 215, 216
Hodge, M. H., 199, 214
Hoedemaeker, M., 182, 193
Hoffman, C., 139, 152
Hoffmeister, R., 127, 133
Holdaway, D., 34, 47
Hollan, J., 35, 37, 47, 49
Holland, S., 218, 235
Holley, P., 18, 24, 25, 26, 27, 28, 30
Holmes-Rovner, M., 145, 153
Holtzblatt, K., 205, 214
Holzman, T., 101, 103, 109
Homes-Rovner, M., 145, 153
Hon, H.-W., 102, 103, 108
Hood, J. N., 150, 154
Hoppin, S., 20, 25, 30
Horn, J. L., 54, 67
Horn, L. J., 72, 90
Hornof, A., 84, 90
Horrey, W. J., 181, 193
Horsky, J., 143, 154
Horton, M., 37, 48
Horwitz, P., 43, 44, 47
Hoselton, R., 84, 90
Houben, G.-J., 104, 107
Hourcade, J. P., 35, 39, 47
Hoysniemi, J., 40, 41, 47
Hu, P. S., 182, 193
Hu, S., 95, 98, 107
Huang, X., 102, 103, 108

SUBJECT INDEX